Quantum Transport in Nanostructures and Molecules

An introduction to molecular electronics

Quantum Transport in Nanostructures and Molecules

An introduction to molecular electronics

Colin John Lambert
Department of Physics, Lancaster University, Lancaster LA1 4YB, United Kingdom

IOP Publishing, Bristol, UK

ISBN 978-0-7503-3639-0 (ebook)
ISBN 978-0-7503-3637-6 (print)
ISBN 978-0-7503-3640-6 (myPrint)
ISBN 978-0-7503-3638-3 (mobi)

DOI 10.1088/978-0-7503-3639-0

Version: 20210301

IOP ebooks

British Library Cataloguing-in-Publication Data: A catalogue record for this book is available from the British Library.

Published by IOP Publishing, wholly owned by The Institute of Physics, London

IOP Publishing, Temple Circus, Temple Way, Bristol, BS1 6HG, UK

US Office: IOP Publishing, Inc., 190 North Independence Mall West, Suite 601, Philadelphia, PA 19106, USA

This book is dedicated to my wife Karen and our children Faye and James.

Contents

Preface **xvi**

Acknowledgements **xvii**

Author biography **xviii**

1 Introduction to molecular-scale electronics **1-1**

1.1 Background 1-1
1.2 The theoretical challenges addressed by this book 1-4
1.3 De Broglie wave patterns 1-5
1.4 The Landauer formula 1-6
1.5 Resonant transport 1-6
1.6 Thermoelectricity 1-8
1.7 Off-resonance transport 1-8
1.8 Intuitive picture of Green's functions 1-11
1.9 Magic ratio theory 1-12
1.10 Controlling quantum interference 1-13
1.11 Summary 1-15
References 1-16

2 Connectivity theory and noisy neighbour equations for quantum transport **2-1**

2.1 Introduction 2-1
2.2 Green's functions for beginners 2-2
2.3 The principle of superposition 2-5
2.4 Destructive versus constructive interference 2-5
2.5 Green's functions produced by noisy neighbours 2-6
2.6 NNEs for transport through more complex coupled quantum structures 2-14
2.7 Identifying quantum interference within the five-component NNE 2-16
2.8 Magic ratio theory 2-18
2.9 Green's function of a molecular core 2-21
2.10 Mid-gap energies versus $E = 0$ 2-28
2.11 Summary 2-30
Appendix A Basic properties of matrices 2-30
References 2-45

3 A beginner's guide to solving the Schrödinger equation 3-1

3.1 Introduction 3-1
3.2 Mathematical aim: an introduction to linear algebra 3-2
3.3 A simple harmonic oscillator 3-2
3.4 The equations of motion of two harmonic oscillators 3-3
3.5 Matrix form of the equations of motion of two harmonic oscillators 3-4
3.6 Combining linear independence with the best choice of basis to solve equation (3.7) 3-8
3.7 Eigenvalues and eigenvectors 3-9
3.8 Solution to the general $N \times N$ dynamical problem 3-11
3.9 Solution to the general $N \times N$ Schrödinger equation 3-12
3.10 How many linearly-independent eigenvectors does a dynamical matrix $\boldsymbol{H}$ possess? 3-13
3.11 A two-level quantum system 3-14
3.12 The Hamiltonian of a two-level quantum system 3-17
3.13 Eigenvalues and eigenvectors of $N \times N$ Hermitian matrices 3-19
3.14 Proof that a $N \times N$ Hermitian matrix possesses N linearly independent eigenvectors 3-21
3.15 Completeness and initial conditions revisited 3-23
3.16 The eigenstates of a two-level quantum system 3-24
3.17 Summary 3-25
Appendix A Basic properties of matrices and complex numbers 3-27

4 Quantum properties of linear chains and simple molecules 4-1

4.1 Introduction 4-1
4.2 The eigenvalues and eigenvectors of a two-level molecule 4-1
4.3 The eigenvalues and eigenvectors of a doubly infinite chain of identical atoms 4-3
4.4 Effective mass and group velocity 4-5
4.5 Degeneracies and the allowed values of k 4-6
4.6 The most general solution to the Schrödinger equation for a linear chain 4-7
4.7 A finite chain of N sites with periodic boundary conditions 4-8
4.8 Group velocity revisited 4-13
4.9 The eigenvalues and eigenvectors of a linear chain of N identical atoms 4-16
4.10 The two-level system revisited 4-19
4.11 A three-level system 4-20

4.12 Normal modes of a vibrating atomic chain 4-20
4.13 Radicals and Huckel's rule 4-23
4.14 Summary of results for the eigenstates and eigenvalues of one-dimensional chains 4-24
4.15 A quantum spider's web 4-24
4.16 Eigenstates of a semi-infinite one-dimensional chain 4-30
4.17 Summary 4-31
Appendix A Relationship to Fourier analysis 4-31
Appendix B Continuity equations 4-35

5 Quantum properties of electrodes in higher dimensions 5-1

5.1 Introduction 5-1
5.2 Peierl's distortion and oligoynes 5-1
5.3 Conduction bands versus valence bands 5-3
5.4 Flexibility in labelling states 5-4
5.5 A finite-width electrodes formed from a linear chain of cells 5-6
5.6 A linear chain of cells, each containing two sites 5-8
5.7 A linear chain of cells, each containing N sites, with an intra-cell Hamiltonian $\boldsymbol{H}_1$ proportional to the unit matrix 5-9
5.8 A linear chain of cells, each containing N sites, with an intra-cell Hamiltonian $\boldsymbol{H}_1$ proportional to the unit matrix and an intra-cell Hamiltonian $\boldsymbol{H}_0$ describing a linear chain with free ends 5-10
5.9 A linear chain of cells, each containing N sites, with an intra-cell Hamiltonian $\boldsymbol{H}_1$ proportional to the unit matrix and an intra-cell Hamiltonian $\boldsymbol{H}_0$ describing a linear chain with periodic boundary conditions 5-11
5.10 A two-dimensional crystal on a square lattice 5-13
5.11 The most general solution to the Schrödinger equation for the periodic structure of figure 5.7, for which $\boldsymbol{H}_1$ is proportional to the unit matrix 5-14
5.12 The equivalence between a finite-width lead and many one-dimensional leads 5-16
5.13 Summary 5-17

6 Scattering theory of electrical conductance and thermopower 6-1

6.1 Introduction 6-1
6.2 A single impurity in one dimension 6-1
6.3 The scattering matrix 6-4
6.4 Bond currents and conservation of probability 6-6

6.5 Current carried by counter-propagating plane waves 6-8
6.6 Unitarity of the scattering matrix 6-9
6.7 Physical meaning of scattering matrix elements 6-10
6.8 Consequences of unitarity of the scattering matrix 6-11
6.9 Bound states 6-12
6.10 The S matrix in higher dimensions 6-13
6.11 The Landauer formula 6-15
6.12 Derivation of the Landauer formula and formulae for thermoelectric properties of a quantum object connected to external electrodes 6-17
6.13 Summary 6-23
Appendix A A general expression for bond currents 6-23
Appendix B Unitarity of the multi-channel S matrix 6-24

7 Thermoelectricity in nanostructures and molecules 7-1

7.1 Introduction 7-1
7.2 A qualitative description of thermoelectricity 7-1
7.3 Linear response formulae for thermoelectric coefficients 7-3
7.4 An expressions for thermoelectric efficiency 7-4
7.5 Relationship between thermoelectric efficiency and ZT 7-5
7.5.1 Efficiency at maximum power 7-5
7.5.2 Maximum efficiency 7-6
7.6 Expressions for thermoelectric coefficients 7-7
7.7 Proof that ZT_e is positive 7-8
7.8 Strategies for maximising thermoelectric performance 7-8
7.9 Summary 7-10

8 A very useful formula (VUF) for the transmission coefficient of an arbitrary scatterer connected to one-dimensional leads and a magic ratio theory for intra-molecular currents 8-1

8.1 Introduction 8-1
8.2 The Schrödinger equation for a scatterer connected to two one-dimensional leads 8-1
8.3 Note about the choice of wave vectors 8-3
8.4 Solution to the Schrödinger equation 8-3
8.5 Expression for the transmission amplitude in terms of the full Green's function $\boldsymbol{G}$ 8-7
8.6 Expression for the transmission amplitude in terms of the Green's function of the isolated scatterer 8-7

8.7 Expression for the reflection amplitude 8-8
8.8 Expression for the wave function inside the scatterer 8-9
8.9 Expressions for bond currents 8-9
8.10 Magic ratio theory for intra-molecular currents 8-11
8.11 Properties of the very useful formula for the transmission coefficient 8-13
8.12 The relationship between magic ratio theory and the VUF 8-14
8.13 The relationship between the Breit–Wigner formula, on-resonance transport and the VUF 8-16
8.14 Summary 8-18
References 8-18

9 A quantum system connected to many scattering channels 9-1

9.1 Introduction 9-1
9.2 Solving the Schrödinger equation 9-1
9.3 Unitarity of the many-channel scattering matrix 9-7
9.4 Expressions for the transmission coefficients 9-8
9.5 Expressions for the reflection coefficients 9-10
9.6 The self-energy matrix 9-10
9.7 Dyson's equation 9-12
9.8 Example: a scatterer connected to two 1-dimensional leads 9-13
9.9 A scatterer attached to M leads, in which each lead connects to only one site of the scatterer 9-14
9.10 Scattering theory in the presence of an open scatterer 9-15
9.11 The effect of closed channels 9-15
9.12 A scatterer connected to finite-dimensional leads 9-16
9.13 Summary 9-16

10 Relationship between Green's functions, wave functions and scattering amplitudes 10-1

10.1 Introduction 10-1
10.2 Green's functions of closed systems 10-2
10.3 Approximations to Green's functions of closed systems 10-3
10.4 Green's functions of a doubly infinite linear chain 10-4
10.5 Relationship between a Green's function of a linear chain and the wave function of a T-shaped junction 10-8
10.6 Green's functions of a semi-infinite linear chain 10-10

10.7 The relationship between Green's functions and wave functions in the presence of a scatterer 10-11
10.8 Green's function of a linear chain with periodic boundary conditions 10-14
10.9 Green's functions of a linear chain with free-end boundary conditions 10-15
10.10 Green's functions of a finite-width nanoribbon 10-18
10.11 Relationships between Green's functions and transmission amplitudes 10-20
10.12 Summary 10-21

11 Connectivity theory revisited: heteroatom substitution, decimation and the Breit–Wigner formula 11-1

11.1 Introduction 11-1
11.2 Recursively describing the effect of changes to a molecular junction 11-2
11.3 Decimation and the effect of pendant groups 11-4
11.4 Illustration of destructive quantum interference due to pendant atoms 11-6
11.5 The effect of a local perturbation 11-7
11.6 The effect of introducing a single heteroatom 11-9
11.7 The effect of introducing two heteroatoms 11-10
11.8 The Green's function of a semi-infinite lead revisited 11-11
11.9 Fusion of substructures via Dyson's equation 11-12
11.10 Constructing a lead-scatterer-lead system using Dyson's equation 11-17
11.11 From Green's functions to transmission functions 11-19
11.12 Simplification of the Green's function of a scatterer in the presence of electrodes 11-21
11.13 Summary of Green's function equations for the transmission coefficient 11-22
11.14 The Breit–Wigner formula for an arbitrary scatterer connected to finite-width leads 11-22
11.15 Derivation of the five-component NNE 11-23
11.16 Summary 11-27
Appendix A Similarity transformations 11-28
Appendix B An alternative derivation of the core Green's function 11-33

12 Linear molecules 12-1

12.1 Introduction 12-1
12.2 Single-site decimation and the variation of conductance with length 12-1
12.3 A negative eigenvalue theorem 12-6

12.4 Recurrence relations via the noisy neighbour equations 12-7
12.5 Decay of conductance with length of an oligoyne and a diatomic chain 12-8
12.6 Quantum circuit rules for the conductance and Seebeck coefficient of linear molecules 12-11
12.7 Summary 12-12
Reference 12-12

13 Quantum interference in molecules with parallel paths and pendant groups 13-1

13.1 Introduction 13-1
13.2 The effect of pendant groups 13-2
13.3 The Green's function of an isolated molecule containing a bridging unit 13-4
13.4 A molecular backbone coupled to two different atoms of a bridging unit 13-6
13.5 A molecule with three moieties in series 13-7
13.6 A molecule with two bridging groups in parallel 13-8
13.7 The electrical conductance of a molecule with weakly coupled bridges 13-12
13.8 The electrical conductance of a molecule with many parallel bridges 13-13
13.9 Summary 13-16

14 Connectivity theory and equations of motion 14-1

14.1 Introduction 14-1
14.2 Stationary states and Green's functions 14-2
14.3 Green's function of a 2-component system revisited 14-4
14.4 Connectivity equations from combinations of active and passive viewpoints 14-6
14.5 Stationary states of vibrating structures 14-10
14.6 Connectivity theory, equality of currents and unitarity of the scattering matrix 14-12
14.7 The limit $\eta \to 0$ 14-13
14.8 The case $\eta \neq 0$ 14-15
14.9 Comparison between advanced and retarded Green's functions 14-15
14.10 Summary 14-16
Appendix A Explicit integration of equation (14.4) 14-16
Appendix B Green's functions of open systems 14-17

Appendix C Green's functions of a double infinite tight-binding chain revisited 14-19

Appendix D Iterative analysis of the 3-component structure of figure 14.3 14-20

Appendix E The Green's function of a 5-component system 14-25

15 Relationship between Green's functions, molecular orbitals and densities of states 15-1

15.1 Introduction 15-1

15.2 The Coulson–Rushbrooke theorem 15-1

15.3 An orbital symmetry rule 15-3

15.4 Densities of states 15-4

15.5 van Hove singularities 15-7

15.6 A more precise form of the density of states 15-8

15.7 Properties of delta functions 15-10

15.8 Examples of delta functions 15-12

15.9 Relationship between Green's functions and densities of states 15-14

15.10 Summary 15-16

References 15-16

16 Solving the time-dependent Schrödinger equation and the theory of representations 16-1

16.1 Introduction 16-1

16.2 The theory of representations 16-2

16.3 Representations involving orthonormal basis functions 16-3

16.4 Representations involving non-orthogonal basis functions 16-6

16.5 Transformations between representations 16-8

16.6 A more compact notation for the transformation equation 16-9

16.7 Atomic orbitals versus pseudo-atomic orbitals 16-10

16.8 Contribution of overlap integrals to nearest-neighbour couplings 16-11

16.9 Dyson's equation in a non-orthogonal basis 16-12

16.10 Summary 16-13

Reference 16-13

17 Scattering theory in the presence of material-specific leads 17-1

17.1 Introduction 17-1

17.2 The most general solution to the Schrödinger equation for an arbitrary crystalline lead 17-3

17.3 Matrix elements between channel eigenvectors 17-5
17.4 Cancellation of off-diagonal contributions to currents 17-7
17.5 Note about degeneracies 17-8
17.6 Analysis of a general scatterer connected to a periodic structure 17-9
17.7 Analysis of currents 17-11
17.8 Expressions for transmission amplitudes 17-12
17.9 Determining the allowed wave vectors for a given energy E 17-16
17.10 The Green's function of an arbitrary material-specific, finite-width, doubly infinite lead 17-18
17.11 The Green's function of an arbitrary finite-width, semi-infinite lead 17-20
17.12 The surface Green's function of an arbitrary finite-width, semi-infinite lead 17-21
17.13 A simple derivation of the surface Green's function of an arbitrary finite-width, semi-infinite lead 17-22
17.14 Dyson's equation for the surface Green's function of an arbitrary finite-width, semi-infinite lead 17-23
17.15 Expressions for transmission and reflection amplitudes of a scatterer connected to arbitrary, material-specific leads 17-26
17.16 Summary 17-27
Appendix A Solving the $\boldsymbol{H}_1$ problem 17-28
Appendix B Translation operators, velocity operators, Green's functions, time reversal symmetry and the transmission coefficient 17-33
Appendix C A note about degeneracies 17-57
References 17-58

Preface

This book originated from a series of tutorials, which I have presented to new PhD students in my group for many years. Chapters 1 and 2 should be of interest to a novice wishing to be exposed to some of the terminology used to describe quantum transport at the molecular scale. Chapters 3–6 are intended to expose the reader to basic theoretical tools and by chapter 8, the reader should have enough knowledge to write their own code to solve for transport properties of arbitrary structures connected to 1d leads. Chapters 6 and 7 provide an introduction to the Landauer formula for electrical conductance and to thermoelectricity in nanostructures. Table 8.1 of chapter 8 summarises analytic formulae for various model Green's functions derived in later chapters. Chapter 8 also presents a new 'magic ratio theory' for intra-molecular bond currents. Chapters 9 and 11 generalise the discussion to structures with many leads or finite-width leads, while chapter 10 provides useful analytic results for Green's functions of linear chains and rings. Analytic solutions for transport properties containing linear chains, pendant group and multiple paths are presented in chapters 11–13. Chapter 14 discusses relationships to time dependent states created by sources, while chapter 15 discusses densities of states and orbital product rules. Finally chapters 16 and 17 discuss technical details needed to describe transport in material specific structures.

Acknowledgements

It is a great pleasure to thank the ~100 PhD students and postdocs, who have been or are currently members of my group, many of whom are now senior academics. Over the years, their enthusiasm, curiosity and multitude of questions have honed the material presented in this book. I would also like to express thanks to many senior theoreticians, with whom it has been a pleasure to collaborate, including Jaime Ferrer (Oviedo), Jozsef Cserti (Budapest), Spiros Evangelou (Ioannina), Stefano Sanvito (Dublin) and John Jefferson (QinetiQ). I would also like to express my appreciation to collaborators in leading experimental groups, whose laboratory demonstrations of molecular-scale quantum interference phenomena underpin many of the concepts discussed in this book. These include the following group leaders and their teams: Martin Bryce (Durham), Richard Nichols and Simon Higgins (Liverpool), Tim Albrecht (Birmingham), Nicholas Long and Lesley Cohen (Imperial College, London), Walther Schwarzacher (Bristol), Wenjing Hong (Xiamen), Paul Low (Perth), Nicolás Agraït (Madrid), Pilar Cea (Zaragoza), Silvio Decurtins, Thomas Wandlowski and Shi-Xia Liu (Bern), Benjamin Robinson and Sam Jarvis (Lancaster), Andrew Briggs, Harry Anderson, Lapo Bogani and Kyriakos Porfyrakis (Oxford), Chris Ford (Cambridge) Dominique Vuillaume (Villeneuve-d'Ascq), Xiangfeng Duan (UCLA), Klaus Müllen (Basel), Nongjian Tao (Arizona), Herre van der Zant (Delft), Marcel Mayor (Basel), Bernd Gotsmann (IBM, Zurich) and Michel Calame (EMPA, Zurich).

Finally I would like to thank my sailing colleagues Roger Pearce, Einar Kristoffersen, Stafford Stuchbury, Norman Whewell, Chris Elleray, Steve Bailey and Jeff Simpkin, all of whom, from time to time, have had to tolerate my enthusiastic outpourings about quantum mechanics and molecular electronics.

Over many years, my research has been supported by a range of funding agencies, including EPSRC and the European Commission. Current research funding is provided by the EU FET Open project 767187-QuIET and the EPSRC Programme QuEEN.

Author biography

Colin John Lambert

Colin Lambert became a Lecturer in theoretical physics at Lancaster University in 1983, a full Professor in 1990 and in 2010 was awarded a Research Professorship. During the period 1992–2013, he led a series of five, 4-year European networks, each involving theory groups from 10 European Universities and since then has provided the theoretical leadership within four European networks, focussing on molecular-scale electronics. He is a member of Academia Europaea, a former President of CECAM (Centre Européen de Calcul Atomique et Moléculaire) and in 2009 was awarded a QinetiQ Fellowship for his 'Distinguished scientific research and leadership'.

The first decade of Lambert's career, spanned 26 orders of magnitude in energy, including predictions for pair production by ultra-high energy ($>10^{19}$ eV) cosmic rays and quasi-particle production in 3He–B at ultra-low temperatures ($<10^{-7}$ eV). The latter study challenged text book assertions that moving objects in superfluid 3He feel no dissipation until the Landau velocity is reached; a concept which continues to be explored today. In the early 1990s, he invented a fundamental scattering theory for describing multi-channel, phase-coherent electron transport in the presence of superconductivity, which underpinned the discovery of unexpected proximity and quantum interference effects in hybrid superconducting nanostructures. This caused a reassessment of more conventional quasi-classical theories, which had missed these important interference effects.

In 1998, he derived the fundamental formulae discussed in chapter 17, describing material-specific quantum transport, which led to the SMEAGOL and GOLLUM codes. During the past two decades, he has collaborated with many experimental groups around the world to identify and exploit room-temperature interference phenomena in molecular junctions and self-assembled monolayers.

https://scholar.google.com/citations?hl=en&user=55vf-xgAAAAJ

IOP Publishing

Quantum Transport in Nanostructures and Molecules
An introduction to molecular electronics
Colin John Lambert

Chapter 1

Introduction to molecular-scale electronics

1.1 Background

When standing on a cliff top overlooking the ocean, one is struck by the contrast between the sea, the land and the coastal region between them. The sea is described in terms of waves, currents and tides, the land is described in terms of farming, the built environment and geographical features, whereas the coast is described as a battle ground between the sea and the land, using terms such as coastal erosion, corrosion and deposition. This separation between land and sea is reflected in the cultural separation between their users, in which fishermen and sailors have few common interests with farmers and builders and both of them prefer to avoid the complexities of the coast.

Similarly when a single molecule is placed between a source and a drain electrode, as shown in figure 1.1, there is a marked contrast between the molecule, the electrodes (sometimes called 'leads') and the coastal region between them. The molecule is described using the language of chemistry, the electrodes are described using the language of solid-state physics and there is a cultural separation between physicists and chemists, with both of them preferring to avoid the complexities of the molecule–electrode interface.

Physicists sometimes describe molecules as analogues of artificial quantum dots and suggest that at a conceptual level, all molecules are the same, because many generic phenomena familiar from mesoscopic physics appear when a molecule is contacted to electrodes, including Kondo physics, Coulomb blockade, Franck–Condon blockade and resonant tunnelling. On the other hand, chemists tend to emphasise the unique features of different molecules and when describing the properties of single-molecule junctions, draw attention to features such as spatial distributions of molecular orbitals, charge distributions within molecules and their precise geometry and chemical makeup. Of course both points of view have their merits, but neither is the whole story. For example, molecules differ from artificial quantum dots, because electrons can be injected into molecular cores with atomic accuracy,

doi:10.1088/978-0-7503-3639-0ch1

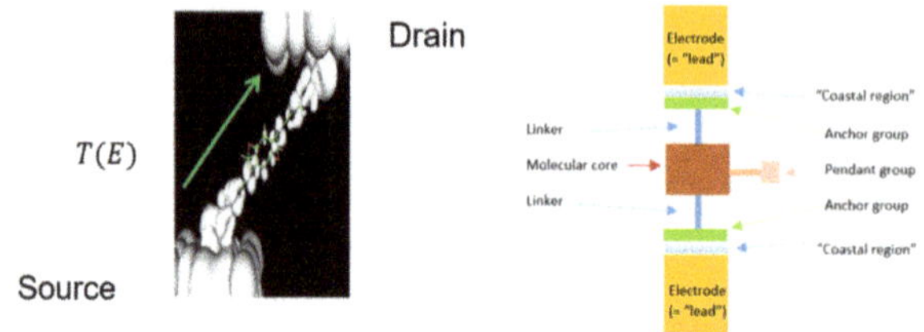

Figure 1.1. (Left) A molecule located between a source and drain electrode [1]. The cloud shows the de Broglie wave pattern formed by an electron entering from the source. $T(E)$ is the transmission probability for an electron of energy E to pass from the source to the drain. (This figure is adapted from [1], with permission of Springer.) (Right) A conceptual picture of a molecule connected to electrodes (i.e. electrical leads) via anchor groups at each end of the molecule. The anchor groups are connected to the central core of the molecule by linkers. Optionally, the central core may possess one or more pendant groups.

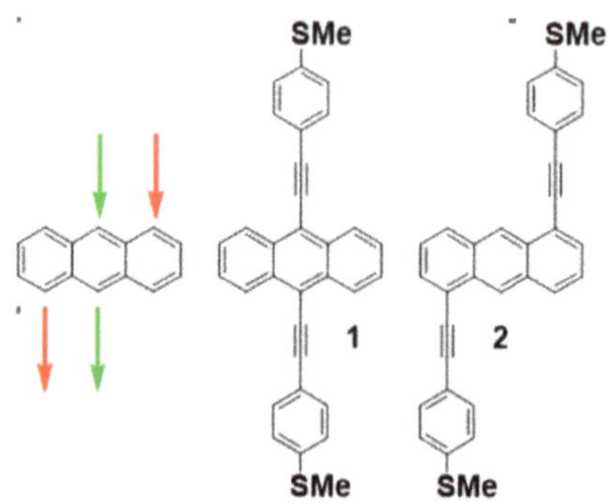

Figure 1.2. A molecular core composed of three six-membered rings, with two different connectivities indicated by the red and green arrows. The connectivities are determined by appropriate design of their chemical synthesis. Electrons enter and leave the core via triple bonds. For the chemically challenged reader, the cores of these molecules are called anthracene. (This figure is reprinted with permission from [153]. Copyright (2020) American Chemical Society.)

engineered by chemical synthesis, as shown in figure 1.2, which means theories of connectivity are particularly germane.

On the other hand transport through single-molecule junctions is not determined by the molecule alone, because the alignment between molecular energy levels and the Fermi energy of the electrodes is not a property of the isolated molecule, and instead is determined by the 'coastal region' between the molecule and the electrodes.

One aim of this book is to try to span this cultural divide by presenting basic concepts from both sciences, which allow us to describe the flow of electricity through molecules and nanostructures. One of the crucial differences between quantum dot physics and transport through single molecules is that in studies of the former, one is primarily interested in on-resonance transport, whereas in the latter, off-resonance phase-coherent tunnelling is of primary interest. This difference arises, because if an energy level of a molecule coincides with the Fermi energy of an electrode, then it becomes partially occupied. Molecules with partially occupied levels are known as radicals. They are reactive and therefore such junctions are usually unstable under ambient conditions. Consequently, studies of on-resonance transport through molecules are usually confined to low temperatures and to three-terminal devices in which the molecular energy levels can be manipulated and moved onto resonance via an electrostatic gate. In contrast, off-resonance transport controls electrical and thermoelectrical properties of molecular-scale materials, such as self-assembled monolayers (SAMs).

The fact that, under ambient conditions, the Fermi energy of an electrode rarely coincides with molecular energy levels leads to certain simplifications, because typically, electrons passing through the molecule from one electrode to the other tunnel through the energy gap between the highest occupied molecular orbital (HOMO) and the lowest unoccupied molecular orbital (LUMO). This gap is typically a few electron volts, which is hundreds of times higher than[1] k_BT at room temperature T. This means that electrons passing through a molecule are protected by the HOMO–LUMO gap and can undergo negligible scattering by other electrons or phonons. Furthermore, off-resonant tunnelling within the gap is much faster than resonant tunnelling through a molecular orbital, which means that if the molecule is sufficiently short, the molecule cannot respond to the passage of an electron through the junction. In other words, transport can be phase coherent and described by a mean-field Hamiltonian, which does not change as an electron passes through the junction. Experiments reveal that at room temperature, this is the case for molecules, which are shorter than a few nanometres. In summary, the key conceptual ingredients in a description of phase-coherent transport through molecules or other phase-coherent structures is a mean-field Hamiltonian H, which describes the atomic and geometrical makeup of the structure, and a source and drain for electrons entering and leaving the structures, whose connectivity to the central core can be varied by careful design.

With the above observations in mind, this book presents a conceptual framework for understanding room-temperature electron and phonon transport through molecules and other quantum objects described by mean-field Hamiltonians, where the absence of electron–electron and electron–phonon scattering means that their electrical and thermoelectrical properties are controlled by quantum interference. The aim is to present material at the level of a final-year undergraduate or first-year postgraduate. On the one hand, the language of solid-state physics is used to describe electrodes and requires an understanding of band structures, densities of states and the Fermi distribution. On the other, the language of chemistry is used to describe the molecules and requires an understanding of how the atomic structure of a molecule determines electronic properties, such as energy level spacings and the spatial distribution of molecular orbitals. After reading this account, it is intended that the reader will have become familiar with the basic concepts of molecular-orbital theory and scattering theory, which underpin mean-field theories of quantum transport. These are the tools and concepts that allow us to confront the

[1] k_B is Boltzmann's constant and at room temperature, $k_BT \approx 0.025$ eV = 0.58 kcal mol^{-1} = 2.4 kJ mol^{-1}.

wave-particle duality of electrons and are therefore also widely applicable within solid-state physics to studies of transport through nanostructures.

Molecular electronics involves the manipulation of molecular orbitals to achieve desired function, such as sensing, switching, electrical or thermal conductance and the thermoelectric conversion of heat into electricity and during the past decade or so, various design strategies have been developed to help to achieve this function. This book will help to rationalise these strategies and help the reader to understand when particular concepts are applicable and when they are not. The remainder of this introductory chapter will summarise some of the key challenges and concepts associated with transport through molecular-scale structures, which of course are equally applicable to any phase-coherent quantum system. To emphasise the generic nature of many of the theoretical concepts, the word 'molecule' will be replaced occasionally by the phrase 'scattering region', or simply 'scatterer'.

1.2 The theoretical challenges addressed by this book

When single-molecule junctions, such as those shown in figures 1.1 and 1.2, are formed by placing a single molecule between a source and a drain electrode, the theoretical challenge is to describe the passage of electrons, as they are injected into the molecule from the source electrode and collected by the drain electrode. As noted above, for junctions in which the distance between the source and drain is less than a few nanometres, it has been demonstrated experimentally that electron transport through the junction is phase coherent, even at room temperature, which means that with a high probability, the energy of the electron is unchanged as it passes through the molecule. Consequently electron transport is described by the Schrödinger equation, which must be solved by treating the source, molecule and drain in a holistic manner. In other words, the electrodes and their coupling to the molecule should be treated on the same footing as the molecule itself, or returning to the analogy at the beginning of this chapter, the land, the sea and the coastal region should be recognised as being equally important.

From a conceptual viewpoint, an isolated molecule (e.g. formed from the green, blue, brown and peach portions of the cartoon in figure 1.1) is a closed system, because there is no possibility for electrons to enter or leave. This means that energy levels of an isolated molecule are quantised and assume a discrete set of values. By applying the Pauli exclusion principle, two electrons (one with spin up and the other with spin down) can occupy each molecular orbital[2]. The HOMO and LUMO, and other orbitals, which are nearby in energy, are often referred to collectively as 'frontier orbitals'. For such a closed system, the mean-field Hamiltonian can be approximated by a matrix of finite size and therefore in principle, such a Hamiltonian can be stored on a computer and the time-independent Schrödinger equation can be solved numerically to obtain the molecular energy levels and molecular orbital wave functions. At a material-specific level, the relevant mean-field Hamiltonian H could be obtained using community codes based on density

[2] In chemistry this is called the Aufbau principle.

functional theory (DFT) or Hartree–Fock theory. For the purpose of developing concepts and describing generic properties, simpler model Hamiltonians are appropriate. Physicists refer to these as tight-binding Hamiltonians, whereas chemists refer to such descriptions as Hückel theory.

In contrast with isolated molecules, electrodes are open systems, which far from the junction are often modelled as crystalline leads, and from a conceptual viewpoint, can be regarded as extending to infinity. This means that at least part of their energy level spectrum forms a continuum, with a vanishing energy level spacing. In this case, after filling all the lowest energy states, each with two electrons, the highest occupied energy level is called the Fermi energy and denoted E_F. The Hamiltonian describing such open systems is infinitely large and cannot be stored on a computer. Therefore, to solve the Schrödinger equation, concepts from solid-state physics, which are appropriate to open systems must be used to reduce the problem to a finite size.

When a molecule is connected to electrodes, the source–molecule–drain system is also open. In this case, to solve for transport properties, a theoretical framework is needed that describes how the quantised levels of the molecule become coupled to the continuous spectrum of the leads. This book is aimed at delivering such a framework. Initially tight-binding/Hückel theory will be used to describe basic concepts, without introducing unnecessary complexity. As the book progresses, the complexity of the structures will be gradually increased. Finally, chapter 17 presents a framework for describing material-specific transport in realistic structures of the kind encountered in the laboratory. This final chapter captures the basic concepts underpinning many quantum-chemistry-based transport codes, which are nowadays widely available.

1.3 De Broglie wave patterns

In the single-molecule junction shown in figure 1.1, the cloud in the right-hand panel depicts the wave function of an electron of energy E as it passes from one electrode to the other. This de Broglie wave can be regarded as a standing wave pattern created by the electron entering from the source and leaving via the drain. As an analogy, one can imagine a singer standing at the entrance to an auditorium and emitting a note of a fixed frequency ω. The emitted sound echoes around the auditorium and sets up a standing wave pattern. Regions of high amplitude are called anti-nodes and correspond to positions in which the echo interferes with itself constructively. A listener standing at such a location would hear a loud sound. Similarly regions of low amplitude are called nodes. They correspond to positions at which the sound waves interfere destructively and a listener standing at such a location would hear a weak sound. Obviously this echo pattern can be manipulated by moving the furniture around the auditorium.

Figure 1.1 shows that an electron of energy E creates a standing de Broglie wave pattern within the molecule, consisting of regions of high and low amplitude. A key quantity that determines the measured transport properties of such a structure, such as electrical conductance or thermopower, is the transmission coefficient $T(E)$, which describes the ease with which an electronic wave of energy E can pass from the source to the drain. If the de Broglie wave pattern has a large amplitude at the drain electrode, then $T(E)$ will be large and this is referred to as constructive quantum interference (CQI), whereas if it has a small amplitude at the drain, then $T(E)$ will be small and this is referred to as destructive quantum interference (DQI). Obviously, this de Broglie wave pattern depends on the molecular 'furniture' and can be manipulated by changing the shape or chemical makeup of the molecule.

1.4 The Landauer formula

This ability to manipulate wave patterns and hence $T(E)$ is central to the field of single-molecule electronics and mesoscopic physics, because it underpins strategies for controlling quantities such as the electrical conductance of nanoscale and molecular-scale devices. Indeed, at a low enough temperature, the electrical conductance σ of such a structure is given by the Landauer formula [2, 3], which is derived in chapter 6:

$$\sigma = G_0 T(E_F) \tag{1.1}$$

In this equation, G_0 is called the quantum of conductance and takes the value $G_0 = \frac{2e^2}{h} = 77\ \mu\text{S}$, where e is the electronic charge and h is Planck's constant. Since G_0 depends only on fundamental constants, it is the natural unit of electrical conductance. The quantity $T(E_F)$ is the transmission coefficient $T(E)$, evaluated at $E = E_F$. This means that an ability to control the wave pattern created by electrons entering a molecule at the Fermi energy E_F of the electrodes, enables control of the single-molecule electrical conductance and other transport properties.

1.5 Resonant transport

To understand why this may be interesting, figure 1.3 shows an example of a transmission coefficient $T(E)$. In this example, the HOMO of the molecule is located at an energy E_H (in this example, $E_H \approx 0.5$ eV) and the spacing between molecule

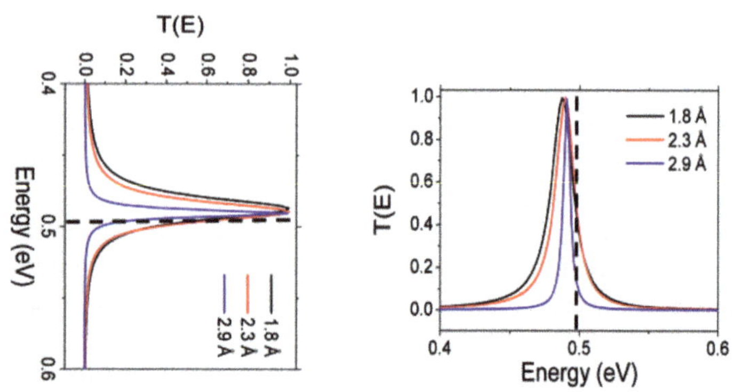

Figure 1.3. A transmission curve versus electron energy E. The left panel shows energy in the vertical direction, whereas the right plot has energy along the horizontal axis. The Fermi energy E_F of the electrodes is indicated by the dashed line.

energy levels is large, so no other energy levels appear in the figure. In figure 1.3, $T(E)$ is plotted over an energy interval containing the HOMO of the molecule and when the energy E of an electron passing through the molecule is close to E_H, $T(E)$ exhibits a resonance and achieves a large value. This is an example of CQI. The energy E of electrons passing through a junction is always close to the Fermi energy E_F of the electrodes indicated by the dashed line. Therefore since the electrical conductance is proportional to the transmission coefficient evaluated at E_F, the electrical conductance is determined by the crossing point between the dashed line and the transmission curve. In figure 1.3, three curves are shown, corresponding to three different separations (1.8, 2.3 and 2.9 Å) between the two ends of the molecule and their respective electrodes. If this separation is large, then the coupling between the molecule and electrodes is weak and the width of the HOMO resonance (denoted Γ_H below) is small, whereas if the separation between the ends of the molecule and the electrodes is small, then the coupling between the molecule electrodes is strong and the width of the resonance is larger. This feature is described by the Breit–Wigner formula [4], derived in chapter 11, which for a symmetric junction takes the form

$$T(E) = \frac{\Gamma_H^2}{(E - \epsilon_H)^2 + \Gamma_H^2} \tag{1.2}$$

A plot of the right-hand side of this equation versus E reproduces the curves shown in figure 1.3, with a full width at half maximum of Γ_H and a peak position of ϵ_H. As explained in chapter 11, the peak position does not exactly coincide with E_H and is given by the equation $\epsilon_H = E_H - \sigma_H$, where σ_H is called the 'self-energy' and describes the fact that the peak position is shifted slightly relative to E_H. As noted above, the field of molecular electronics is interdisciplinary and involves chemists and physicists (and other disciplines) working and communicating together. This involves overcoming many cultural differences between the fields, one of which is shown in figure 1.3; namely chemists prefer energy to be plotted along the vertical axis, whereas physicists prefer energy to be plotted along the horizontal axis.

In equation (1.2), the 'coastal parameters' Γ_H and σ_H are determined by a mixture of physics and chemistry, and depend on the shape of the HOMO orbital, the density of electronic states in the electrodes and the coupling between the molecule and the electrodes. The formula is valid, provided E is sufficiently close to ϵ_H and provided Γ_H is much smaller than the spacing δ between the resonating orbital (in this case the HOMO) and the nearest frontier orbitals. Clearly, by controlling the parameter Γ_H, the electrical conductance can be changed significantly, because small changes in the width of the resonance lead to large changes in $T(E_F)$. Similarly, since the energy of the HOMO determines the peak position ϵ_H of the resonance, any mechanism, which shifts the HOMO energy E_H, will lead to large changes in conductance. Such changes in conductance have many potential applications, ranging from transistor-like switching to DNA sequencing and to the sensing of changes in the local environment, which could be caused by the presence of drugs, explosives, light, electric fields or magnetic fields.

1.6 Thermoelectricity

The large slope in $T(E)$ near a resonance is also of interest for thermoelectricity. As discussed in chapter 7, if a temperature difference $T_{\text{source}} - T_{\text{drain}}$ is applied to a single-molecule junction, then this creates a voltage difference $V_{\text{source}} - V_{\text{drain}}$, given by

$$V_{\text{source}} - V_{\text{drain}} = -S(T_{\text{source}} - T_{\text{drain}}), \tag{1.3}$$

where S is the Seebeck coefficient. If the temperature difference is generated by coupling the junction to a waste heat source, then this provides a mechanism for generating electricity from waste heat, in a silent and environmentally friendly manner. Conversely, if a voltage difference is applied to a molecular junction, then this creates a temperature difference, which could be used to cool objects at the nanoscale via the Peltier effect. For this reason, one of the goals of the field of molecular-scale electronics is to design molecules and self-assembled monolayers (SAMs) of molecules, which possess high values of S.

To understand how this might be achievable, note that if the slope of $T(E)$ does not change significantly on the scale of k_BT, then the Seebeck coefficient is obtained from the transmission coefficient via the formula

$$S(E_{\text{F}}) \approx -S_0\left(\frac{dlnT(E)}{dE}\right)_{E=E_{\text{F}}} \tag{1.4}$$

The above formula is derived in chapter 7. In SI units, $S_0 = \alpha\, |e|\, T$, where $\alpha = \left(\frac{k_{\text{B}}}{e}\right)^2 \frac{\pi^2}{3} = 2.44 \times 10\text{–}8\ \text{W}\Omega\text{K}^{-2}$ is the Lorentz number. On the other hand, if E is measured in units of electron volts, then $S_0 = \alpha T$, which at $T = 300$ K, takes the value $S_0 = 7.3\ \mu\text{V}\ K^{-1}$. This value sets the scale for room-temperature, single-molecule thermoelectricity. From equation (1.4), if E_{F} is known, then from the slope of $\ln T(E)$ at E_{F}, one can predict the Seebeck coefficient. Clearly a large slope in $T(E)$ near the Fermi energy is highly desirable, since this leads to a high Seebeck coefficient.

1.7 Off-resonance transport

As mentioned above, unless the frontier orbitals of a molecular junction are controlled electrostatically, electrochemically or by some other means, when a molecule is placed in a junction, charge transfer between the electrodes and the molecule causes the molecular energy levels to adjust themselves, such that the Fermi energy lies within the HOMO–LUMO gap and not close to a resonance. In other words, if the HOMO energy is E_{H} and the LUMO energy is E_{L}, then E_{F} tends to lie closer to the mid-gap energy $E_{\text{HL}} = (E_{\text{H}} + E_{\text{L}})/2$, as shown in figure 1.4. The precise location of the orbital energies relative to E_{F} depends on a combination of many factors, including the nature of the electrodes and the anchor groups. For example, in a junction formed from gold electrodes and a molecule with thiol anchor groups, the Au-thiol bond attracts electronic charge and this build up of negative charge acts like an electrostatic gate, which repels the HOMO and LUMO, thereby

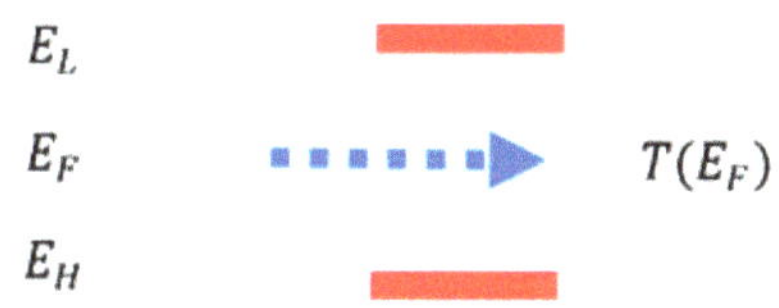

Figure 1.4. Two frontier energy levels $E_{\rm H}$, $E_{\rm L}$, with the Fermi energy $E_{\rm F}$ located between them.

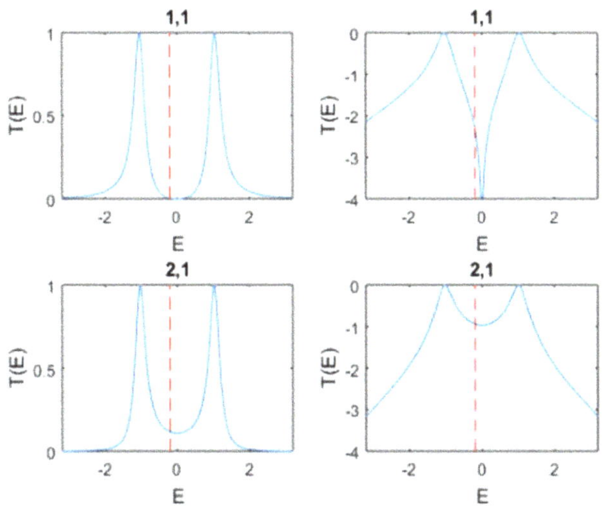

Figure 1.5. The transmission coefficients $T(E)$ of two examples of a two-level molecule, labelled 1,1 and 2,1. The vertical scales of the left-most figures are linear, whereas the vertical scales of the right-most figures are logarithmic.

increasing their energies and moving the HOMO closer to $E_{\rm F}$. On the other hand pyridyl anchor groups tend to decrease the values of $E_{\rm H}$ and $E_{\rm L}$. Since $E_{\rm F}$ lies between $E_{\rm H}$ and $E_{\rm L}$, this moves $E_{\rm L}$ closer to $E_{\rm F}$.

Equation (1.2) is not valid when $|E-\epsilon_{\rm H}|$ is of the order of the spacing ($\delta = E_{\rm L} - E_{\rm H}$) between the LUMO and the HOMO and cannot be used when E is located in the vicinity of the middle of the HOMO–LUMO gap (i.e. when $E \approx E_{\rm HL}$), because near the mid gap all frontier orbitals (such as the HOMO, HOMO-1, HOMO-2, … and LUMO, LUMO+1, LUMO+2, ….) make non-negligible contributions to the transmission coefficient. In this case an alternative theory of mid-gap electron transport is needed.

To understand the distinction between mid-gap transport and resonant transport, consider the fictitious molecule with two molecular orbitals (i.e. only a HOMO and a LUMO), shown in figure 1.4. Two examples of transmission coefficients for such a two-level system are shown in figure 1.5.

Since figure 1.5 possesses two resonances (near $E_{\rm H}$ and $E_{\rm L}$), one might hope that it could be described by an equation of the form (1.5), which is simply a sum of two Breit–Wigner resonances.

$$T(E) = \frac{\Gamma_{\rm H}^2}{(E - \epsilon_{\rm H})^2 + \Gamma_{\rm H}^2} + \frac{\Gamma_{\rm L}^2}{(E - \epsilon_{\rm L})^2 + \Gamma_{\rm L}^2} \qquad \text{(wrong!)} \tag{1.5}$$

However this expression is not correct, because it misses important quantum interference effects that occur within the core of a molecule, when E lies within the HOMO–LUMO gap. A more accurate description of mid-gap transport is obtained by noting that equation (1.2) can be written in the form

$$T(E) = c|g_{\rm H}|^2, \qquad \text{where} \quad g_{\rm H} = \frac{A_{\rm H}}{(E - \epsilon_{\rm H}) + {\rm i}\Gamma_{\rm H}} \tag{1.6}$$

and $\Gamma_{\rm H} = cA_{\rm H}^2$. In this expression, c is a constant describing the strength of the contacts to the electrodes, $i = \sqrt{-1}$ and $A_{\rm H}$ is a constant that describes the amplitude of the HOMO wave function in the regions of the contacts to the left and right the electrodes. Using this notation, equation (1.5) is of the form

$$T(E) = c[|g_{\rm H}(E)|^2 + |g_{\rm L}(E)|^2] \qquad \text{(wrong!)} \tag{1.7}$$

In fact for a fictitious molecule with two energy levels, a more accurate generalisation of equation (1.2) is

$$T(E) = c|g_{\rm H} + g_{\rm L}|^2, \tag{1.8}$$

where the contribution from the LUMO is captured by the term

$$g_{\rm L} = \frac{A_{\rm L}}{(E - \epsilon_{\rm L}) + {\rm i}\Gamma_{\rm L}}, \tag{1.9}$$

and the contribution $g_{\rm H}$ from the HOMO is given by equation (1.6). Since $|g_{\rm H} + g_{\rm L}|^2 = (g_{\rm H} + g_{\rm L})(g_{\rm H}^* + g_{\rm L}^*)$, equation (1.8) yields

$$T(E) = T(E) = c\left[|g_{\rm H}|^2 + |g_{\rm L}|^2 + g_{\rm L}^* g_{\rm H} + g_{\rm L} g_{\rm H}^*\right] \tag{1.10}$$

This differs from equation (1.7) due to the presence of the term $g_{\rm L}^* g_{\rm H} + g_{\rm L} g_{\rm H}^*$, which describes the effect of quantum interference between the HOMO and LUMO. This term is important, because it leads to the phenomenon of destructive quantum interference (DQI) within the HOMO–LUMO gap. To recognise this, note that if either or both of $g_{\rm H}$ and $g_{\rm L}$ are non-zero, then $|g_{\rm H}|^2 + |g_{\rm L}|^2$ is necessarily positive and therefore the right-hand side of equation (1.7) cannot vanish. On the other hand, in equation (1.10), the interference term $g_{\rm L}^* g_{\rm H} + g_{\rm L} g_{\rm H}^*$ could be either positive or negative and in the latter case, could exactly cancel the term $|g_{\rm H}|^2 + |g_{\rm L}|^2$ leading to $T(E) = 0$. The occurrence of DQI is a crucial difference between on-resonant transport described by the Breit–Wigner formula (1.2) and off-resonant electron transport within the gap. Such a DQI feature is shown in the top row of figure 1.4 (labelled 1,1).

Unfortunately, when describing mid-gap transport, equation (1.8) is not the whole story, because as noted above, real molecules possess many orbitals and they all make non-negligible contributions to the transmission coefficient. To demonstrate this, note that for a molecule possessing a HOMO-1, a HOMO, a LUMO and a LUMO+1, and other orbitals, equation (1.7) becomes

$$T(E) = c|\ldots\ \ldots + g_{\mathrm{H}-1} + g_{\mathrm{H}} + g_{\mathrm{L}} + g_{\mathrm{L}+1} + \ldots\cdots|^2 \tag{1.11}$$

It turns out (see section 10.3) that the contributions from other frontier orbitals, such as g_{H-1} and g_{L+1} etc can be almost as big as those from g_{H} and g_{L} and therefore although a simple description based on equation (1.8) is appealing and can provide qualitative insight, a theory which ignores the contributions from other orbitals is usually inaccurate.

1.8 Intuitive picture of Green's functions

At first sight, the prospect of including contributions from all orbitals seems to be rather daunting, but fortunately, when transport takes places near the middle of the HOMO–LUMO gap, certain simplifications are possible, as described in chapter 2 and later chapters. To understand these simplifications, one should recognise that the amplitude of a wave pattern at some location is space r, depends on the location r' of the source that created the wave. For example, in figure 1.6, a pebble thrown into a pond at r' creates a wave at location r of instantaneous amplitude $g(r, r')$. These wave patterns are referred to as Green's fucntions.

Similarly, the sound wave created by a singer standing at the entrance to an auditorium depends on the location of the entrance and the de Broglie wave pattern within a molecule created by a source injecting an electron into a particular atom, depends on the choice of atom. The latter is illustrated in figure 1.7, which shows the de Broglie wave pattern $g(r, r')$ at all points r in space, created inside anthracene-based (upper panel) and anthraquinone-based (lower panel) molecules, when the source injects mid-gap electrons of energy $E = E_{\mathrm{HL}}$ at a source point r', indicated by the tiny red dot near the bottom right-hand corner of the pictures. These wave patterns are also referred to as Green's functions. They are different from molecular orbitals, because Green's functions require a source, whereas molecular orbitals do not.

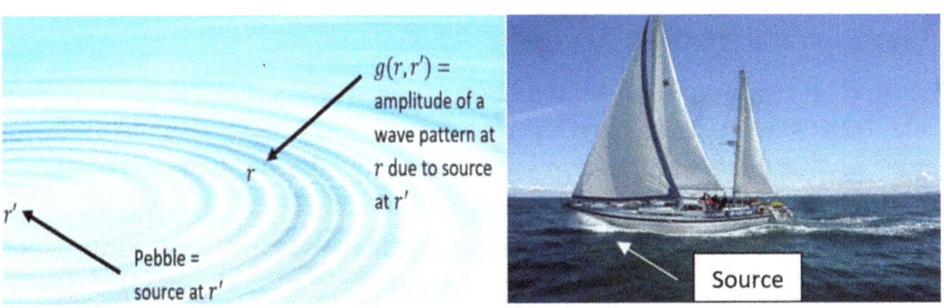

Figure 1.6. The wave pattern on the left is created by a pebble hitting the surface of a pond at position r'. The sailing boat on the right acts as a source of additional waves, which are superposed on a background wave pattern created by the wind and tide.

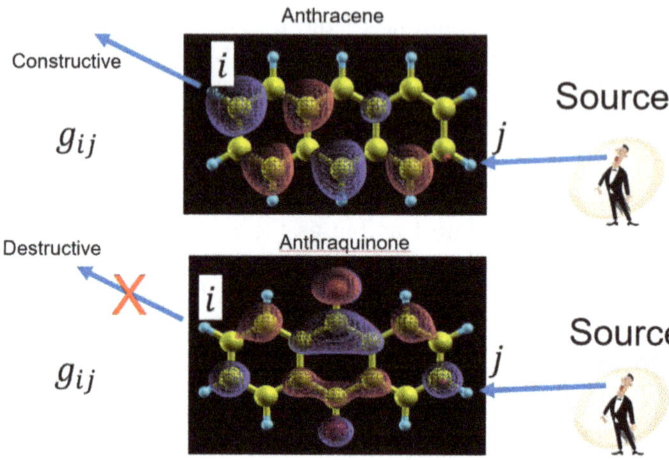

Figure 1.7. de Broglie wave patterns $g(r, r')$ created by mid-gap electrons injected into anthracene (upper panel) and anthraquinone (lower panel) molecules. When the source injects electrons into the atom labelled j, the resulting amplitude on atom i is $g(r_i, r_j)$ and to simplify the notation is denoted g_{ij}. (This figure was created by Songjun Hou and Qingqing Wu, within the Lancaster Theory of Molecular-scale Transport group.)

When an electron is injected into an atom labelled j, (which is attached to the source electrode) then to simplify the notation, the amplitude $g(r_i, r_j)$ of the wave pattern on some other atom i (attached to the drain electrode) will be denoted g_{ij}. If g_{ij} is large, as in the upper panel of figure 1.7, then the transmission coefficient will be large and the molecule exhibits CQI, whereas if g_{ij} is small, the transmission coefficient will be small and the molecule exhibits DQI, as in the lower panel of figure 1.7. In fact, as discussed in chapter 2, if the molecule only weakly couples to the electrodes, then the transmission coefficient is proportional to $|g_{ij}|^2$. Therefore if the corresponding transmission coefficient is denoted $T_{ij}(E_F)$ and the corresponding electrical conductance is $\sigma_{ij} = G_0 T_{ij}(E_F)$, one finds

$$\sigma_{ij} = A\,|g_{ij}|^2 \tag{1.12}$$

where A is a constant, which does not depend on the labels i, j.

1.9 Magic ratio theory

As discussed in chapter 2, when the source and drain are weakly coupled to the molecule, these wave patterns are a property of the isolated molecule alone and for the simplest case of molecular cores formed from alternant polyaromatic hydrocarbons, such as the pyrene cores in figure 1.8, the amplitudes g_{ij} are proportional to 'magic numbers' denoted M_{ij}. More precisely, the magic numbers are defined by $M_{ij} = Bg_{ij}$, where B is a constant of proportionality, which is independent of the connectivity i, j. From the magic number table of pyrene, the magic number corresponding to 5,1′ connectivity of the blue pyrene core is $M_{51'} = -1$ and the

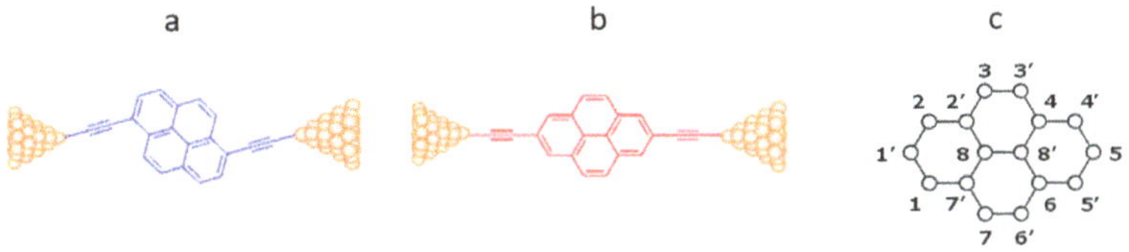

Figure 1.8. (a) and (b): Pyrene cores formed from four 6-membered rings, with different connectivities to electrodes. (c) An example of a numbering system used to label atoms. The triple bonds act as weak couplings to the electrodes. (This figure is reprinted with permission from [36]. Copyright (2015) American Chemical Society.)

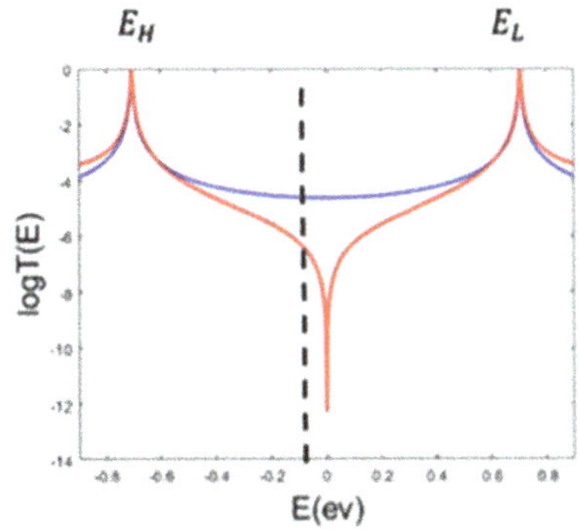

Figure 1.9. The red curve shows $T(E)$ versus E for a connectivity corresponding to DQI and an anti-resonance at the mid-gap. The blue curve shows $T(E)$ versus E for a connectivity corresponding to CQI. The dashed line indicates a possible location for the Fermi energy. E_{H} and E_{L} are the energies of the HOMO and LUMO resonances respectively.

magic number corresponding to 14′ connectivity of the red pyrene core is $M_{14'} = -3$. Consequently from equation (1.12), the ratio of their conductances is predicted to be

$$\frac{G_{14'}}{G_{51'}} = \frac{(-3)^2}{(-1)^2} = 9$$

This demonstrates that ratios of conductances corresponding to different connectivities are a reflection of the differences between the wave function patterns created by mid-gap electrons injected at different source points. This is good news from the point of view of a synthetic chemist seeking structure–function relationships, because apart from E_{F} (which enters equation (1.12)) properties of the electrodes do not appear in such ratios.

For the first time, an extension of magic ratio theory to intra-molecular bond currents is presented in chapter 8.

1.10 Controlling quantum interference

The ability to locate instances of DQI and to control their energetic location is useful, because as shown by the red curve in figure 1.9, DQI creates anti-resonances in the transmission coefficient $T(E)$, leading to steep slopes, whereas as shown by the blue curve, connectivities leading to mid-gap CQI lead to rather featureless transmission coefficients, with smaller slopes near the mid-gap. As noted above, if the Fermi energy is located at the energy indicated by the dashed line in figure 1.9, the Seebeck coefficient will be high, due to the steep slope of $T(E_F)$. Furthermore the electrical conductance will be sensitive to the location of the DQI feature, which means that connectivities leading to DQI could be useful for switching and sensing. As mentioned above, when a molecule is placed between source and drain electrodes, charge transfer on or off the molecule usually leads to the Fermi energy being located somewhere within the HOMO–LUMO gap, which is typically the region of energy where anti-resonances are located.

In view of the importance of controlling quantum interference effects, chemists have designed and synthesised families of molecules with a variety of structures and connectivities. From a conceptual viewpoint, these take the form shown in figure 1.10. Specific examples can be found in the many articles cited at the end of this chapter. The theoretical tools described in this book will allow the reader to deconstruct such molecules conceptually and thereby understand how their overall quantum interference properties arise from their component parts. The main features of these structures can be understood using 'connectivity theory,' which describes how a complex structure can be assembled from smaller substructures and how wave functions and interference patterns arise from the interactions between the substructures. An intuitive introduction to connectivity theory is presented in chapter 2. Of course to be able to utilise these ideas, the theoretical tools described in chapters 3 to 17 are needed. After reading these later chapters, the reader should

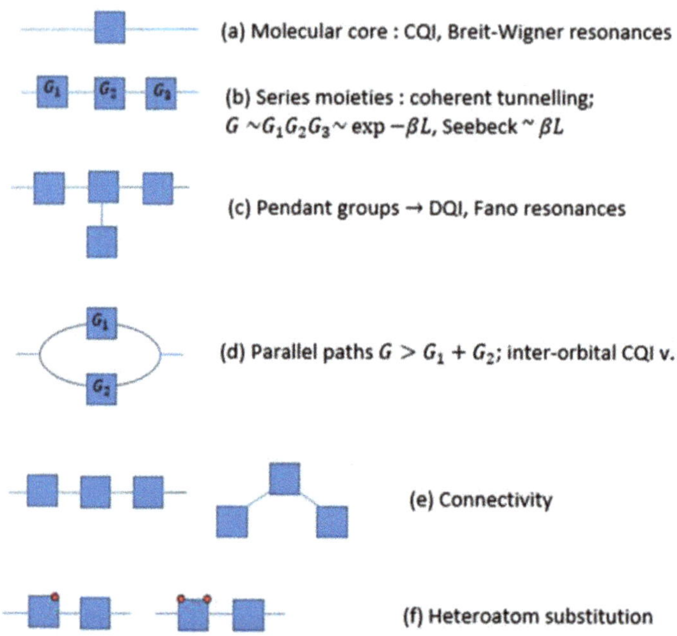

Figure 1.10. Molecular architectures used to control molecular-scale quantum interference, along with indicators of key concepts used to interpret their transport properties.

have acquired the concepts and research-level skills needed to model phase-coherent electron and phonon transport through molecules and nanostructures.

1.11 Summary

In summary, the primary aim of this book is to present the underlying concepts and mathematics, which underpin theories of quantum transport through phase-coherent structures. What sets this book apart from other texts, is the focus on electron transport from the very beginning, starting with 'noisy neighbour equations' (NNEs) in chapter 2, which contain many concepts of quantum interference and provides an easy introduction to Green's functions and quantum transport for the novice reader. The intended reader is a final-year undergraduate, a Master's level student or a first-year PhD student, who would like to see detailed derivations of all equations. Although it is assumed that the reader has an elementary knowledge of quantum mechanics and a rudimentary understanding of matrices, introductory chapters describing both are included. At each stage, the aim is to develop intuitive pictures, based on rigorous mathematics and exact solutions of model systems. It is hoped that the reader will complete the book with an intuitive understanding of scattering theory and Green's functions, and their applications to molecular electronics and phase-coherent transport in molecular-scale, nanoscale and mesoscopic systems. While developing this understanding, the formulae presented above and their generalisations will be derived from first principles and used to illustrate various mechanisms for controlling transport properties of molecules. The concepts outlined in this book form the foundations of quantum transport codes such as SMEAGOL, TransSIESTA and GOLLUM, which enable material-specific studies of transport properties, by interfacing the scattering theory described here, with mean-field Hamiltonians obtained from density functional theory.

For those readers new to the field, the following (non-comprehensive) list of references shows a selection of papers addressing room-temperature quantum interference effects in molecules. The possibility of using DQI to control electron transport through single molecules was first proposed in [5] where the role of Fano resonances due to cross-conjugated oxygens and pendant bipyridyl units was highlighted. This was followed by a joint experimental and theoretical study [6] of molecules, which possess different pendant substituents, including a cross-conjugated oxygen and a theoretical prediction that quantum interference could be used to enhance the thermoelectric performance of single molecules [7]. Interesting theoretical discussions were also presented in [8–13]. The first joint experimental and theoretical study, designed to find evidence of DQI associated with cross-conjugated pendant oxygens in anthraquinones, was presented in [14] and rapidly followed by related experiments in [15–18]. In the same year, an early prediction regarding CQI in molecules with parallel paths [19] was investigated experimentally [20]. Following these early investigations, during the past decade, there has been an outpouring of work aimed at probing and controlling QI in single molecules, a selection of which can be found in [21–107]. In particular, a growing number of studies have identified quantum interference effects in the thermoelectric properties of single molecules [108–145]. More recently, it has been demonstrated that these single-molecule QI effects can be translated into self-assembled monolayers [146–158], thereby creating two-dimensional materials, whose electronic and thermoelectric properties are controlled by room-temperature quantum interference. Interest in molecular-scale thermoelectricity has also stimulated studies of thermal transport through single molecules [159–171], where room-temperature phonon interference may provide a route to suppressing thermal conductance and increasing the thermoelectric performance of molecular-scale devices and materials. Single-molecule thermoelectricity and phonon transport are discussed in chapter 7. A very recent demonstration of room-temperature QI effects in single molecules is a tendency for the electrical conductance of certain conjugated molecules to be length independent or even increase with length [172–176]. This quantum effect is discussed in chapter 12. Of course single molecule junctions exhibit a wide range of phenomena besides quantum interference. For a broad overview of the many issues associated with the development of single-molecule electronic devices, the following review articles are recommended [177–190].

References

[1] Haiss W *et al* 2006 Precision control of single-molecule electrical junctions *Nat. Mater.* **5** 995

[2] Landauer R 1957 Spatial variation of currents and fields due to localized scatterers in metallic conduction *IBM J. Res. Dev.* **1** 223

[3] Büttiker M, Imry Y, Landauer R and Pinhas S Generalized many-channel conductance formula with application to small rings *Phys. Rev.* B **31** 6207

[4] Breit G and Wigner E 1936 Capture of slow neutrons *Phys. Rev.* **49** 519

[5] Papadopoulos T A, Grace I M and Lambert C J 2006 Control of electron transport through Fano resonances in molecular wires *Phys. Rev.* B **74** 193306

[6] Wang C, Bryce M R, Gigon J, Ashwell G J, Grace I and Lambert C J 2008 Synthesis and properties of functionalized 4 nm scale molecular wires with thiolated termini for self-assembly onto metal surfaces *J. Org. Chem.* **73** 4810

[7] Finch C M, García-Suárez V M and Lambert C J 2009 Giant thermopower and figure of merit in single-molecule devices *Phys. Rev.* B **79** 033405

[8] Bergfield J P, Solis M and Stafford C A 2010 Giant thermoelectric effect from transmission supernodes *ACS Nano.* **4** 5314

[9] Ricks A B, Solomon G C, Colvin M T, Scott A M, Chen K, Ratner M A and Wasielewski M R 2010 Controlling electron transfer in donor–bridge–acceptor molecules using cross-conjugated bridges *J. Am. Chem. Soc.* **132** 15427

[10] Markussen T, Schiötz J and Thygesen K S 2010 Electrochemical control of quantum interference in anthraquinone-based molecular switches *J. Chem. Phys.* **132** 224104

[11] Li X, Staykov A and Yoshizawa K 2010 Orbital views of the electron transport through polycyclic aromatic hydrocarbons with different molecular sizes and edge type structures *J. Phys. Chem.* C **114** 9997

[12] Sparks R E, García-Suárez V M and Lambert C J 2011 Quantum interference in single molecule electronic systems *Phys. Rev.* B **83** 075437

[13] Solomon G C, Bergfield J P, Stafford C A and Ratner M A 2011 When 'small' terms matter: coupled interference features in the transport properties of cross-conjugated molecules *Beilstein J. Nanotechnol.* **2** 862
[14] Hong W, Valkenier H, Mészáros G, Manrique D Z, Mishchenko A, Putz A, García P M, Lambert C J, Hummelen J C and Wandlowski T 2011 *Beilstein J. Nanotechnol.* **2** 699
[15] Sedghi G, Esdaile L J, Anderson H L, García-Suárez V M, Lambert C J, Martin S, Bethell D, Higgins S J and Nichols R J 2011 Long-range electron tunnelling in oligo-porphyrin molecular wires *Nat. Nano.* **6** 517
[16] Ballmann S, Härtle R, Coto P B, Elbing M, Mayor M, Bryce M R, Thoss M and Weber H B 2012 Experimental evidence for quantum interference and vibrationally induced decoherence in single-molecule junctions *Phys. Rev. Lett.* **109** 056801
[17] Kaliginedi V, Moreno-García P, Valkenier H, Hong W, García-Suárez V M, Buiter P, Otten J L H, Hummelen J C, Lambert C J and Wandlowski T 2012 Correlations between molecular structure and single-junction conductance: a case study with oligo(phenylene-ethynylene)-type wires *J. Am. Chem. Soc.* **134** 5262
[18] Guedon C M, Valkenier H, Markussen T, Thygesen K S, Hummelen J C and van der Molen S J 2012 *Nat. Nano.* **7** 304
[19] Magoga M and Joachim C 1999 Conductance of molecular wires connected or bonded in parallel *Phys. Rev.* B **59** 1601
[20] Vazquez H, Skouta R, Schneebeli S, Kamenetska M, Breslow R, Venkataraman L and Hybertsen M S 2012 Probing the superposition law *Nat. Nano.* **7** 663
[21] Hong W, Manrique D Z, Moreno-Garcia P, Gulcur M, Mishchenko A, Lambert C J, Bryce M R and Wandlowski T 2012 Single molecular conductance of tolanes: experimental and theoretical study on the junction evolution dependent on the anchoring group *J. Am. Chem. Soc.* **134** 2292
[22] Yoshizawa K 2012 An orbital rule for electron transport in molecules *Acc. Chem. Res.* **45** 1612–21
[23] Arroyo C R, Tarkuc S, Frisenda R, Seldenthuis J S, Woerde C H M, Eelkema R, Grozema F C and van der Zant H S J 2013 Signatures of quantum interference effects on charge transport through a single benzene ring *Angew. Chem.* **125** 3234
[24] García-Suárez V M, Ferradás R and Ferrer J 2013 Impact of Fano and Breit–Wigner resonances in the thermoelectric properties of nanoscale junctions *Phys. Rev.* B **88** 235417
[25] Ferradás R and García-Suárez V M 2013 Symmetry-induced quantum interference effects in metalloporphyrin wires *J. Phys.: Condens. Matt.* **25** 325501
[26] Arroyo C R, Risenda R, Moth-Poulsen K, Seldenthuis J S, Bjørnholm T and van der Zant H S J 2013 Quantum interference effects at room temperature in OPV-based single-molecule junctions *Nanoscale Res. Lett.* **8** 234
[27] Arroyo C R *et al* 2013 Signatures of quantum interference effects on charge transport through a single benzene ring *Angew. Chem. Int. Ed.* **52** 3152
[28] Tsuji Y and Hoffmann R 2014 Frontier orbital control of molecular conductance and its switching *Angew. Chem.* **126** 4177
[29] Chen W, Li H, Widawsky J R, Appayee C, Venkataraman L and Breslow R 2014 Aromaticity decreases single-molecule junction conductance *J. Am. Chem. Soc.* **136** 918
[30] Lykkebo J, Gagliardi A, Pecchia A and Solomon G C 2014 IETS and quantum interference: propensity rules in the presence of an interference feature *J. Chem. Phys.* **141** 124119

[31] Markussen T and Thygesen K S 2014 Temperature effects on quantum interference in molecular junctions *Phys. Rev.* B **89** 085420

[32] Valkenier H, Guédon C M, Markussen T, Thygesen K S, van der Molen S J and Hummelen J C 2014 Cross-conjugation and quantum interference: a general correlation? *Phys. Chem. Chem. Phys.* **16** 653

[33] Sun L, Diaz-Fernandez Y A, Gschneidtner T A, Westerlund F, Lara-Avila S and Moth-Poulsen K 2014 Single-molecule electronics: from chemical design to functional devices *Chem. Soc. Rev.* **43** 7378

[34] Batra A, Meisner J S, Darancet P, Chen Q, Steigerwald M L, Nuckolls C and Venkataraman L 2014 Molecular diodes enabled by quantum interference *Faraday Discuss.* **174** 79

[35] Pedersen K G L *et al* 2014 Quantum interference in off-resonant transport through single molecules *Phys. Rev.* B **90** 125413

[36] Sangtarash S, Huang C, Sadeghi H, Sorohhov G, Hauser J, Wandlowski T, Hong W, Decurtins S, Liu S-X and Lambert C J 2015 Searching the hearts of graphene-like molecules for simplicity, sensitivity, and logic *J. Am. Chem. Soc.* **137** 11425

[37] Geng Y, Sangtarash S, Huang C, Sadeghi H, Fu Y, Hong W, Wandlowski T, Decurtins S, Lambert C J and Liu S-X 2015 Magic ratios for connectivity-driven electrical conductance of graphene-like molecules *J. Am. Chem. Soc.* **137** 4469

[38] Solomon G C 2015 Interfering with interference *Nat. Chem.* **7** 621

[39] Manrique D Z *et al* 2015 A quantum circuit rule for interference effects in single-molecule electrical junctions *Nat. Commun.* **6** 6389

[40] Berritta M, Manrique D Z and Lambert C J 2015 Interplay between quantum interference and conformational fluctuations in single-molecule break junctions *Nanoscale* **7** 1096

[41] Vacek J, Chocholoušová J V, Stará I G, Starý I and Dubi Y 2015 Mechanical tuning of conductance and thermopower in helicene molecular junctions *Nanoscale* **7** 8793

[42] Miguel D *et al* 2015 Toward multiple conductance pathways with heterocycle-based oligo (phenyleneethynylene) derivatives *J. Am. Chem. Soc.* **137** 13818

[43] Tada T and Yoshizawa K 2015 Molecular design of electron transport with orbital rule: toward conductance-decay free molecular junctions *Phys. Chem. Chem. Phys.* **17** 32099

[44] Pedersen K G L, Borges A, Hedegård P, Solomon G C and Strange M 2015 Illusory connection between cross-conjugation and quantum interference *J. Phys. Chem.* C **119** 26919

[45] Sadeghi H, Mol J, Lau C, Briggs G A D, Warner J and Lambert C J 2015 Conductance enlargement in pico-scale electro-burnt graphene nanojunctions *Proc. Natl. Acad. Sci.* **9** 2658–63

[46] Vezzoli A *et al* 2015 Gating of single molecule junction conductance by charge transfer complex formation *Nanoscale* **7** 18949

[47] Tsuji Y, Hoffmann R, Strange M and Solomon G C 2016 Close relation between quantum interference in molecular conductance and diradical existence *Proc. Natl. Acad. Sci.* **113** E413

[48] Sangtarash S, Sadeghi H and Lambert C J 2016 Exploring quantum interference in heteroatom-substituted graphene-like molecules *Nanoscale* **8** 13199

[49] Su T A *et al* 2016 Tuning conductance in π–σ–π single-molecule wires *J. Am. Chem. Soc.* **138** 7791

[50] Borges A, Fung E-D, Ng F, Venkataraman L and Solomon G C 2016 Probing the conductance of the σ-system of bipyridine using destructive interference *J. Phys. Chem. Lett.* **7** 4825
[51] Prasongkit J and Rocha A R 2016 Quantum interference effects in biphenyl dithiol for gas detection *RSC Adv.* **6** 59299
[52] Garner M H, Solomon G C and Strange M 2016 Tuning conductance in aromatic molecules: constructive and counteractive substituent effects *J. Phys. Chem.* C **120** 9097
[53] Nozaki D and Toher C 2017 Is the antiresonance in meta-contacted benzene due to the destructive superposition of waves traveling two different routes around the benzene ring? *J. Phys. Chem.* C **121** 11739
[54] Borges A and Solomon G C 2017 Effects of aromaticity and connectivity on the conductance of five-membered rings *J. Phys. Chem.* C **121** 8272
[55] Tsuji Y and Yoshizawa K 2017 Frontier orbital perspective for quantum interference in alternant and nonalternant hydrocarbons *J. Phys. Chem.* C **121** 9621
[56] Yin X *et al* 2017 A reversible single-molecule switch based on activated antiaromaticity *Sci. Adv.* **3** eaao2615
[57] Borges A, Xia J, Liu S H, Venkataraman L and Solomon G C 2017 The role of through-space interactions in modulating constructive and destructive interference effects in benzene *Nano Lett.* **17** 4436
[58] Klöckner J C, Cuevas J C and Pauly F 2017 Tuning the thermal conductance of molecular junctions with interference effects *Phys. Rev.* B **96** 245419
[59] Liu X, Sangtarash S, Reber D, Sadeghi H, Hong W, Lambert C J and Liu S-X 2017 Gating of quantum interference in molecular junctions by heteroatom substitution, *Angew. Chem.* **56** 173
[60] Jingxian J Y, Horsley J R and Abell A D 2017 Exploiting the interplay of quantum interference and backbone rigidity on electronic transport in peptides: a step towards bio-inspired quantum interferometers *Mol. Syst. Design Eng.* **2** 67
[61] Ismael A K, Grace I M and Lambert C J 2017 Connectivity dependence of Fano resonances in single molecules *Phys. Chem. Chem. Phys.* **19** 6416
[62] Richert S *et al* 2017 Constructive quantum interference in a bis-copper six-porphyrin nanoring *Mat. Comms.* **8** 14842
[63] Gantenbein M, Wang L, Al-Jobory A A, Ismael A K, Lambert C J, Hong W and Bryce M R 2017 Quantum interference and heteroaromaticity of para- and meta-linked bridged biphenyl units in single molecular conductance measurements *Sci. Rep.* **7** 1794 7
[64] Zhao X, Kastlunger G and Stadler R 2017 Quantum interference in coherent tunneling through branched molecular junctions containing ferrocene centers *Phys. Rev.* B **96** 085421
[65] Burkle M *et al* 2017 The orbital selection rule for molecular conductance as manifested in tetraphenyl-based molecular junctions *J. Am. Chem. Soc.* **139** 2989
[66] Ismael A K *et al* 2017 Side group-mediated mechanical conductance switching in molecular junctions, *Angew. Chem.* **129** 15580
[67] Yang Yang *et al* 2018 Heteroatom-induced molecular asymmetry tunes quantum interference in charge transport through single-molecule junctions *J. Phys. Chem.* C **122** 14965
[68] Garner M H *et al* 2018 Comprehensive suppression of single-molecule conductance using destructive σ-interference *Nature* **558** 415

[69] Yang G, Wu H, Wei J, Zheng J, Chen Z, Liu J, Shi J, Yang Y and Hong W 2018 Quantum interference effect in the charge transport through single-molecule benzene dithiol junction at room temperature: an experimental investigation *Chin. Chem. Lett.* **29** 147

[70] Carlotti M *et al* 2018 Systematic experimental study of quantum interference effects in anthraquinoid molecular wires *Nanoscale Adv.* **1** 2018

[71] Huang B *et al* 2018 Controlling and observing sharp-valleyed quantum interference effect in single molecular junctions *J. Am. Chem. Soc.* **140** 17685

[72] Garner M H *et al* 2018 Comprehensive suppression of single-molecule conductance using destructive σ-interference *Nature* **558** 415

[73] Frisenda R, Stefani D and van der Zant H S J 2018 Quantum transport through a single conjugated rigid molecule, a mechanical break junction study *Acc. Chem. Res.* **51** 1359

[74] Zhang Y *et al* 2018 Distinguishing diketopyrrolopyrrole isomers in single-molecule junctions via reversible stimuli-responsive quantum interference *J. Am. Chem. Soc.* **140** 6531

[75] Li X, Tan Z, Huang X, Bai J, Liu J and Hong W 2019 Experimental investigation of quantum interference in charge transport through molecular architectures *J. Mat. Chem.* C **7** 12790

[76] Tang J-H *et al* 2019 Single-molecule level control of host-guest interactions in metallocycle-C60 complexes *Nat. Comms.* **10** 4599

[77] Gantenbein M *et al* 2019 Exploring antiaromaticity in single-molecule junctions formed from biphenylene derivatives *Nanoscale* **11** 20659

[78] Li Y *et al* 2019 Gate controlling of quantum interference and direct observation of anti-resonances in single molecule charge transport *Nat. Mater.* **18** 357

[79] Alanazy A *et al* 2019 Cross conjugation increases the conductance of meta-connected fluorenones. *Nanoscale* **11** 13720

[80] Tan Z *et al* 2019 Atomically defined angstrom-scale all-carbon junctions *Nat. Comms.* **10** 1748

[81] Li D, Banerjee R, Mondal S, Maliyov I, Romanova M, Dappe Y J and Smogunov A 2019 Symmetry aspects of spin filtering in molecular junctions: hybridization and quantum interference effects *Phys. Rev.* B **99** 115403

[82] Mijbil Z Y 2019 Quantum interference independence of the heteroatom position *Chem. Phys. Lett.* **716** 69

[83] Li S, Yu H, Schwieter K, Chen K, Li B, Liu Y, Moore J S and Schroeder C M 2019 Charge transport and quantum interference effects in oxazole-terminated conjugated oligomers *J. Am. Chem. Soc.* **141** 16079

[84] Li Y, Yu X, Zhen Y, Dong H and Hu W 2019 Two-pathway viewpoint to interpret quantum interference in molecules containing five-membered heterocycles: thienoacenes as examples *J. Phys. Chem.* C **123** 15977

[85] Liu J, Huang X, Wang F and Hong W 2019 Quantum interference effects in charge transport through single-molecule junctions: detection, manipulation, and application *Acc. Chem. Res.* **52** 151

[86] Rakyta P *et al* 2019 Magic number theory of superconducting proximity effects and Wigner delay times in graphene-like molecules *J. Phys. Chem.* C **123** 6812

[87] Ulčakar L *et al* 2019 On the resilience of magic number theory for conductance ratios of aromatic molecules *Sci. Rep.* **9** 3478

[88] Zheng H *et al* 2019 Room-temperature quantum interference in single perovskite quantum dot junctions. *Nat. Commun.* **10** 5458
[89] Bai J *et al* 2019 Anti-resonance features of destructive quantum interference in single-molecule thiophene junctions achieved by electrochemical gating *Nat. Mater.* **18** 364
[90] Ismael A K and Lambert C J 2019 Single-molecule conductance oscillations in alkane rings *J. Mat. Chem.* C **7** 6578
[91] Jiang F *et al* 2019 Turning the tap: conformational control of quantum interference to modulate single molecule conductance *Angew. Chem.* **131** 19163
[92] Naghibi S *et al* 2019 Synthetic control of quantum interference by regulating charge on a single atom in heteroaromatic molecular junctions *J. Phys. Chem. Lett.* **10** 6419
[93] Alanazy A *et al* 2019 Cross conjugation increases the conductance of meta-connected fluorenones *Nanoscale* **11** 13720
[94] Gryn'ova G and Corminboeuf C 2019 Topology-driven single-molecule conductance of carbon nanothreads *J. Phys. Chem. Lett.* **10** 825
[95] Wang K *et al* 2019 Charge transfer complexation boosts molecular conductance through Fermi level pinning *Chem. Sci.* **10** 2396
[96] Liu J *et al* 2019 Quantum interference effects in charge transport through single-molecule junctions: detection, manipulation, and application *Acc. Chem. Res.* **52** 151
[97] Ramakrishnan R 2020 Charge-transfer selectivity and quantum interference in real-time electron dynamics: gaining insights from time-dependent configuration interaction simulations *J. Chem. Phys.* **152** 194111
[98] Garner M H and Solomon G C 2020 Simultaneous suppression of π-and σ-transmission in π-conjugated molecules *J. Phys. Chem. Lett.* **11** 7400
[99] Wang K, Meyhofer E and Reddy P 2020 Thermal and thermoelectric properties of molecular junctions *Adv. Funct. Mat.* **30** 1904534
[100] Garner M H and Solomon G C 2020 Simultaneous suppression of π- and σ-transmission in π-conjugated molecules. *J. Phys. Chem. Lett.* **11** 7400–6
[101] Okazawa K, Yuta Y T and Yoshizawa K 2020 Understanding single-molecule parallel circuits on the basis of frontier orbital theory *J. Phys. Chem.* C **124** 3322
[102] Baghernejad M *et al* 2020 Constructive quantum interference in single-molecule benzodichalcogenophene junctions *Chem. Eur. J.* **26** 5264
[103] Chen H L *et al* 2020 Giant conductance enhancement of intramolecular circuits through interchannel gating *Matter* **2** 378–89
[104] Tang C *et al* 2020 Electric-field-induced connectivity switching in single-molecule junctions *Iscience* **23** 100770
[105] Chen L *et al* 2020 Non-additive transport in multi-channel single-molecule circuits *Small* **16** 02808
[106] Camarasa-Gómez M *et al* 2020 Mechanically-tunable quantum interference in ferrocene-based single-molecule junctions *Nano Lett.* **20** 6381
[107] Gunasekaran S, Greenwald J E and Venkataraman L 2020 Visualizing quantum interference in molecular junctions *Nano Lett.* **20** 2843
[108] Pauly F, Viljas J K and Cuevas J C 2008 Length-dependent conductance and thermopower in single-molecule junctions of dithiolated oligophenylene derivatives: a density functional study *Phys. Rev.* B **78** 035315

[109] Nozaki D, Sevinçli H, Li W, Gutiérrez R and Cuniberti G 2010 Engineering the figure of merit and thermopower in single-molecule devices connected to semiconducting electrodes *Phy. Rev.* B **81** 235406
[110] Bürkle M *et al* 2012 *Ab initio* study of the thermopower of biphenyl-based single-molecule junctions *Phys. Rev.* B **86** 115304
[111] Evangeli C, Gillemot K, Leary E, González M T, Rubio-Bollinger G, Lambert C J and Agraït N 2013 Engineering the thermopower of C60 molecular junctions *Nano Lett.* **13** 2141
[112] Guo S, Zhou G and Tao N 2013 Single molecule conductance, thermopower, and transition voltage *Nano Lett.* **13** 4326
[113] Karlström O, Strange M and Solomon G C 2014 Understanding the length dependence of molecular junction thermopower *J. Chem. Phys.* **140** 044315
[114] Péterfalvi C G, Grace I, Manrique D Z and Lambert C J 2014 Thermoelectric performance of various benzo-difuran wires *J. Chem. Phys.* **140** 174711
[115] García-Suárez V M, Lambert C J, Manrique D Z and Wandlowski T 2014 Redox control of thermopower and figure of merit in phase-coherent molecular wires *Nanotechnology* **25** 205402
[116] Ismael A K, Grace I and Lambert C J 2015 Increasing the thermopower of crown-ether-bridged anthraquinones *Nanoscale* **7** 17338
[117] Lambert C J, Sadeghi H and Sangtarash S 2015 Oligoyne molecular junctions for efficient room temperature thermoelectric power generation *Nano Lett.* **15** 7467
[118] Bürkle M, Hellmuth T J, Pauly F and Asai Y 2015 First-principles calculation of the thermoelectric figure of merit for [2, 2] paracyclophane-based single-molecule junctions *Phys. Rev.* B **91** 165419
[119] Rincón-García L, Ismael A K, Evangeli C, Grace I, Rubio-Bollinger G, Porfyrakis K, Agraït N and Lambert C J 2016 Molecular design and control of fullerene-based bi-thermoelectric materials *Nat. Mater.* **15** 289
[120] Perroni C A, Ninno D and Cataudella V 2016 Thermoelectric efficiency of molecular junctions *J. Phys.: Condens. Matt.* **28** 373001
[121] Manrique D Z, Al-Galiby Q, Hong W and Lambert C J 2016 A new approach to materials discovery for electronic and thermoelectric properties of single-molecule junctions *Nano Lett.* **16** 1308
[122] Lambert C J, Sadeghi H and Al-Galiby Q H 2016 Quantum-interference-enhanced thermoelectricity in single molecules and molecular films *C. R. Phys.* **17** 1084
[123] Li Y, Xiang L, Palma J L, Asai Y and Tao N 2016 Thermoelectric effect and its dependence on molecular length and sequence in single DNA molecules *Nat. Comms.* **7** 11294
[124] Wu Q, Sadeghi H, Garcia-Suarez V, Ferrer J and Lambert C J 2017 Thermoelectricity in vertical graphene-C60-graphene architectures *Sci. Rep.* **7** 11680
[125] Gehring P *et al* 2017 Field-effect control of graphene-fullerene thermoelectric nanodevices *Nano Lett.* **17** 7055
[126] Klöckner J C, Siebler R, Cuevas J C and Pauly F 2017 Thermal conductance and thermoelectric figure of merit of -based single-molecule junctions: electrons, phonons, and photons *Phys. Rev.* B **95** 245404
[127] Noori M, Sadeghi H, Al-Galiby Q, Bailey S W D and Lambert C J 2017 High cross-plane thermoelectric performance of metallo-porphyrin molecular junctions *Phys. Chem. Chem. Phys.* **19** 17356

[128] Al-Galiby Q, Sadeghi H, Manrique D Z and Lambert C J 2017 Tuning the Seebeck coefficient of naphthalenediimide by electrochemical gating and doping *Nanoscale* **9** 4819–25
[129] Sangtarash S, Sadeghi H and Lambert C J 2018 Connectivity-driven bi-thermoelectricity in heteroatom-substituted molecular junctions *Phys. Chem. Chem. Phys.* **20** 9630–7
[130] Harzheim A 2018 Mater thermoelectricity in single-molecule devices *Sci. Tech.* **31** 1275
[131] Noori M, Sadeghi H and Lambert C J 2018 Stable-radicals increase the conductance and Seebeck coefficient of graphene nanoconstrictions *Nanoscale* **10** 19220
[132] Algharagholy L A A, Pope T and Lambert C J 2018 Strain-induced bi-thermoelectricity in tapered carbon nanotubes *J. Phys.: Condens. Matt.* **30** 105304
[133] Yzambart G, Rincón-García L, Al-Jobory A A, Ismael A K, Rubio-Bollinger G, Lambert C J, Agraït N and Bryce M R 2018 Thermoelectric properties of 2,7-dipyridylfluorene derivatives in single-molecule junctions *J. Phys. Chem.* C **122** 27198
[134] Al-Khaykanee M K, Grace I and Lambert C J 2018 Oscillating seebeck coefficient in π-stacked molecular junctions *RSC Adv.* **8** 24711
[135] Miao R *et al* 2018 Influence of quantum interference on the thermoelectric properties of molecular junctions *Nano Lett.* **18** 5666
[136] Cui L *et al* 2018 Peltier cooling in molecular junctions *Nat. Nanotechnol.* **13** 122
[137] Famili M, Grace I M, Al-Galiby Q, Sadeghi H and Lambert C J 2018 Towards high thermoelectric performance of thiophene and ethylenedioxythiophene (EDOT) molecular wires *Adv. Funct. Mat.* **28** 1703135
[138] Khalatbari H *et al* 2019 Effect of dopant nitrogen on the thermoelectric properties of C-20 and C-60 fullerene in graphene nanoribbon junction *Phys.* E **108** 372
[139] Hang Chen *et al* 2020 Exploring thermoelectric properties of oligophenylene-ethynylene derivatives *Nanoscale* **12** 15150
[140] Dekkiche H *et al* 2020 Electronic conductance and thermopower of single-molecule junctions of oligo (phenyleneethynylene) derivatives *Nanoscale* **12** 18908
[141] Ismael A K and Lambert C J 2020 Molecular-scale thermoelectricity: a worst-case scenario *Nanoscale Horiz.* **5** 1073
[142] Dekkiche H *et al* 2020 Electronic conductance and thermopower of single-molecule junctions of oligo (phenyleneethynylene) derivatives *Nanoscale* **12** 18908
[143] Grace I M, Olsen G, Hurtado-Gallego J, Rincón-García L, Rubio- Bollinger G, Bryce M R, Agrait N and Lambert C J 2020 Connectivity dependent thermopower of bridged biphenyl molecules in single-molecule junctions *Nanoscale* **12** 14682
[144] Ismael A K and Lambert C J 2020 Molecular-scale thermoelectricity: a worst-case scenario *Nanoscale Horiz.* **5** 1073
[145] Fracasso D, Valkenier H, Hummelen J C, Solomon G C and Chiechi R C 2011 Evidence for quantum interference in SAMs of arylethynylene thiolates in tunneling junctions with eutectic Ga–In (EGaIn) top-contacts *J. Am. Chem. Soc.* **133** 9556
[146] Kumar K S, Pasula R R, Lim S and Nijhuis C A 2016 Long-range tunneling processes across ferritin-based junctions *Adv. Mater.* **28** 1824
[147] Jia C *et al* 2018 Quantum interference mediated vertical molecular tunnel transistors *Sci. Adv.* **4** 8237
[148] Li B, Famili M, Pensa E, Grace I, Long N J, Lambert C J, Albrecht T and Cohen L F 2018 Cross-plane conductance through a graphene—molecular-monolayer—gold sandwich *Nanoscale* **10** 19791

[149] Zhang Y *et al* 2018 Tunneling probability increases with distance in junctions comprising self-assembled monolayers of oligothiophenes *J. Am. Chem. Soc.* **140** 15048
[150] Zhang Y *et al* 2018 Controlling destructive quantum interference in tunneling junctions comprising self-assembled monolayers via bond topology and functional groups *Chem. Sci.* **9** 4414
[151] Famili M *et al* 2019 Self-assembled molecular-electronic films controlled by room temperature quantum interference *Chem.* **5** 474
[152] Herrer L *et al* 2019 Single molecule vs large area design of molecular electronic devices incorporating an efficient 2-aminepyridine double anchoring group *Nanoscale* **11** 15871–80
[153] Wang X *et al* 2020 Scale-up of room-temperature constructive quantum interference from single molecules to self-assembled molecular-electronic films *J. Am. Chem. Soc.* **142** 8555
[154] Jia C *et al* 2020 Redox control of charge transport in vertical ferrocene molecular tunnel junctions *Chem.* **6** 968
[155] Soni S, Ye G, Zheng J, Zhang Y, Asyuda A, Zharnikov M, Hong W and Chiechi R C 2020 Understanding role of parallel pathways via *in situ* switching of quantum interference in molecular tunneling junctions *Angew. Chem. Int. Ed.* **59** 14308
[156] Shiqiang Zhao *et al* 2020 Cross-plane transport in a single two-dimensional van der Waals heterojunction *Sci. Adv.* **6** 6714
[157] Ismael A, Wanga X, Bennett T L R, Wilkinson L A, Robinson B J, Long N J, Cohen L F and Lambert C J 2020 Tuning the thermoelectrical properties of anthracene-based self-assembled monolayers *Chem. Sci.* **11** 6836
[158] Huang Z, Xu B, Chen Y, Ventra M D and Tao N 2006 Measurement of current-induced local heating in a single molecule junction *Nano Lett.* **6** 1240
[159] Lee W, Kim K, Jeong W, Zotti L A, Pauly F, Cuevas J C and Reddy P 2013 Heat dissipation in atomic-scale junctions *Nature* **498** 209
[160] Kiršanskas G, Li Q, Flensberg K, Solomon G C and Leijnse M 2014 Designing π-stacked molecular structures to control heat transport through molecular junctions *Appl. Phys. Lett.* **105** 233102
[161] Zotti L A *et al* 2014 Heat dissipation and its relation to thermopower in single-molecule junctions *New J. Phys.* **16** 015004
[162] Li Q, Strange M, Duchemin I, Donadio D and Solomon G C 2017 A strategy to suppress phonon transport in molecular junctions using π-stacked systems *J. Phys. Chem.* C **121** 7175
[163] Meier T, Menges F, Nirmalraj P, Hölscher H, Riel H and Gotsmann B 2014 Length-dependent thermal transport along molecular chains *Phys. Rev. Lett.* **113** 060801
[164] Han H *et al* 2016 Functionalization mediates heat transport in graphene nanoflakes *Nat. Commun.* **7** 11281
[165] Klöckner J C, Bürkle M, Cuevas J C and Pauly F 2016 Length dependence of the thermal conductance of alkane-based single-molecule junctions: an *ab initio* study *Phys. Rev.* B **94** 205425
[166] Klöckner J C, Siebler R, Cuevas J C and Pauly F 2017 Thermal conductance and thermoelectric figure of merit of -based single-molecule junctions: electrons, phonons, and photons *Phys. Rev.* B **95** 245404
[167] Famili M, Grace I, Sadeghi H and Lambert C J 2017 Suppression of phonon transport in molecular Christmas trees *Chem. Phys. Chem.* **18** 1234
[168] Cui L *et al* 2017 Quantized thermal transport in single-atom junctions *Science* **355** 1192

[169] Miao R *et al* 2018 Influence of quantum interference on the thermoelectric properties of molecular junctions *Nano Lett.* **18** 5666
[170] Mosso N, Sadeghi H, Gemma A, Sangtarash S, Lambert C and Gotsmann B 2019 Thermal transport through single molecule junctions *Nanoletters* **19** 7614
[171] Cui L *et al* 2019 Thermal conductance of single-molecule junctions *Nature* **572** 628
[172] Algethami N, Sadeghi H, Sangtarash S and Lambert C J 2018 The conductance of porphyrin-based molecular nanowires increases with length *Nano Lett.* **18** 4482
[173] Leary E *et al* 2018 Bias-driven conductance increase with length in porphyrin tapes *J. Am. Chem. Soc.* **140** 12877
[174] Gunasekaran S *et al* 2018 Near length-independent conductance in polymethine molecular wires *Nano Lett.* **18** 6387
[175] Xu W *et al* 2019 Unusual length-dependence of conductance in cumulene molecular wires *Angew. Chem.* **131** 8466
[176] Zang Y, Fu T, Zou Q, Ng F, Li H, Steigerwald M L, Nuckolls C and Venkataraman L 2020 Cumulene wires display increasing conductance with increasing length *Nano Lett.* **20** 8415
[177] van der Molen S J *et al* 2013 Visions for a molecular future *Nat. Nano* **8** 385
[178] Sun L, Diaz-Fernandeza Y A, Gschneidtner T A, Westerlund F, Lara-Avila S and Moth-Poulsen K R 2014 Single-molecule electronics: from chemical design to functional devices *Chem. Soc. Rev.* **43** 7378
[179] Lambert C J 2015 Basic concepts of quantum interference and electron transport in single-molecule electronics *Chem. Soc. Rev.* **44** 875
[180] Perrin M L, Burzurí E and van der Zant H S J 2015 Single-molecule transistors *Chem. Soc. Rev.* **44** 902
[181] Guldi D M, Nishihara H and Venkataraman L 2015 Molecular wires *Chem. Soc. Rev.* **44** 842
[182] Perroni C A, Ninno D and Cataudella V 2016 Thermoelectric efficiency of molecular of junctions *J. Phys.: Condens. Matter* **28** 373001
[183] Rincon-Garcia L, Evangeli C, Rubio-Bollinger G and Agraıt N 2016 Thermopower measurements in molecular junctions *Chem. Soc. Rev.* **45** 4285
[184] Su T A, Neupane M, Steigerwald M I L, Venkataraman L and Nuckolls C 2016 Chemical principles of single-molecule electronics *Nat. Rev.* **1** 16002
[185] Xiang D, Wang X, Jia C, Lee T and Guo X 2016 Molecular-scale electronics: from concept to function *Chem. Rev.* **116** 4318
[186] Wang K and Xu B 2017 Modulation and control of charge transport through single-molecule junctions *Top. Curr. Chem.* **375** 17
[187] Cui L, Miao R, Jiang C, Meyhofer E and Reddy P 2017 Perspective: thermal and thermoelectric transport in molecular junctions *J. Chem. Phys.* **146** 092201
[188] Thoss P M and Evers F 2018 Perspective: theory of quantum transport in molecular junctions *J. Chem. Phys.* **148** 030901
[189] Gehring P, Thijssen J M and van der Zant H S J 2019 Single-molecule quantum-transport phenomena in break junctions *Nat. Rev. Phys.* **1** 381
[190] Evers F, Korytár R, Tewari S and van Ruitenbeek J M 2020 Advances and challenges in single-molecule electron transport *Rev. Mod. Phys.* **92** 035001

Chapter 2

Connectivity theory and noisy neighbour equations for quantum transport

2.1 Introduction

The aim of this chapter is to put meat on the bones of the ideas introduced in chapter 1, by presenting a rather intuitive 'Connectivity Theory' of quantum interference. Connectivity theory is a generic theory, which describes wave-carrying structures, created by joining together smaller substructures. It is based the recognition that 'a wave is a wave is a wave' and that interference phenomena do not derive from any particular wave equation. Instead they are generic properties of waves and are a consequence of the principle of superposition. For this reason, connectivity theory can be used to describe interference phenomena in de Broglie waves, electromagnetic waves, sound waves, spin waves, water waves and many others.

The material in this chapter would normally be presented towards the end of an advanced course about scattering theory and can only be appreciated fully, after reading chapters 3–17. It is presented here to show that the fundamental equations used to describe quantum interference can be obtained without using complicated mathematics and arise from basic properties of waves. This chapter is intended as an introduction to the basic equations used to describe interference phenomena and as a revision chapter for those students reading the later technical chapters. It aims to introduce wave propagation and interference from an intuitive viewpoint, but the novice reader may find it bewildering and may prefer to move directly to chapter 3 and return to this material after reaching chapter 13. Concepts will be illustrated using vibrating structures and de Broglie waves as examples. In section 2.6, fundamental and extremely versatile 'noisy neighbour equations' (NNEs) will be derived, which capture almost every idea discussed in chapter 1. An introduction to NNEs can by found in the following YouTube video: https://www.youtube.com/watch?v=OObmMRApbnk. In sections 2.8–2.10, their use will be illustrated using 'magic ratio theory'. Magic ratio

doi:10.1088/978-0-7503-3639-0ch2

theory is a simple example of connectivity theory and describes changes to transport properties of molecules, when the connectivity of a molecular core to electrodes is changed with atomic precision. The ability to achieve this by chemical synthesis makes molecular electronics distinct from more physics-oriented studies of artificial quantum dots, where such a level of control is not possible. In chapter 8, an extension of magic ratio theory to intra-molecular bond currents is presented.

2.2 Green's functions for beginners

In chapter 1, the idea of producing a wave from a source was introduced and it was noted that the resulting wave is known as a Green's function. To make this idea more precise, consider how the information about the snapshot of the wave in figure 1.6 could be stored on a computer. One possibility would be to sample the height of the wave $g(r, r')$ at each point r and then store the resulting measurements in a table. Before doing this, consider a much simple problem of two masses connected by a spring, which are constrained to vibrate in the vertical direction only, as shown in figure 2.1. Initially, they are at rest, in equilibrium positions, corresponding to no stretching or compression of the spring. Imagine a tuning fork vibrating with a single frequency ω, which interacts only with mass 1 and has the effect of applying a driving force to mass 1. This tuning-fork 'source' is a source of energy that will cause both masses to vibrate and after some time, the pair of masses will become synchronised with the source and vibrate with the same frequency ω.

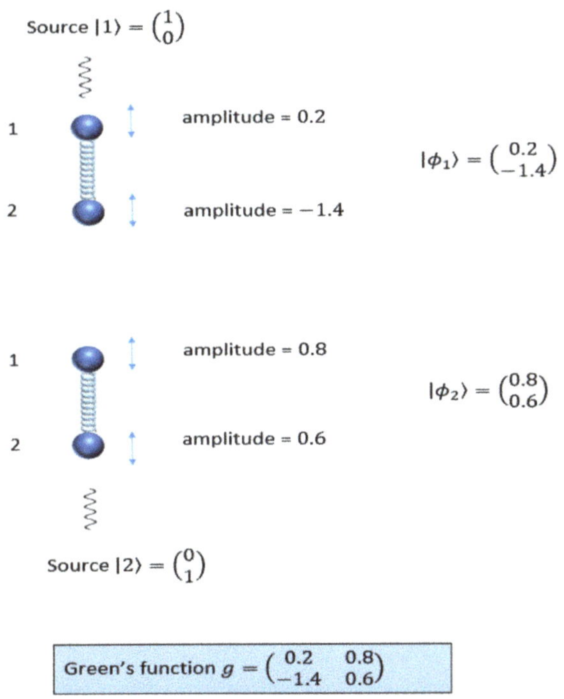

Figure 2.1. Two masses connected by a spring, which are constrained to vibrate in the vertical direction only.

In what follows, it is assumed that the tuning fork is 'controlled', such that the driving force it imposes is not affected by the motion of the masses. Such a source will be called a 'controlled source'. At some instant of time, take a snapshot of the system and note the displacements of the masses from their equilibrium positions. Obviously these will depend on the size of the driving force and the instant at which the snapshot is taken. To eliminate these dependences, divide the instantaneous equilibrium positions by the instantaneous driving force imposed on mass[1] 1. The resulting numbers are the amplitudes of oscillation per unit driving force, or simply 'amplitudes' and since the motion of the masses is synchronised with the source, these ratios do not depend on time.

For the structure shown in figure 2.1, during an experiment in which the source is applied to mass 1, the amplitudes of mass 1 and mass 2 are measured to be 0.2 and −1.4 respectively and as mentioned above, it is convenient to store these in a table. If the table is denoted $|\phi_1\rangle$, then the result of the experiment is $|\phi_1\rangle = \begin{pmatrix} 0.2 \\ -1.4 \end{pmatrix}$. (Note that the symbols $|\rangle$ is called a 'ket' and $|\phi_1\rangle$ is pronounced 'ket phi one'. Throughout this book, kets will be used to denote columns of numbers.) To record the fact that these amplitudes arise from a unit source applied to mass 1, the corresponding source will be denoted $|1\rangle = \begin{pmatrix} 1 \\ 0 \end{pmatrix}$. As shown in figure 2.1, if the experiment is repeated, with a source $|2\rangle = \begin{pmatrix} 0 \\ 1 \end{pmatrix}$ applied to mass 2, then the amplitudes are measured to be $|\phi_2\rangle = \begin{pmatrix} 0.8 \\ 0.6 \end{pmatrix}$. To record the results of both experiments, it is useful to introduce a table with two columns. The resulting table will be denoted g, as shown in equation (2.1).

$$g = \begin{pmatrix} 0.2 & 0.8 \\ -1.4 & 0.6 \end{pmatrix} \tag{2.1}$$

Clearly the table g is a 2×2 matrix of numbers. It contains two columns, because there are two distinct possibilities for applying the source and each column has two rows, because there are two amplitudes of vibration. The matrix g is the Green's function of the two-mass structure.

In the case of the wave in figure 1.6, imagine introducing a fine grid of N positions located across the pond. If a tiny tuning fork creates a force at some grid point r', then the wave amplitudes at all points r would be stored in a column of N numbers. Furthermore, if the experiment is repeated for each of the N possible source points defined by the grid, then the results could be stored in a table g with N columns and N rows. Again, the resulting table is the Green's function of the pond. The points on such a grid will be referred to as 'sites'. They are the locations about which the water oscillates. Similarly in figure 2.1, the equilibrium positions of the masses are referred to as 'sites'. They are the locations about which each of the two masses oscillate.

---

[1] This idea is discussed in more detail in chapter 14, where it is noted that the vibrating masses will possess in phase and out of phase components. For simplicity, here we are discussing in-phase components only.

In a quantum system, de Broglie waves also have amplitudes on sites. The sites are sometimes chosen to correspond to a mesh of a large number of points in real space, as was chosen for the Green's function of the pond in figure 1.6. However when describing de Broglie waves of electrons in molecules or solids, one is often more interested in the amplitude of a de Broglie wave on particular atomic orbitals, such as the π orbitals of benzene, shown in figure 2.2. The π orbital located on carbon atom i (for benzene $i = 1, 2, \ldots 6$) is denoted $\pi_i(r)$ and has a known functional dependence on position r, which can be found using a quick Google search. As indicated in the thought experiments shown in figure 2.2, a source applied to orbital π_1 is measured to create a wave $|\phi_1\rangle$, with amplitudes 1, −1, 1 on orbitals π_2, π_4 and π_6 respectively. This means that in real space, the wave function would have the form $\phi(r) = (1).\ \pi_2(r) + (-1).\ \pi_4(r) + (1).\ \pi_6(r)$. From the known real-space functional forms of the π orbitals $\pi_2(r)$, $\pi_4(r)$ and $\pi_6(r)$, this equation allows one to plot the wave function $\phi(r)$ at all points r in space.

Clearly, this is a very efficient way of storing the details of a de Broglie wave, since a column matrix with only 6 entries is needed, from which the wave function could be plotted at many points (r) in space. In this example, the 'sites' are the six π orbitals and a de Broglie wave is specified by noting the six amplitudes on each site. As indicated in figure 2.2, a source of a given frequency ω could be applied to any of the six sites, each leading to six distinct de Broglie waves. If the amplitudes of each wave are stored in the columns of a 6×6 table of numbers, then the resulting table is the Green's function of the π system of benzene. In quantum mechanics, ω is related to the energy of the electronic de Broglie wave by $E = \hbar\omega$, where $\hbar$ is Planck's constant divided by 2π.

Since tables (i.e. matrices) of numbers arise naturally when describing wave amplitudes on a grid of sites, the reader is advised to note a few elementary properties of matrices, which are summarised in appendix A.

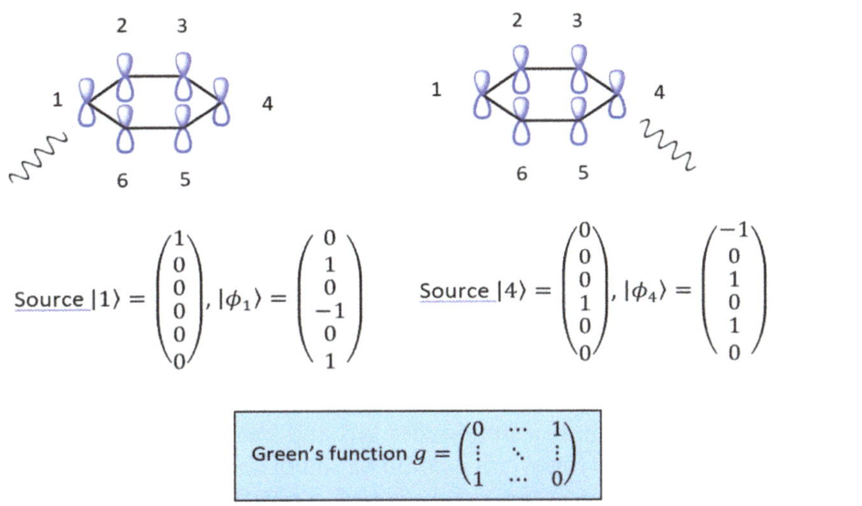

Figure 2.2. Six π orbitals located on the carbon atoms of benzene. A source injecting a de Broglie wave at a chosen site will create a wave function with amplitudes on each of the six π orbitals.

2.3 The principle of superposition

It is interesting to note that if the matrix g of equation (2.1) is multiplied into the source $|1\rangle = \begin{pmatrix} 1 \\ 0 \end{pmatrix}$, then the result is $|\phi_1\rangle$. i.e.

$$\begin{pmatrix} 0.2 & 0.8 \\ -1.4 & 0.6 \end{pmatrix}\begin{pmatrix} 1 \\ 0 \end{pmatrix} = \begin{pmatrix} 0.2 \\ -1.4 \end{pmatrix}$$

Similarly, when g is multiplied into the source $|2\rangle$ one obtains

$$\begin{pmatrix} 0.2 & 0.8 \\ -1.4 & 0.6 \end{pmatrix}\begin{pmatrix} 0 \\ 1 \end{pmatrix} = \begin{pmatrix} 0.8 \\ 0.6 \end{pmatrix}$$

In other words,

$$|\phi_1\rangle = g|1\rangle \text{ and } |\phi_2\rangle = g|2\rangle \tag{2.2}$$

Equation (2.2) shows that for a given source, the amplitudes of the two masses can be obtained by multiplying g into the source. Equation (2.2) also provides the amplitudes that arise when a superposition of sources of the same frequency are imposed simultaneously on the two masses. For example if the source is

$$|S\rangle = a|1\rangle + b|2\rangle \tag{2.3}$$

Then the resulting amplitudes are

$$|\phi\rangle = g|S\rangle \tag{2.4}$$

Writing this out in full yields

$$|\phi\rangle = g(a|1\rangle + b|2\rangle) = ag|1\rangle + bg|2\rangle$$

i.e.

$$|\phi\rangle = a|\phi_1\rangle + b|\phi_2\rangle \tag{2.5}$$

This is an example of the principle of superposition, because it shows that a superposition of sources such as $|S\rangle = a|1\rangle + b|2\rangle$ produces a corresponding superposition of amplitudes $|\phi\rangle = a|\phi_1\rangle + b|\phi_2\rangle$. Clearly the same principle applies to any system, whose amplitudes are related to sources by equation (2.4). In particular, it applies to de Broglie wave patterns such as those in figure 1.7 and to the six-component de Broglie waves in figure 2.2.

2.4 Destructive versus constructive interference

The principle of superposition immediately leads to interference phenomena. For example, the two columns of amplitudes $|\phi_1\rangle$ and $|\phi_2\rangle$ in figure 2.1 have non-zero values on both masses. However if a combination of sources is applied, then it could be arranged that the resulting amplitude of oscillation on mass 2 is zero. For example if the source $|S\rangle$ in equation (2.3) is chosen such that $a = 3$ and $b = 7$, then equation (2.4) shows that the resulting amplitudes of oscillation are

$$|\phi\rangle = 3|\phi_1\rangle + 7|\phi_2\rangle = 3\begin{pmatrix} 0.2 \\ -1.4 \end{pmatrix} + 7\begin{pmatrix} 0.8 \\ 0.6 \end{pmatrix} = \begin{pmatrix} 6.2 \\ 0 \end{pmatrix}$$

Since the amplitude of oscillation of mass 2 is zero, this corresponds to perfect destructive interference for mass 2 and since mass 1 has a non-zero amplitude, it corresponds to constructive interference for mass 1. Similarly, choosing $a = 4$ and $b = -1$ would lead to perfect destructive interference for mass 1 and constructive interference for mass 2.

Clearly in the case of benzene, one could arrange a combination of sources to create a wave with zero amplitude on any chosen site, corresponding to perfect destructive quantum interference on that site. This is why varying the connectivity of a molecular core to electrodes is an effective strategy for probing quantum interference effects in molecules, because it is a method of varying the source. More generally, quantum interference effects can be revealed in any quantum structure by applying combinations of sources and observing or calculating the resulting wave amplitudes.

2.5 Green's functions produced by noisy neighbours

The above discussion can be applied to any wave created by a source that satisfies equation (2.4). In this section, starting from equation (2.4), fundamental equations of connectivity theory are derived, which describe the wave amplitudes in a structure formed by connecting together two or more substructures, such as the two pyridine moieties shown in figure 2.3(a). As noted above, the resulting equations could describe de Broglie waves in a double quantum dot structure, formed by coupling together two quantum dots, de Broglie waves in a larger molecule formed by connecting together two smaller moieties, or it could describe sound waves in room created by a noisy neighbour playing music in a next door room, as shown in figure 2.3(b). For this reason, the derived equations will be called 'noisy neighbour equations (NNEs)'. They provide an answer to the question 'What are the wave patterns produced in a substructures 1 and 2 (such as the rooms in figure 2.3(b)), due to a controlled source which acts only on substructure 2 (such as room 2 in figure 2.3(b))?'

As a first step towards solving this 'noisy neighbour problem,' imagine a thought experiment in which each neighbour could measure the Green's function of their isolated rooms by applying all possible sources and measuring the generated wave amplitudes. During these experiments, all other disturbances due to outside sources are excluded and no sound is allowed to pass between the rooms. This would be the case if the wall between the rooms were infinitely rigid. As indicated in figure 2.3(c) and following equation (2.4), the Green's function g_{11} of isolated room 1 is defined such that the wave pattern created by a source $|S_1\rangle$ acting on the room is

$$|\phi_1\rangle = g_{11}|S_1\rangle \qquad (2.6)$$

As discussed above, g_{11} is a square matrix, whose columns can be constructed by measuring the resulting wave amplitudes from sources applied in turn to a grid of

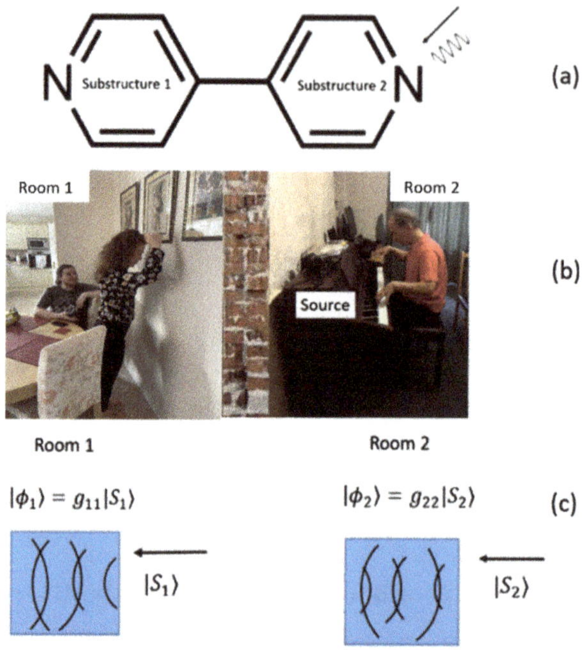

Figure 2.3. (a) A source acting on substructure 2 creates a de Broglie wave in substructure 1. (b) A source in room 1 disturbs the neighbours in room 2. (c) The occupant of each room in isolation (i.e. when the wall is infinitely rigid) characterises their rooms by measuring the wave patterns produced by all possible sources.

sites in the room. In equation (2.5), by analogy with equation (2.3), $|S_1\rangle$ could be chosen to be any linear superposition of such sources.

Similarly, if an arbitrary source $|S_2\rangle$ is applied to isolated room 2, the resulting wave pattern $|\phi_2\rangle$ created by the source is

$$|\phi_2\rangle = g_{22}|S_2\rangle \tag{2.7}$$

In equations (2.6) and (2.7), g_{11} and g_{22} are the Green's functions of the rooms when they are in isolation, such that a source in one room does not disturb the other.

In the presence of a physical wall, the rooms interact with each other, so that when the occupant of room 2 plays music and creates a wave $|\phi_2\rangle$, this wave creates a noise source $|S_1\rangle$ acting on room 1, which annoys the neighbours in room 1. To solve the noisy neighbour problem, instead of becoming angry, the neighbours agree to cooperate and explore the properties of the wall. The sound source acting on room 1 will not simply be $|\phi_2\rangle$, because vibrations of the wall will depend on how sound from room 2 is diminished and distorted by the wall. To explore this feature, an experiment could be carried out in which a wave of unit amplitude is created at each position (i.e. site) in room 2 and the sound source created in room 1 is measured at each site in room 1. As indicated in 2.4(b), these measurements could be used to create a table H_{12} which relates the source $|S_1\rangle$ acting on each site in room 1, due to the wave amplitudes $|\phi_2\rangle$ created in room 2 via the equation

$$|S_1\rangle = H_{12}|\phi_2\rangle \tag{2.8}$$

Clearly H_{12} is matrix of numbers, which describes the fact that the two rooms are coupled together and therefore it is called the 'coupling matrix'. Similarly, by measuring sources created in room 2 by waves in room 1, one obtains the relation

$$|S_2\rangle = H_{21}|\phi_1\rangle \tag{2.9}$$

The two rooms are said to be isolated from each other if $H_{12} = H_{21} = 0$.

As an example (figure 2.5), if room 1 is a substructure containing two masses and room 2 is a substructure containing 3 masses, then H_{12} would have three columns and each column would contain two numbers. The first column would contain the source felt by masses 1 and 2 of substructure 1, when mass 1 of room 2 is vibrating with unit amplitude and the other masses are stationary. The second column of H_{12} would contain the source felt by masses 1 and 2 when mass 2 of room 2 is vibrating with unit amplitude and the other masses are stationary, etc. This illustrates the fact that when the substructures have different numbers of sites, H_{12} is a rectangular matrix. For such a structure, to carry out the experiments leading to equations (2.6) and (2.7), room 2 could be isolated from room 1 by increasing the masses in room 1 to infinity, in which case vibrating masses in room 2 would not create a wave in room 1 and one would find $H_{12} = 0$. After measuring the Green's function g_{22} of isolated room 2, one could then start again from the structure in figure 2.5, increase the masses in room 2 to infinity and measure the Green's function g_{11} of isolated substructure 1.

For the interacting rooms, combining equations (2.6) and (2.8) yields

$$|\phi_1\rangle = g_{11}H_{12}|\phi_2\rangle \tag{2.10}$$

This equation yields the wave pattern $|\phi_1\rangle$ in substructure 1, created by a known wave pattern $|\phi_2\rangle$ in substructure 2 and makes use of the tables g_{11} and H_{12} obtained experimentally. In the case of a quantum structure such as a molecule, the same equation relates the de Broglie wave in one part of a molecule to the de Broglie wave in another part of the molecule. For a mesoscopic structure such as two interacting quantum dots, it would relate the wave function in one dot to the wave function in the other. More generally, equation (2.10) relates the wave pattern in an arbitrary substructure 1 to a known wave pattern in a neighbouring substructure 2.

To solve the noisy neighbour problem, consider the case where the waves $|\phi_1\rangle$ and $|\phi_2\rangle$ are created by a controlled source $|S_2'\rangle$ acting on substructure 2 alone and furthermore, the only way this source can be detected in substructure 1 is if the sound passes to substructure 1 via substructure 2, as shown in figure 2.4(c). In the presence of such a source, the occupant of room 2 notices that the wave in room 2 is no longer simply $|\phi_2\rangle = g_{22}|S_2'\rangle$, because the table g_{22} was measured when room 2 was isolated from room 1, whereas now, waves pass from room 1 to room 2 and part of these are reflected back into room 1, thereby creating an additional source acting on room 2. This new source is

$$|S_2''\rangle = H_{21}|\phi_1\rangle \tag{2.11}$$

as indicated in figure 2.4(c). Therefore to solve the noisy neighbour problem experimentally, neighbour 2 creates a new table G_{22}, whose columns contain the wave amplitudes in room 2 due to all possible controlled sources $|S_2'\rangle$ acting on room 2. This satisfies and

$$|\phi_2\rangle = G_{22}|S_2'\rangle \tag{2.12}$$

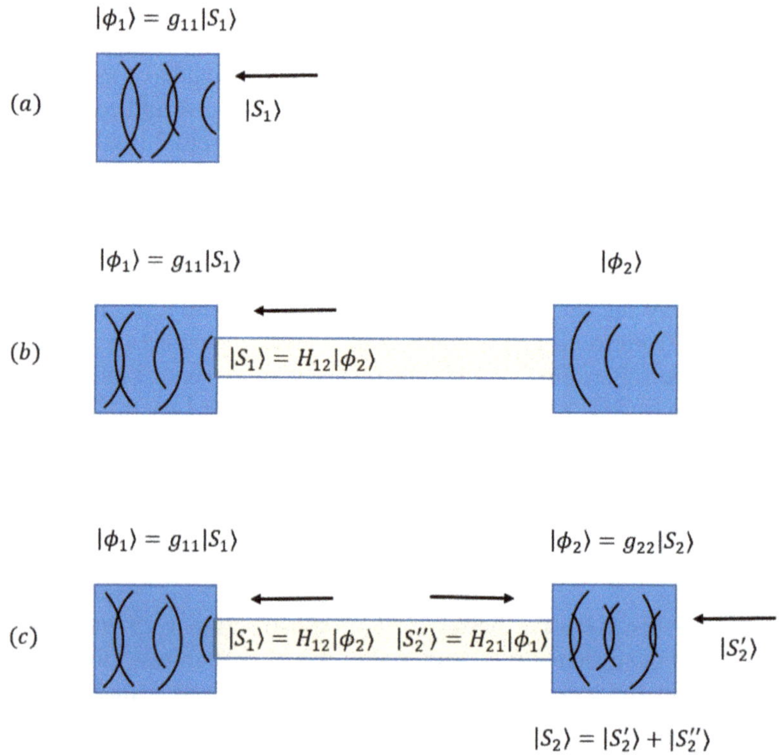

Figure 2.4. (a) The wave pattern $|\phi_1\rangle$ arising in an isolated room 1, due to a controlled source $|S_1\rangle$ applied to the room. (b) The wave pattern $|\phi_1\rangle$ arising in room 1, when a wave function $|\phi_2\rangle$ is present in a neighbouring room 2. (c) If the wave functions in the rooms arise from a controlled source $|S_2'\rangle$ applied to room 2 only, then the wave function $|\phi_2\rangle$ in room 2 arises from $|S_2'\rangle$ and a source $|S_2''\rangle$ due to the reflected wave from room 1.

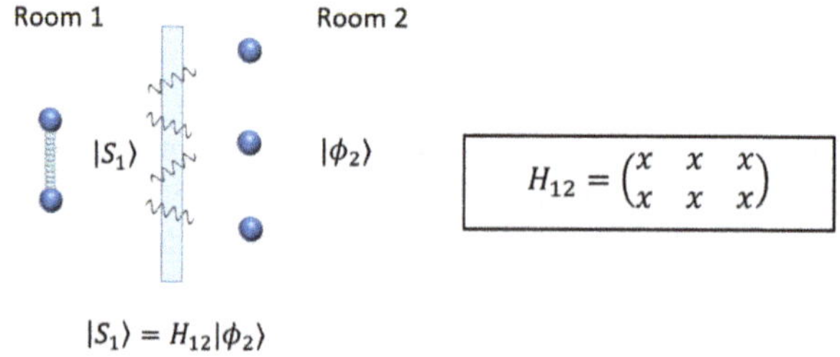

Figure 2.5. Oscillations of the masses in substructure 2 creates a source $|S_1\rangle = H_{12}|\phi_2\rangle$ in substructure 1.

At the same time, the occupants of room 1 could measure the noise in room 2 and created a table G_{12}, whose columns contain the corresponding wave amplitudes in room 1. This satisfies

$$|\phi_1\rangle = G_{12}|S_2'\rangle \tag{2.13}$$

The measured tables G_{22} and G_{12} are the experimental solution to the noisy neighbour problem.

On the other hand, the occupants of room 1 may be lazy and not wish to undertake this new experiment to measure the noise in room 1. Instead, they may prefer to play a passive role and take advantage of equations (2.10) and (2.12), which combine to yield equation (2.13), with

$$G_{12} = g_{11}H_{12}G_{22} \text{ (passive solution)} \tag{2.14}$$

From the viewpoint of the occupants of room 1, this is the passive solution to the noisy neighbour problem and takes advantage of the fact that the occupants of room 2 carried out an extra experiment to determine G_{22}.

Alternatively, if the occupant of room 2 is not willing to measure G_{22}, then one could seek a theoretical solution, by predicting the Green's functions G_{22} and G_{12} from the known tables g_{11}, g_{22}, H_{12} and H_{21}. This is obtained by noting that equation (2.7) yields

$$|\phi_2\rangle = g_{22}\left[|S_2'\rangle + |S_2''\rangle\right] \tag{2.15}$$

Substituting equation (2.11) into equation (2.14) yields

$$|\phi_2\rangle = g_{22}\left[|S_2'\rangle + H_{21}|\phi_1\rangle\right] \tag{2.16}$$

and substituting equation (2.10) into equation (2.15) yields

$$|\phi_2\rangle = g_{22}\left[|S_2'\rangle + H_{21}g_{11}H_{12}|\phi_2\rangle\right] \tag{2.17}$$

i.e.

$$|\phi_2\rangle = g_{22}\left[|S_2'\rangle + \Sigma_2|\phi_2\rangle\right] \tag{2.18}$$

where

$$\Sigma_2 = H_{21}g_{11}H_{12} \tag{2.19}$$

The quantity Σ_2 is called a self-energy and by reading the right-hand side of equation (2.19) from right to left, it describes how a wave from room 2 passes into room 1 (via H_{12}) and creates a wave in room 1 (via g_{11}), which then reflects back into room 2, thereby creating an additional source $|S_2''\rangle$.

To solve equation (2.18), multiply from the left by the inverse matrix g_{22}^{-1} satisfying $g_{22}^{-1}g_{22} = I$, where I is the unit matrix. This yields

$$g_{22}^{-1}|\phi_2\rangle = |S_2'\rangle + \Sigma_2|\phi_2\rangle \tag{2.20}$$

i.e.

$$\left[g_{22}^{-1} - \Sigma_2\right]|\phi_2\rangle = |S_2'\rangle \tag{2.21}$$

Finally, multiply both side from the left by the inverse matrix $\left[g_{22}^{-1} - \Sigma_2\right]^{-1}$ to yield

$$|\phi_2\rangle = \left[g_{22}^{-1} - \Sigma_2\right]^{-1}|S_2'\rangle \tag{2.22}$$

Comparing this with equation (2.13) yields

$$G_{22} = \left[g_{22}^{-1} - \Sigma_2\right]^{-1} \tag{2.23}$$

The table G_{22} is the Green's function of room 2 in the presence of room 1 and differs from the Green's function g_{22} due to the presence of the self-energy in equation (2.23). When equation (2.23) is substituted into equation (2.14), this completes the theoretical solution to the noisy neighbour problem. In what follows, equation (2.14) will be referred to as a noisy neighbour equation (NNE).

The above expression for G_{12} is not unique. As shown below, an alternative expression is

$$G_{12} = G_{11}H_{12}g_{22} \text{ (active)} \tag{2.24}$$

This is referred to as the 'active' NNE, because it requires the occupant of room 1 to take a more active role and carry out an extra experiment to measure the wave pattern G_{11} in room 1 due to all controlled sources acting on room 1, in the presence of room 2. To obtain this alternative NNE substitute equation (2.16) into equation (2.10) to yield

$$|\phi_1\rangle = g_{11}H_{12}g_{22}\left[|S_2'\rangle + H_{21}|\phi_1\rangle\right] \tag{2.25}$$

Multiplying from the left by g_{11}^{-1} yields

$$g_{11}^{-1}|\phi_1\rangle = H_{12}g_{22}\left[|S_2'\rangle + H_{21}|\phi_1\rangle\right] \tag{2.26}$$

and therefore

$$\left[g_{11}^{-1} - H_{12}g_{22}H_{21}\right]|\phi_1\rangle = H_{12}g_{22}|S_2'\rangle \tag{2.27}$$

Multiplying from the left by the inverse matrix

$$G_{11} = \left[g_{11}^{-1} - H_{12}g_{22}H_{21}\right]^{-1} \tag{2.28}$$

again yields equation (2.13), where G_{12} is given by the active NNE (2.24).

In the above equation

$$G_{11} = \left[g_{11}^{-1} - \Sigma_1\right]^{-1} \tag{2.29}$$

where the self-energy Σ_1 is given by

$$\Sigma_1 = H_{12}g_{22}H_{21} \tag{2.30}$$

Comparing equation (2.29) with equation (2.23) reveals that G_{11} is indeed the Green's function of room 1 in the presence of room 2. The wave amplitudes in room 1 due to an arbitrary controlled source $|S_1'\rangle$ acting on room 1 are therefore

$$|\phi_1\rangle = G_{11}|S_1'\rangle \tag{2.31}$$

In summary, the following 'noisy neighbour equations (NNEs)' have been obtained:

$$G_{11} = [g_{11}^{-1} - \Sigma_1]^{-1} \tag{2.32}$$

$$G_{22} = [g_{22}^{-1} - \Sigma_2]^{-1} \tag{2.33}$$

$$|\phi_1\rangle = G_{12}|S_2'\rangle \tag{2.34}$$

where

$$G_{12} = G_{11}H_{12}g_{22} = g_{11}H_{12}G_{22} \tag{2.35}$$

In the above equations, G_{11} determines the wave pattern in room 1 due to a controlled source acting on room 1, taking into account the fact that a wave from room 1 enters room 2 and reflects back into room 1 to create an additional source. Similarly G_{22} determines the wave pattern in room 2 due to a controlled source acting on room 2, taking into account the fact that a wave from room 2 enters room 1 and reflects back into room 2. In other words, G_{11} is the Green's function of room 1 in the presence of room 2 and G_{22} is the Green's function of room 2 in the presence of room 1. In the literature, they are sometimes called renormalized Green's functions, whereas g_{11} and g_{22} are called bare Green's functions or Green's functions of the isolated rooms.

In the NNEs (2.35), G_{12} is a table, which determines the wave pattern in substructure 1 due to any controlled source acting on substructure 2. The equation should be read from right to left, taking note of the indices. For example in the experiment described by the active equation (2.26), a controlled source $|S_2'\rangle$ creates a partial wave $g_{22}|S_2'\rangle$ in substructure 2, which passes though the wall to create a partial source $H_{12}g_{22}|S_2'\rangle$, and in turn this creates a wave $G_{11}H_{12}g_{22}|S_2'\rangle$ in substructure 1 via the renormalized Green's function G_{11}. In the experiment described by the passive equation (2.14), the controlled source $|S_2'\rangle$ creates a total wave $G_{22}|S_2'\rangle$ in substructure 2 via the renormalized Green's function G_{22}, which passes though the wall to create the complete source $H_{12}G_{22}|S_2'\rangle$. In turn this creates the wave function $g_{11}H_{12}G_{22}|S_2'\rangle$ via the bare Green's function. The NNEs (2.35) are fundamental equations of connectivity theory and by applying them iteratively, they allow one to describe wave amplitudes in a complex quantum structure by joining together a series of simpler substructures.

In section A.14 of appendix A, it is noted that equations (2.26) and (2.28) combine to yield

$$G_{11} = g_{11} + g_{11}H_{12}G_{22}H_{21}g_{11} \tag{2.36a}$$

and similarly

$$G_{22} = g_{22} + g_{22}H_{21}G_{11}H_{12}g_{22} \tag{2.36b}$$

Furthermore in section A.16 of appendix A, it is demonstrated that the NNEs imply that

$$\begin{pmatrix} G_{11} & G_{12} \\ G_{21} & G_{22} \end{pmatrix} = \begin{pmatrix} g_{11}^{-1} & -H_{12} \\ -H_{21} & g_{22}^{-1} \end{pmatrix}^{-1} \tag{2.37}$$

Equation (2.36) is called Dyson's equation and is a direct consequence of the NNEs. The above fundamental equations, which underpin connectivity theory, are collected together in the text boxes below, which summarise the relationships between sources and waves.

$$|\phi_2\rangle = G_{22}|S_2'\rangle \tag{2.12}$$

$$|\phi_1\rangle = G_{12}\,|S_2'\rangle \tag{2.34}$$

$$|\phi_1\rangle = g_{11}H_{12}G_{22}|S_2'\rangle \text{(passive viewpoint)} \tag{2.14}$$

$$|\phi_1\rangle = G_{11}H_{12}g_{22}|S_2'\rangle \text{(active viewpoint)} \tag{2.26}$$

$$G_{12} = g_{11}H_{12}G_{22} = G_{11}H_{12}g_{22} \tag{2.35}$$

$$G_{11} = (g_{11}^{-1} - \Sigma_1)^{-1} \tag{2.27}$$

$$G_{11} = g_{11} + g_{11}H_{12}G_{22}H_{21}g_{11} \tag{2.36a}$$

$$G_{22} = (g_{22}^{-1} - \Sigma_2)^{-1} \tag{2.23}$$

$$G_{22} = g_{22} + g_{22}H_{21}G_{11}H_{12}g_{22} \tag{2.36b}$$

$$\Sigma_1 = H_{12}g_{22}H_{21} \tag{2.28}$$

$$\Sigma_2 = H_{21}g_{11}H_{12} \tag{2.19}$$

$$\begin{pmatrix} G_{11} & G_{12} \\ G_{21} & G_{22} \end{pmatrix} = \begin{pmatrix} g_{11}^{-1} & -H_{12} \\ -H_{21} & g_{22}^{-1} \end{pmatrix}^{-1} \tag{2.37}$$

The equivalence of the two NNEs for G_{12} in equation (3.34) is proved directly in section A.12 of appendix A.

The above equations show how Green's functions (such as g_{11} and g_{22}) of isolated substructures combine to yield the Green's functions (G_{11}, G_{22} and G_{12}) of a more complex structure and are independent of the physical nature of the Green's functions. It is clear that the above equations are general properties of any wave satisfying equations (2.6) and (2.8), in which the former shows how a source creates a wave, whereas the latter describes how a wave creates a source. However they do not provide a method for computing the Green's functions g_{11} and g_{22}. To compute these, one needs to specify a wave equation, which describes the underlying physical dynamics. In later chapters, the above equations will be re-derived by solving the Schrödinger equation, which also provides methods for computing the Green's functions g_{11} and g_{22} associated with de Broglie waves. Examples of such methods are introduced section 2.8, but before discussing these examples, it is of interest to iterate the NNEs to provide equations for Green's functions that are relevant to molecular-scale electron transport.

2.6 NNEs for transport through more complex coupled quantum structures

By applying the two-component NNEs (2.35) iteratively, NNEs for multi-component structures can be obtained from the structure of figure 2.3(c), as indicated in 2.6. These iterative procedures are discussed in detail in chapters 11 and 14. They are mentioned here simply to illustrate the fundamental nature of the two-component NNEs (2.35) and to show how they lead to quantum interference effects, which are observed experimentally in molecular-scale junctions.

As indicated in figure 2.6, a three-component NNE can be obtained by starting from two interacting substructures labelled 1 and 2′ and considering the case where substructure 2′ is composed of two substructures, labelled 2 and 3. In this case, if the occupant of substructure 2 takes an active viewpoint, then the wave amplitudes in substructure 2 are given by $|\phi_2\rangle = G_{22}H_{23}g_{33}|S\rangle$, where G_{22} is the Green's function of

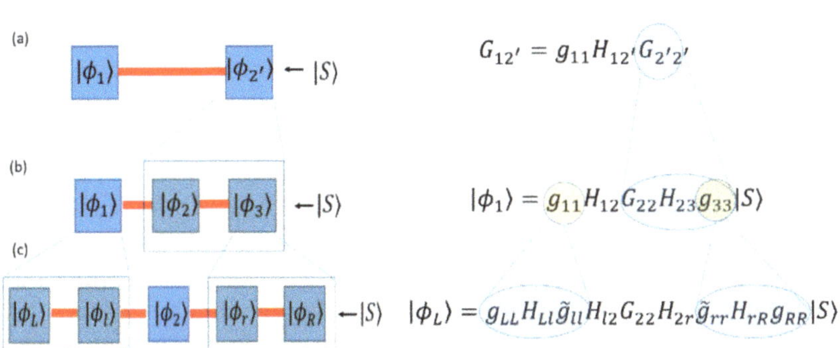

Figure 2.6. (a)–(c) By sub-dividing rooms, the two-component NNEs can be applied iteratively to yield multi-component NNEs.

substructure 2 in the presence of substructures 1 and 3. The wave amplitudes in substructure 1 are therefore

$$|\phi_1\rangle = g_{11}H_{12}|\phi_2\rangle = g_{11}H_{12}G_{22}H_{23}g_{33}|S\rangle \tag{2.38}$$

In the three-component NNE (2.37), g_{11} and g_{33} are the Green's functions of the isolated substructures 1 and 3. If the bare Green's function of substructure 2 is g_{22}, then the renormalized Green's function is $G_{22} = \left(g_{22}^{-1} - \Sigma_2\right)^{-1}$, where the self-energy Σ_2 takes into account reflections from substructure 2 to substructures 1 and 3 and back to substructure 2. A more detailed derivation of equation (2.37) is provided in section 11.10. Furthermore in section 14.4, it is noted that families of multi-component NNEs can be obtained by performing different sequences of active and passive iterations of the two-component NNE. The resulting NNEs allow one to view quantum interference in multi-component structures from a variety of different viewpoints.

As indicated in figure 2.6, and discussed in more detail in section 11.5, the five-component NNE is obtained by dividing substructure 3 of figure 2.6 into sub-structures labelled r and R and dividing substructure 1 into two substructures labelled L and l. As an example of the five-component NNE shown in figure 2.6, consider the case where the coupling matrix H_{l2} connects only to a subset of sites labelled a in substructure 2 and the coupling matrix H_{2r} connects to only a subset of sites labelled b in substructure 2. If the sub-matrix of G_{22} involving these sites is denoted $(G_M)_{ab}$, then the five-component NNE takes the form

$$G_{LR} = g_{LL}H_{Ll}\tilde{g}_{ll}H_{la}(G_M)_{ab}H_{br}\tilde{g}_{rr}H_{rR}g_{RR} \tag{2.39a}$$

This equation describes the five-component structure shown in figure 2.7, comprising a molecular core M (brown) connected by linkers (blue) to anchor groups (green) on the left and right, which in turn are connected via left and right contacts (faded blue) to left and right electrodes (gold). This means that a source $|S\rangle$ applied to the right-hand electrode creates a wave function $|\phi_L\rangle$ in the left electrode given by

$$|\phi_L\rangle = G_{LR}|S\rangle \tag{2.39b}$$

The five-component NNE (2.39a) is derived in detail in section 11.15.

The precise nature of the different terms in equation (2.39a) cannot be appreciated fully, until the material in chapters 3–14 is assimilated. Nevertheless, the qualitative features can be understood by reading the equation from right to left, taking note of the indices. From equations (2.39a) and (2.39b), one can see that a controlled source $|S\rangle$ on the right creates a partial wave $g_{RR}|S\rangle$ in the right electrode, which produces a partial source $H_{rR}g_{RR}|S\rangle$ acting on the right anchor. In turn this creates a partial wave $\tilde{g}_{rr}H_{rR}g_{RR}|S\rangle$ in the right anchor due to the controlled source, where $\tilde{g}_{rr}$ is the renormalized Green's function of the right anchor, in the presence of the right electrode, which takes into account reflections between the anchor and the electrode, but not from the molecular core. The wave function in the right anchor produces a partial source $H_{br}\tilde{g}_{rr}H_{rR}g_{RR}|S\rangle$ acting on the molecular core, which in turn

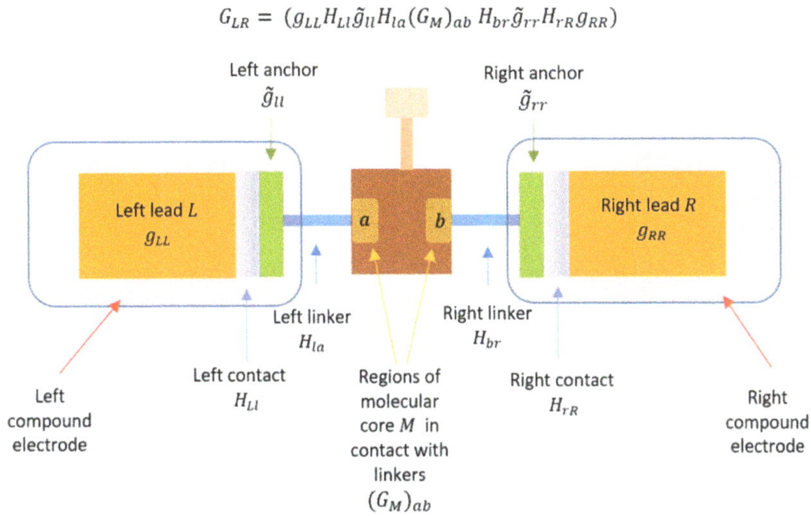

Figure 2.7. A structure described by the five-component NNE. A molecular core M (brown) connected by linkers (blue) to anchor groups (green) on the left and right, which in turn are connected to left and right electrodes (gold).

produces a total wave function $(G_M)_{ab}H_{br}\tilde{g}_{rr}H_{rR}g_{RR}|S\rangle$ within the core. This is described by the renormalized Green's function of the core, which takes into account reflections from all other components of the molecular junction, including the anchors and electrodes. Next, the wave function in the molecular core creates a complete source $H_{la}(G_M)_{ab}H_{br}\tilde{g}_{rr}H_{rR}g_{RR}|S\rangle$ acting on the left anchor group, which in turn creates a wave function $\tilde{g}_{ll}H_{la}(G_M)_{ab}H_{br}\tilde{g}_{rr}H_{rR}g_{RR}|S\rangle$. Finally, this creates a complete source $H_{Ll}\tilde{g}_{ll}H_{la}(G_M)_{ab}H_{br}\tilde{g}_{rr}H_{rR}g_{RR}|S\rangle$, which produces the wave function $|\phi_{\text{left}}\rangle$ in the left electrode. Not surprisingly, the five-component NNE is able to describe many quantum interference phenomena, which arise in molecules and mesoscopic structures. Fundamentally, they are all a consequence of the principle of superposition captured by equations (2.3) and (2.4), which in turn lead to the NNEs (2.35).

2.7 Identifying quantum interference within the five-component NNE

The above discussion shows how basic ideas of connectivity theory can combine to yield wave amplitudes in rather complex structures. As a simple example of the five-component NNE, consider the case where each linker H_{la} and H_{br} connects to only single sites (labelled a and b) in the molecular core, in which case $(G_M)_{ab}$ is the wave function amplitude at point a due to a source at point b. That is $(G_M)_{ab}$ is simply a number and the five-component NNE reduces to

$$G_{LR} = R_{LR}(G_M)_{ab} \tag{2.40}$$

In equation (2.40), R_{LR} is a matrix describing wave amplitudes in the left and right compound electrodes, formed from the electrodes, anchors and linkers. Clearly,

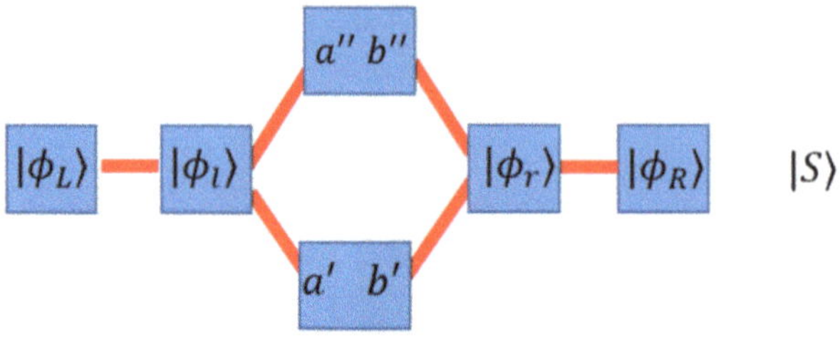

Figure 2.8. A structure with two parallel paths.

equation (2.40) describes how quantum interference takes place within different parts of a molecular core. For example, if for any reason $(G_M)_{ab} = 0$, then G_{LR} will vanish and perfect DQI is said to occur within the molecular core. On the other hand, if for some reason the magnitude of $(G_M)_{ab}$ is large, then CQI is said to occur within the molecular core. Clearly these are mere examples, since as shown in equation (2.39a) interference can occur in all other parts of the five-component structure.

In chapter 10, the relationship between Green's functions and transmission functions of molecular junctions is discussed, which in turn relates Green's functions to measurable properties such as electrical conductance and thermopower. For the above example in which compound electrodes connect to only single sites a, b of the core, the corresponding transmission coefficient T_{ab} is proportional to $|(G_M)_{ab}|^2$ and therefore from the Landauer formula (1.1) the electrical conductance is proportional to $|(G_M)_{ab}|^2$ evaluated at the Fermi energy E_F of the electrodes. In this case, the frequency ω of the injected wave is given by $\hbar\omega = E_F$.

As another example, consider the case where in equation (2.39a), H_{la} connects to two sites in 2, labelled a'', a' and H_{br} connect to two sites in 2 labelled b'', b'. Furthermore, consider the case where $(G_{22})_{a''b'} = (G_{22})_{a'b''} = 0$. This corresponds to the structure in 2.8, containing two parallel branches. If $H_{la''} = H_{la'}$ and $H_{b''r} = H_{b'r}$, then equation (2.40) is again obtained, with $(G_M)_{ab}$ replaced by

$$G_{LR} = R_{LR}[(G_M)_{a''b''} + (G_M)_{a'b'}] \tag{2.41a}$$

If $(G_M)_{ab}$ and $(G_M)_{a''b''}$ have the same sign, then they interfere constructively. If they have opposite signs, then they interfere destructively.

The electrical conductance of such a molecule is proportional to $|(G_M)_{a''b''} + (G_M)_{a'b'}|^2$ Therefore, as noted by Magoga and Joachim [1], if the branches are identical and $(G_M)_{a''b''} = (G_M)_{a'b'}$, then $|(G_M)_{a''b''} + (G_M)_{a'b'}|^2 = 4\,|(G_M)_{a'b'}|^2$. If the coupling between the anchors and the core is weak, so that the effect of the self-energies can be neglected then the electrical conductance of a single branch would be proportional to $|(G_M)_{a'b'}|^2$. Therefore, provided the coupling to the anchors is sufficiently weak[2], one expects the electrical conductance of cores with

[2] If the coupling is strong, then since the self-energies in the presence of two branches are not the same as the self-energy in the presence of a single branch, $(G_M)_{a'b'}$ is not the Green's function of a single-branch molecule and the factor of 4 is modified.

two parallel identical branches to be approximately 4 times the conductance of a single branch.

In equation (2.41a), the two contributions $(G_M)_{a''b''}$ and $(G_M)_{a'b'}$ belong to physical branches of the molecule, which provide physically separated paths, along which waves can travel. In general, one can always divide the Green's function $(G_M)_{ab}$ of the core into a sum of two contributions

$$(G_M)_{ab} = (G_H)_{ab} + (G_L)_{ab} \tag{2.41b}$$

where $(G_H)_{ab}$ is the contribution from all molecular orbitals with energies less than E_F and $(G_L)_{ab}$ is the contribution from all orbitals with energies greater than E_F. In this case, the paths occupy the same physical space, but are distinguished by their MO energies. As discussed in section 1.7, for the fictitious two-level molecule of figure 1.4, $(G_H)_{ab}$ would be the contribution from the HOMO and $(G_L)_{ab}$ would be the contribution from the LUMO. More generally, as noted in equation (1.11) and discussed in section 15.3, contributions from all orbitals should be included.

As discussed in chapter 12, the iterative procedure leading to the three- and five-component NNEs can be implemented for even longer molecules and the result for G_{LR} involves a product of more Green's functions. When the Fermi energy lies within the HOMO–LUMO gap, products of such matrices decay exponentially with the number of matrices and therefore the electrical conductance of the corresponding molecule will decay exponentially with length. This is a common feature of single-molecule conductances, but as discussed in chapter 12, this behaviour is altered if the HOMO–LUMO gap decreases with increasing length.

These consequences of NNEs are simple examples of the ubiquitous phenomenon of interference. As shown by the above discussion of connectivity theory, they are generic manifestations of wave interference, whose origins lie in the principle of superposition and equations (2.6) and (2.8). They are independent of the physical origin of the waves of interest. The fact that they are observed experimentally in single-molecule junctions, even at room temperature, is clear evidence that quantum interference controls electron transport through molecules. When applying NNEs to real physical systems, it is necessary to compute the Green's functions themselves. In the following sections, simple examples of Green's functions are presented within the context of magic ratio theory.

2.8 Magic ratio theory

The 'elephant in the room' when studying electron transport through single molecules is the presence of fluctuations. One of the selling points of molecular-scale electronics is that all molecules of a given kind are identical and can be synthesised in huge quantities. Therefore, if these can be wired into molecular-scale circuits, it should be possible to continue Moore's law to sub-10 nanometre dimensions. Unfortunately, although all molecules of a given kind may be identical, the electrodes, their couplings to the electrodes and their local environments are not. Even small atomic-scale changes in the electrode configuration and the local environment can cause huge changes in the electrical conductance of a single-molecule junction and therefore to obtain meaningful results, in the absence of atomic-scale control over the 'coastal region' between a molecule and the electrodes, many thousands of measurements should be carried out. This is why the most popular techniques for measuring single-molecule conductances, include mechanically-controlled break junctions or various scanning-tunnelling-microscopy-based methods, which allow rapid accumulation of such data. Since the measured conductances exhibit a large degree of randomness, typically spanning many orders of magnitude, the data is usually used to form histograms of the logarithms of conductances. The most probable value of the logarithmic conductance is then obtained from with the peak of the histogram, as shown in 2.9, where the dashed lines indicate the most-probable conductances for molecules **1** and **2**. Therefore a theory of single-molecule conductances should attempt to compute this most-probable value.

Since environmental fluctuations play such an important role, this leads to the question: 'what is the contribution of the molecule to the most probable logarithmic conductance?' The answer to this question lies in 'magic ratio theory,' which does not attempt to compute conductances, but instead computes ratios of the most-probable conductances of different molecules, such as the two anthanthrene-based molecules on the right-hand side of figure 2.9. The phrase 'magic ratio' is used, because the theory is rather simple and it seems 'like magic' that such a simple theory can predict the outcomes of real experiments. Magic ratio theory was developed recently in a series of papers [2–7]. Reasons why such a simple theory applies to molecules in the presence of coulomb interactions are discussed in [5] and an extension to superconducting proximity effects in molecular-scale superconducting junctions is discussed in [6, 7].

Magic ratio theory is a simple example of connectivity theory, which applies when the Fermi energy E_F of electrodes lies within the HOMO–LUMO gap of measured molecules and molecular cores are weakly coupled to electrodes. It starts by noting that the cores of the molecules such as the anthanthrenes in figure 2.9 are attached to pyridyl anchor groups by triple bonds and in turn the anchor groups are attached to gold electrodes. Triple bonds act as tunnel barriers, which only weakly couple the core to the anchor groups and therefore, as in the five-component NNE, it is fruitful to regard the anchor groups and gold electrodes as 'compound electrodes,' which are weakly coupled to the central core. These compound electrodes are indicated by the

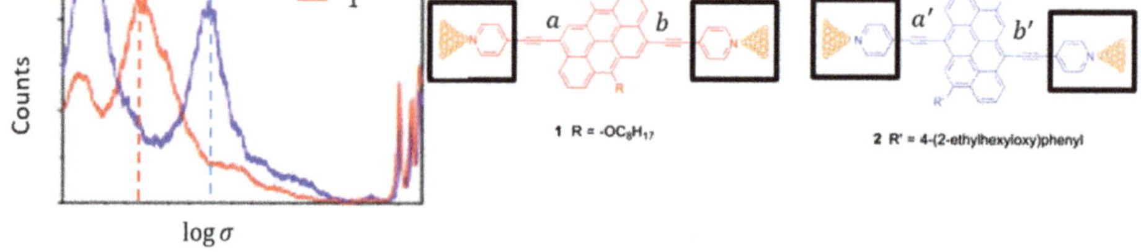

Figure 2.9. Histograms of the logarithm of measured conductances for the red and blue molecules **1** and **2**, with graphene-like anthanthrene cores. (This figure is adapted from [2].)

rectangular boxes in figure 2.9. For molecule **1**, the compound electrodes attach to atoms labelled a and b, whereas for **2**, they are attached to atoms a' and b'. In chapter 11, it will be shown that due to the weak coupling to the compound electrodes, provided the Fermi energy E_F of the electrodes lies within the HOMO–LUMO gap of the molecule, the self-energies in equations (2.38) and (2.39) can be ignored and the electrical conductance σ_{ba} of a molecule such as **1**, with connectivity a, b can be written

$$\sigma_{ba} = R[g_{ba}(E_F)]^2 \tag{2.42}$$

where R is associated with the compound electrodes, whereas $g_{ba}(E_F)$ is the Green's function of the isolated molecular core, obtained by setting $\Sigma_2 = 0$ in the renormalized core Green's function $(G_M)_{ab}$; i.e. $g_{ab}(E_F)$ is the amplitude of the wave function of the isolated core at site a due to a source at site b. For such a molecule, the sites may be π orbitals, such those shown in figure 2.2. For simplicity, it will be assumed that the Hamiltonian of the molecule is real, so that $g_{ba}(E_F)$ is a real number. In practice the shape of the gold electrodes and the coupling between the pyridyl anchors and the gold electrodes varies randomly from measurement to measurement, whereas the core of the molecule does not change. Consequently R is a random number, whereas $g_{ba}(E_F)$ has a definite value. Taking the logarithm of equation (2.41) yields

$$\log\sigma_{ba} = \log R + \log[g_{ba}(E_F)]^2 \tag{2.43}$$

This shows that the most probable value of $\log\sigma_{ba}$ is the most probable value of $\log R$ plus the non-random number $\log[g_{ba}(E_F)]^2$. Under the assumption that the statistical properties of R do not depend on connectivity, the conductance of molecule 2 of figure 2.9 is given by

$$\log\sigma_{b'a'} = \log R + \log[g_{b'a'}(E_F)]^2 \tag{2.44}$$

Hence, subtracting equation (2.44) from (2.43) yields for the most-probable conductances

$$\log\sigma_{ba} - \log\sigma_{b'a'} = \log[g_{ba}(E_F)]^2 - \log[g_{b'a'}(E_F)]^2$$

This is an equation for the difference between the two histogram peaks shown in figure 2.9. It contains no information about their absolute positions, since these are determined by the properties of the random numbers R.

Since $\log\sigma_{ba} - \log\sigma_{b'a'} = \log\frac{\sigma_{ba}}{\sigma_{b'a'}}$, magic ratio theory predicts that the ratio of the most-probable conductances of a molecular core, with different connectivities a, b and a', b' to the electrodes, is given by

$$\boxed{\frac{\sigma_{ba}}{\sigma_{b'a'}} = \frac{[g_{ba}(E_F)]^2}{[g_{b'a'}(E_F)]^2}} \tag{2.45}$$

Equation (2.44) is useful, because the 'interference pattern' $g_{ba}(E)$ is a property of the molecular core alone. The only property associated with the electrodes is the Fermi energy E_F.

2.9 Green's function of a molecular core

Before discussing the Green's function $g_{ba}(E)$ in more detail, it is helpful to recall that when describing electrons in molecules, one is interested in predicting the probability P_j that an electron is located on a particular site j of the molecule. If the electron is definitely on the molecule, then these probabilities must sum to unity. So if the molecule is composed of N sites, the following equation is satisfied

$$\sum_{j=1}^{N} P_j = 1 \tag{2.46}$$

Within a quantum mechanical description, P_j is determined by first obtaining the amplitude ψ_j of the wave function of the electron on site j. The probability P_j is then given by

$$P_j = |\psi_j|^2 \tag{2.47}$$

To simplify the discussion, consider molecules built from fictitious atoms, which possess a single atomic-like π orbital, which plays the role of a 'site'[3]. Each atom is assumed to possess a single electron, which can reside within the molecule. As an example, consider a molecule with only a single atom, possessing a π atomic orbital of energy ϵ_0. In this case, $\psi_1 = 1$ and we say that the energy E of an electron sitting on this site is $E = \epsilon_0$, or simply 'the site energy is ϵ_0'. A cartoon representing this site is shown in figure 2.10.

Now consider a diatomic molecule formed from two such atoms. In this case there will be two molecular orbitals (MOs); the HOMO of energy E_H and the LUMO of energy E_L, where $E_L > E_H$. In view of the Pauli exclusion principle, at most two electrons can occupy each MO and therefore the lowest energy state of the molecule occurs when both electrons occupy the HOMO, with a total energy of $2E_H$. To introduce the notion of a chemical bond, consider the case where the atoms are

Figure 2.10. An isolated site with a site energy ϵ_0.

[3] To simplify the discussion, we consider molecules built from atoms which possess a single pseudo-atomic orbital containing a single electron. These atomic-like orbitals are assumed to be linear combinations of atomic orbitals, as discussed in chapter 16, so that pseudo-atomic orbitals on neighbouring sites are orthogonal.

infinitely far apart. In this case, the individual sites are represented in figure 2.11 and $E_L = E_H = \epsilon_0$.

On the other hand, as already noted, if the atoms are moved closer together, a chemical bond forms and $E_L > E_H$. In this case, the average of the two MO energies is $\frac{E_L + E_H}{2} = \epsilon_0$ and their difference will be denoted 2γ i.e. $E_L - E_H = 2\gamma$. From the above discussion, $\gamma = 0$ when the atoms are infinitely far apart and increases to a non-zero value as the distance between them decreases. With this notation, $E_H = \epsilon_0 - \gamma$ and $E_L = \epsilon_0 + \gamma$ and one can say that a chemical bond of energy $-\gamma$ is formed, as depicted in figure 2.12. In the language of tight-binding theory, $-\gamma$ is called a nearest neighbour coupling. In many cases, γ turns out to be positive, so the nearest neighbour coupling is negative.

For the HOMO of such a molecule, the MO amplitudes are $\psi_1 = \psi_2 = c$, where c is a constant chosen such that equation (2.46) is satisfied (with $N = 2$ for our diatomic molecule). That is $c = 1/\sqrt{2}$. For the LUMO, one finds $\psi_1 = -\psi_2 = c$. Often it is convenient to write these amplitudes in the form of a column of numbers, in which case the HOMO would be written

$$\text{HOMO} = \begin{pmatrix} \psi_1 \\ \psi_2 \end{pmatrix} = \begin{pmatrix} c \\ c \end{pmatrix} \tag{2.48}$$

and the LUMO would be written

$$\text{LUMO} = \begin{pmatrix} \psi_1 \\ \psi_2 \end{pmatrix} = \begin{pmatrix} c \\ -c \end{pmatrix} \tag{2.49}$$

The above results could have been obtained by noting that the Hamiltonian H describing such a molecule is a 2×2 matrix of the form

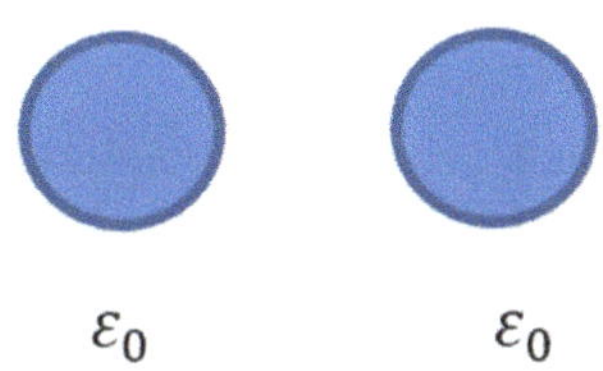

Figure 2.11. Two non-interacting sites.

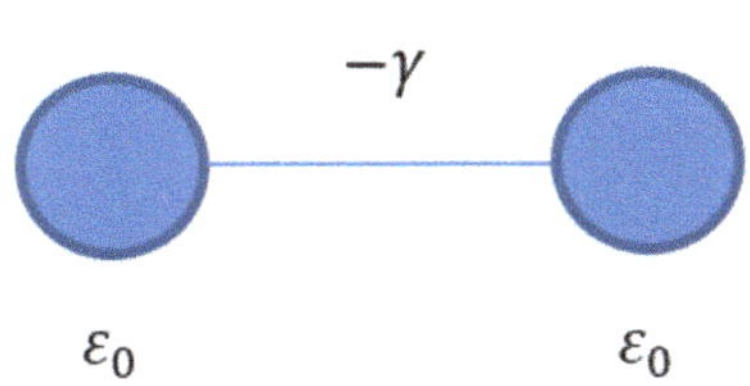

Figure 2.12. A representation of a diatomic molecule formed from atoms with site energies ε_0 and a nearest neighbour coupling $-\gamma$.

$$H = \begin{pmatrix} \varepsilon_0 & -\gamma \\ -\gamma & \varepsilon_0 \end{pmatrix} \tag{2.50}$$

and that MOs are solutions to the time-independent Schrödinger equation

$$\begin{pmatrix} \varepsilon_0 & -\gamma \\ -\gamma & \varepsilon_0 \end{pmatrix}\begin{pmatrix} \psi_1 \\ \psi_2 \end{pmatrix} = E\begin{pmatrix} \psi_1 \\ \psi_2 \end{pmatrix} \tag{2.51}$$

To check that the HOMO of equation (2.48) is indeed a solution of this equation, substitute equation (2.48) into the left-hand side of equation (2.51) to obtain the following series of steps

$$H\begin{pmatrix} \psi_1 \\ \psi_2 \end{pmatrix} = \begin{pmatrix} \varepsilon_0 & -\gamma \\ -\gamma & \varepsilon_0 \end{pmatrix}\begin{pmatrix} \psi_1 \\ \psi_2 \end{pmatrix} = \begin{pmatrix} \varepsilon_0 & -\gamma \\ -\gamma & \varepsilon_0 \end{pmatrix}\begin{pmatrix} c \\ c \end{pmatrix} = \begin{pmatrix} c[\epsilon_0 - \gamma] \\ c[\epsilon_0 - \gamma] \end{pmatrix} = [\epsilon_0 - \gamma]\begin{pmatrix} c \\ c \end{pmatrix} = E_{\mathrm{H}}\begin{pmatrix} \psi_1 \\ \psi_2 \end{pmatrix}$$

This proves that the HOMO $\begin{pmatrix} c \\ c \end{pmatrix}$ is indeed a solution of (2.51), with energy $E_{\mathrm{H}} = \epsilon_0 - \gamma$. Similarly by substituting (2.49) into (2.51), one can prove that the LUMO $\begin{pmatrix} c \\ -c \end{pmatrix}$ satisfies equation (2.51), with energy $E_{\mathrm{L}} = \epsilon_0 + \gamma$.

If the column matrix $\begin{pmatrix} \psi_1 \\ \psi_2 \end{pmatrix}$ is denoted $|\psi\rangle = \begin{pmatrix} \psi_1 \\ \psi_2 \end{pmatrix}$, equation (2.51) is written

$$H|\psi\rangle = E|\psi\rangle \tag{2.52}$$

If the HOMO is denoted $|\psi_{\mathrm{H}}\rangle$ and the LUMO denoted $|\psi_{\mathrm{L}}\rangle$, then $|\psi_{\mathrm{H}}\rangle = \begin{pmatrix} c \\ c \end{pmatrix}$ and $|\psi_{\mathrm{L}}\rangle = \begin{pmatrix} c \\ -c \end{pmatrix}$. To emphasise the differences between the HOMO and the LUMO, it is sometimes convenient to represent the amplitudes of the MOs on each site by a circle, coloured blue if the amplitude is positive and red if it is negative. If the radius of the circle is chosen to be proportional to the magnitude of the amplitude, then one obtains the representation of MOs shown in figure 2.13, in which each circle has the same radius $(1/\sqrt{2})$.

To describe molecules which are more complicated than the above diatomic molecule, one should start from the appropriate Hamiltonian matrix H and then solve equation (2.52) to obtain the associated MOs and their energies. Examples of more complicated molecules are discussed in the later chapters, where it is shown that the Green's function matrix of the core is defined by

$$g(E) = (E - H)^{-1} \tag{2.53}$$

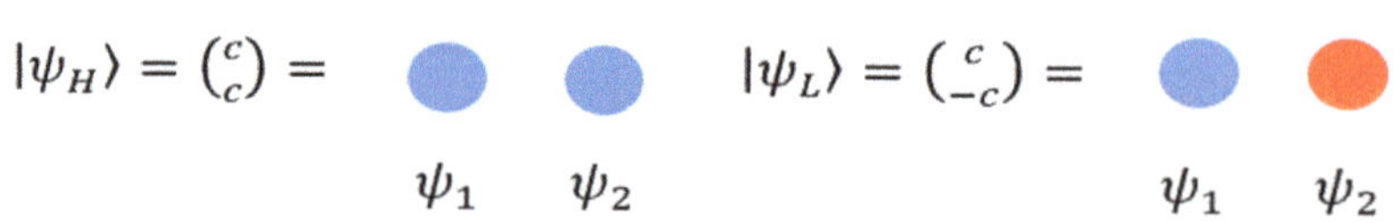

Figure 2.13. A representation of the HOMO and LUMO amplitudes of a diatomic molecule.

and $g_{ba}(E)$ is the b, a th element of this matrix. Obviously the size of H increases with the number of sites on a molecule and therefore the matrix inversion on the right-hand side of equation (2.53) can become a challenge.

To highlight the simplicity of magic ratio theory, it is convenient to take equation (2.45) one step further by defining magic numbers $M_{ba}(E_F)$ via the relation

$$M_{ba}(E_F) = Bg_{ba}(E_F) \tag{2.54}$$

where the constant B is independent of a, b and is chosen for convenience, as discussed below. Clearly B cancels in the ratio of equation (2.45), so one obtains finally

$$\boxed{\frac{\sigma_{ba}}{\sigma_{b'a'}} = \frac{|M_{ba}(E_{\mathrm{F}})|^2}{|M_{b'a'}(E_{\mathrm{F}})|^2)}} \tag{2.55}$$

where

$$M = B(E - H)^{-1} \tag{2.56}$$

This final step is of course totally trivial, but with an appropriate choice of B, it leads to simplifications, which reveal the simple form of interference patterns within certain molecules. Indeed when describing molecules using a simple tight-binding Hamiltonian, M sometimes reduces to a table of integers. In what follows, M is called a 'magic number table' or simply a 'M-table'.

To illustrate this simplification, it should be noted that in chemistry, as in physics, only energy differences matter and therefore the energy origin can be chosen for convenience. Furthermore, to avoid unnecessary detail, it can be useful to choose a convenient energy scale. In the above two-atom molecule, a natural energy origin is the mid-gap energy $(E_{\mathrm{H}} + E_{\mathrm{L}})/2$, which is simply the site energy ϵ_0. In other words, if the site energy is chosen to be $\epsilon_0 = 0$, the quantities $E_{\mathrm{H}} = -\gamma$ and $E_{\mathrm{L}} = +\gamma$ are the HOMO and LUMO energies measured relative to the middle of the HOMO–LUMO gap. Similarly, since $\frac{E_{\mathrm{H}}}{\gamma} = -1$ and $\frac{E_{\mathrm{L}}}{\gamma} = +1$ the natural energy scale is γ. With this choice of energy scale, one finds that relative to the middle of the HOMO–LUMO gap, and in units of 'half the energy gap', $E_{\mathrm{H}} = -1$ and $E_{\mathrm{L}} = +1$. Using these units, the Hamiltonian of equation (2.50) becomes

$$H = \begin{pmatrix} 0 & -1 \\ -1 & 0 \end{pmatrix} \tag{2.57}$$

or equivalently $H = -C$, where $C = \begin{pmatrix} 0 & 1 \\ 1 & 0 \end{pmatrix}$. The matrix C is simply a connectivity table, obtained by inserting the number 1 when two atoms are connected and zero otherwise. To emphasise this point, it is convenient to write C as in the form shown in table 2.1.

As mentioned in chapter 1, whenever a molecule is placed between two electrodes, the MO energies adjust themselves, such that the Fermi energy lies in the vicinity of

Table 2.1. The connectivity table of a diatomic molecule.

C	1	2
1	0	1
2	1	0

Table 2.2. The M-table of a diatomic molecule.

M	1	2
1	0	1
2	1	0

the middle of the HOMO–LUMO gap. Therefore if no further information about E_F is available, then it is of interest to investigate mid-gap transport properties, which arise when E_F coincides with the middle of the HOMO–LUMO gap. In the case of the above diatomic molecule, this means that $E_F = 0$ and from equation (2.53), $M(E_F) = B(E_F - H)^{-1} = BC^{-1}$. For the matrix $C = \begin{pmatrix} 0 & 1 \\ 1 & 0 \end{pmatrix}$, it is easy to show (see section A.7 of appendix A) that $C^{-1} = \begin{pmatrix} 0 & 1 \\ 1 & 0 \end{pmatrix}$, so choosing $B = 1$ yields $M = \begin{pmatrix} 0 & 1 \\ 1 & 0 \end{pmatrix}$. When written in the style of a table, this takes the form of table 2.2.

(Note that $MC = \begin{pmatrix} 0 & 1 \\ 1 & 0 \end{pmatrix}\begin{pmatrix} 0 & 1 \\ 1 & 0 \end{pmatrix} = \begin{pmatrix} 1 & 0 \\ 0 & 1 \end{pmatrix} = I$, where I is the unit matrix, which proves that M is indeed equal to the inverse matrix of C.)

Based on this magic number table and equation (2.55), the mid-gap conductances of this diatomic molecule are predicted to satisfy $\frac{\sigma_{11}}{\sigma_{12}} = 0$. Of course for a real molecule, whose atoms possess more than single π atomic orbital, this is an approximate result. In practice it means that if the electrodes are connected to the same atom (e.g. if the connectivity is 1,1), the electrical conductance will be much lower than that obtained when the electrodes connect to the different atoms (i.e. when the connectivity is 1,2). These two connectivities are shown in figure 2.14. The fact that the conductance of the 1,1 connectivity (on the left) is lower than that of the 1,2 connectivity (on the right) may appear to be counterintuitive, because the distance between the source and drain is greater for the latter. Nevertheless, it is a natural consequence of the mid-gap destructive quantum interference (DQI) that is present for the 1,1 connectivity.

As a less trivial example, consider the lattice of sites, which captures the connectivity of naphthalene, shown in figure 2.15.

In this case the connectivity table C and magic number table M are shown in table 2.3.

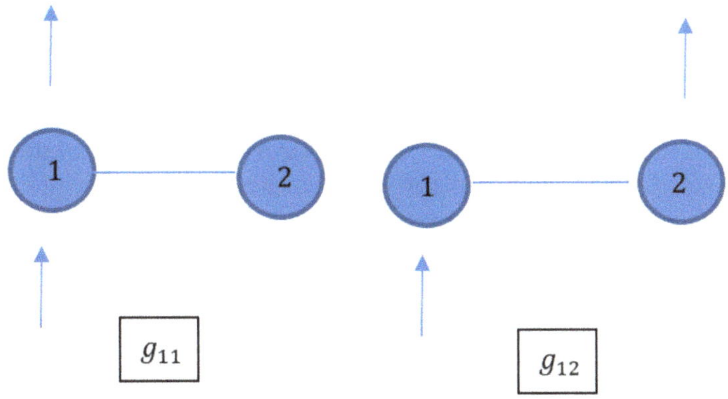

Figure 2.14. A two-site molecule with two possible connectivities.

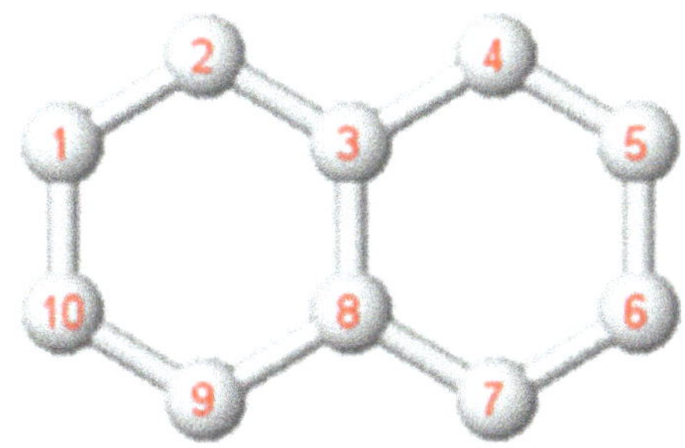

Figure 2.15. A lattice of sites representing the connectivity of naphthalene.

Table 2.3. The M-table of anthracene.

C	1	3	5	7	9	2	4	6	8	10
1	0	0	0	0	0	1	0	0	0	1
3	0	0	0	0	0	1	1	0	1	0
5	0	0	0	0	0	0	1	1	0	0
7	0	0	0	0	0	0	0	1	1	0
9	0	0	0	0	0	0	0	0	1	1
2	1	1	0	0	0	0	0	0	0	0
4	0	1	1	0	0	0	0	0	0	0
6	0	0	1	1	0	0	0	0	0	0
8	0	1	0	1	1	0	0	0	0	0
10	1	0	0	0	1	0	0	0	0	0

M	1	3	5	7	9	2	4	6	8	10
1	0	0	0	0	0	-2	1	-1	1	-1
3	0	0	0	0	0	-1	-1	1	-1	1
5	0	0	0	0	0	1	-2	-1	1	-1
7	0	0	0	0	0	-1	2	-2	-1	1
9	0	0	0	0	0	2	-1	1	-1	-2
2	-2	-1	1	-1	2	0	0	0	0	0
4	1	-1	-2	2	-1	0	0	0	0	0
6	-1	1	-1	-2	1	0	0	0	0	0
8	1	-1	1	-1	-1	0	0	0	0	0
10	-1	1	-1	1	-2	0	0	0	0	0

One can verify by direct multiplication that the product of the two matrices in table 2.3 is $MC = (-3)I$, where I is the 10×10 unit matrix. This proves that M is proportional to C^{-1}. (i.e. $M = BC^{-1}$, with a constant of proportionality $B = -3$.)

From the magic number table 2.3, it predicted that the electrical conductances σ_{ab} between odd-numbered sites and odd-numbered sites, or even-numbered sites and even-numbered sites are zero. These are the connectivities associated with the yellow blocks of the M-table. The only non-zero conductances are those associated with the blue blocks of the M-table, which occur between odd-numbered sites and even-numbered sites. This M-table structure occurs, because the connectivity table C is block off-diagonal, with non-zero numbers in the blue blocks only. This in turn arises because in the anthracene lattice, odd-numbered sites are connected to even-numbered sites only and vice versa. Such a lattice is called a bipartite lattice and always leads to a block off-diagonal connectivity table and a corresponding mid-gap M-table $M = BC^{-1}$, which is also block off-diagonal. The zeros contained in the diagonal blocks of M are inevitable and therefore referred to as trivial zeros. In what follows, to save space and avoid writing sub-matrices containing only zeros, the upper right-hand block of the connectivity table will be denoted $\bar{C}$ and the lower left-hand block of M by $\bar{M}$, and only these will be displayed. Clearly $\bar{M} = B\bar{C}^{-1}$.

The connectivities corresponding to zeros in the M-tables are examples of gap-centre destructive quantum interference (DQI). The connectivities in the blue blocks correspond to non-zero magic numbers and are examples of gap-centre constructive quantum interference (CQI). By inspecting this table, one can identify high-conductance connectivities. For example, the conductance $G_{51'}$ is expected to be $2^2 = 4$ times higher than the conductance $G_{52'}$.

Often zeros also occur in the off-diagonal blocks of M-tables. These are called non-trivial zeros. Examples of non-trivial zeros occurring in the off-diagonal blocks $\bar{M}$ of an M-table are encountered in the M-table of pyrene shown in table 2.4.

Here the connectivities highlighted in yellow correspond to non-trivial occurrences of gap-centre DQI. This table allows one to identify the highest-conductance connectivities as 15,5 and 12,11, which are predicted to be $5^2 = 25$ times higher than e.g. the 12,1 connectivity.

It is perhaps worth mentioning that such M-tables are easy to obtain using MATLAB. Starting from a connectivity table C, the M-table corresponding to $E = 0$ is given by the following three lines (in which the constant B in equation (2.56) is $B = \det(C)/x$, where $\det C$ is the determinant of the matrix C):

Table 2.4. The M-table of pyrene.

2	3	3	-3	3	-3	3	-3	0
4	-1	1	5	-3	3	-1	1	-2
6	1	-1	1	3	-3	1	-1	2
8	-1	1	-1	3	3	-1	1	-2
10	1	-1	1	-3	3	1	-1	2
12	-1	1	-1	3	-3	5	1	-2
14	3	-3	3	-3	3	-3	3	0
16	-2	2	-2	0	0	-2	2	2

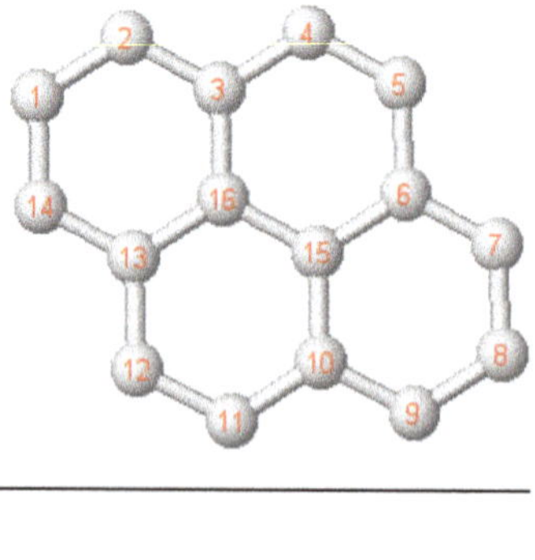

$$A = \det(\mathrm{C}) \times \mathrm{inv}(\mathrm{C});\ x = \min(\mathrm{A}(A > 0));\ M = A/x$$

This means that the problem of identifying zeros in the $E = 0$ Green's function is solved by implementing the above three lines of code. The reader is encouraged to construct his/her own connectivity tables for other interesting molecules and to obtain the corresponding M-tables. Alternatively, analytic counting rules can be used to locate DQI at $E = 0$ in tight-binding Hamiltonians formed from connectivity matrices, as described in [8, 9].

2.10 Mid-gap energies versus $E = 0$

For bipartite lattices such as naphthalene and pyrene, when the Hamiltonian is a simple connectivity table, with $\epsilon_0 = 0$ and $\gamma = 1$, the middle of the HOMO–LUMO gap occurs at $E = 0$. Therefore a theory of $g_{ba}(E_{\mathrm{F}})$ at $E_{\mathrm{F}} = 0$ is a theory of mid-gap transport. On the other hand, if the lattice is not bipartite, then the middle of the HOMO–LUMO gap does not necessarily correspond to $E = 0$ and therefore a theory of $g_{ba}(0)$ is not a theory of mid-gap transport. In this case the energy $E = 0$ has no particular physical significance. If the middle of the HOMO–LUMO gap is denoted $E_{\mathrm{HL}} = (E_{\mathrm{H}} + E_{\mathrm{L}})/2$, then a theory of mid-gap transport should consider $g_{ba}(E_{\mathrm{HL}})$, rather than $g_{ba}(0)$. In this case, even if the Hamiltonian is a simple connectivity matrix, the magic number table $M(E_{\mathrm{HL}}) = B(E_{\mathrm{HL}} - C)^{-1}$ will no longer contain simple integers, due to the presence of the energy E_{HL} on the diagonal. On the other hand the $E = 0$ M-table $M_{ba}(0) = Bg_{ba}(0) = B(-C)^{-1}$ is a table of integers. This means that the $E = 0$ M-table is particularly simple and appealing, even though it is of no particular physical significance. This situation occurs for azulene, which is not a bipartite lattice, because as shown in figure 2.16, the odd-numbered site 1 is connected to the odd-numbered site 7 (table 2.5).

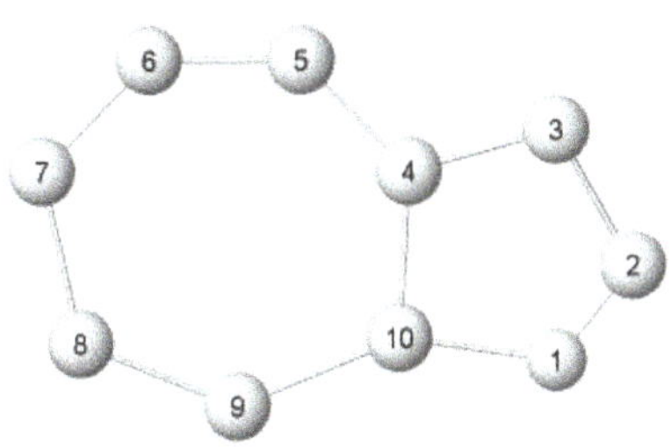

C	1	2	3	4	5	6	7	8	9	10
1	0	1	0	0	0	0	0	0	0	1
2	1	0	1	0	0	0	0	0	0	0
3	0	1	0	1	0	0	0	0	0	0
4	0	0	1	0	1	0	0	0	0	1
5	0	0	0	1	0	1	0	0	0	0
6	0	0	0	0	1	0	1	0	0	0
7	0	0	0	0	0	1	0	1	0	0
8	0	0	0	0	0	0	1	0	1	0
9	0	0	0	0	0	0	0	1	0	1
10	1	0	0	1	0	0	0	0	1	0

Figure 2.16. A lattice of sites representing azulene. The connectivity table of azulene.

Table 2.5. (Left) The mid-gap ($E = E_{HL}$) M-table of azulene. (Right) The $E = 0$ M-table of azulene.

M	1	2	3	4	5	6	7	8	9	10
1	1.14	1.00	-1.10	-1.04	0.02	1.05	0.02	-1.04	-0.06	1.04
2	1.00	0.09	1.00	-0.05	-0.95	0.02	0.95	0.02	-0.95	-0.05
3	-1.10	1.00	1.14	1.04	-0.06	-1.04	0.02	1.05	0.02	-1.04
4	-1.04	-0.05	1.04	0.09	0.95	-0.06	-0.95	0.02	0.95	0.01
5	0.02	-0.95	-0.06	0.95	-0.86	1.02	0.89	-0.99	-0.93	0.95
6	1.05	0.02	-1.04	-0.06	1.02	0.10	0.98	-0.06	-0.99	0.02
7	0.02	0.95	0.02	-0.95	0.89	0.98	-0.86	0.98	0.89	-0.95
8	-1.04	0.02	1.05	0.02	-0.99	-0.06	0.98	0.10	1.02	-0.06
9	-0.06	-0.95	0.02	0.95	-0.93	-0.99	0.89	1.02	-0.86	0.95
10	1.04	-0.05	-1.04	0.01	0.95	0.02	-0.95	-0.06	0.95	0.09

M	1	2	3	4	5	6	7	8	9	10
1	1	1	-1	-1	0	1	0	-1	0	1
2	1	0	1	0	-1	0	1	0	-1	0
3	-1	1	1	1	0	-1	0	1	0	-1
4	-1	0	1	0	1	0	-1	0	1	0
5	0	-1	0	1	-1	1	1	-1	-1	1
6	1	0	-1	0	1	0	1	0	-1	0
7	0	1	0	-1	1	1	-1	1	1	-1
8	-1	0	1	0	-1	0	1	0	1	0
9	0	-1	0	1	-1	-1	1	1	-1	1
10	1	0	-1	0	1	0	-1	0	1	0

Table 2.6. (Left) The connectivity table for a linear chain of 3 sites. (Middle) The connectivity table for a chain of 4 sites with periodic boundary conditions. (Right) A minimal Hamiltonian for a chain of 4 sites with periodic boundary conditions and alternating bonds α and β.

C	1	2	3
1	0	1	0
2	1	0	1
3	0	1	0

C	1	2	3	4
1	0	1	0	1
2	1	0	1	0
3	0	1	0	1
4	1	0	1	0

H	1	2	3	4
1	0	α	0	β
2	α	0	β	0
3	0	β	0	α
4	β	0	α	0

When constructing M-tables, it will be found that in some cases the inverse of C diverges, because the determinant of C vanishes (see section A.7 of appendix A). This is a sign that the corresponding molecule does not exist or is unstable. This happens, for example, for a linear chain with free ends containing an odd number of sites, or a chain with period boundary conditions containing $4n$ sites, where n is an integer. Table 2.6 shows connectivity tables for a linear chain of 3 sites with free ends and a ring of 4 sites with periodic boundary conditions. The inverses of both of these matrices diverge. When approximating Hamiltonians by simple connectivity matrices, such instabilities can be avoided by only considering rings, which satisfy Huckel's rule, i.e. rings containing $4n + 2$ sites or chains containing $2n$ sites. For other structures, a simple connectivity matrix is not adequate, because it assumes equal bond lengths between sites. Cyclic molecules such as cyclobutadiene ($n = 4$), which do not satisfy Huckel's rule, undergo distortions to stabilise themselves, such that bond lengths are no longer equal. Therefore the Hamiltonian should be modified to account for these alternating bond strengths. If these alternating nearest neighbour couplings are denoted α and β, then a more appropriate model Hamiltonian is shown on the right of table 2.6, whose inverse H^{-1} is well behaved. With an appropriate choice of α and β this provides a minimal model of cyclobutadiene.

2.11 Summary

In summary, magic ratio theory is a theory of most-probable-conductance ratios based on equation (2.45). It applies when molecules are weakly coupled to electrodes and transport takes place within the HOMO–LUMO gap. It states that if the core Green's function $g_{ba}(E_F)$ is known, then ratios of most-probable conductances can be predicted. The final step of multiplying $g_{ba}(E_F)$ by some convenient constant B, as in equation (2.54), is of course trivial and not needed. It is merely a useful trick for highlighting the simple structure of $g_{ba}(E_F)$, when the Hamiltonian is a simple connectivity matrix. More generally, for an arbitrary Hamiltonian H, $g_{ba}(E_F)$ should be obtained by evaluating the matrix equation $g(E_F) = (E_F - H)^{-1}$, where H is the Hamiltonian describing the core of the molecule. On the other hand, as discussed in section 11.14, when E_F is close to the HOMO or LUMO, the self-energy (2.42) cannot be neglected, even if the core is weakly coupled to the compound electrodes. In this case, the five-component NNE predicts that transport resonances occur and as discussed in chapter 11, if the coupling is weak, it reproduces the Breit–Wigner formula.

Appendix A Basic properties of matrices

As discussed in section 2.2 and shown in equation (2.1), matrices are simply tables of numbers and are a convenient way of storing experimental data. In order to understand the consequences of such experiments, it is necessary to manipulate matrices of data. In this appendix, the rules of matrix manipulation are summarised.

A.1 Matrix addition and subtraction

Consider the matrix $M = \begin{pmatrix} 1 & 2 & 5 \\ 3 & 4 & 6 \end{pmatrix}$. This matrix has two rows and three columns. The number 1 belongs to the first row and the first column. If the number belonging to the i th row and j th column is denoted m_{ij}, then one can say that $m_{11} = 1, m_{21} = 3$, $m_{23} = 6$ etc. Here, m_{ij} is called a matrix element of the matrix M or simply an element of M and the integers i, j are called indices or labels of the matrix element m_{ij}. The first index i always refers to a row and the second index j always refers to a column.

Now consider two square matrices A, B such that $A = \begin{pmatrix} 1 & 2 \\ 3 & 4 \end{pmatrix}$ and $B = \begin{pmatrix} 5 & 6 \\ 7 & 8 \end{pmatrix}$. These matrices have two rows and two columns and the numbers contained in such matrices are identified by their row and column. Then the sum of A and B is another matrix C, such that

$$C = A + B = \begin{pmatrix} 1 & 2 \\ 3 & 4 \end{pmatrix} + \begin{pmatrix} 5 & 6 \\ 7 & 8 \end{pmatrix} = \begin{pmatrix} 1+5 & 2+6 \\ 3+7 & 4+8 \end{pmatrix} = \begin{pmatrix} 6 & 8 \\ 10 & 12 \end{pmatrix}$$

That is the top left element of A is added to the top left element of B to yield the top left element of C, etc. In general, if $A = \begin{pmatrix} a_{11} & a_{12} \\ a_{21} & a_{22} \end{pmatrix}$, $B = \begin{pmatrix} b_{11} & b_{12} \\ b_{21} & b_{22} \end{pmatrix}$ and $C = \begin{pmatrix} c_{11} & c_{12} \\ c_{21} & c_{22} \end{pmatrix}$, then

$$C = \begin{pmatrix} c_{11} & c_{12} \\ c_{21} & c_{22} \end{pmatrix} = \begin{pmatrix} a_{11} + b_{11} & a_{12} + b_{12} \\ a_{21} + b_{21} & a_{22} + b_{22} \end{pmatrix}$$

An alternative way of expressing this rule is to say that matrix elements of C are given by

$$c_{ij} = a_{ij} + b_{ij},$$

where the indices i, j can take values 1 or 2.

Similarly if $C = A - B$, then $c_{ij} = a_{ij} - b_{ij}$.

Note that for these general addition and subtraction formulae to make sense, only matrices with the same number of rows and columns can be added or subtracted. For example, expressions such as $3 - \begin{pmatrix} 1 & 2 \\ 3 & 4 \end{pmatrix}$ or $\begin{pmatrix} 1 & 2 \\ 3 & 4 \end{pmatrix} + \begin{pmatrix} 5 \\ 6 \end{pmatrix}$ are nonsense.

A.2 Matrix multiplication

Consider the column matrix $|\varphi\rangle = \begin{pmatrix} 5 \\ 6 \end{pmatrix}$. The left product of A and $|\varphi\rangle$ is another column matrix $|\psi\rangle$, given by

$$|\psi\rangle = A|\varphi\rangle = \begin{pmatrix} 1 & 2 \\ 3 & 4 \end{pmatrix}\begin{pmatrix} 5 \\ 6 \end{pmatrix} = \begin{pmatrix} 3.5 + 2.6 \\ 3.5 + 4.6 \end{pmatrix} = \begin{pmatrix} 27 \\ 39 \end{pmatrix}$$

Hence one finds that $A|\varphi\rangle = |\psi\rangle$, where $|\psi\rangle = \begin{pmatrix} 27 \\ 39 \end{pmatrix}$. Similarly the product of A and B is a matric C, given by

$$C = AB = \begin{pmatrix} 1 & 2 \\ 3 & 4 \end{pmatrix}\begin{pmatrix} 5 & 6 \\ 7 & 8 \end{pmatrix} = \begin{pmatrix} 3.5 + 2.7 & 3.6 + 2.8 \\ 3.5 + 4.7 & 3.6 + 4.8 \end{pmatrix} = \begin{pmatrix} 19 & 22 \\ 43 & 50 \end{pmatrix}$$

Hence $AB = C$, where $C = \begin{pmatrix} 19 & 22 \\ 43 & 50 \end{pmatrix}$.

By combining the above rules, one also finds $(A + B)C = AC + BC$.

Note that $A + B = B + A$, but in general $AB \neq BA$. If two matrices A and B have the property $AB = BA$, then they are said to commute. If $AB \neq BA$ then they do not commute. (For the above matrices, the reader should evaluate the product BA and show that indeed it is not equal to the product AB.) The fact that matrices do not usually commute means that when manipulating products of matrices, the order of the matrices must be preserved, because $ABC \neq BAC \neq CAB$ etc.

A.3 Multiplication of a matrix by a number

For example if $A = \begin{pmatrix} 1 & 2 \\ 3 & 4 \end{pmatrix}$ and $|\varphi\rangle = \begin{pmatrix} 5 \\ 6 \end{pmatrix}$, then $2 \times A = \begin{pmatrix} 2 & 4 \\ 6 & 8 \end{pmatrix}$ and $(-3)x|\varphi\rangle = \begin{pmatrix} -15 \\ -18 \end{pmatrix}$. In general, multiplying a matrix by a number means that every element of the matrix is multiplied by the number.

A.4 Expressions for individual matrix elements of a product

Consider the matrices $A = \begin{pmatrix} a_{11} & a_{12} \\ a_{21} & a_{22} \end{pmatrix}$, $B = \begin{pmatrix} b_{11} & b_{12} \\ b_{21} & b_{22} \end{pmatrix}$, $C = \begin{pmatrix} c_{11} & c_{12} \\ c_{21} & c_{22} \end{pmatrix}$, $|\psi\rangle = \begin{pmatrix} \psi_1 \\ \psi_2 \end{pmatrix}$ and $|\varphi\rangle = \begin{pmatrix} \varphi_1 \\ \varphi_2 \end{pmatrix}$. If $C = AB$, then

$$\begin{pmatrix} c_{11} & c_{12} \\ c_{21} & c_{22} \end{pmatrix} = \begin{pmatrix} a_{11} & a_{12} \\ a_{21} & a_{22} \end{pmatrix}\begin{pmatrix} b_{11} & b_{12} \\ b_{21} & b_{22} \end{pmatrix} = \begin{pmatrix} a_{11}b_{11} + a_{12}b_{21} & a_{11}b_{12} + a_{12}b_{22} \\ a_{21}b_{11} + a_{22}b_{21} & a_{21}b_{12} + a_{22}b_{22} \end{pmatrix} \tag{A.1}$$

Hence $c_{ij} = a_{i1}b_{1j} + a_{i2}b_{2j} = \sum_{k=1}^{2} a_{ik}b_{kj}$.

More generally, if A has N columns and M rows and B has N rows and P columns, and $C = AB$, then the i, j th element of C is given by

$$c_{ij} = \sum_{k=1}^{N} a_{ik}b_{kj} \tag{A.2}$$

Since i labels rows of A and j labels columns of B, C is a matrix with M rows and P colums.

Similarly if $|\psi\rangle = A|\varphi\rangle$, then $\begin{pmatrix} \psi_1 \\ \psi_2 \end{pmatrix} = \begin{pmatrix} a_{11} & a_{12} \\ a_{21} & a_{22} \end{pmatrix}\begin{pmatrix} \varphi_1 \\ \varphi_2 \end{pmatrix} = \begin{pmatrix} a_{11}\varphi_1 + a_{12}\varphi_2 \\ a_{21}\varphi_1 + a_{22}\varphi_2 \end{pmatrix}$.

Hence $\psi_i = \sum_{k=1}^{2} a_{ik}\varphi_k$. In this example, A has 2 rows and 2 columns, whereas $|\varphi\rangle$ has 2 rows and 1 column, therefore as expected, $|\psi\rangle$ has 2 rows and 1 column. More generally, if $|\varphi\rangle$ is an N-component column matrix and A is an $N \times N$ matrix, then the elements of the N-component vector $|\psi\rangle = A|\varphi\rangle$ are given by

$$\psi_i = \sum_{k=1}^{N} a_{ik}\varphi_k \tag{A.3}$$

A.5 The unit matrix

A $N \times N$ square matrix with the number 1 on the diagonal and zeros elsewhere is called the $N \times N$ unit matrix and denoted by the symbols I. For example the 2×2 unit matrix is

$$I = \begin{pmatrix} 1 & 0 \\ 0 & 1 \end{pmatrix}$$

This has the property that $I\begin{pmatrix} a_{11} & a_{12} \\ a_{21} & a_{22} \end{pmatrix} = \begin{pmatrix} a_{11} & a_{12} \\ a_{21} & a_{22} \end{pmatrix} I = \begin{pmatrix} a_{11} & a_{12} \\ a_{21} & a_{22} \end{pmatrix}$ and $I\begin{pmatrix} \varphi_1 \\ \varphi_2 \end{pmatrix} = \begin{pmatrix} \varphi_1 \\ \varphi_2 \end{pmatrix}$.

(Note however that $\begin{pmatrix} \varphi_1 \\ \varphi_2 \end{pmatrix} I$ is nonsense.)

In general if I is the $N \times N$ unit matrix and A has N rows, then $IA = A$. Similarly if B has N columns, then $BI = B$.

Similarly if H is a $N \times N$ matrix and E is a scalar (i.e. a number or equivalently a 1×1 matrix) then the expression $E - H$ is nonsense. This expression appears many times in text books and is shorthand for the expression $EI - H$, where I is a $N \times N$ unit matrix. E.g. if $H = \begin{pmatrix} 1 & 2 \\ 2 & 4 \end{pmatrix}$ then

$$E - H = EI - H = E\begin{pmatrix} 1 & 0 \\ 0 & 1 \end{pmatrix} - \begin{pmatrix} 1 & 2 \\ 2 & 4 \end{pmatrix} = \begin{pmatrix} E & 0 \\ 0 & E \end{pmatrix} - \begin{pmatrix} 1 & 2 \\ 2 & 4 \end{pmatrix} = \begin{pmatrix} E - 1 & -2 \\ -2 & E - 4 \end{pmatrix}$$

A.6 The determinant of a matrix

$$A = \begin{pmatrix} a_{11} & a_{12} \\ a_{21} & a_{22} \end{pmatrix}$$

Then the determinant of A is denoted $\det A$ and defined to be

$$\det A = a_{11}a_{22} - a_{12}a_{21} \tag{A.4}$$

For example if $A = \begin{pmatrix} 1 & 2 \\ 3 & 4 \end{pmatrix}$, then $\det A = 1.4 - 2.3 = -2$.

In general, the determinant of a product of matrices is the product of their determinants. That is if $C = AB$, then $\det C = (\det A)(\det B)$. Methods for computing determinants of larger matrices can be found in any book about linear algebra, but these will rarely be needed in this book.

A.7 Inverse of a matrix

It makes no sense to try to divide one matrix by another. However, given a square matrix A, it may be possible to find another matrix B, such that $BA = I$, where I is the unit matrix. In this case, B is said to be the inverse of A and denotedA^{-1}. For example, if

$A = \begin{pmatrix} 1 & 2 \\ 3 & 4 \end{pmatrix}$ and $B = \frac{1}{d}\begin{pmatrix} 4 & -2 \\ -3 & 1 \end{pmatrix}$, where $d = \det A = -2$, then

$BA = \frac{1}{d}\begin{pmatrix} 4 & -2 \\ -3 & 1 \end{pmatrix}\begin{pmatrix} 1 & 2 \\ 3 & 4 \end{pmatrix} = \frac{1}{d}\begin{pmatrix} -2 & 0 \\ 0 & -2 \end{pmatrix} = \begin{pmatrix} 1 & 0 \\ 0 & 1 \end{pmatrix} = I$, where I is the unit matrix.

Hence $B = A^{-1}$.

I.e., $A^{-1} = \frac{1}{d}\begin{pmatrix} 4 & -2 \\ -3 & 1 \end{pmatrix}$. More generally, if

$$A = \begin{pmatrix} a_{11} & a_{12} \\ a_{21} & a_{22} \end{pmatrix}$$

Then

$$A^{-1} = \frac{1}{d}\begin{pmatrix} a_{22} & -a_{12} \\ -a_{21} & a_{11} \end{pmatrix} \tag{A.5}$$

where $d = \det A = a_{11}a_{22} - a_{12}a_{21}$.

Clearly, if $A^{-1}A = I$, then $AA^{-1} = I$, which can be checked by direct multiplication of the two matrices.

A.8 Book keeping the Green's function of a structure composed of two substructures

In many cases, it is fruitful to divide matrices into sub-matrices, because submatrices may be associated with particular substructures of a more complex structure.

For a substructure comprising two sites labelled 1 and 2, equation (2.1) shows that the Green's function

$\begin{pmatrix} 0.2 & 0.8 \\ -1.4 & 0.6 \end{pmatrix}$ is a 2x2 table of numbers. The first column contains the wave amplitudes obtained by applying the source $|S_1\rangle = \begin{pmatrix} 1 \\ 0 \end{pmatrix}$ applied to site 1 and the second columns contained the wave amplitudes due to a source $|S_2\rangle = \begin{pmatrix} 0 \\ 1 \end{pmatrix}$ applied to site 2.

Now consider a second substructure (which does not interact with the first substructure) with two sites labelled 3 and 4, whose Green's function is

$$\begin{pmatrix} 0.7 & -0.4 \\ 0.3 & 0.5 \end{pmatrix}$$

Both 2x2 Green's function matrices could be stored in a single 4x4 Green's function matrix. Since the two substructures do not interact, the Green's function for the four-site combined structure is a 4x4 matrix of the form

$$g = \begin{pmatrix} 0.2 & 0.8 & 0 & 0 \\ -1.4 & 0.6 & 0 & 0 \\ 0 & 0 & 0.7 & -0.4 \\ 0 & 0 & 0.3 & 0.5 \end{pmatrix} \tag{A.6}$$

where the yellow sub-matrix shows the amplitudes on sites 1 and 2, while the grey sub-matrix shows amplitudes on sites 3 and 4.

In this case, the source applied to mass 1 would be denoted

$$|\bar{1}\rangle = \begin{pmatrix} 1 \\ 0 \\ 0 \\ 0 \end{pmatrix} \tag{A.7}$$

and the resulting amplitudes would be denoted

$$|\bar{\phi}_1\rangle = \begin{pmatrix} 0.2 \\ -1.4 \\ 0 \\ 0 \end{pmatrix}$$

So that equation (2.2) remains valid; i.e. $|\phi_1\rangle = g|1\rangle$ and $|\phi_2\rangle = g|2\rangle$. Similarly the source applied to mass 3 would be denoted

$$|\bar{3}\rangle = \begin{pmatrix} 0 \\ 0 \\ 1 \\ 0 \end{pmatrix}$$

and the resulting amplitudes would be denoted $|\bar{\phi}_3\rangle = \begin{pmatrix} 0 \\ 0 \\ 0.7 \\ 0.3 \end{pmatrix}$. These satisfy $|\bar{\phi}_3\rangle = g|\bar{3}\rangle$.

In view of the sub-block structure of g in equation (A.6), it is natural to write g in the form

$$g = \begin{pmatrix} g_{11} & g_{12} \\ g_{21} & g_{22} \end{pmatrix} \tag{A.9}$$

where g_{ij} are 2×2 matrices given by

$g_{11} = \begin{pmatrix} 0.2 & 0.8 \\ -1.4 & 0.6 \end{pmatrix}$, $g_{22} = \begin{pmatrix} 0.7 & -0.4 \\ 0.3 & 0.5 \end{pmatrix}$ and $g_{12} = g_{21} = \begin{pmatrix} 0 & 0 \\ 0 & 0 \end{pmatrix}$. The fact that g_{12} and g_{21} contain only zeroes is a reflection of the fact that the two structures are non-interacting, and guarantees that when g is multiplied into sources $|S_1\rangle$ and $|S_2\rangle$, the amplitudes on sites 3 and 4 are zero.

Similarly, it is natural to write the 4-component column matrices representing sources and waves in the form

$$|\bar{1}\rangle = \begin{pmatrix} 1 \\ 0 \\ 0 \\ 0 \end{pmatrix} = \begin{pmatrix} |1\rangle \\ |0\rangle \end{pmatrix}, |\bar{2}\rangle = \begin{pmatrix} 0 \\ 2 \\ 0 \\ 0 \end{pmatrix} = \begin{pmatrix} |2\rangle \\ |0\rangle \end{pmatrix}, |\bar{3}\rangle = \begin{pmatrix} 0 \\ 0 \\ 1 \\ 0 \end{pmatrix} = \begin{pmatrix} |0\rangle \\ |1\rangle \end{pmatrix}, |\bar{4}\rangle = \begin{pmatrix} 0 \\ 0 \\ 0 \\ 1 \end{pmatrix} = \begin{pmatrix} |0\rangle \\ |2\rangle \end{pmatrix}$$

where $|0\rangle = \begin{pmatrix} 0 \\ 0 \end{pmatrix}$, $|1\rangle = \begin{pmatrix} 1 \\ 0 \end{pmatrix}$ and $|2\rangle = \begin{pmatrix} 0 \\ 1 \end{pmatrix}$.

Similarly

$$|\bar{\phi}_1\rangle = \begin{pmatrix} 0.2 \\ -1.4 \\ 0 \\ 0 \end{pmatrix} = \begin{pmatrix} |\phi_1\rangle \\ |0\rangle \end{pmatrix}, |\bar{\phi}_2\rangle = \begin{pmatrix} 0.8 \\ 0.6 \\ 0 \\ 0 \end{pmatrix} = \begin{pmatrix} |\phi_2\rangle \\ |0\rangle \end{pmatrix}, |\bar{\phi}_3\rangle = \begin{pmatrix} 0 \\ 0 \\ 0.7 \\ 0.3 \end{pmatrix} = \begin{pmatrix} |0\rangle \\ |\phi_3\rangle \end{pmatrix} \text{ etc.}$$

With this notation, an arbitrary source $|\bar{S}\rangle$ acting on any of the 4 sites can be written

$$|\bar{S}\rangle = \begin{pmatrix} |S_1\rangle \\ |S_2\rangle \end{pmatrix} \quad \text{(A.10)}$$

If the wave amplitudes $|\bar{\phi}\rangle = g|\bar{S}\rangle$ created by the source are denoted

$$|\bar{\phi}\rangle = \begin{pmatrix} |\phi_1\rangle \\ |\phi_2\rangle \end{pmatrix} \quad \text{(A.11)}$$

Then since matrices can be multiplied sub-block by sub-block (see section A.9 below), one finds

$$|\bar{\phi}\rangle = g|\bar{S}\rangle = \begin{pmatrix} g_{11} & 0 \\ 0 & g_{22} \end{pmatrix} \begin{pmatrix} |S_1\rangle \\ |S_2\rangle \end{pmatrix} = \begin{pmatrix} g_{11}|S_1\rangle \\ g_{22}|S_2\rangle \end{pmatrix} \quad \text{(A.12)}$$

Comparing equations (A.11) and (A.12) shows that

$$|\phi_1\rangle = g_{11}|S_1\rangle \quad \text{and} \quad |\phi_2\rangle = g_{22}|S_2\rangle \quad \text{(A.13)}$$

in agreement with equations (2.5) and (2.7).

The above separation of matrices into sub-blocks makes it easy to keep track of which amplitudes and sources belong to which substructure.

If the two substructures are allowed to interact with each other, then the Green's function G describing the whole structure will have the form

$$G = \begin{pmatrix} G_{11} & G_{12} \\ G_{21} & G_{22} \end{pmatrix} \quad \text{(A.14)}$$

where the off-diagonal sub-blocks G_{12} and G_{21} are no longer zero.

In this case the wave amplitudes created by a source $|\bar{S}\rangle$ are $|\bar{\phi}\rangle = G|\bar{S}\rangle$ and equation (A.12) is replaced by

$$|\bar{\phi}\rangle = \begin{pmatrix} |\phi_1\rangle \\ |\phi_2\rangle \end{pmatrix} = G|\bar{S}\rangle = \begin{pmatrix} G_{11} & G_{12} \\ G_{21} & G_{22} \end{pmatrix} \begin{pmatrix} |S_1\rangle \\ |S_2\rangle \end{pmatrix} = \begin{pmatrix} G_{11}|S_1\rangle + G_{12}|S_2\rangle \\ G_{21}|S_1\rangle + G_{22}|S_2\rangle \end{pmatrix} \quad \text{(A.15)}$$

i.e.

$$|\phi_1\rangle = G_{11}|S_1\rangle + G_{12}|S_2\rangle \quad \text{and} \quad |\phi_2\rangle = G_{21}|S_1\rangle + G_{22}|S_2\rangle \quad \text{(A.16)}$$

For example, if the source only acts on sites 3 and 4, so that $|\bar{S}\rangle = \begin{pmatrix} 0 \\ |S_2\rangle \end{pmatrix}$, then

$$|\phi_1\rangle = G_{12}|S_2\rangle \tag{A.17}$$

and

$$|\phi_2\rangle = G_{22}|S_2\rangle \tag{A.18}$$

A.9 Multiplying matrices sub-matrix by sub-matrix

Often it is convenient to multiply matrices sub-matrix by sub-matrix. For example the following matrix

$$A = \begin{pmatrix} 1 & 2 & a & b \\ 3 & 4 & c & d \\ e & f & 5 & 6 \\ g & h & 7 & 8 \end{pmatrix} \text{ can be written } \qquad A = \begin{pmatrix} a_{11} & a_{12} \\ a_{21} & a_{22} \end{pmatrix}, \tag{A.19}$$

where

$$a_{11} = \begin{pmatrix} 1 & 2 \\ 3 & 4 \end{pmatrix}, a_{12} = \begin{pmatrix} a & b \\ c & d \end{pmatrix}, a_{21} = \begin{pmatrix} e & f \\ g & h \end{pmatrix} \text{ and } a_{22} = \begin{pmatrix} 5 & 6 \\ 7 & 8 \end{pmatrix}$$

Similarly the matrix $B = \begin{pmatrix} i & j & 5 & 6 \\ k & l & 9 & 4 \\ x & y & 4 & 3 \\ p & q & 2 & 1 \end{pmatrix}$ can be written $B = \begin{pmatrix} b_{11} & b_{12} \\ b_{21} & b_{22} \end{pmatrix}$, where

$$b_{11} = \begin{pmatrix} i & j \\ k & l \end{pmatrix}, b_{12} = \begin{pmatrix} 5 & 6 \\ 9 & 4 \end{pmatrix}, b_{21} = \begin{pmatrix} x & y \\ p & q \end{pmatrix} \text{ and } b_{22} = \begin{pmatrix} 4 & 3 \\ 2 & 1 \end{pmatrix}$$

Then if $C = AB$, the 4×4 matrix C can be written

$$C = \begin{pmatrix} c_{11} & c_{12} \\ c_{21} & c_{22} \end{pmatrix}$$

where the 2×2 sub-matrices c_{ij} are given by (cf equation (A.1)).

$c_{11} = a_{11}b_{11} + a_{12}b_{21}$, $c_{12} = a_{11}b_{12} + a_{12}b_{22}$, etc, or in general.
Hence

$$c_{ij} = a_{i1}b_{1j} + a_{i2}b_{2j} \tag{A.20}$$

The crucial difference between equation (A.20) and equation (A.1) is that in (A.20) the order of the 2×2 matrices must be preserved, because for example, $a_{11}b_{12} \neq b_{12}a_{11}$.

Similarly, if $|\bar{S}\rangle = \begin{pmatrix} k \\ l \\ m \\ n \end{pmatrix}$, then the product $|\bar{\phi}\rangle = A|\bar{S}\rangle = \begin{pmatrix} 1 & 2 & a & b \\ 3 & 4 & c & d \\ e & f & 5 & 6 \\ g & h & 7 & 8 \end{pmatrix}\begin{pmatrix} k \\ l \\ m \\ n \end{pmatrix}$ could be evaluated by multiplying each row of the 4×4 matrix A into the 4-component column matrix $|\bar{S}\rangle$ to obtain each element of the column matrix $|\bar{\phi}\rangle$. Alternatively, one could write A in terms of four sub-matrices as in equation (A.19). Similarly one could write $|\bar{S}\rangle$ as two sub-columns, as shown below

$$|\bar{S}\rangle = \begin{pmatrix} k \\ l \\ m \\ n \end{pmatrix} = \begin{pmatrix} |S_1\rangle \\ |S_2\rangle \end{pmatrix} \tag{A.21}$$

where $|S_1\rangle = \begin{pmatrix} k \\ l \end{pmatrix}$ and $|S_2\rangle = \begin{pmatrix} m \\ n \end{pmatrix}$. The result of the product of the two matrices could then be written $|\bar{\phi}\rangle = \begin{pmatrix} |\phi_1\rangle \\ |\phi_2\rangle \end{pmatrix}$. Then by direct multiplication of the 2×2 sub-matrices into 2 component sub-columns, one finds

$$|\bar{\phi}\rangle = A|\bar{S}\rangle = \begin{pmatrix} |\phi_1\rangle \\ |\phi_2\rangle \end{pmatrix} = A\begin{pmatrix} |S_1\rangle \\ |S_2\rangle \end{pmatrix} = \begin{pmatrix} a_{11}|S_1\rangle + a_{12}S_2\rangle \\ a_{21}|S_1\rangle + a_{22}|S_2\rangle \end{pmatrix} \tag{A.22}$$

i.e.

$$|\phi_1\rangle = a_{11}|S_1\rangle + a_{12}S_2\rangle \text{ and } |\phi_2\rangle = a_{21}|S_1\rangle + a_{22}|S_2\rangle \tag{A.23}$$

In general, if a_{11}, b_{11}, c_{11} are $N \times N$ square matrices and a_{22}, b_{22}, c_{22} are $M \times M$ square matrices, then a_{12}, b_{12}, c_{12} would be rectangular matrices with N rows and M columns, whereas a_{21}, b_{21}, c_{21} would have $= M$ rows and N columns. Breaking larger matrices into sub-matrices is a convenient method of keeping track of the contributions to products from different substructures.

A.10 Products of three matrices

If $C = hGH$, where h, G, H, C are square matrices, then equation (A.2) generalises to

$$c_{ij} = \sum_{k,l=1}^{N} h_{ik} G_{kl} H_{lj} \tag{A.24}$$

In many cases, the matrices are sparse, in which case this simplifies. For example, if the only non-zero matrix elements of h are h_{ia} and h_{ib}, then this simplifies to

$$c_{ij} = \sum_{l=1}^{N} (h_{ia} G_{al} + h_{ib} G_{bl}) H_{lj} \tag{A.25}$$

If the only non-zero matrix elements of H are $H_{a'j}$ and $H_{b'j}$, then this simplifies further to

$$c_{ij} = h_{ia} G_{aa'} H_{a'j} + h_{ia} G_{ab'} H_{b'j} + h_{ib} G_{ba'} H_{a'j} + h_{ib} G_{bb'} H_{b'j} \tag{A.26}$$

For the case where $G_{ba'} = 0$ and $G_{ab'} = 0$ this simplifies even further to

$$c_{ij} = h_{ia}G_{aa'}H_{a'j} + h_{ib}G_{bb'}H_{b'j} \tag{A.27}$$

Equation (A.27) is a simple example of the more general case, where h, G and H have the following sub-matrix structures $h = \begin{pmatrix} h_{11} & h_{12} \\ 0 & 0 \end{pmatrix}$, $G = \begin{pmatrix} G_{11} & 0 \\ 0 & G_{22} \end{pmatrix}$ and $H = \begin{pmatrix} H_{11} & 0 \\ H_{21} & 0 \end{pmatrix}$, in which case, the product of the three matrices becomes

$$hGH = \begin{pmatrix} h_{11} & h_{12} \\ 0 & 0 \end{pmatrix}\begin{pmatrix} G_{11} & 0 \\ 0 & G_{22} \end{pmatrix}\begin{pmatrix} H_{11} & 0 \\ H_{21} & 0 \end{pmatrix} = \begin{pmatrix} h_{11}G_{11}H_{11} + h_{12}G_{22}H_{21} & 0 \\ 0 & 0 \end{pmatrix} \tag{A.28}$$

If the matrix elements of G_{11} are $G_{aa'}$ and the matrix elements of G_{22} are $G_{bb'}$, then if the only non-zero matrix elements of h are h_{ia} and h_{ib} and if the only non-zero matrix elements of H are $H_{a'j}$ and $H_{b'j}$, then this simplifies to equation (A.27).

A.11 Products of inverses of matrices

Consider the product $C = AB$. Being careful to preserve the order of matrix products, note that

$$B^{-1}A^{-1}AB = B^{-1}B = I$$

Hence $C^{-1} = B^{-1}A^{-1}$. In other words, the inverse of a product of matrices such as AB is equal to the product of their inverses in reverse order. That is

$$B^{-1}A^{-1} = (AB)^{-1} \tag{A.29}$$

In the case where $A = \left(g_{11}^{-1} - \Sigma_1\right)$ and $B = g_{11}$, this yields

$$g_{11}^{-1}\left(g_{11}^{-1} - \Sigma_1\right)^{-1} = \left[\left(g_{11}^{-1} - \Sigma_1\right)g_{11}\right]^{-1} = [I - \Sigma_1 g_{11}]^{-1} \tag{A.30}$$

Similarly, if $G_{11} = \left(g_{11}^{-1} - \Sigma_1\right)^{-1}$, then

$$G_{11} = \left(g_{11}^{-1} - \Sigma_1\right)^{-1} = \left[(I - \Sigma_1 g_{11})g_{11}^{-1}\right]^{-1} = g_{11}(I - \Sigma_1 g_{11})^{-1} \tag{A.31}$$

A.12 Proof of the equivalence of the NNEs $G_{11}H_{12}g_{22} = g_{11}H_{12}G_{22}$

Multiply both equations from the left by G_{11}^{-1} and from the right by G_{22}^{-1} to yield

$$H_{12}g_{22}G_{22}^{-1} = G_{11}^{-1}g_{11}H_{12}$$

That is from equations (2.34) and (2.36) this can be written

$$H_{12}g_{22}\left(g_{22}^{-1} - \Sigma_2\right) = \left(g_{11}^{-1} - \Sigma_1\right)g_{11}H_{12}$$

i.e.

$$(H_{12} - H_{12}g_{22}\Sigma_2) = (H_{12} - \Sigma_1 g_{11}H_{12})$$

i.e.

$$H_{12}g_{22}\Sigma_2 = \Sigma_1 g_{11}H_{12}$$

From equations (2.30) and (2.25) this becomes

$$H_{12}g_{22}H_{21}g_{11}H_{12} = H_{12}g_{22}H_{21}g_{11}H_{12}$$

which is indeed correct.

A.13 Proof that the equation $G_{11} = \left(g_{11}^{-1} - \Sigma_1\right)^{-1}$ implies $G_{11} = g_{11} + g_{11}H_{12}G_{22}H_{21}g_{11}$

From equation (A.31), G_{11} can be written

$$G_{11} = g_{11}(I - \Sigma_1 g_{11})^{-1} \tag{A.32}$$

For any matrix X, since $I = I - X + X$ and $(I - X)^{-1}(I - X) = I$,

$$(I - X)^{-1} = (I - X)^{-1}(I - X + X) = I + (I - X)^{-1}X = I + X(I - X)^{-1} \tag{A.33}$$

Choosing $X = \Sigma_1 g_{11}$ yields for equation (A.32),

$$G_{11} = g_{11} + g_{11}(I - \Sigma_1 g_{11})^{-1}\Sigma_1 g_{11}$$

From equation (A.31), this can be written

$$G_{11} = g_{11} + G_{11}\Sigma_1 g_{11}$$

i.e.

$$G_{11} = g_{11} + G_{11}H_{12}g_{22}H_{21}g_{11}$$

Using the NNE $G_{11}H_{12}g_{22} = g_{11}H_{12}G_{22}$ yields

$$G_{11} = g_{11} + g_{11}H_{12}G_{22}H_{21}g_{11} \tag{A.34}$$

A.14 Dealing with large (possibly infinite) matrices

In the discussion of section 2.5, if substructure 1 has N_1 sites and substructure 2 has N_2 sites, then H_{12} is a matrix with N_1 rows and N_2 columns, while H_{21} is a matrix with N_2 rows and N_1 columns. If N_1 and/or N_2 are large, perhaps infinitely large, then it is important to keep track of the sizes of matrices and avoid the ill-defined and perhaps impossible task of inverting matrices of infinite dimensions. Even if N_1 and N_2 are finite, it is helpful to save effort by inverting matrices of the smallest possible size. In many structures, the coupling matrices H_{12} and H_{21} do not couple all sites in one substructure to all sites in the other substructure. If H_{12} couples only N_a sites in substructure 1 to N_b sites in substructure 2, where $N_a < N_1$ and $N_b < N_2$, then it is convenient to write H_{12} and H_{21} in the form

$$H_{12} = \begin{pmatrix} h_{ab} & 0 \\ 0 & 0 \end{pmatrix} \text{ and } H_{21} = \begin{pmatrix} h_{ba} & 0 \\ 0 & 0 \end{pmatrix} \tag{A.35}$$

where h_{ab} is a matrix with N_a rows and N_b columns, while H_{ba} is a matrix with N_b rows and N_a columns. Similarly it is convenient to write

$$|\phi_1\rangle = \begin{pmatrix} |\phi_a\rangle \\ |\phi_{a'}\rangle \end{pmatrix} \text{ and } |\phi_2\rangle = \begin{pmatrix} |\phi_b\rangle \\ |\phi_{b'}\rangle \end{pmatrix}, \tag{A.36}$$

where $|\phi_a\rangle$ is a column of N_a amplitudes on the N_a sites of substructure 1 coupled to substructure 2 by h_{ab} and $|\phi_{a'}\rangle$ is a column of the remaining $N_1 - N_a$ amplitudes of substructure 1. Similarly, $|\phi_b\rangle$ is a column of N_b amplitudes on the N_b sites of substructure 2 coupled to substructure 1 by h_{ab} and $|\phi_{b'}\rangle$ is a column of the remaining $N_2 - N_b$ amplitudes.

Using the above notation, equation (2.8) takes the form $|S_1\rangle = H_{12}|\phi_2\rangle = \begin{pmatrix} h_{ab} & 0 \\ 0 & 0 \end{pmatrix}\begin{pmatrix} |\phi_b\rangle \\ |\phi_{b'}\rangle \end{pmatrix} = \begin{pmatrix} h_{ab}|\phi_b\rangle \\ 0 \end{pmatrix}$ and therefore it is convenient to write

$$|S_1\rangle = \begin{pmatrix} |S_a\rangle \\ 0 \end{pmatrix}, \text{ where } |S_a\rangle = h_{ab}|\phi_b\rangle \tag{A.37}$$

Similarly the sum of sources on the right-hand side of equation (2.16) is of the form

$$|S_2\rangle = H_{21}|\phi_1\rangle + |S_2'\rangle = \begin{pmatrix} h_{ba} & 0 \\ 0 & 0 \end{pmatrix}\begin{pmatrix} |\phi_a\rangle \\ |\phi_{a'}\rangle \end{pmatrix} + |S_2'\rangle = \begin{pmatrix} h_{ba}|\phi_a\rangle \\ 0 \end{pmatrix} + |S_2'\rangle$$

Writing

$$|S_2\rangle = \begin{pmatrix} |S_b\rangle \\ |S_{b'}\rangle \end{pmatrix} \text{ and } |S_2'\rangle = \begin{pmatrix} |S_b'\rangle \\ |S_{b'}'\rangle \end{pmatrix} \tag{A.38}$$

yields

$$|S_b\rangle = h_{ba}|\phi_a\rangle + |S_b'\rangle \text{ and } |S_{b'}\rangle = |S_{b'}'\rangle \tag{A.39}$$

Using the above notation, equation (2.5) becomes $|\phi_1\rangle = \begin{pmatrix} |\phi_a\rangle \\ |\phi_{a'}\rangle \end{pmatrix} = g_{11}\begin{pmatrix} h_{ab}|\phi_b\rangle \\ 0 \end{pmatrix}$ and therefore it is convenient to write g_{11} in terms of sub-matrices as follows

$$g_{11} = \begin{pmatrix} g_{aa} & g_{aa'} \\ g_{a'a} & g_{a'a'} \end{pmatrix} \tag{A.40}$$

This yields

$$|\phi_1\rangle = \begin{pmatrix} |\phi_a\rangle \\ |\phi_{a'}\rangle \end{pmatrix} = \begin{pmatrix} g_{aa}h_{ab}|\phi_b\rangle \\ g_{a'a}h_{ab}|\phi_b\rangle \end{pmatrix} \tag{A.41}$$

Similarly, since $|\phi_2\rangle = g_{22}h_{21}|\phi_1\rangle$,

$$|\phi_2\rangle = \begin{pmatrix} |\phi_b\rangle \\ |\phi_{b'}\rangle \end{pmatrix} = g_{22}\begin{pmatrix} h_{ba} & 0 \\ 0 & 0 \end{pmatrix}\begin{pmatrix} |\phi_a\rangle \\ |\phi_{a'}\rangle \end{pmatrix} + g_{22}\begin{pmatrix} |S_b'\rangle \\ |S_{b'}'\rangle \end{pmatrix} = g_{22}\begin{pmatrix} h_{ba}|\phi_a\rangle \\ 0 \end{pmatrix} + g_{22}\begin{pmatrix} |S_b'\rangle \\ |S_{b'}'\rangle \end{pmatrix}$$

Therefore it is convenient to write

$$g_{22} = \begin{pmatrix} g_{bb} & g_{bb'} \\ g_{b'b} & g_{b'b'} \end{pmatrix} \tag{A.42}$$

This yields

$$\begin{pmatrix} |\phi_b\rangle \\ |\phi_{b'}\rangle \end{pmatrix} = \begin{pmatrix} g_{bb}[h_{ba}|\phi_a\rangle + |S_b'\rangle] + g_{bb'}|S_{b'}'\rangle \\ g_{b'b}[h_{ba}|\phi_a\rangle + |S_b'\rangle] + g_{b'b'}|S_{b'}'\rangle \end{pmatrix}$$

That is

$$\begin{pmatrix} |\phi_b\rangle \\ |\phi_{b'}\rangle \end{pmatrix} = \begin{pmatrix} g_{bb}[h_{ba}|\phi_a\rangle + |\bar{S}_b'\rangle] \\ g_{b'b}[h_{ba}|\phi_a\rangle + |\bar{S}_b'\rangle] \end{pmatrix} \tag{A.43}$$

where

$$|\bar{S}_b'\rangle = |S_b'\rangle + g_{bb}^{-1}g_{bb'}|S_{b'}'\rangle \tag{A.44}$$

The top row of equation (A.43) is $|\phi_b\rangle = g_{bb}[h_{ba}|\phi_a\rangle + |\bar{S}_b'\rangle]$. Substituting the top row of equation (A.41) into this yields

$$|\phi_b\rangle = g_{bb}[h_{ba}g_{aa}h_{ab}|\phi_b\rangle + |\bar{S}_b'\rangle]$$

That is

$$g_{bb}^{-1}|\phi_b\rangle = [\Sigma_b|\phi_b\rangle + |\bar{S}_b'\rangle]$$

where

$$\Sigma_b = [h_{ba}g_{aa}h_{ab}] \tag{A.45}$$

Hence

$$|\phi_b\rangle = G_{bb}|\bar{S}_b'\rangle \tag{A.46}$$

where

$$G_{bb} = \left(g_{bb}^{-1} - \Sigma_b\right)^{-1} \tag{A.47}$$

Substituting equation (A.45) into equation (A.41) yields

$$|\phi_1\rangle = \begin{pmatrix} |\phi_a\rangle \\ |\phi_{a'}\rangle \end{pmatrix} = \begin{pmatrix} G_{ab}|\bar{S}_b'\rangle \\ G_{a'b}|\bar{S}_b'\rangle \end{pmatrix} \tag{A.48}$$

where

$$G_{ab} = g_{aa}h_{ab}G_{bb} \text{ and } G_{a'b} = g_{a'a}h_{ab}G_{bb} \tag{A.49}$$

Note that following section A.12, one can also write

$$G_{ab} = g_{aa}h_{ab}G_{bb} = G_{aa}h_{ab}g_{bb} \tag{A.50}$$

where

$$G_{aa} = \left(g_{aa}^{-1} - \Sigma_a\right)^{-1} \text{ and } \Sigma_{aa} = h_{ab}g_{bb}h_{ba} \tag{A.51}$$

Finally, substituting equation (A.48) into the bottom row of equation (A.65) yields

$$|\phi_{b'}\rangle = g_{b'b}[h_{ba}G_{ab} + 1]|\bar{S}_b'\rangle] = g_{b'b}[\Sigma_b G_{bb} + 1]|\bar{S}_b'\rangle]$$

That is

$$|\phi_{b'}\rangle = g_{b'b}(I - \Sigma_b g_{bb})^{-1}|\bar{S}_b'\rangle \tag{A.52}$$

In obtaining the last line, it was noted that replacing 1 by 2 in equation (A.32) yields, $\Sigma_2 G_{22} = \Sigma_2 g_{22}(I - \Sigma_2 g_{22})^{-1}$. So choosing $X = \Sigma_2 g_{22}$ in equation (A.52) yields $+\Sigma_2 G_{22} = (I - \Sigma_2 g_{22})^{-1}$.

The crucial point in the above analysis is that only finite matrices are inverted, so the analysis is valid even if N_1 or N_2 are infinite. In this case, $g_{a'a}$ and/or $g_{b'b}$ would be matrices with an infinite number of rows, but their inverse is not required. Since $g_{a'a}$ and $g_{b'b}$ are presumed to be known, they can be used to evaluate the amplitudes $|\phi_{a'}\rangle$ and $|\phi_{b'}\rangle$. The text box below summarises the main results.

In summary, when a controlled source is applied to substructure 2 of the form

$$|S_2'\rangle = \begin{pmatrix} |S_b'\rangle \\ |S_{b'}'\rangle \end{pmatrix}$$

and only sites labelled a in 1 are coupled to sites labelled b in 2, then the resulting amplitudes in substructures 1 and 2 are $|\phi_1\rangle = \begin{pmatrix} |\phi_a\rangle \\ |\phi_{a'}\rangle \end{pmatrix}$ and $|\phi_2\rangle = \begin{pmatrix} |\phi_b\rangle \\ |\phi_{b'}\rangle \end{pmatrix}$, where

$$|\phi_b\rangle = G_{bb}|\bar{S}_b'\rangle \tag{A.46}$$

$$|\phi_a\rangle = G_{ab}|\bar{S}_b'\rangle \tag{A.48a}$$

$$G_{ab} = g_{aa}h_{ab}G_{bb} = G_{aa}h_{ab}g_{bb} \tag{A.50}$$

$$G_{bb} = \left(g_{bb}^{-1} - \Sigma_b\right)^{-1} \text{ and } \Sigma_b = h_{ba} g_{aa} h_{ab} \tag{A.47}$$

$$G_{aa} = \left(g_{aa}^{-1} - \Sigma_a\right)^{-1} \text{ and } \Sigma_{aa} = h_{ab} g_{bb} h_{ba} \tag{A.51}$$

$$|\bar{S}_b'\rangle = |S_b'\rangle + g_{bb}^{-1} g_{bb'} |S_{b'}'\rangle \tag{A.44}$$

$$|\phi_{a'}\rangle = g_{a'a} h_{ab} G_{bb} |\bar{S}_b'\rangle \tag{A.48b}$$

$$|\phi_{b'}\rangle = g_{b'b}(I - \Sigma_b g_{bb})^{-1} |\bar{S}_b'\rangle \tag{A.52}$$

Note that if the source is applied only to sites in 2, which are not coupled to sites in 1, then

$|S_{b'}'\rangle = g_{bb}^{-1} g_{bb'} |S_{b'}'\rangle$ and equations (A.48a) and the second of equations (A.50) yield

$$|\phi_a\rangle = G_{aa} h_{ab} g_{bb'} |\bar{S}_b'\rangle \tag{A.53}$$

A.15 Proof that the NNEs imply $\begin{pmatrix} G_{11} & G_{12} \\ G_{21} & G_{22} \end{pmatrix} = \begin{pmatrix} g_{11}^{-1} & -H_{12} \\ -H_{21} & g_{22}^{-1} \end{pmatrix}^{-1}$

This will be proved by demonstrating that

$$\begin{pmatrix} G_{11} & G_{12} \\ G_{21} & G_{22} \end{pmatrix} \begin{pmatrix} g_{11}^{-1} & -H_{12} \\ -H_{21} & g_{22}^{-1} \end{pmatrix} = \begin{pmatrix} I & 0 \\ 0 & I \end{pmatrix}$$

Proof:

$$\begin{pmatrix} G_{11} & G_{12} \\ G_{21} & G_{22} \end{pmatrix} \begin{pmatrix} g_{11}^{-1} & -H_{12} \\ -H_{21} & g_{22}^{-1} \end{pmatrix} = \begin{pmatrix} G_{11} g_{11}^{-1} - G_{12} H_{21} & -G_{11} H_{12} + G_{12} g_{22}^{-1} \\ G_{21} g_{11}^{-1} - G_{22} H_{21} & -G_{21} H_{12} + G_{22} g_{22}^{-1} \end{pmatrix}$$

From the NNE $G_{12} = G_{11} H_{12} g_{22}$, where $G_{11} = \left(g_{11}^{-1} - \Sigma_1\right)^{-1}$, the top left element of the product is

$$G_{11} g_{11}^{-1} - G_{12} H_{21} = G_{11}\left(g_{11}^{-1} - H_{12} g_{22} H_{21}\right) = G_{11}\left(g_{11}^{-1} - \Sigma_1\right) = I$$

Similarly, from the NNE $G_{21} = G_{22} H_{21} g_{11}$ the bottom right element is

$$-G_{21} H_{12} + G_{22} g_{22}^{-1} = G_{22}\left(-H_{21} g_{11} H_{12} + g_{22}^{-1}\right) = G_{22}\left(g_{22}^{-1} - \Sigma_2\right) = I$$

Also, the top right element is

$$-G_{11} H_{12} + G_{12} g_{22}^{-1} = G_{11}(-H_{12} + H_{12}) = 0$$

Similarly, the bottom left element also vanishes.

A.16 Note regarding the notation $\begin{pmatrix} G_{11} & G_{12} \\ G_{21} & G_{22} \end{pmatrix} = \begin{pmatrix} E - H_{11} & -H_{12} \\ -H_{21} & E - H_{22} \end{pmatrix}^{-1}$

In section A.15, the matrices g_{11}^{-1} and g_{22}^{-1} can be used to define two other matrices, H_{11} and H_{22} respectively, via the equations

$$H_{11} = E_{11} - g_{11}^{-1} \text{ and } H_{22} = E_{22} - g_{22}^{-1} \tag{A.54}$$

where E_{11} and E_{22} are square matrices, chosen for convenience.

As an example, one could choose $E_{11} = E_{22} = EI$, where E is an arbitrary constant, in which case

$H_{11} = EI - g_{11}^{-1}$ and $H_{22} = EI - g_{22}^{-1}$. Since the matrices H_{11} and H_{22} are obtained from the inverses of g_{11} and g_{22}, they will be referred to as 'inverse Green's functions'. As discussed in later chapters, they may or may not be Hermitian, depending on whether or not the substructures 1 and 2 are open or closed. Using the above notation,

$$G = (EI - H)^{-1} \tag{A.55}$$

where $G = \begin{pmatrix} G_{11} & G_{12} \\ G_{21} & G_{22} \end{pmatrix}$ and $H = \begin{pmatrix} H_{11} & H_{12} \\ H_{21} & H_{22} \end{pmatrix}$.

For isolated quantum substructures, such a 1 and 2, the matrices H_{11} and H_{22} are called Hamiltonians of the isolated substructures. For the structure formed by coupling substructures 1 and 2, H is the Hamiltonian of the whole structure. One can also define a matrix $\bar{H}_{11}$ via the equation

$$\bar{H}_{11} = EI - G_{11}^{-1} \tag{A.56}$$

From equation (2.27), this has the form

$$\bar{H}_{11} = H_{11} + \Sigma_1 \tag{A.57}$$

$\bar{H}_{11}$ is called a 'renormalized' Hamiltonian of substructure 1, in the presence of substructure 2. In terms of $\bar{H}_{11}$, the Green's function $G_{11} = (EI - \bar{H}_{11})^{-1}$. This describes waves in 1 created by a source in 1, in the presence of 2. The above concepts will be discussed in more detail in chapter 11.

References

[1] Magoga M and Joachim C 1999 Conductance of molecular wires connected or bonded in parallel *Phys. Rev.* B **59** 1601

[2] Lambert C J, Liu S-X and Magic A 2018 Ratio rule for beginners: a chemist's guide to quantum interference in molecules *Chem. A Eur. J.* **24** 4193

[3] Sangtarash S, Huang C, Sadeghi H, Sorohhov G, Hauser J, Wandlowski T, Hong W, Decurtins S, Liu S-X and Lambert C J 2015 Searching the hearts of graphene-like molecules for simplicity, sensitivity, and logic *J. Am. Chem. Soc.* **137** 11425

[4] Geng Y, Sangtarash S, Huang C, Sadeghi H, Fu Y, Hong W, Wandlowski T, Decurtins S, Lambert C J and Liu S-X 2015 Magic ratios for connectivity-driven electrical conductance of graphene-like molecules *J. Am. Chem. Soc.* **137** 4469

[5] Ulčakar L, Rejec T, Kokalj J, Sangtarash S, Sadeghi H, Ramšak A, Jefferson J H and Lambert C J 2019 On the resilience of magic number theory for conductance ratios of aromatic molecules *Sci. Rep.* **9** 3478
[6] Rakyta P, Alanazy A, Kormányos A, Tajkov Z, Kukucska G, Koltai J, Sangtarash S, Sadeghi H, Cserti J and Lambert C J 2019 Magic number theory of superconducting proximity effects and Wigner delay times in graphene-like molecules *J. Phys. Chem.* C **123** 6812
[7] Plaszkó N L, Rakyta P, Cserti J, Kormányos A and Lambert C J 2020 Quantum interference and non-equilibrium Josephson current in molecular Andreev interferometers *Nanomaterials* **10** 1033
[8] Markussen T, Stadler R and Thygesen K S 2010 The relation between structure and quantum interference in single molecule junctions *Nano Lett.* **10** 4260
[9] Markussen T, Stadler R and Thygesen K S 2011 Graphical prediction of quantum interference-induced transmission nodes in functionalized organic molecules *Phys. Chem. Chem. Phys.* **13** 14311

IOP Publishing

Quantum Transport in Nanostructures and Molecules
An introduction to molecular electronics
Colin John Lambert

Chapter 3

A beginner's guide to solving the Schrödinger equation

3.1 Introduction

The discussion in chapters 1 and 2 is intended to be accessible to the general reader, who may be interested in the broad concepts used to understand quantum interference in molecules and other phase-coherent structures. The chapters which follow are aimed at graduate students who would like to become proficient in the theoretical techniques and technical language used to describe phase-coherent transport.

In this chapter, the analogy between vibrations and de Broglie waves will be further developed by discussing similarities between the Schrödinger equation and Newton's laws of motion applied to harmonic oscillators. This enables the development of a common approach to solving electron and phonon transport problems in phase-coherent nanostructures. Even though the physical meaning of the symbols in the two problems are completely different, their equations of motion have much in common and therefore the steps needed to solve one are almost identical to the steps needed to solve the other.

As demonstrated below, the equation of motion of a system of classical oscillating masses connected by Hooke springs can be written

$$-\frac{d^2}{dt^2}|x(t)\rangle = H|x(t)\rangle \tag{3.1}$$

whereas the time-dependent Schrödinger equation can be written

$$i\hbar\frac{d}{dt}|\varphi(t)\rangle = H|\varphi(t)\rangle \tag{3.2}$$

In both of these equations H is a square matrix, with N rows and N columns. In equation (3.1) H is called the dynamical matrix, whereas in equation (3.2) H is called the Hamiltonian. In equation (3.1) the quantity $|x(t)\rangle$ is a column of numbers

doi:10.1088/978-0-7503-3639-0ch3

representing the instantaneous displacement from equilibrium of the oscillating masses, whereas in equation (3.2), $|\varphi(t)\rangle$ represents the instantaneous quantum mechanical wave function, evaluated at each site[1] of the quantum system. Chapter 16 describes how to obtain the matrix equation (3.2) from the differential-equation form of the Schrödinger equation, but this detail is not needed at this stage.

3.2 Mathematical aim: an introduction to linear algebra

A common approach to the solution of equations (3.1) and (3.2) exists, because they are both linear equations. For example in equation (3.1) only $|x(t)\rangle$ appears, whereas squares or cubes of displacements do not appear. Similarly, the connectivity theory discussed in chapter 2 is a consequence of the linear relationship between sources and waves. For this reason, this section begins by discussing the ideas and techniques of linear algebra, which underpin the language needed to solve these problems. This will involve developing an understanding of the following mathematical terms and how they allow the formal solutions of the above equations, to be written down almost by inspection. These key words are:

Boundary conditions and initial conditions, linearity, linear superposition, vectors, basis vectors, linear dependence, linear independence, eigenvalue, eigenvector, representation, completeness, bra vectors, ket vectors, normalisation, orthogonality, orthonormality, Hermitian, degenerate and non-degenerate.

The above key words and concepts form the language of matrices and will be used to describe the solutions to equations such as (3.1) and (3.2). By working through the following examples, it is intended that the reader will develop an intuitive understanding of their meaning and how to apply them when solving the equations of motion of linear dynamical systems.

These ideas will be developed by first considering the vibrational dynamics of harmonic oscillators and then moving on to consider the simple two-level quantum mechanical system introduced in section 2.9. Initially, a single oscillator will be considered and then two coupled oscillators will be discussed, whose equations of motion are a simple example of equation (3.1).

3.3 A simple harmonic oscillator

As a first example, consider a mass m connected to a rigid immovable wall by a Hooke spring of stiffness α, as shown in figure 3.1. When $x = 0$, the spring is neither stretched nor compressed and the system is in equilibrium. If the mass is displaced from equilibrium by a distance x, then according to Hooke's law, the restoring force is $F = -\alpha x$ and therefore from Newton's second law, one obtains the following equation of motion

$$m\frac{d^2x}{dt^2} = -\alpha x \tag{3.3}$$

[1] Here a 'site' is a pseudo atomic orbital (PAO), as discussed in section 16.8, and can be regarded as an atomic orbital, with a little leakage onto neighbouring atoms. For simplicity, PAOs are assumed to be orthogonal.

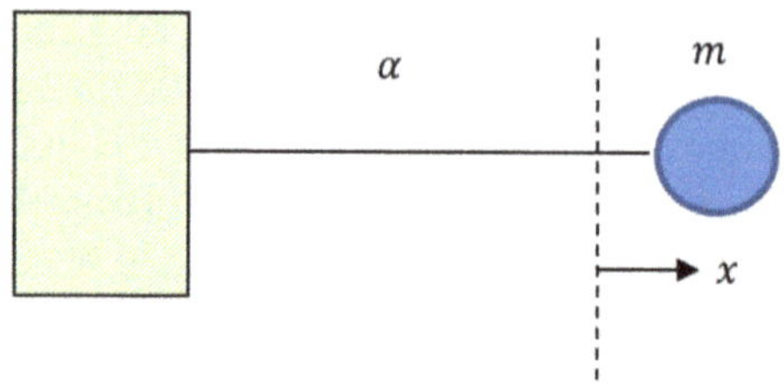

Figure 3.1. A mass m (blue) connected to an immovable wall (green) by a Hooke spring of stiffness α. The spring is denoted by a straight horizontal line. The dashed vertical line marks the equilibrium site of the mass, at which it experiences no force. x denotes the displacement from equilibrium.

Since this is a second order differential equation, the most general solution involves two arbitrary constants of integration and can be written

$$x(t) = A\cos(\omega t + B)$$

where A and B are the constants of integration. Since $\cos(\theta + 2\pi) = \cos\theta$, $x(t) = x(t + 2\pi/\omega)$, so $x(t)$ varies periodically in time, with period $2\pi/\omega$. Since the motion involves only a single angular frequency ω, it is referred to as simple harmonic motion. Since $\frac{dx}{dt} = -\omega A\sin(\omega t + B)$ and $\frac{d^2x}{dt^2} = -\omega^2 A\cos(\omega t + B)$, substituting this solution into equation (3.3) yields $\omega^2 = \alpha/m$. In other words, the (angular) frequency of oscillation is $\omega = \sqrt{\alpha/m}$.

Key words: Boundary conditions and initial conditions. Note that the constants of integration are determined by initial conditions, not by the equation of motion. For example, if at $t = 0$ the initial displacement is $x(0) = 1$ and the velocity $\dot{x}(t) = dx(t)/dt$ at $t = 0$ is $\dot{x}(0) = 0$, then $A\cos(B) = 1$ and $-A\omega\sin(B) = 0$, which can be solved to yield $A = 1$ and $B = 0$. As a second example, if the initial displacement is $x(0) = A\cos(B) = 0$ and the initial velocity is $\dot{x}(0) = -A\omega\sin(B) = 1$, then $B = \pi/2$ and $A = -1/\omega$. The generic phrase for such initial conditions is 'boundary conditions'. The use of boundary conditions to determine arbitrary constants is always encountered when solving dynamical problems.

3.4 The equations of motion of two harmonic oscillators

So far, the language of linear algebra has not been needed. This only becomes necessary when attempting to describe the motion of two or more masses. To demonstrate this, consider two masses, connected to each other by a Hooke spring β and to rigid immovable walls by springs of strength α. Figure 3.2 shows a representation of such a system, where the springs are denoted by horizontal straight lines.

If the displacements from equilibrium of the two masses are denoted x_1 and x_2, then the force on mass 1 is $F_1 = -\alpha x_1 - \beta(x_1 - x_2)$ and the force on mass 2 is $F_2 = -\alpha x_2 + \beta(x_1 - x_2)$. Note that forces to the right are defined to be positive and forces to the left are defined to be negative. As an example, to obtain the correct signs for F_1, note that if x_1 is positive, then the left-most spring α is stretched and the force it exerts on mass 1 is negative. Similarly if $(x_1 - x_2)$ is positive, then the spring

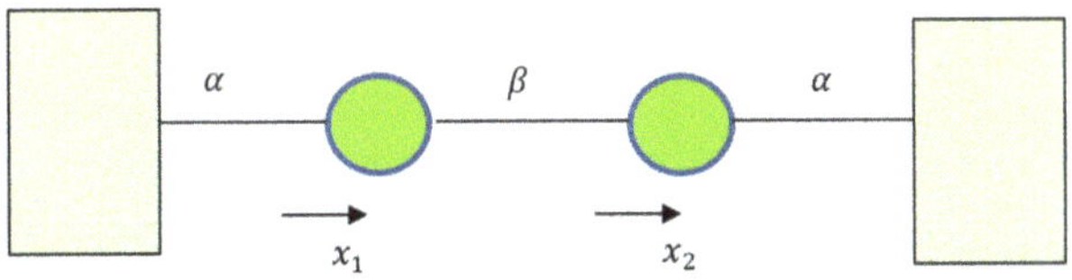

Figure 3.2. Two masses connected by harmonic springs to each other and to immovable walls.

β is compressed, so the force it exerts on mass 1 is negative and the force on mass 2 is positive. Hence from Newton's second law,

$$m\frac{d^2x_1}{dt^2} = -\alpha x_1 - \beta(x_1 - x_2) \text{ and } m\frac{d^2x_2}{dt^2} = -\alpha x_2 + \beta(x_1 - x_2)$$

Key words: Linearity and linear superposition. To see how the above equations illustrate the key words linearity and linear superposition, recall that the equation for a straight line is $y = mx + c$. That is $y = mx^1 + cx^0$ The right-hand side of this equation involves x raised to the power 1 or 0 only. It does not involve terms like x^2, x^3, ... etc. For this reason, any equation involving only quantities raised at most to the power 1 is called a linear equation. So the above two equations of motion for x_1 and x_2 are linear equations. Furthermore expressions such as $3x_1 + 4x_2$ are called linear superpositions of x_1 and x_2. Clearly the right-hand sides of the above equations involve linear superpositions of x_1 and x_2.

3.5 Matrix form of the equations of motion of two harmonic oscillators

To understand why the language of linear algebra is needed to describe the solution to the above equations of motion, note that they can be written in matrix form:

$$-\frac{d^2}{dt^2}\begin{pmatrix} x_1 \\ x_2 \end{pmatrix} = \begin{pmatrix} \varepsilon_0 & -\gamma \\ -\gamma & \varepsilon_0 \end{pmatrix}\begin{pmatrix} x_1 \\ x_2 \end{pmatrix} \tag{3.4}$$

where $\varepsilon_0 = (\alpha + \beta)/m$ and $\gamma = \beta/m$.

As discussed in chapter 2, it is convenient to store numbers such as x_1 and x_2 as a column of numbers $|x(t)\rangle$ of the form

$$|x(t)\rangle = \begin{pmatrix} x_1(t) \\ x_2(t) \end{pmatrix} \tag{3.5}$$

Similarly it is convenient to define the 2 × 2 dynamical matrix

$$H = \begin{pmatrix} \varepsilon_0 & -\gamma \\ -\gamma & \varepsilon_0 \end{pmatrix} \tag{3.6}$$

in which case equation (3.4) can be written

$$-\frac{d^2}{dt^2}|x(t)\rangle = H|x(t)\rangle \tag{3.7}$$

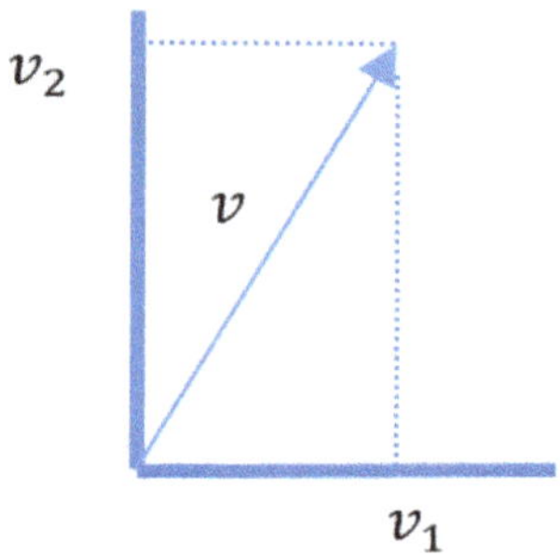

Figure 3.3. A vector $\underline{v}$ in two dimensions.

This is an example of equation (3.1). It is common to refer to the numbers x_1 and x_2 as 'displacements from equilibrium' or 'degrees of freedom'. For a more complicated system involving N degrees of freedom, H would be a $N \times N$ matrix and $|x(t)\rangle$ would be a N-component column of displacements. For the above system with two degrees of freedom, our goal is to compute the time-dependent displacements $x_1(\mathrm{t})$ and $x_2(t)$. Before proceeding, it is useful to examine the column of numbers $|x(t)\rangle$ from different viewpoints and to understand how it can be manipulated mathematically.

Key words: Vectors and basis vectors. Clearly the first component of $|x\rangle$ is x_1 and the second component of $|x\rangle$ is x_2. This reminds us of the notation used for ordinary vectors. For example in 2 dimensions, a vector $\underline{v}$ has components v_1 in the horizontal direction and v_2 in the vertical direction, as shown in figure 3.3.

The vector $\underline{v}$ could also be written as a column vector $|v\rangle$:

$$|v\rangle = \begin{pmatrix} v_1 \\ v_2 \end{pmatrix} \tag{3.8}$$

This is equivalent to writing $|v\rangle = v_1\binom{1}{0} + v_2\binom{0}{1}$ or equivalently

$$|v\rangle = v_1|1\rangle + v_2|2\rangle \tag{3.9}$$

where $|1\rangle = \binom{1}{0}$ is a unit vector along the horizontal axis and $|2\rangle = \binom{0}{1}$ is a unit vector along the vertical axis, as shown in figure 3.4.

In the language of linear algebra, the vectors $|1\rangle$ and $|2\rangle$ are called 'basis vectors'. Equation (3.9) and figure 3.4 show that $|v\rangle$ can be written as a linear superposition of the basis vectors $|1\rangle$ and $|2\rangle$, with amplitudes v_1 and v_2 respectively. Of course this way of representing $|v\rangle$ is not unique. For example equation (3.8) could be written

$$|v\rangle = \begin{pmatrix} \dfrac{v_1 + v_2}{2} + \dfrac{v_1 - v_2}{2} \\ \dfrac{v_1 + v_2}{2} - \dfrac{v_1 - v_2}{2} \end{pmatrix} = \frac{v_1 + v_2}{2}\begin{pmatrix} 1 \\ 1 \end{pmatrix} + \frac{v_1 - v_2}{2}\begin{pmatrix} 1 \\ -1 \end{pmatrix} \tag{3.10}$$

That is

$$|v\rangle = a_1|\psi_1\rangle + a_2|\psi_2\rangle \tag{3.11}$$

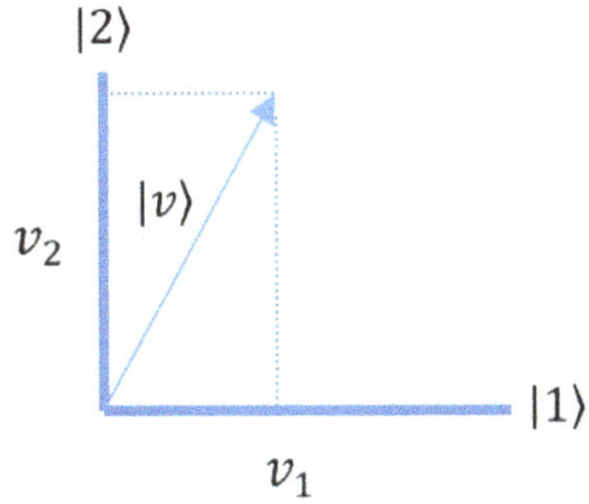

Figure 3.4. A vector $|v\rangle$ in two dimensions.

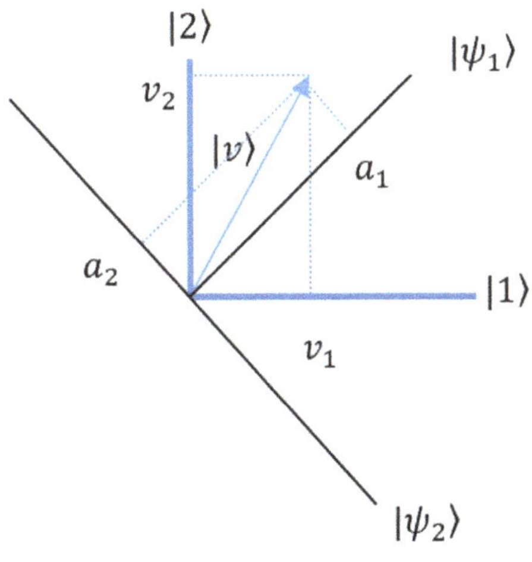

Figure 3.5. A vector $|v\rangle$ in two dimensions.

where $|\psi_1\rangle = \binom{1}{1}$, $a_1 = \frac{v_1 + v_2}{2}$, $|\psi_2\rangle = \binom{1}{-1}$ and $a_2 = \frac{v_1 - v_2}{2}$. This shows that the vectors $|\psi_1\rangle$ and $|\psi_2\rangle$ form a new pair of basis vectors and that $|v\rangle$ can be written as a linear superposition of these basis vectors, with amplitudes a_1 and a_2 respectively, as shown in figure 3.5.

In equations (3.9) and (3.11), $|v\rangle$ is the same vector in both cases. However the amplitudes of the vector depend on the chosen basis. In the basis $|1\rangle, |2\rangle$ the amplitudes are v_1, v_2, whereas in the basis $|\psi_1\rangle, |\psi_2\rangle$ the amplitudes are a_1, a_2. This demonstrates that to specify a vector such as $|v\rangle$, it is necessary to specify both the basis vectors and the amplitudes.

Key words: Linear independence, linear dependence and completeness. Of course one can think of many other choices for the basis vectors $|\psi_1\rangle$ and $|\psi_2\rangle$, because the 2-dimensional vector $|v\rangle$ can be written as a linear superposition of any two vectors $|\psi_1\rangle$ and $|\psi_2\rangle$, provided $|\psi_1\rangle$ and $|\psi_2\rangle$ are not parallel.

Vectors which are not parallel are said to be **linearly independent**, so in two dimensions, any two linearly-independent vectors form a basis. The idea of linear independence is useful, because as shown below, it helps to solve equation (3.1).

Clearly if two vectors $|\psi_1\rangle$ and $|\psi_2\rangle$ are linearly independent, then if c is a constant, one cannot write $b|\psi_1\rangle = c|\psi_2\rangle$, because this would imply that $|\psi_1\rangle$ is parallel to $|\psi_2\rangle$ which contradicts the fact that they are linearly independent (i.e. not parallel). This means that if $|\psi_1\rangle$ and $|\psi_2\rangle$ are linearly independent, then the only way the equation $b|\psi_1\rangle = c|\psi_2\rangle$ can be satisfied is if $c = b = 0$. Similarly, if $|\psi_1\rangle$ and $|\psi_2\rangle$ are linearly independent, then the only way the equation $a_1|\psi_1\rangle + a_2|\psi_2\rangle = 0$ can be satisfied is if $a_1 = a_2 = 0$. This argument means that if $|\psi_1\rangle$ and $|\psi_2\rangle$ are linearly independent and if the following equation is satisfied,

$$a_1|\psi_1\rangle + a_2|\psi_2\rangle = 0 \tag{3.12a}$$

then one can conclude that

$$a_1 = 0 \text{ and } a_2 = 0 \tag{3.12b}$$

Linear independence is important, because as already mentioned and shown in figure 3.6, if $|\psi_1\rangle$ and $|\psi_2\rangle$ are linearly independent, then any other vector $|v\rangle$ which lies in the 2-dimensional plane containing $|\psi_1\rangle$ and $|\psi_2\rangle$ can be written as a linear superposition $|v\rangle = a|\psi_1\rangle + b|\psi_2\rangle$, where a and b are constants. The set of two vectors $|\psi_1\rangle$ and $|\psi_2\rangle$ is said to be '**complete**', because for any choice of the two-dimensional vector $|v\rangle$, no other vectors are needed on the right-hand side of equation (3.11).

A vector such as $|v\rangle$, which can be written as a linear superposition of two other vectors $|\psi_1\rangle$ and $|\psi_2\rangle$ is said to be **linearly dependent** on $|\psi_1\rangle$ and $|\psi_2\rangle$. On the other hand, if $|v\rangle$ points outside of the 2-dimensional plane containing $|\psi_1\rangle$ and $|\psi_2\rangle$ then $|v\rangle$ cannot be equal to $a|\psi_1\rangle + b|\psi_2\rangle$, where a and b are constants, because $a|\psi_1\rangle+b|\psi_2\rangle$ lies in the plane containing $|\psi_1\rangle$ and $|\psi_2\rangle$, whereas $|v\rangle$ does not. In this case, $|v\rangle$, $|\psi_1\rangle$ and $|\psi_2\rangle$ are obviously linearly independent. The discussion leading to equations (3.12a) and (3.12b) can now be repeated, because if c is a constant and $|v\rangle$, $|\psi_1\rangle$ and $|\psi_2\rangle$ are linearly independent, then the vector $-c|v\rangle$ does not lie in the plane containing $|\psi_1\rangle$ and $|\psi_2\rangle$. Hence $-c|v\rangle \neq a|\psi_1\rangle + b|\psi_2\rangle$ i.e. $a|\psi_1\rangle + b|\psi_2\rangle + c|v\rangle \neq 0$ In other words, adding three linearly-independent vectors cannot result in zero. Alternatively one can state that if $|v\rangle$, $|\psi_1\rangle$ and $|\psi_2\rangle$ are linearly independent, then the only way the equation $a|\psi_1\rangle + b|\psi_2\rangle + c|v\rangle = 0$ can be satisfied is if $a = 0$, $b = 0$ and $c = 0$. This is a more democratic statement about linear dependence, because it treats the vectors $|v\rangle$, $|\psi_1\rangle$ and $|\psi_2\rangle$ on an equal footing.

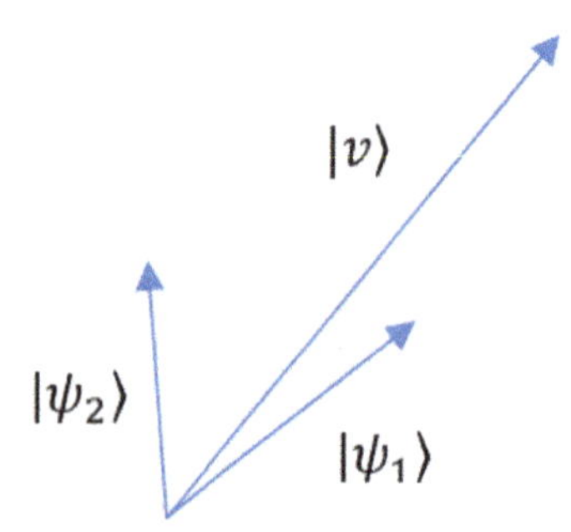

Figure 3.6. Three vectors $|v\rangle$, $|\psi_1\rangle$, $|\psi_2\rangle$ in two dimensions. Any one of them is linearly dependent on the other two.

3.6 Combining linear independence with the best choice of basis to solve equation (3.7)

Equations (3.12a) and (3.12b) will now be used to solve equation (3.7). The first step is to express the solution $|x(t)\rangle$, at any instant of time t, as a linear superposition of two linearly independent basis vectors $|\psi_1\rangle$ and $|\psi_2\rangle$, by writing

$$|x(t)\rangle = a_1(t)|\psi_1\rangle + a_2(t)|\psi_2\rangle \tag{3.13}$$

As mentioned above, the choice of basis vectors $|\psi_1\rangle$ and $|\psi_2\rangle$ is not unique, so later a convenient choice will be made, which helps to find a solution to equation (3.7). Once $|\psi_1\rangle$ and $|\psi_2\rangle$ have been chosen, the task is to compute the amplitudes $a_1(t)$ and $a_2(t)$.

Since the basis vectors are time independent, whereas $|x(t)\rangle$ is time dependent, the amplitudes $a_1(t)$ and $a_2(t)$ are time dependent. Substituting equation (3.13) into equation (3.7) yields

$$-\frac{d^2}{dt^2}a_1(t)|\psi_1\rangle + -\frac{d^2}{dt^2}a_2(t)|\psi_2\rangle = a_1(t)H|\psi_1\rangle + a_2(t)H|\psi_2\rangle \tag{3.14}$$

Since any choice of linearly-independent basis vectors is allowed, it is convenient to choose $|\psi_1\rangle$ and $|\psi_2\rangle$ to satisfy

$$H|\psi_1\rangle = E_1|\psi_1\rangle \tag{3.15}$$

$$H|\psi_2\rangle = E_2|\psi_2\rangle \tag{3.16}$$

where E_1 and E_2 are numbers. Equations (3.15) and (3.16) define the best choice of basis vectors, because with such a choice, equation (3.14) becomes easy to solve. The vectors $|\psi_1\rangle$ and $|\psi_2\rangle$ which satisfy equations (3.15) and (3.16) will be discussed in the next section, where it will be shown that $|\psi_1\rangle$ and $|\psi_2\rangle$ are indeed linearly independent. To see why this choice leads to a solution, substitute equations (3.15) and (3.16) into (3.14) to obtain

$$-\frac{d^2}{dt^2}a_1(t)|\psi_1\rangle + -\frac{d^2}{dt^2}a_2(t)|\psi_2\rangle = a_1(t)E_1|\psi_1\rangle + a_2(t)E_2|\psi_2\rangle \tag{3.17}$$

or alternatively

$$b_1(t)|\psi_1\rangle + b_2(t)|\psi_2\rangle = 0 \tag{3.18}$$

where

$$b_1(t) = \frac{d^2}{dt^2}a_1(t) + a_1(t)E_1 \text{ and } b_2(t) = \frac{d^2}{dt^2}a_2(t) + a_2(t)E_2 \tag{3.19}$$

Since, as shown below, $|\psi_1\rangle$ and $|\psi_2\rangle$ are linearly independent, comparing equation (3.19) with equations (3.12a) and (3.12b) leads to the conclusion that at any instant of time t, $b_1(t) = 0$ and $b_2(t) = 0$.

Hence, the amplitudes $a_1(t)$ and $a_2(t)$ are given by

$$-\frac{d^2}{dt^2}a_1(t) = a_1(t)E_1 \text{ and } -\frac{d^2}{dt^2}a_2(t) = a_2(t)E_2 \tag{3.20}$$

The solution to these equations (Note. Check this by differentiating equations (3.21) twice) is

$$a_1(t) = A\cos(\omega_1 t + B) \text{ and } a_2(t) = C\cos(\omega_2 t + D) \tag{3.21}$$

where $\omega_1^2 = E_1$, $\omega_2^2 = E_2$ and A, B, C, D are constants of integration, which are determined by initial conditions. For example if the boundary conditions are such that $a_1(0) = a_2(0) = 1$ and $\dot{a}_1(0) = \dot{a}_2(0) = 0$, then $A = C = 1$ and $B = D = 0$. Hence $a_1(t) = \cos(\omega_1 t)$ and $a_2(t) = \cos(\omega_2 t)$.

Since the amplitudes $a_1(t)$ and $a_2(t)$ in equation (3.13) are now known, the problem of finding $|x(t)\rangle$ is solved, provided the linearly-independent basis vectors $|\psi_1\rangle$ and $|\psi_2\rangle$ can be obtained.

3.7 Eigenvalues and eigenvectors

The vectors $|\psi_1\rangle$ and $|\psi_2\rangle$ are solutions of equations (3.15) and (3.16), where H is a square, 2×2 matrix. In general, if H is a known $N \times N$ square matrix, E is a number and $|\psi\rangle$ is a N-dimensional vector, then the problem of solving an equation of the form

$$H|\psi\rangle = E|\psi\rangle \tag{3.22}$$

is called an eigenvalue problem. The vector $|\psi\rangle$ is called an eigenvector and the number E is called an eigenvalue. Hence to solve equations (3.15) and (3.16), there is a need to be able to find the eigenvectors and eigenvalues of the matrix H.

To obtain the eigenvectors of H, note that

$$H\begin{pmatrix}1\\1\end{pmatrix} = \begin{pmatrix}\varepsilon_0 & -\gamma\\ -\gamma & \varepsilon_0\end{pmatrix}\begin{pmatrix}1\\1\end{pmatrix} = \begin{pmatrix}\varepsilon_0 - \gamma\\ \varepsilon_0 - \gamma\end{pmatrix} = (\varepsilon_0 - \gamma)\begin{pmatrix}1\\1\end{pmatrix}$$

Hence the vector $|\psi_1\rangle = \begin{pmatrix}1\\1\end{pmatrix}$ satisfies equation (3.15) with eigenvalue $E_1 = \varepsilon_0 - \gamma$. Similarly one finds that the vector $|\psi_2\rangle = \begin{pmatrix}1\\-1\end{pmatrix}$ satisfies equation (3.16) with eigenvalue $E_2 = \varepsilon_0 + \gamma$. (Note. Prove this by evaluating the product $\begin{pmatrix}1\\-1\end{pmatrix}$.)

Hence from equation (3.13)

$$|x(t)\rangle = \begin{pmatrix}x_1(t)\\x_2(t)\end{pmatrix} = a_1(t)|\psi_1\rangle + a_2(t)|\psi_2\rangle = \begin{pmatrix}a_1(t) + a_2(t)\\a_1(t) - a_2(t)\end{pmatrix} \tag{3.23}$$

As an example, if the initial conditions are such that $a_1(t) = \cos(\omega_1 t)$ and $a_2(t) = \cos(\omega_2 t)$, then the above solution for the time dependence of $x_1(t)$ and $x_2(t)$ is

$$\begin{pmatrix}x_1(t)\\x_2(t)\end{pmatrix} = \begin{pmatrix}\cos(\omega_1 t) + \cos(\omega_2 t)\\ \cos(\omega_1 t) - \cos(\omega_2 t)\end{pmatrix} \tag{3.24}$$

In summary, the solution to equation (3.7) can be written in the form

$$|x(t)\rangle = x_1(t)|1\rangle + x_2(t)|2\rangle \tag{3.25}$$

where $x_1(t) = \cos(\omega_1 t) + \cos(\omega_2 t)$, $x_2(t) = \cos(\omega_1 t) - \cos(\omega_2 t)$, $|1\rangle = \binom{1}{0}$ and $|2\rangle = \binom{0}{1}$.

Alternatively one can write

$$|x(t)\rangle = a_1(t)|\psi_1\rangle + a_2(t)|\psi_2\rangle \tag{3.26}$$

where $a_1(t) = \cos(\omega_1 t)$, $a_2(t) = \cos(\omega_2 t)$, $|\psi_1\rangle = \binom{1}{1}$ and $|\psi_2\rangle = \binom{1}{-1}$.

Key word: Representations. Equations (3.25) and (3.26) are different representations of the left-hand side of equation (3.24). In equation (3.25), the amplitudes $x_1(t)$ and $x_2(t)$ are relatively complicated (because they are sums of cosines with two different angular frequencies ω_1 and ω_2) and the vectors $|1\rangle$ and $|2\rangle$ are simple, whereas in equation (3.26), the amplitudes $a_1(t)$ and $a_2(t)$ are simple, but the vectors $|\psi_1\rangle$ and $|\psi_2\rangle$ are more complicated. The choice of which representation to use is simply a matter of taste, because the left-hand side of equation (3.24) is the same in both cases. However it is important to be able to switch between the two representations, because sometimes one representation is more convenient than the other. For example, although the most general motion of the masses involves a sum of two frequencies, equation (3.26) reveals that there are two special types of motion that involve only single frequencies, namely

$$|x(t)\rangle = a_1(t)|\psi_1\rangle \tag{3.27}$$

or

$$|x(t)\rangle = a_2(t)|\psi_2\rangle \tag{3.28}$$

where $a_1(t)$ and $a_2(t)$ are given by equation (3.21).

Key word: Normal modes. The two types of motion (3.27) and (3.28) are called 'normal modes' of vibration and the representation (3.26) tells that that the most general motion is a linear superposition of these two normal modes. Note that for normal modes of vibration, the ratio $x_1(t)/x_2(t)$ is stationary; i.e. it does not depend on time. Therefore normal modes are examples of 'stationary states,' discussed in chapter 2. In equation (3.28), since $|\psi_2\rangle = \binom{1}{-1}$, the masses are moving in opposite directions at any instant, whereas for the normal mode (3.27), since $|\psi_1\rangle = \binom{1}{1}$, the masses are moving in the same direction at any instant of time. As shown in figure 3.7(a), for the latter motion, the spring β is not stretched, and therefore the frequency ω_1 does not involve β. (Note. $\omega_1{}^2 = E_1 = \varepsilon_0 - \gamma = \alpha/m$.)

The right-hand sides of figures 3.7(a) and 3.7(b) show an alternative method of depicting these normal modes, in which blue circles represent displacements of one sign (say positive) and red circles represent displacements of the opposite sign (say negative). So $\binom{1}{1}$ is represented by two blue circles and $\binom{1}{-1}$ is represented by a blue

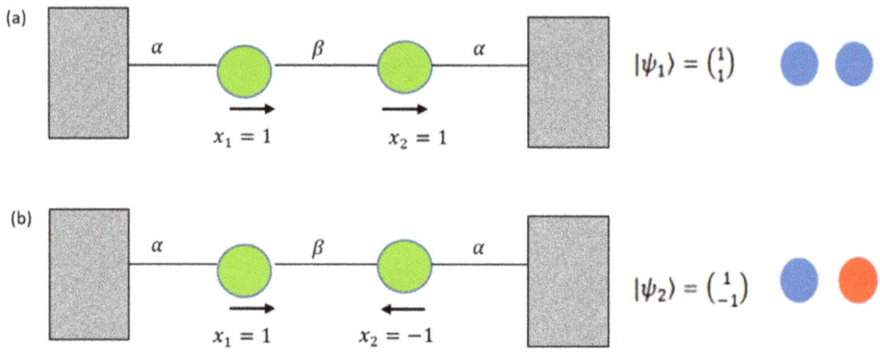

Figure 3.7. (a) The normal mode $|\psi_1\rangle = \begin{pmatrix} 1 \\ 1 \end{pmatrix}$ in which spring β is not stretched. (b) The normal mode $|\psi_2\rangle = \begin{pmatrix} 1 \\ -1 \end{pmatrix}$, in which spring β is periodically stretched and compressed.

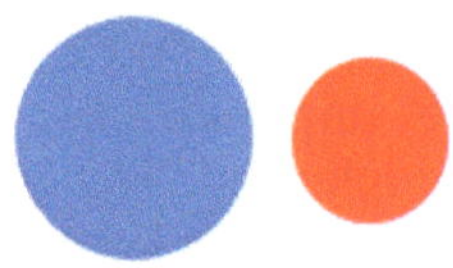

Figure 3.8. A pictorial representation of the normal mode $= \begin{pmatrix} 2 \\ -1 \end{pmatrix}$.

and a red circle. If the radius of the circles is proportional to displacement of a particular mass, then for example the state $\begin{pmatrix} 2 \\ -1 \end{pmatrix}$ would be represented by figure 3.8.

Note that interchanging the two blue circles representing $|\psi_1\rangle$ in figure 3.7(a) changes nothing and therefore this lowest frequency mode is a symmetric mode. Similarly interchanging the two circles representing $|\psi_2\rangle$ yields a state $-|\psi_2\rangle = \begin{pmatrix} -1 \\ 1 \end{pmatrix}$ and therefore the highest frequency mode $|\psi_2\rangle$ is said to be anti-symmetric. As shown in figure 2.13, when solving for the energy levels of a two-atom molecule, it was found that the lowest energy molecular orbital is also symmetric and the highest energy orbital is anti-symmetric.

3.8 Solution to the general $N \times N$ dynamical problem

The above problem of two oscillating masses demonstrates that eigenvectors such as $|\psi_1\rangle$ and $|\psi_2\rangle$ are useful vectors for solving dynamical problems. In general, for more complicated problems, the underlying equation of motion will involve a $N \times N$ square matrix H and as shown in section 3.14, there are guaranteed to be N linearly-independent solutions to the eigenvalue equation (3.22). To clarify how these eigenvectors can be used to solve the general dynamical problems (3.1) and (3.2), various properties of eigenvectors and eigenvalues of $N \times N$ square matrices will now be highlighted.

Consider first a system of masses described by a $N \times N$ dynamical matrix H and suppose that H possesses N linearly independent eigenvectors $|\psi_n\rangle$, where $n = 1,2, \ldots N$. These satisfy $H|\psi_n\rangle = E_n|\psi_n\rangle$, so the solution to equation (3.1) can be written as a linear superposition of the form (cf equation (3.13))

$$|x(t)\rangle = \sum_{n=1}^{N} a_n(t)|\psi_n\rangle \tag{3.29}$$

Substituting into equation (3.1) yields

$$\sum_{n=1}^{N} -\frac{d^2}{dt^2} a_n(t)|\psi_n\rangle = \sum_{n=1}^{N} a_n(t)H|\psi_n\rangle = \sum_{n=1}^{N} a_n(t)E_n|\psi_n\rangle \tag{3.30}$$

Equation (3.30) can be written (cf equation (3.18))

$$\sum_{n=1}^{N} b_n(t)|\psi_n\rangle = 0 \tag{3.31}$$

where $b_n(t) = \frac{d^2}{dt^2}a_n(t) + a_n(t)E_n$. Since the vectors $|\psi_n\rangle$ are linearly independent, $b_n(t) = 0$ for all n. Consequently $\frac{d^2}{dt^2}a_n(t) + a_n(t)E_n = 0$, which means that (cf equation (3.21))

$$a_n(t) = A_n \cos(\omega_n t + B_n) \tag{3.32}$$

where $\omega_n^2 = E_n$ and A_n, B_n are constants of integration, which are determined by initial conditions. Equation (3.29) reveals that the general solution is a linear superposition of N normal modes of vibration, each with its own normal mode frequency ω_n and amplitude $a_n(t)$, where the latter contains constants of integration, which are determined by initial conditions.

3.9 Solution to the general $N \times N$ Schrödinger equation

Exactly the same steps lead to the solution of equation (3.2), in which $\hbar$ is Planck's constant divided by 2π and $i^2 = -1$. Instead of equation (3.29) write

$$|\varphi(t)\rangle = \sum_{n=1}^{N} a_n(t)|\psi_n\rangle \tag{3.33}$$

and substitute into equation (3.2) to again obtain equation (3.31), where in this case, $b_n(t) = -i\hbar\frac{d}{dt}a_n(t) + a_n(t)E_n$. Since the vectors $|\psi_n\rangle$ are linearly independent, $b_n(t) = 0$. That is $\frac{d}{dt}a_n(t) = -\frac{i}{\hbar}E_n a_n(t)$. The solution to this equation is

$$a_n(t) = A_n \exp(-i\omega_n t)$$

where $\hbar\omega_n = E_n$ and A_n is a constant of integration, which is determined by initial conditions.

In summary, the general solution to equation (3.2) is

$$|\varphi(t)\rangle = \sum_{n=1}^{N} A_n \exp(-i\omega_n t)|\psi_n\rangle \tag{3.34}$$

where the constants of integration A_n are determined by boundary (i.e. initial) conditions.

The story is almost complete, provided one can prove that H has N linearly-independent eigenvectors. For the dynamical problems described by equations (3.29) and (3.33), a general method of determining the constants A_n and B_n is also needed. For example, to describe the motion that arises when the initial displacements are known to be $x_1(0)$, $x_2(0)$, ... $x_N(0)$ and the initial velocities are known to be $\dot{x}_1(0)$, $\dot{x}_2(0)$, ... $\dot{x}_N(0)$, there is a need to determine the constants A_n and B_n, which correspond to this initial condition. Similarly for a given choice of the initial vector $|\varphi(0)\rangle$ in equation (3.34), the constants A_n must be determined. To address these issues, it is first useful to discuss general properties of eigenvectors and eigenvalues.

3.10 How many linearly-independent eigenvectors does a dynamical matrix H possess?

The above analysis demonstrates that to solve the equation of motion of a system possessing a $N \times N$ dynamical matrix H, it is first necessary to find N linearly independent eigenvectors $|\psi_n\rangle$, satisfying

$$H|\psi_n\rangle = E_n|\psi_n\rangle \tag{3.35}$$

If such a set of N such vectors can be found, then they are said to be 'complete,' because any other N-dimensional vector $|\varphi\rangle$ can be written as a linear superposition:

$$|\varphi\rangle = \sum_{n=1}^{N} a_n|\psi_n\rangle \tag{3.36}$$

In other words, the vectors $|\psi_n\rangle$ form a representation. On the other hand, if only $N - 1$ solutions to equation (3.35) could be found, then they would not be complete and in general, it would not be correct to write equation (3.36). Therefore this raises the question: how do we know that there are N linearly-independent solutions to equation (3.35)?

From the discussion of two harmonic oscillators, it is clear that there can be no more than two linearly-independent vectors in two dimensions and similarly, there can be no more than N linearly-independent vectors is N dimensions. Therefore a $N \times N$ matrix cannot have more than N linearly-independent eigenvectors. However it may have less than N.

As an example, consider the matrix shown below

$$A = \begin{pmatrix} 0 & 1 \\ 0 & 0 \end{pmatrix} \tag{3.37}$$

This satisfies $A|\psi\rangle = E|\psi\rangle$, where the eigenvalue is $E = 0$ and the eigenvector is $|\psi\rangle = \begin{pmatrix} 1 \\ 0 \end{pmatrix}$. However there are no other linearly-independent eigenvectors. In other words, the eigenvectors of A do not form a complete set. (Amusingly MATLAB mistakenly claims that this matrix has two linearly independent eigenvectors, but close inspection of MATLAB's two vectors reveals that they are identical to within machine accuracy. i.e. although MATLAB claims to have found two linearly independent vectors, it has only found one.)

Fortunately, the dynamical matrices and Hamiltonians encountered in nature are not of the form of the matrix A. Instead, they belong to a class of matrices known as Hermitian matrices and as shown below, $N \times N$ Hermitain matrices are guaranteed to possess N linearly independent eigenvectors. In other words, the eigenvectors of a Hermitian matrix form a complete set. Before proving this property, we first digress a little and discuss the quantum analogue of the above two-oscillator problem.

3.11 A two-level quantum system

In this section, ideas which underpin the language of quantum mechanics will be recalled, starting from a simple example of a 'two-site' quantum system possessing wave amplitudes ψ_1, ψ_2 located on sites labelled 1 and 2. The physical meaning of these amplitudes is not important. As an example, as discussed in section 2.9, for a molecule such as that shown in figure 2.12, they could be regarded as amplitudes of wave functions on π-like atomic orbitals. If a measurement is made, the probability that the system is found to reside on site 1 is proportional to $|\psi_1|^2$ and the probability that the system is found to reside on site 2 is proportional to $|\psi_2|^2$.

Figure 3.9 shows a cartoon representation of such a two-level system.

As discussed in chapter 2, it is convenient to store the two numbers ψ_1 and ψ_2 as a two-component vector. For example, one could define the two-component column vector $|\psi\rangle$ as

$$|\psi\rangle = \begin{pmatrix} \psi_1 \\ \psi_2 \end{pmatrix} \tag{3.38}$$

The main difference between this vector and those encountered so far is that ψ_1 and ψ_2 could be complex numbers. This means that one must be careful when defining the length of such a vector. To illustrate this feature, consider a two-dimensional vector with components v_1 in the x direction and v_2 in the y direction, as shown in figure 3.10.

As discussed above, the vector $\underline{v}$ could be represented by a column vector $|v\rangle$:

$$|v\rangle = \begin{pmatrix} v_1 \\ v_2 \end{pmatrix}$$

Alternatively $\underline{v}$ could be represented by as a row vector $\langle v|$, defined by

$$\langle v| = (v_1 \quad v_2)$$

Figure 3.9. A cartoon of a two-level system, where the levels are represented by circles.

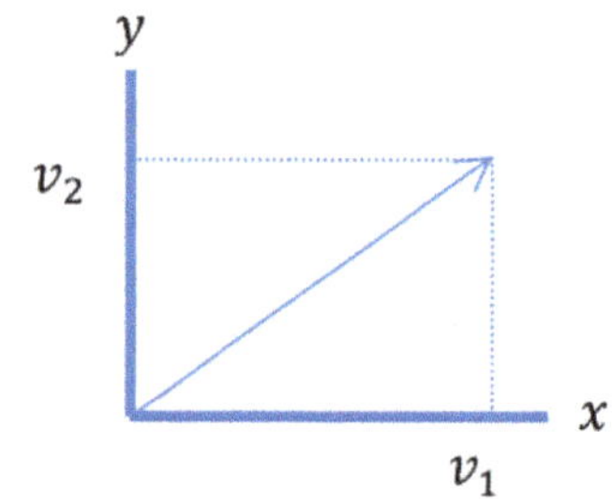

Figure 3.10. A vector in two dimensions, with components v_1 in the x direction and v_2 in the y direction.

Key words: The length of a vector. Clearly, the product of the row vector with the column vector, takes the form

$$\langle v|v\rangle = (v_1 \;\; v_2)\begin{pmatrix} v_1 \\ v_2 \end{pmatrix} = (v_1)^2 + (v_2)^2$$

which according to Pythogoras' theorem, is the squared length $|v|^2$ of the vector.

In other words, $\langle v|v\rangle = |v|^2 = (v_1)^2 + (v_2)^2$. For this reason, in the literature, the fact that the product of $\langle v|$ with $|v\rangle$ yields the squared length $\langle v|v\rangle$ has led to $|v\rangle$ being called a 'ket' vector and $|v\rangle$ being called a 'bra' vector, so that, the quantity $\langle v|v\rangle$ could be called a bracket (i.e. bra(c)ket). By analogy, it is tempting to define the squared length of the vector $|\psi\rangle$ to be $(\psi_1)^2 + (\psi_2)^2$. However there is a slight complication, because in quantum mechanics, ψ_1 and ψ_2 could be complex numbers, whereas to make sense, the squared length of a vector must be a positive real number. For this reason, the squared length of a complex vector is defined to be $|\psi_1|^2 + |\psi_2|^2$.

Key words: Bra and ket vectors. To allow $|\psi_1|^2 + |\psi_2|^2$ to be equated to a bra(c)ket, to each ket vector such as

$$|\psi\rangle = \begin{pmatrix} \psi_1 \\ \psi_2 \end{pmatrix}$$

the corresponding bra vector is defined to be

$$\langle\psi| = (\psi_1^* \;\; \psi_2^*)$$

and then the squared length of $|\psi\rangle$ can be written

$$\langle\psi|\psi\rangle = (\psi_1^* \;\; \psi_2^*)\begin{pmatrix} \psi_1 \\ \psi_2 \end{pmatrix} = \psi_1^*\psi_1 + \psi_2^*\psi_2 = |\psi_1|^2 + |\psi_2|^2$$

Since the modulus squared of a complex number is real and positive, the sum $|\psi_1|^2$ and $|\psi_2|^2$ is real and positive, so it is reasonable to define $\langle\psi|\psi\rangle$ to be the squared length of $|\psi\rangle$.

More generally, for a N-component ket vector

$$|\psi\rangle = \begin{pmatrix} \psi_1 \\ \psi_2 \\ \vdots \\ \psi_N \end{pmatrix}$$

the corresponding bra vector is

$$\langle\psi| = \begin{pmatrix} \psi_1^* & \psi_2^* \cdots \psi_N^* \end{pmatrix}$$

and the squared length is

$$\langle\psi|\psi\rangle = \sum_{i=1}^{N} |\psi_i|^2$$

Key word: Normalisation. If $\langle\psi|\psi\rangle = 1$, then the length of $|\psi\rangle$ is 1 and $|\psi\rangle$ is said to be 'normalised'. Equivalently, one can say that $|\psi\rangle$ is a 'unit vector'. If $|\psi\rangle$ is normalised, then the components ψ_i satisfy $\sum_{i=1}^{N} |\psi_i|^2 = 1$.

Key words: Orthogonality and orthonormality. If two real Cartesian vectors $|v\rangle = \begin{pmatrix} v_1 \\ v_2 \end{pmatrix}$ and $|w\rangle = \begin{pmatrix} w_1 \\ w_2 \end{pmatrix}$ are at 90° to each other, then they are said to be orthogonal and one finds $\langle w|v\rangle = 0$. In other words

$$\langle w|v\rangle = (w_1 \;\; w_2)\begin{pmatrix} v_1 \\ v_2 \end{pmatrix} = w_1 v_1 + w_2 v_2 = 0$$

For example if $|v\rangle = \begin{pmatrix} 1 \\ 1 \end{pmatrix}$ and $|w\rangle = \begin{pmatrix} 1 \\ -1 \end{pmatrix}$ then $\langle w|v\rangle = (1 \; -1)\begin{pmatrix} 1 \\ 1 \end{pmatrix} = 1 - 1 = 0$

Similarly, if two complex vectors $|\psi\rangle$ and $|\varphi\rangle$ satisfy $\langle\psi|\varphi\rangle = \langle\varphi|\psi\rangle^* = 0$ then they are defined to be orthogonal.

Note that linearly independent vectors must not be parallel, but they are not necessarily orthogonal. On the other hand orthogonal vectors are necessarily not parallel and are therefore linearly independent. To prove mathematically that two orthogonal vectors are linearly independent, one simply needs to prove that the equation $a|\psi\rangle + b|\varphi\rangle = 0$ can only be satisfied if both $a = 0$ and $b = 0$. To prove this result, multiply both sides by $\langle\psi|$ to yield $a\langle\psi|\psi\rangle + b\langle\psi|\varphi\rangle = 0$ Since $\langle\psi|\varphi\rangle = 0$ and $\langle\psi|\psi\rangle \neq 0$, this yields $a = 0$. Similarly multiply both sides by $\langle\varphi|$ yields $b = 0$.

Note that vectors can be both normalised and orthogonal, in which case they are described as being orthonormal. That is, if the set of vectors $|\psi_n\rangle$, $n = 1,2, \ldots N$ satisfy $\langle\psi_m|\psi_n\rangle = 0$ for $n \neq m$ and $\langle\psi_n|\psi_n\rangle = 1$, then they are orthonormal. In that case one often writes

$$\langle\psi_m|\psi_n\rangle = \delta_{nm} \tag{3.39}$$

where δ_{nm} is the Kronecker delta, which satisfies $\delta_{nm} = 1$ for $n = m$ and $\delta_{nm} = 0$ for $n \neq m$.

3.12 The Hamiltonian of a two-level quantum system

As mentioned in section 3.1, the dynamics of a quantum system are determined by the Hamiltonian. The Hamiltonian H is a square matrix and contains information about the energies of the the system and how the components of the system are coupled together. For a two-level system, if an energy ε_1 is assigned to orbital 1 and ε_2 to orbital 2 and the coupling between these orbitals is denoted Δ, then instead of figure 3.9, the picture representing this two-site system would be as shown in figure 3.11.

The Hamiltonian H of this system is a 2×2 square matrix of the form

$$H = \begin{pmatrix} \varepsilon_1 & \Delta \\ \Delta^* & \varepsilon_2 \end{pmatrix} \tag{3.40}$$

In this equation, ε_1 and ε_2 are real energies and Δ is a complex number.

The Hamiltonian of equation (3.40) is an example of the following 2×2 matrix

$$H = \begin{pmatrix} H_{11} & H_{12} \\ H_{21} & H_{22} \end{pmatrix}$$

with the property that H_{11}, H_{12} are real and H_{21}, H_{22} are complex numbers satisfying $H_{21} = (H_{12})^*$. Such a matrix is said to be Hermitian.

Key words: Complex conjugates, transposes and Hermitian conjugates. To define more precisely the term 'Hermitian', it is timely to recall a few facts about complex conjugates, transposes and Hermitian conjugates. If $i = \sqrt{-1}$, then $3 + i5$ is a complex number, whose real part is 3 and whose imaginary part is 5. As another example, if $z = e^{ikj}$ where k and j are real, then the real and imaginary parts of z are obtained using de Moivre's theorem, which states $e^{ikj} = \cos kj + i \sin kj$. Hence the real part of z is $\cos kj$ and the imaginary part of z is $\sin kj$. The complex conjugate of a complex number is obtained by replacing every occurrence of i by $-i$. Hence, for example, the complex conjugate of z is $z^* = e^{-ikj} = \cos kj - i \sin kj$ and the complex conjugate of $z = 3 + i5$ is $z^* = 3 - i5$.

Similarly the complex conjugate of the vector $|\psi\rangle = \begin{pmatrix} \psi_1 \\ \psi_2 \end{pmatrix}$ is $|\psi\rangle^* = \begin{pmatrix} \psi_1^* \\ \psi_2^* \end{pmatrix}$ and the complex conjugate of the Hamiltonian $H = \begin{pmatrix} H_{11} & H_{12} \\ H_{21} & H_{22} \end{pmatrix}$ is $H^* = \begin{pmatrix} H_{11}{}^* & H_{12}{}^* \\ H_{21}{}^* & H_{22}{}^* \end{pmatrix}$. In the case of matrices and vectors, their transposes and Hermitian conjugates are also of interest. The transpose of a matrix or a vector is obtained by replacing every row by a column and vice versa. For example if $|\psi\rangle = \begin{pmatrix} \psi_1 \\ \psi_2 \end{pmatrix}$, then the transpose of $|\psi\rangle$ is

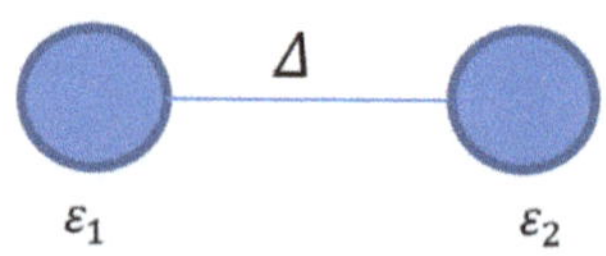

Figure 3.11. A representation of a two-level system, with orbital energies ε_1 and ε_2, coupled by Δ.

denoted $|\psi\rangle^t$ and is given by $|\psi\rangle^t = (\psi_1 \;\; \psi_2)$. Similarly if $H = \begin{pmatrix} H_{11} & H_{12} \\ H_{21} & H_{22} \end{pmatrix}$, then the transpose of H is denoted H^t and given by $H^t = \begin{pmatrix} H_{11} & H_{21} \\ H_{12} & H_{22} \end{pmatrix}$. This shows that H^t is obtained by reflecting H about the diagonal.

Finally the Hermitian conjugate of a matrix is defined to be the complex conjugate of the transpose and is denoted by the symbol †. This means that $H^\dagger = H^{t*}$. For example if $H = \begin{pmatrix} H_{11} & H_{12} \\ H_{21} & H_{22} \end{pmatrix}$, then

$$H^\dagger = \begin{pmatrix} H_{11}^* & H_{21}^* \\ H_{12}^* & H_{22}^* \end{pmatrix} \tag{3.41}$$

Similarly if $|\psi\rangle = \begin{pmatrix} \psi_1 \\ \psi_2 \end{pmatrix}$, then $|\psi\rangle^\dagger$ is given by

$$|\psi\rangle^\dagger = (\psi_1^* \;\; \psi_2^*) \tag{3.42}$$

Hence the Hermitian conjugate of a ket vector is the corresponding bra vector and vice versa. That is

$$|\psi\rangle^\dagger = \langle\psi| \text{ and } \langle\psi|^\dagger = |\psi\rangle \tag{3.43}$$

Finally it should be noted that the transpose of a product of two matrices A and B is equal to the product of their transposes in reverse order: i.e. $(AB)^t = B^tA^t$ and consequently, Hermitian conjugates of a product of two matrices A and B is equal to the product of their Hermitian conjugates in reverse order. That is

$$(AB)^\dagger = B^\dagger A^\dagger \tag{3.44}$$

As an example of this equation, it is useful to note that

$$(H|\psi\rangle)^\dagger = \langle\psi|H^\dagger \tag{3.45}$$

Key words: Hermitian matrices. Finally note that if a matrix H satisfies $H^\dagger = H$, then H is said to be Hermitian. In the above example, H is Hermitian if $\begin{pmatrix} H_{11} & H_{12} \\ H_{21} & H_{22} \end{pmatrix} = \begin{pmatrix} H_{11}^* & H_{21}^* \\ H_{12}^* & H_{22}^* \end{pmatrix}$, which means that H_{11}, H_{12} are real and H_{21}, H_{12} are complex numbers satisfying $H_{21} = (H_{12})^*$. Clearly, the matrix of equation (3.40) is Hermitian, provided ε_1 and ε_2 are real. Similarly the dynamical matrix of the two-oscillator problem of equation (3.6) is also Hermitian.

In quantum mechanics, all Hamiltonians are Hermitian, so in the above example, it is a requirement of quantum mechanics that ε_1 and ε_2 are real. More generally, for a $N \times N$ Hamiltonian H, with matrix elements H_{ij}, the requirement $H^\dagger = H$ implies that $H_{ij} = H_{ji}^*$, so the diagonal elements H_{ii} are real.

3.13 Eigenvalues and eigenvectors of $N \times N$ Hermitian matrices

We are almost in a position to determine how many linearly-independent eigenvectors a dynamical matrix H possesses. Since the only dynamical matrices of interest are Hermitian, we start by examining the eigenvalues E_n and eigenvectors $|\psi_n\rangle$ of Hermitian matrices, which are obtained by solving

$$H|\psi_n\rangle = E_n|\psi_n\rangle \tag{3.46}$$

In this equation, n is an integer, which labels the eigenvectors and their corresponding eigenvalues.

The problem posed by equation (3.46) is to determine all linearly-independent solutions, which have a non-zero length. Consider two such solutions $|\psi_n\rangle$ and $|\psi_m\rangle$, where $m \neq n$. Since $|\psi_m\rangle$ satisfies $H|\psi_m\rangle = E_m|\psi_m\rangle$, taking the Hermitian conjugate and noting equation (3.45) yields

$$\langle\psi_m|H^\dagger = \langle\psi_m|E_m^* \tag{3.47}$$

Multiply both sides of equation (3.46) from the left by $\langle\psi_m|$ and both sides of equation (3.47) from the right by $|\psi_n\rangle$ to yield $\langle\psi_m|H|\psi_n\rangle = E_n\langle\psi_m|\psi_n\rangle$ and $\langle\psi_m|H^\dagger|\psi_n\rangle = E_m^*\langle\psi_m|\psi_n\rangle$ respectively. Subtracting one equation from the other yields

$$\langle\psi_m|H^\dagger - H|\psi_n\rangle = (E_m^* - E_n)\langle\psi_m|\psi_n\rangle \tag{3.48}$$

So far nothing has been assumed about the Hamiltonian H. However if H is Hermitian, then $H^\dagger = H$ and the left-hand side of equation (3.48) vanishes. Hence

$$(E_m^* - E_n)\langle\psi_m|\psi_n\rangle = 0 \tag{3.49}$$

From this equation, the following conclusions are obtained:

A. Choosing $m = n$ and noting that the squared length $\langle\psi_n|\psi_n\rangle \neq 0$, equation (3.49) yields $E_n^* = E_n$ Hence all eigenvalues of Hermitian matrices are real.
   Consequently equation (3.49) can be written

$$(E_m - E_n)\langle\psi_m|\psi_n\rangle = 0 \tag{3.50}$$

B. Clearly if $E_m \neq E_n$, then equation (3.50) yields $\langle\psi_m|\psi_n\rangle = 0$. Hence the eigenvectors belonging to different eigenvalues are necessarily orthogonal.

C. If $E_m = E_n$, then no conclusions can be drawn about the orthogonality properties of $|\psi_n\rangle$ and $|\psi_m\rangle$. Hence the linearly-independent eigenvectors of a Hermitian matrix possessing the same eigenvalue are not necessarily orthogonal.

Key word: Degeneracy. If two linearly-independent eigenvectors $|\psi_n\rangle$ and $|\psi_m\rangle$ are found to possess the same eigenvalue E then the eigenvalue E is said to be doubly degenerate. If three linearly-independent eigenvectors possess the same eigenvalue E, then E is said to be triply degenerate etc. Degeneracies are common in nature and it is important to understand their consequences.

As an example, consider the matrix $H = \begin{pmatrix} 8 & 0 \\ 0 & 8 \end{pmatrix}$, which satisfies $\begin{pmatrix} 8 & 0 \\ 0 & 8 \end{pmatrix}\begin{pmatrix} 1 \\ 0 \end{pmatrix} = 8\begin{pmatrix} 1 \\ 0 \end{pmatrix}$ and $\begin{pmatrix} 8 & 0 \\ 0 & 8 \end{pmatrix}\begin{pmatrix} 0 \\ 1 \end{pmatrix} = 8\begin{pmatrix} 0 \\ 1 \end{pmatrix}$. Hence there are two linearly-independent eigenvectors $\begin{pmatrix} 1 \\ 0 \end{pmatrix}$ and $\begin{pmatrix} 0 \\ 1 \end{pmatrix}$ which possess the same eigenvalue 8. On the other hand, $\begin{pmatrix} 8 & 0 \\ 0 & 8 \end{pmatrix}\begin{pmatrix} 1 \\ 1 \end{pmatrix} = 8\begin{pmatrix} 1 \\ 1 \end{pmatrix}$.

Hence $\begin{pmatrix} 1 \\ 1 \end{pmatrix}$ is also another eigenvector with eigenvalue 8. However $\begin{pmatrix} 1 \\ 1 \end{pmatrix} = \begin{pmatrix} 1 \\ 0 \end{pmatrix} + \begin{pmatrix} 0 \\ 1 \end{pmatrix}$ and therefore $\begin{pmatrix} 1 \\ 1 \end{pmatrix}$ is linearly dependent on $\begin{pmatrix} 1 \\ 0 \end{pmatrix}$ and $\begin{pmatrix} 0 \\ 1 \end{pmatrix}$. Since the task is to find all linearly independent eigenvectors, only two of the three should be retained. One could keep $\begin{pmatrix} 1 \\ 0 \end{pmatrix}$ and $\begin{pmatrix} 0 \\ 1 \end{pmatrix}$, but not $\begin{pmatrix} 1 \\ 1 \end{pmatrix}$. Alternatively, one could keep $\begin{pmatrix} 1 \\ 0 \end{pmatrix}$ and $\begin{pmatrix} 1 \\ 1 \end{pmatrix}$, but not $\begin{pmatrix} 0 \\ 1 \end{pmatrix}$. More generally, for any non-zero a and b, $\begin{pmatrix} 8 & 0 \\ 0 & 8 \end{pmatrix}\begin{pmatrix} a \\ b \end{pmatrix} = 8\begin{pmatrix} a \\ b \end{pmatrix}$. Hence for this simple example, any vector $\begin{pmatrix} a \\ b \end{pmatrix}$ is an eigenvector. Since $\begin{pmatrix} a \\ b \end{pmatrix} = a\begin{pmatrix} 1 \\ 0 \end{pmatrix} + b\begin{pmatrix} 0 \\ 1 \end{pmatrix}$, $\begin{pmatrix} a \\ b \end{pmatrix}$ is again linearly dependent on $\begin{pmatrix} 1 \\ 0 \end{pmatrix}$ and $\begin{pmatrix} 0 \\ 1 \end{pmatrix}$, so one cannot keep all three. Since any two linearly-independent combinations of $\begin{pmatrix} 1 \\ 0 \end{pmatrix}$ and $\begin{pmatrix} 0 \\ 1 \end{pmatrix}$ are eigenvectors, one could choose $\begin{pmatrix} 1 \\ 2 \end{pmatrix}$ and $\begin{pmatrix} 19 \\ 7 \end{pmatrix}$. These are clearly linearly independent and are perfectly acceptable solutions.

Given the above freedom of choice, it is natural to ask: what is the best choice of the degenerate eigenstates of a matrix such as $\begin{pmatrix} 8 & 0 \\ 0 & 8 \end{pmatrix}$? The answer to this question depends on taste and convenience. For example if $H_0 = \begin{pmatrix} 8 & 0 \\ 0 & 8 \end{pmatrix}$ and $H_1 = \begin{pmatrix} 1 & 0 \\ 0 & -1 \end{pmatrix}$, then one may be interested in the eigenvectors of a related matrix such as $H(\eta) = H_0 + \eta H_1 = \begin{pmatrix} 8+\eta & 0 \\ 0 & 8-\eta \end{pmatrix}$, in which case one may wish to choose the eigenstates of H_0 to be also eigenstates of $H(\eta)$. The only possible choice of eigenvectors of $H(\eta)$ is $\begin{pmatrix} 1 \\ 0 \end{pmatrix}$ and $\begin{pmatrix} 0 \\ 1 \end{pmatrix}$, with non-degenerate eigenvalues $8+\eta$ and $8-\eta$ respectively. When $\eta = 0$, any linear combination is acceptable, but when $\eta \neq 0$ only this choice is possible, no matter how small η.

As a second example, if $H_0 = \begin{pmatrix} 8 & 0 \\ 0 & 8 \end{pmatrix}$, $H_1 = \begin{pmatrix} 0 & 1 \\ 1 & 0 \end{pmatrix}$ and $H(\eta) = H_0 + \eta H_1 = \begin{pmatrix} 8 & \eta \\ \eta & 8 \end{pmatrix}$. The only possible choice of eigenvectors of $H(\eta)$ is $\begin{pmatrix} 1 \\ 1 \end{pmatrix}$ and $\begin{pmatrix} 1 \\ -1 \end{pmatrix}$, with non-degenerate eigenvalues $8+\eta$ and $8-\eta$ respectively. Again this choice is necessary no matter how small η, provided η is non-zero.

Key words. Non-degenerate eigenstates. It is important to recognise that apart from a multiplicative normalisation constant, there is no freedom of choice for eigenstates belonging to non-degenerate eigenvalues. For example, consider the matrix $H = \begin{pmatrix} 8 & 0 \\ 0 & 7 \end{pmatrix}$, which satisfies $\begin{pmatrix} 8 & 0 \\ 0 & 7 \end{pmatrix}\begin{pmatrix} 1 \\ 0 \end{pmatrix} = 8\begin{pmatrix} 1 \\ 0 \end{pmatrix}$ and $\begin{pmatrix} 8 & 0 \\ 0 & 7 \end{pmatrix}\begin{pmatrix} 0 \\ 1 \end{pmatrix} = 7\begin{pmatrix} 0 \\ 1 \end{pmatrix}$. Hence the matrix H possesses two linearly-independent eigenvectors $\begin{pmatrix} 1 \\ 0 \end{pmatrix}$ and $\begin{pmatrix} 0 \\ 1 \end{pmatrix}$. On the other hand $\begin{pmatrix} 8 & 0 \\ 0 & 7 \end{pmatrix}\begin{pmatrix} 1 \\ 1 \end{pmatrix} = 8\begin{pmatrix} 1 \\ 7/8 \end{pmatrix}$. Since the right-hand side of this equation is not proportional to $\begin{pmatrix} 1 \\ 1 \end{pmatrix}$, this means that $\begin{pmatrix} 1 \\ 1 \end{pmatrix}$ is not an eigenvector. If an eigenvalue such as 7 has only one linearly independent eigenvector $\begin{pmatrix} 0 \\ 1 \end{pmatrix}$, then the only freedom of choice is the normalisation of the vector. Clearly for any constant A,

$$\begin{pmatrix} 8 & 0 \\ 0 & 7 \end{pmatrix}\begin{pmatrix} 0 \\ A \end{pmatrix} = 7\begin{pmatrix} 0 \\ A \end{pmatrix}$$

which means that the vector $A\begin{pmatrix} 0 \\ 1 \end{pmatrix}$ is an eigenvector with eigenvalue 7. However this vector is simply proportional to $\begin{pmatrix} 0 \\ 1 \end{pmatrix}$ and therefore is linearly dependent on it. This illustrates the general property that eigenvectors are only determined up to a constant of proportionality A and the choice of A is completely arbitrary. This is obvious from the equation $H|\psi_n\rangle = E_n|\psi_n\rangle$, because multiplying both sides by A yields $H(A|\psi_n\rangle) = E_n\,(A|\psi_n\rangle)$. Hence if $|\psi_n\rangle$ is an eigenvector with eigenvalue E_n, then so is $A|\psi_n\rangle$. In other words, the constant A (and hence the length of $|\psi_n\rangle$) is not determined by the eigenvalue equation.

3.14 Proof that a $N \times N$ Hermitian matrix possesses N linearly independent eigenvectors

Consider now an arbitrary $N \times N$ matrix A. The inverse of A is the matrix M, which satisfies $MA = I$, where I is the unit matrix.

$$I = \begin{pmatrix} 1 & 0 & 0 & 0 \\ 0 & 1 & 0 & 0 \\ 0 & 0 & \ddots & 0 \\ 0 & 0 & 0 & 1 \end{pmatrix}$$

Conventionally, M is denoted A^{-1} and given by

$$A^{-1} = \frac{1}{\det A} C^t$$

where C is the cofactor matrix of A, C^t is the transpose of the cofactor matrix and $\det A$ is the determinant of A. For example if $A = \begin{pmatrix} a & b \\ c & d \end{pmatrix}$, then $A^{-1} = \frac{1}{\det A}\begin{pmatrix} d & -b \\ -c & a \end{pmatrix}$, where $\det A = ad - bc$. **Check:** By multiplying the two matrices it is clear that A^{-1} satisfies $A^{-1}A = AA^{-1} = I$.

Now consider the equation $H|\psi\rangle = E|\psi\rangle = EI|\psi\rangle$. This can be written $(EI - H)|\psi\rangle = 0$, which implies $|\psi\rangle = (EI - H)^{-1}0$. That is $|\psi\rangle = \frac{1}{\det(EI - H)}C^t . 0$, where C is the cofactor matrix of $H - EI$. Hence $|\psi\rangle$ must vanish unless

$$\det(EI - H) = 0 \tag{3.51}$$

This equation is called the secular equation and is the equation that determines the eigenvalues of H. As an example, for the Hamiltonian (3.40),

$$EI - H = \begin{pmatrix} \mathrm{E} - \varepsilon_1 & -\Delta \\ -\Delta^* & \mathrm{E} - \varepsilon_2 \end{pmatrix} \tag{3.52}$$

Hence $\det(EI - H) = (\mathrm{E} - \varepsilon_1)(\mathrm{E} - \varepsilon_2) - |\Delta|^2$ and the secular equation is

$$(\mathrm{E} - \varepsilon_1)(\mathrm{E} - \varepsilon_2) - |\Delta|^2 = 0 \tag{3.53}$$

Notice that this is a quadratic equation for E of the form

$$p_2(E) = 0 \tag{3.54}$$

where $p_2 = (\mathrm{E} - \varepsilon_1)(\mathrm{E} - \varepsilon_2) - |\Delta|^2$. Furthermore the polynomial p_2 can be written in the form

$$p_2(E) = (E - E_1)(E - E_2), \tag{3.55}$$

where

$$E_1 = \frac{\varepsilon_1 + \varepsilon_2}{2} + R \text{ and } E_2 = \frac{\varepsilon_1 + \varepsilon_2}{2} - R, \text{ with } R = \sqrt{\left(\frac{\varepsilon_1 - \varepsilon_2}{2}\right)^2 + |\Delta|^2} \tag{3.56}$$

Clearly E_1 and E_2 are the two eigenvalues of H.

To prove that a $N \times N$ Hermitian matrix H possesses N linearly independent eigenvectors, note that the secular equation for a $N \times N$ Hamiltonian is an Nth order polynomial equation for E of the form

$$p_N(E) = 0 \tag{3.57}$$

Furthermore the fundamental theorem of algebra (see appendix A) tells us that $p_N(E)$ can be written

$$p_N(E) = (E - E_1)(E - E_2)\ldots(E - E_N) \tag{3.58}$$

where E_1, E_2 etc are called roots of the polynomial. The N roots of the polynomial $p_N(E)$ are the solutions of the secular equation (3.57) and are not necessarily distinct. If they are distinct, then H possesses N non-degenerate eigenvalues and since eigenstates of a Hermitian matrix belonging to different eigenvalues are orthogonal, H possesses N orthogonal eigenstates, which (see the end of section 11) are clearly linearly independent. On the other hand if the solutions of the secular equation (3.57) are not distinct, then it is convenient to consider the Hamiltonian $H(\eta) = H + \eta H_1$, where H_1 is a random Hermitian matrix, such that for any value of η, the eigenvalues of $H(\eta)$ are non-degenerate. Since $H(\eta)$ necessarily possesses N orthogonal eigenstates, and since this property persists even in the limit that η tends to zero, H also possesses N orthogonal eigenstates. In other words, even though some of the eigenvalues of $H(\eta)$ become degenerate as η tends to zero, their eigenstates remain orthogonal and are therefore linearly independent. (Examples of this property are shown at the end of the discussion of the key word 3.14 'degeneracy'.)

Note that the above argument does not work for a non-Hermitian matrix A, because the matrix $A(\eta) = A + \eta H_1$ is non-Hermitian and eigenvectors belonging to different eigenvalues are not necessarily orthogonal. Therefore eigenvectors which are non-parallel for $\eta \neq 0$ can become parallel in the limit η tends to zero. For the example of equation (3.37), it is easy to see that if $(\eta) = \begin{pmatrix} 0 & 1 \\ \eta & 0 \end{pmatrix}$, then the eigenstates are $\begin{pmatrix} 1 \\ \pm\eta^{1/2} \end{pmatrix}$, which both tend to the same vector $\begin{pmatrix} 1 \\ 0 \end{pmatrix}$ as η tends to zero.

3.15 Completeness and initial conditions revisited

Knowing that a $N \times N$ Hermitian matrix H possesses N orthogonal eigenvectors $|\psi_n\rangle$ means that the eigenvectors of a $N \times N$ Hermitian matrix H form a complete set and any other N-dimensional vector $|\varphi\rangle$ can be written as a linear superposition:

$$|\varphi\rangle = \sum_{n=1}^{N} a_n|\psi_n\rangle. \tag{3.59}$$

This means that equations (3.29) and (3.33) are indeed valid.

If the eigenvectors are chosen to be orthonormal, then multiplying equation (3.59) from the left by $\langle\psi_m|$ and using $\langle\psi_m|\psi_n\rangle = \delta_{nm}$ yields

$$\langle\psi_m|\varphi\rangle = \sum_{n=1}^{N} a_n\langle\psi_m|\psi_n\rangle = \sum_{n=1}^{N} a_n\delta_{nm} = a_m \tag{3.60}$$

Substituting this back into equation (3.59) yields $|\varphi\rangle = \sum_{n=1}^{N}|\psi_n\rangle\langle\psi_n|\varphi\rangle = \left\{\sum_{n=1}^{N}|\psi_n\rangle\langle\psi_n|\right\}|\varphi\rangle$. Since this is valid for any vector $|\varphi\rangle$, it shows that any complete orthonormal set of vectors satisfies

$$\sum_{n=1}^{N}|\psi_n\rangle\langle\psi_n| = I \tag{3.61}$$

where I is the $N \times N$ unit matrix.

Furthermore, for given boundary conditions, equation (3.60) makes it straightforward to determine the constants A_n and B_n in equations (3.32) and (3.34).

To determine these constants for a given set of vibrating masses, consider the case where the known initial displacements are $x_1(0), x_2(0), \ldots x_N(0)$ and known the initial velocities are $\dot{x}_1(0), \dot{x}_2(0), \ldots \dot{x}_N(0)$. That is consider the case where the initial conditions are

$$|x(0)\rangle = \begin{pmatrix} x_1(0) \\ x_2(0) \\ \vdots \\ x_N(0) \end{pmatrix} \text{ and } |\dot{x}(0)\rangle = \begin{pmatrix} \dot{x}_1(0) \\ \dot{x}_2(0) \\ \vdots \\ \dot{x}_N(0) \end{pmatrix}$$

From equations (3.29) and (3.32),

$$|x(t)\rangle = \sum_{n=1}^{N} a_n(t)|\psi_n\rangle \tag{3.62}$$

where $a_n(t) = A_n \cos(\omega_n t + B_n)$.

Hence $|x(0)\rangle = \sum_{n=1}^{N} a_n(0)|\psi_n\rangle$ and $|\dot{x}(0)\rangle = \sum_{n=1}^{N} \dot{a}_n(0)|\psi_n\rangle$. From equation (3.60) A_n and B_n must be chosen such that

$$a_n(0) = \langle\psi_n|x(0)\rangle \text{ and } \dot{a}_n(0) = \langle\psi_n|\dot{x}(0)\rangle \tag{3.63}$$

In other words, A_n and B_n are given by $A_n \cos(B_n) = \langle\psi_n|x(0)\rangle$ and $-A_n\omega_n \sin(B_n) = \langle\psi_n|\dot{x}(0)\rangle$.

Similarly for a given choice of the initial vector $|\varphi(0)\rangle$ in the solution to the Schrödinger equation (3.34), the constants A_n are given by

$$A_n = \langle\psi_n|\varphi(0)\rangle \tag{3.64}$$

3.16 The eigenstates of a two-level quantum system

The two-level quantum system shown in figure 3.11, whose Hamiltonian is

$$H = \begin{pmatrix} \varepsilon_1 & \Delta \\ \Delta^* & \varepsilon_2 \end{pmatrix} \tag{3.65}$$

is a classic example of the above discussion.

In this equation, ε_1 and ε_2 are real energies, Δ is a complex number and therefore H is Hermitian.

Notice that $H = \begin{pmatrix} c + \varepsilon & \Delta \\ \Delta^* & c - \varepsilon \end{pmatrix}$ where $c = \frac{\varepsilon_1 + \varepsilon_2}{2}$ and $\varepsilon = \frac{\varepsilon_1 - \varepsilon_2}{2}$. Hence $H = cI + h$, where I is the unit matrix and $h = \begin{pmatrix} \varepsilon & \Delta \\ \Delta^* & -\varepsilon \end{pmatrix}$. Since H and h differ only by a constant, they share the same eigenvectors and the eigenvalues of H are obtained by simply adding c to the eigenvalues of h. For this reason, it is only necessary to determine the eigenvalues and eigenvectors of h, which are obtained by solving

$$\begin{pmatrix} \varepsilon & \Delta \\ \Delta^* & -\varepsilon \end{pmatrix}\begin{pmatrix} u \\ v \end{pmatrix} = E\begin{pmatrix} u \\ v \end{pmatrix} \tag{3.66}$$

Multiplying the two matrices on the left-hand side yields

$$\begin{pmatrix} \varepsilon u + \Delta v \\ \Delta^* u - \varepsilon v \end{pmatrix} = \begin{pmatrix} Eu \\ Ev \end{pmatrix}$$

Hence $\varepsilon u + \Delta v = Eu$ and $\Delta^* u - \varepsilon v = Ev$; i.e.

$$\frac{u}{v} = \frac{\Delta}{E - \varepsilon} = \frac{E + \varepsilon}{\Delta^*} \tag{3.67}$$

The last two equations combine to yield the secular equation $(E - \varepsilon)(E + \varepsilon) - |\Delta|^2 = 0$, from which one obtains

$$E = \pm R \tag{3.68}$$

where $R = \sqrt{\varepsilon^2 + |\Delta|^2}$. Hence, as expected there are two linearly-independent eigenstates
$|\psi_+\rangle = \begin{pmatrix} u_+ \\ v_+ \end{pmatrix}$ and $|\psi_-\rangle = \begin{pmatrix} u_- \\ v_- \end{pmatrix}$, where

$$\frac{u_\pm}{v_\pm} = \frac{\Delta}{\pm R - \varepsilon} = \frac{\pm R + \varepsilon}{\Delta^*} \tag{3.69}$$

Clearly the eigenvalue equation (3.66) determines only the ratio $\frac{u}{v}$, so as expected, the magnitude of the eigenvectors is not determined. Optionally, they can be chosen to be normalised by insisting that $|u_\pm|^2 + |v_\pm|^2 = 1$. It is easy to show that this is satisfied if

$$|u_\pm|^2 = \frac{1}{2}\left(1 + \frac{\varepsilon}{\pm R}\right) \text{ and } |v_\pm|^2 = \frac{1}{2}\left(1 - \frac{\varepsilon}{\pm R}\right) \tag{3.70}$$

Equations (3.68)–(3.70) are the complete solution to the two-level quantum eigenvalue problem. It is well worth memorising these equations, because they are encountered frequently in quantum mechanics.

Note that from equation (3.69), $\frac{u_+^* u_-}{v_+^* v_-} = \frac{\Delta^*}{R - \varepsilon}\frac{-R + \varepsilon}{\Delta^*} = -3$. Hence $\langle\psi_+|\psi_-\rangle = (u_+^* \; v_+^*)\begin{pmatrix} u_- \\ v_- \end{pmatrix} = u_+^* u_- + v_+^* v_- = v_+^* v_-\left[\frac{u_+^* u_-}{v_+^* v_-} + 1\right] = 0$, i.e. $|\psi_+\rangle$ and $|\psi_-\rangle$ are orthogonal, as expected.

3.17 Summary

In summary, the equations of motion $-\frac{d^2}{dt^2}|x(t)\rangle = H|x(t)\rangle$ and $i\hbar\frac{d}{dt}|\varphi(t)\rangle = H|\varphi(t)\rangle$ can be solved by noting that the $N \times N$ Hermitian matrix H possesses a complete set of N linearly independent eigenvectors $|\psi_n\rangle$, satisfying $H|\psi_n\rangle = E_n|\psi_n\rangle$, which can be chosen to be orthonormal. The corresponding eigenvalues E_n are necessarily real.

The solution to the equations of motion are obtained by writing the vectors $|x(t)\rangle$ or $|\varphi(t)\rangle$ as linear superposition of these eigenvectors.

For the classical systems of harmonic oscillators, $|x(t)\rangle = \sum_{n=1}^{N} a_n(t)|\psi_n\rangle$, where $a_n(t) = A_n \cos(\omega_n t + B_n)$, with $\omega_n^2 = E_n$. For the quantum system, $|\varphi(t)\rangle = \sum_{n=1}^{N} a_n(t)|\psi_n\rangle$, where $a_n(t) = A_n \exp(-i\omega_n t)$, with $\hbar\omega_n = E_n$. In the latter equation, if the eigenvectors $|\psi_n\rangle$ are chosen to be orthonormal, then $A_n = \langle\psi_n|\varphi(0)\rangle$. Hence the general solution to the time-dependent Schrodinger equation is

$$|\varphi(t)\rangle = \sum_{n=1}^{N} \exp(-i\omega_n t)|\psi_n\rangle\langle\psi_n|\varphi(0)\rangle \tag{3.71}$$

Note that in this equation, $|\varphi(t)\rangle = \begin{pmatrix} \varphi_1(t) \\ \varphi_2(t) \\ \vdots \\ \varphi_N(t) \end{pmatrix}$ and $|\psi_n\rangle = \begin{pmatrix} \psi_1^{(n)} \\ \psi_2^{(n)} \\ \vdots \\ \psi_N^{(n)} \end{pmatrix}$, where $\psi_j^{(n)}$ is the time-independent amplitude of eigenvector $|\psi_n\rangle$ on atomic orbital j and $\varphi_j(t)$ is the time-dependent amplitude of the wave function on atomic orbital j at time t. Hence equation (3.71) yields for the wave function on atomic orbital j at time t,

$$\varphi_j(t) = \sum_{n=1}^{N} A_n \psi_j^{(n)} \exp(-i\omega_n t) \tag{3.72}$$

Taking advantage of completeness, solutions to equations (3.1) and (3.2) can now be written down by inspection. For example the solution to equation (3.2) is

$$|\varphi(t)\rangle = \exp -\frac{iHt}{\hbar}|\varphi(0)\rangle \tag{3.73}$$

as can be verified by differentiating both sides with respect to t. Using the completeness condition $\sum_{n=1}^{N}|\psi_n\rangle\langle\psi_n| = I$ allows this to be written as

$$\begin{aligned}|\varphi(t)\rangle &= \exp -\tfrac{iHt}{\hbar}|\varphi(0)\rangle \\ &= \exp -\tfrac{iHt}{\hbar} I|\varphi(0)\rangle \\ &= \sum_{n=1}^{N} \exp -\tfrac{iHt}{\hbar}|\psi_n\rangle\langle\psi_n|\varphi(0)\rangle \\ &= \sum_{n=1}^{N} \exp -i\omega_n t|\psi_n\rangle\langle\psi_n|\varphi(0)\rangle = \sum_{n=1}^{N} \exp -i\omega_n t|\psi_n\rangle A_n,\end{aligned}$$

which is identical to equation (3.71). In the penultimate step above, it was noted (see appendix A.1) that when a function of a matrix acts on an eigenvector, the result is the same function of the corresponding eigenvalue multiplied by the eigenvector. In the final step, the expression $A_n = \langle \psi_n | \varphi(0) \rangle$ was used.

Similarly the solution to equation (3.1) can be written $|x(t)\rangle = \cos H^{\frac{1}{2}} t |x(0)\rangle + H^{-\frac{1}{2}} \sin H^{\frac{1}{2}} t |\dot{x}(0)\rangle$, where $H^{\frac{1}{2}}.\ H^{\frac{1}{2}} = H$. Again using $\sum_{n=1}^{N} |\psi_n\rangle\langle\psi_n| = I$ yields $|x(t)\rangle = \sum_{n=1}^{N} a_n(t)|\psi_n\rangle$, where $a_n(t) = [\langle\psi_n|x(0)\rangle]\cos\omega_n t + \left[\frac{\langle\psi_n \mid \dot{x}(0)\rangle}{\omega_n}\right]\sin\omega_n t$.

Appendix A Basic properties of matrices and complex numbers

In addition to the properties of matrices noted in appendix A in chapter 2, the following matrix properties are useful.

A.1 Functions of matrices

Consider a function such as $f(x) = 1 + x + x^2$. If H is a $N \times N$ square matrix, then the corresponding function of H is defined to be $f(H) = 1 + H + H^2$, where $H^2 = H.\ H$. Similarly, since $e^x = 1 + x + \frac{x^2}{2!} + \frac{x^3}{3!} + \ldots$, the corresponding function of H is

$$e^H = 1 + x + \frac{H^2}{2!} + \frac{H^3}{3!} + \ldots$$

More generally, if a function $f(x)$ is analytic and possesses a Taylor expansion of the form $f(x) = a_0 + a_1 x + a_2 x^2 + \ldots$. Then $f(H)$ is defined to be

$$f(H) = a_0 + a_1 H + a_2 H^2 + \ldots \tag{A.1}$$

Note that if $|\psi\rangle$ is an eigenvector of H satisfying $H|\psi\rangle = \lambda|\psi\rangle$, then $H^2|\psi\rangle = H\lambda|\psi\rangle = \lambda^2|\psi\rangle$ and $H^n|\psi\rangle = \lambda^n|\psi\rangle$. Hence from equation (A.1)

$$f(H)|\psi\rangle = (a_0 + a_1 H + a_2 H^2 + \ldots)|\psi\rangle = (a_0 + a_1\lambda + a_2\lambda^2 + \ldots)|\psi\rangle$$

That is

$$f(H)|\psi\rangle = f(\lambda)|\psi\rangle \tag{A.2}$$

This proves that any analytic function of a matrix acting on an eigenvector is the same function of the eigenvalues multiplied into the vector.

A.2 Complex numbers

In this section, a few basic properties of complex number will be presented. De Moivres theorem

$$e^{i\theta} = \cos\theta + i\sin\theta \tag{A.3}$$

yields

$$\cos\theta = \frac{e^{i\theta} + e^{-i\theta}}{2} \tag{A.4}$$

and

$$\sin\theta = \frac{e^{i\theta} - e^{-i\theta}}{2i} \tag{A.5}$$

These can be used to prove a variety of formulae such as the following:

$$[e^{i\theta}]^n = e^{in\theta} = \cos n\theta + i\sin n\theta \tag{A.6}$$

$$\cos(iy) = \frac{e^{i(iy)} + e^{-i(iy)}}{2} = \frac{e^{-y} + e^{y}}{2} = \cos hy \tag{A.7}$$

As a further example, since $e^{iA}e^{iB} = e^{i(A+B)}$, one finds

$$[\cos A + i\sin A][\cos B + i\sin B] = \cos(A+B) + i\sin(A+B)$$

Equating real and imaginary parts of each side of this equation yields

$$\cos(A+B) = \cos(A)\cos(B) - \sin(A)\sin(B) \tag{A.8}$$

$$\sin(A+B) = \sin(A)\cos(B) + \cos(A)\sin(B) \tag{A.9}$$

Consider the complex number $z = x + iy$, where $i = \sqrt{-1}$ and x, y are real. The real part of z, is denoted Re z and the imaginary part of z denoted Im z. That is Re $z = x$ and Im $z = y$. The complex conjugate of a complex number is denoted $z*$ and obtained by replacing i by $-i$ everywhere. That is $z* = x - iy$. As a more complicated example, if $u = \frac{x + e^{ip}}{a + (b + ic)^2}$, where x, p, a, b, c are real, then $u* = \frac{x + e^{-ip}}{a + (b - ic)^2}$. The squared length of z is $zz* = (x + iy)(x - iy) = x^2 + y^2$. If this is denoted r^2, then if we define $\tan\theta = \frac{x}{y}$, $x = r\cos\theta$ and $y = r\sin\theta$. Hence $z = x + iy = r\cos\theta + ir\sin\theta = r(\cos\theta + i\sin\theta) = re^{i\theta}$, where the last step follows from de Moivre's theorem. This shows that $x + iy$ and $re^{i\theta}$ are simply two different ways of writing the same complex number.

When discussing Green's functions, the following complex number is encountered

$$g = \frac{1}{E + i\eta}$$

where E and η are real. Multiplying top and bottom by $E - i\eta$ yields

$$g = \frac{E - i\eta}{(E + i\eta)(E - i\eta)} = \frac{E - i\eta}{E^2 + \eta^2}$$

Hence the real and imaginary parts of g are

$$\text{Re}\, g = \frac{E}{E^2 + \eta^2} \quad \text{and} \quad \text{Im}\, g = \frac{-\eta}{E^2 + \eta^2}$$

respectively.

A.3 The fundamental theorem of algebra

This states that every polynomial equation of the form

$$p(z) = a_0 + a_1 z + a_2 z^2 + \dots a_n z^n = 0,$$

where $n \geqslant 1$ and $a_n \neq 0$, has exactly n roots.

Proof. If $p(z)$ has no roots, then $f(z) = 1/p(z)$ is bounded and analytic for all z. Then by Liouville's theorem, $f(z)$ and therefore $p(z)$ is a constant. Clearly $p(z)$ is not a constant, therefore it must have at least one root, which we denote α. That is $p(\alpha) = 0$.

Hence

$$p(z) - p(\alpha) = a_0 + a_1 z + a_2 z^2 + \dots a_n z^n - (a_0 + a_1 \alpha + a_2 \alpha^2 + \dots a_n \alpha^n)$$

That is

$$p(z) - p(\alpha) = a_1(z - \alpha) + a_2(z^2 - \alpha^2) + \dots a_n(z^n - \alpha^n) = (z - \alpha)Q(z)$$

where $Q(z)$ is a polynomial of degree $n - 3$. Since $p(\alpha) = 0$, this tells us that

$$p(z) = (z - \alpha)Q(z)$$

Since $Q(z)$ must also have one root, which we denote β, we find

$$Q(z) = (z - \beta)R(z),$$

where $R(z)$ is a polynomial of degree $n - 2$ etc.

IOP Publishing

Quantum Transport in Nanostructures and Molecules
An introduction to molecular electronics
Colin John Lambert

Chapter 4

Quantum properties of linear chains and simple molecules

4.1 Introduction

Atoms possess orbitals and from a quantum mechanical viewpoint, one is interested in how these orbitals combine to form molecular orbitals in molecules, or band structures in solids. An individual atom can possess many orbitals, but for the purpose of illustrating the key concepts, it is useful to start by considering fictitious atoms possessing only a single orbital, which will be referred to as a site. For simplicity, we shall also assume that atoms interact with nearest neighbour atoms only. Once the concepts are established, these restrictions can be relaxed.

4.2 The eigenvalues and eigenvectors of a two-level molecule

As a prelude to solving the Schrödinger equation describing many atoms, it is useful to recall the discussion of a two-level system from chapter 3, by starting from two sites, which are sufficiently well separated such that they do not interact. The energy of an electron located on site orbital 1, is ε_1, while the energy of an electron located on site orbital 2, is ε_2. If the atoms are identical, then $\varepsilon_1 = \varepsilon_2$. In what follows, the site orbital belonging to atom 1 will be denoted $|1\rangle = \begin{pmatrix} 1 \\ 0 \end{pmatrix}$ and the site orbital belonging to atom 2 will be denoted $|2\rangle = \begin{pmatrix} 0 \\ 1 \end{pmatrix}$ (figure 4.1).

The Hamiltonian describing such a system is

$$H = \begin{pmatrix} \varepsilon_1 & 0 \\ 0 & \varepsilon_2 \end{pmatrix} \tag{4.1}$$

Figure 4.1. Two non-interacting atoms, with energy levels ε_1 and ε_2.

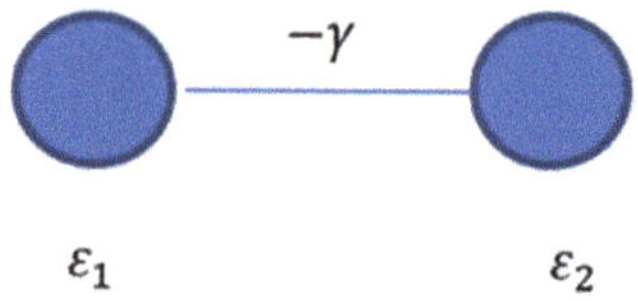

Figure 4.2. A representation of a two-level system, with orbital energies ε_1 and ε_2, coupled by $-\gamma$.

and (as can be verified by direct multiplication) satisfies $H|1\rangle = \varepsilon_1|1\rangle$ and $H|2\rangle = \varepsilon_2|2\rangle$. In other words, the individual site orbitals are eigenstates of the Hamiltonian describing the non-interacting atoms.

As noted in chapter 3, if the atoms are moved closer together, then the main effect is to introduce an off-diagonal matrix element into the Hamiltonian. This matrix element is often real and negative and therefore it will be denoted $-\gamma$, where γ is assumed to be positive and real. Again, this restriction can be relaxed later, where necessary (figure 4.2).

The new Hamiltonian is therefore a real Hermitian (and therefore symmetric) matrix of the form

$$H = \begin{pmatrix} \varepsilon_1 & -\gamma \\ -\gamma & \varepsilon_2 \end{pmatrix} \tag{4.2}$$

In equation (4.2), $\gamma = 0$ when the atoms are far apart, and is non-zero when they are closer together. The molecular orbitals $|\psi_1\rangle$ and $|\psi_2\rangle$ of this two-atom molecule are obtained by solving

$$H|\psi_n\rangle = E_n|\psi_n\rangle \tag{4.3}$$

This two-level problem was solved in chapter 3. For example, if the two site orbitals have the same energy, which we denote ε_0 (i.e. $\varepsilon_1 = \varepsilon_2 = \varepsilon_0$), then one obtains

$$|\psi_1\rangle = \begin{pmatrix} 1 \\ 1 \end{pmatrix}, \; E_1 = \varepsilon_0 - \gamma \; \text{ and } \; |\psi_2\rangle = \begin{pmatrix} 1 \\ -1 \end{pmatrix}, \; E_2 = \varepsilon_0 + \gamma \tag{4.4}$$

In equation (3.69) of chapter 3 the eigenvectors were denoted $|\psi_+\rangle = \begin{pmatrix} u_+ \\ v_+ \end{pmatrix}$ and $|\psi_-\rangle = \begin{pmatrix} u_- \\ v_- \end{pmatrix}$. Here it is convenient to change to a notation that can be generalised to $N \times N$ matrices and denote the eigenvectors by $|\psi_n\rangle = \begin{pmatrix} \psi_1^{(n)} \\ \psi_2^{(n)} \end{pmatrix}$. In other words, instead of denoting the first eigenvector by $|\psi_+\rangle$, it will be denoted $|\psi_1\rangle$ and instead of denoting the amplitudes on sites 1 and 2 by u_+, v_+, they will be denoted $\psi_1^{(1)}, \psi_2^{(1)}$ respectively. Then the probability that an electron that resides in molecular orbital $|\psi_n\rangle$ is located on site 1 is proportional to $|\psi_1^{(n)}|^2$ and the probability that an electron in molecular orbital $|\psi_n\rangle$ is located on site 2 is proportional[1] to $|\psi_2^{(n)}|^2$. Clearly for both $|\psi_1\rangle$ and $|\psi_2\rangle$, equation (4.4) reveals that the electron is found on both sites with equal probability, which is intuitively reasonable, because both sites are identical. On the other hand, in the case of $|\psi_2\rangle$, the amplitudes $\psi_1^{(2)}$ and $\psi_2^{(2)}$ have opposite signs, which is necessary, because eigenstates of Hermitian matrices belonging to different eigenvalues must be orthogonal. Note that for positive γ, the symmetric molecular orbital $\begin{pmatrix}1\\1\end{pmatrix}$ has a lower energy than ε_0. In other words, the presence of the coupling $-\gamma$ allows the two electrons in the molecule to lower their energy by residing in the symmetric orbital, thereby creating a chemical bond. For this reason the molecular orbital $\begin{pmatrix}1\\1\end{pmatrix}$ is called the 'bonding state'. Similarly, the higher-energy molecular orbital $\begin{pmatrix}1\\-1\end{pmatrix}$ is called the 'anti-bonding state'.

4.3 The eigenvalues and eigenvectors of a doubly infinite chain of identical atoms

Consider the doubly infinite chain of sites shown below, in which each site or site possesses a single orbital of energy ε_0, with nearest neighbour couplings $-\gamma$ (figure 4.3).

Our aim is to find all linearly-independent vectors $|\psi\rangle$, which satisfy the equation

$$H|\psi\rangle = E|\psi\rangle \tag{4.5}$$

When written out in full, this takes the form of equation (4.6).

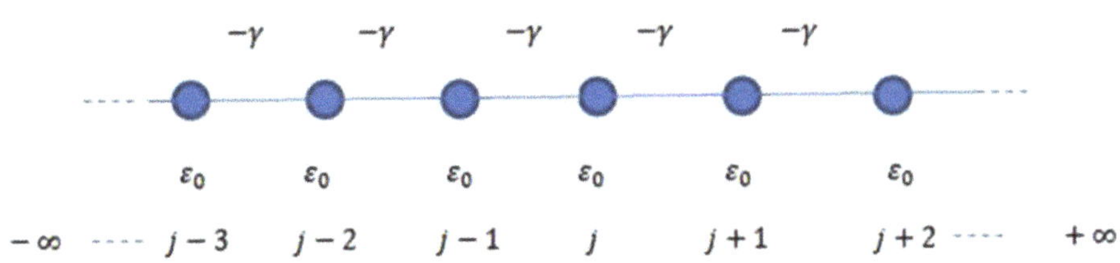

Figure 4.3. A doubly infinite chain of atoms labelled j.

[1] If the eigenstates are normalised by multiplying them by $1/\sqrt{2}$, then the probabilities would equal $|\psi_1^{(n)}|^2$ and $|\psi_2^{(n)}|^2$.

$$\begin{pmatrix} \cdot & \cdot & \cdot & \cdot & \cdot & \cdot & \cdot & \cdot & \cdot & \cdot \\ \cdot & \cdot & \varepsilon_o & -\gamma & 0 & 0 & 0 & \cdot & \cdot \\ \cdot & \cdot & -\gamma & \varepsilon_o & -\gamma & 0 & 0 & \cdot & \cdot \\ \cdot & \cdot & 0 & -\gamma & \varepsilon_o & -\gamma & 0 & \cdot & \cdot \\ \cdot & \cdot & 0 & 0 & -\gamma & \varepsilon_o & -\gamma & \cdot & \cdot \\ \cdot & \cdot & 0 & 0 & 0 & -\gamma & \varepsilon_o & \cdot & \cdot \\ \cdot & \cdot & \cdot & \cdot & \cdot & \cdot & \cdot & \cdot & \cdot \end{pmatrix} \begin{pmatrix} \cdot \\ \cdot \\ \psi_{j-2} \\ \psi_{j-1} \\ \psi_j \\ \psi_{j+1} \\ \psi_{j+2} \\ \cdot \\ \cdot \end{pmatrix} = \mathrm{E} \begin{pmatrix} \cdot \\ \cdot \\ \psi_{j-2} \\ \psi_{j-1} \\ \psi_j \\ \psi_{j+1} \\ \psi_{j+2} \\ \cdot \\ \cdot \end{pmatrix} \quad (4.6)$$

Equivalently, the equation for each row j is

$$\sum_{l=-\infty}^{\infty} H_{jl}\psi_l = E\psi_j \quad (4.7)$$

For a particular choice of j, $H_{jj} = \varepsilon_o$, $H_{jj+1} = H_{jj-1} = -\gamma$ and all other matrix elements are zero. Hence in the sum on the left-hand side of equation (4.7) there are only three non-zero terms and the equation reduces to

$$\varepsilon_o\psi_j - \gamma\psi_{j-1} - \gamma\psi_{j+1} = E\psi_j \quad (4.8)$$

One solution to this equation is

$$\psi_j = e^{ikj} \quad (4.9)$$

To verify this solution, substitute equation (4.9) into equation (4.8) to obtain

$$\varepsilon_o e^{ikj} - \gamma e^{ik(j-1)} - \gamma e^{ik(j+1)} = Ee^{ikj}$$

i.e.

$$\varepsilon_o e^{ikj} - \gamma e^{ikj}.\, e^{-ik} - \gamma e^{ikj}.\, e^{ik} = Ee^{ikj}$$

After cancelling e^{ikj} on both sides, this yields

$$\varepsilon_o - \gamma e^{ik} - \gamma e^{-ik} = E$$

Since $\cos\theta = \frac{(e^{i\theta} + e^{-i\theta})}{2}$, this shows that equation (4.9) is indeed a solution, provided $E = E(k)$, where

$$E(k) = \varepsilon_o - 2\gamma\cos k \quad (4.10)$$

This equation is called a 'dispersion relation' or alternatively a 'band structure' and shows that when plotted as a function of k, the eigenvalues $E(k)$ take the form shown in figure 4.4.

In the literature k is called a wave number or a wave vector. Since k is a continuous variable, the eigenvalues of H form a continuous band of energies $E(k)$ in the range

$$E_{\min} \leqslant E \leqslant E_{\max} \quad (4.11)$$

where $E_{\max} = \varepsilon_0 + 2\gamma$ and $E_{\min} = \varepsilon_0 - 2\gamma$. The width of this energy band is

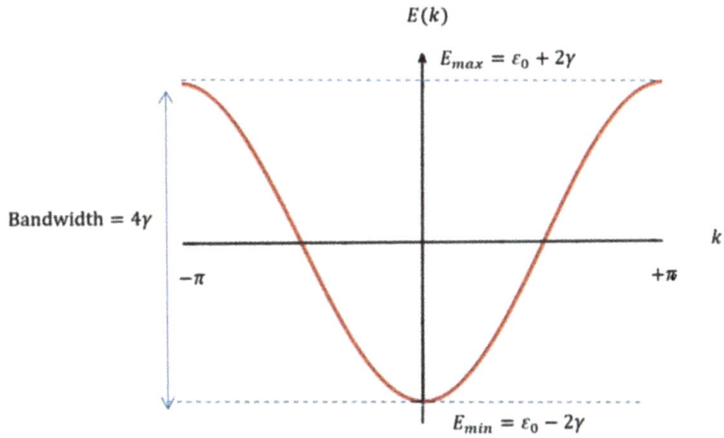

Figure 4.4. The dispersion relation of a linear chain.

$$W = E_{\max} - E_{\min} = 4\gamma \tag{4.12}$$

Note that the function e^{ikj} is often called a 'plane wave' and the fact that it is an eigenstate of the crystalline chain shown in figure 4.3 is a consequence of Bloch's theorem, which states that eigenstates of any crystalline system can be chosen to be proportional to plane waves. These have the property that $|e^{ikj}|^2 = 1$ for all j, which means that an electron in such a state is found to be on any atom with equal probability.

4.4 Effective mass and group velocity

The dimensionless variable k is simply a label used to distinguish between different eigenvectors and eigenvalues. It is useful, because it can also be used to label other properties. For example, as shown in section 4.7 below, electrons in the eigenstate $\psi_j = e^{ikj}$ can be assigned a velocity (usually called the 'group velocity') given by

$$v = \frac{1}{\hbar}\frac{dE}{dk} = \frac{2\gamma}{\hbar}\sin k \tag{4.13}$$

where the units of v are 'atomic spacings per unit time'. Furthermore, for small k, noting that $\cos k \approx 1 - k^2/2$ yields from equation (4.10),

$$E - E_{\min} \approx \gamma k^2 \tag{4.14}$$

This is reminiscent of the equation $E = \hbar^2 k^2/2m$, which describes an electron in free space, where m is the mass of the electron. To make equation (4.14) look like the equation for a free electron, one can write $\gamma = \hbar^2/2m^*$ and call m^* the effective mass of the electron in the linear chain. With this notation, the effective mass is related to the nearest neighbour coupling by

$$m^* = \frac{\hbar^2}{2\gamma} = 2\hbar^2/W \tag{4.15}$$

Note that since γ and the band width W tend to zero when the atoms are far apart, the effective mass m^* becomes very large when the atoms are far apart and the band is narrow. In other words, electrons in narrow bands have a large effective mass and can loosely be regarded as 'heavy electrons'. This is reasonable, since one expects electrons to find it difficult to move when the atoms are far apart.

4.5 Degeneracies and the allowed values of *k*

To complete the solution to equation (4.6), it is necessary to determine the allowed values of k. First it should be stressed that the answer to this question is not contained in equation (4.10). Instead, the answer is obtained by recalling that the problem posed by equation (4.5) is to find only linearly-independent eigenstates $|\psi\rangle$. Since

$$|\psi\rangle = \begin{pmatrix} \vdots \\ \psi_{j-2} \\ \psi_{j-1} \\ \psi_j \\ \psi_{j+1} \\ \psi_{j+2} \\ \vdots \end{pmatrix} \qquad (4.16)$$

where $\psi_j = e^{ikj}$, the allowed values of k are the set of all values for which these vectors are linearly independent. To emphasise that k labels the eigenstates, it is convenient to write $\psi_j^{(k)} = e^{ikj}$. Since $\psi_j^{(k+2\pi)} = e^{i(k+2\pi)j} = e^{ikj}e^{i2\pi j}$ and furthermore, for any integer j, $e^{i2\pi j} = 1$, one finds $\psi_j^{(k+2\pi)} = \psi_j^{(k)}$. Hence $\psi_j^{(k+2\pi)}$ and $\psi_j^{(k)}$ are not linearly independent and therefore if $\psi_j^{(k)}$ is retained, then $\psi_j^{(k+2\pi)}$ is not allowed. This means that k is restricted to an interval of size 2π.

For example, one could choose $0 \leqslant k < 2\pi$ or $-\pi \leqslant k < \pi$. Alternatively one could choose $a \leqslant k < a + 2\pi$, where a is any number. The precise choice is a matter of taste, provided values of k which differ by 2π are not retained. In figure 4.4, the choice $-\pi \leqslant k < \pi$ has been made. (Note that since $k = -\pi$ is retained, $k = +\pi$ is not allowed.) This choice is convenient, because (provided γ is positive) from equation (4.13), positive values of k in this range have positive group velocities v and negative values of k in this range have negative group velocities v, which makes it easy to remember the sign of v. On the other hand if one chooses $0 \leqslant k < 2\pi$, then positive group velocities correspond to $0 \leqslant k < \pi$ and negative group velocities correspond to $\pi < k < 2\pi$.

Notice that the two non-degenerate states corresponding to $k = 0$ and $k = -\pi$ are real, because $e^{i0} = 1$ and $e^{-i\pi j} = (-1)^j$. This is reasonable, because if a real matrix H is presented to a numerical eigenvalue solver such as MATLAB, then the resulting eigenstates will be real. On the other hand for other values of k, the eigenstate $\psi_j^{(k)} = e^{ikj}$ is complex and therefore this would not coincide with a solution presented by MATLAB. The answer to this conundrum lies in the presence of degeneracies. From equation (4.10), since $\cos(k) = \cos(-k)$, one finds $E(k) = E(-k)$ and therefore apart from $k = -\pi$ and $k = 0$, all eigenvalues are doubly degenerate. As discussed in the previous chapter, for the doubly degenerate level $E(k) = E(-k)$, any linear combination of the two states e^{ikj} and e^{-ikj} is also an eigenstate. For example $\cos kj = [e^{ikj} + e^{-ikj}]/2$ and $\sin kj = [e^{ikj} - e^{-ikj}]/2i$ are also eigenstates. Therefore instead of e^{ikj} and e^{-ikj}, if real solutions are preferred, then one could choose $\cos kj$ and $\sin kj$. Obviously one cannot gain eigenstates by making this alternative choice. If $\cos kj$ and $\sin kj$ are the chosen states, then to avoid double counting, one must revisit the question of the allowed values of k. Since $\cos kj = \cos(-k)j$, these states are linearly dependent, so if positive values of k are chosen, then negative values are not allowed. Similarly $\sin kj = -\sin(-k)j$, so again if positive values of k are chosen, then negative values are not allowed.

Note that for $k = 0$, $\cos 0j = 1$ for all, while for $k = \pi$, $= \cos \pi j = (-1)^j$. Hence for these eigenfunctions, $k = 0$ and $k = \pi$ are allowed. On the other hand, $\sin 0j = 0$ for all j and $\sin \pi j = 0$ for all j. Hence for these eigenfunctions, $k = 0$ and $k = \pi$ are not allowed.

In summary, one could either choose the single set of functions

$$\psi_j = e^{ikj}, \quad \text{with} \quad -\pi \leqslant k < \pi \tag{4.17}$$

or one could choose the pair of functions

$$\psi_j = \cos kj \ \text{ with } \ 0 \leqslant k \leqslant \pi \ \text{ and } \ \psi_j = \sin kj \ \text{ with } \ 0 < k < \pi \tag{4.18}$$

More generally, any linear combination could be chosen of the form

$$\psi_j = Ae^{ikj} + Be^{-ikj}, \tag{4.19}$$

where A and B are arbitrary constants. In this case, one should be careful to choose those values of A, B and k for which all linearly-independent solutions are found, without including any linearly-dependent solutions.

4.6 The most general solution to the Schrödinger equation for a linear chain

It is interesting to prove that equation (4.19), which contains two arbitrary constants A, B, is the most general solution to equation (4.8) by noting that the latter can be written

$$\psi_{j+1} = \frac{(\varepsilon_o - E)}{\gamma}\psi_j - \psi_{j-1} \tag{4.20a}$$

This is a two-point recurrence relation. Starting from any two values such as ψ_1 and ψ_2, substituting these into the right hand side gives ψ_3. Then substituting ψ_2 and ψ_3 into the right hand side gives ψ_4 etc. This means that all values of ψ_j for $j > 2$ are determined by the two numbers ψ_1 and ψ_2. Rewriting equation (4.20a) in the form

$$\psi_{j-1} = \frac{(\varepsilon_o - E)}{\gamma}\psi_j - \psi_{j+1} \tag{4.20b}$$

This shows that all values of ψ_j for $j < 1$ are determined by the two numbers ψ_1 and ψ_2. Hence for a given E the most general solution contains two arbitrary constants. Equation (4.19) contains two arbitrary constants A and B and therefore it is the most general solution to equation (4.8).

4.7 A finite chain of N sites with periodic boundary conditions

For the doubly infinite chain of figure 4.3, it was found that k is a continuous label. It will now be shown that for a finite chain, k becomes a discrete variable. Consider the linear chain of N atoms shown in figure 4.5, in which each atom possesses a single site of energy ε_0, with nearest neighbour couplings $-\gamma$.

This differs from the system of figure 4.3, because it has a finite number of N sites. Furthermore atom N is a nearest neighbour of atom 1. A system with the latter property is said to have periodic boundary conditions.

The eigenvalue equation for this structure is

$$H|\psi\rangle = E|\psi\rangle \tag{4.21}$$

or equivalently

$$\sum\nolimits_{l=1}^{N} H_{jl}\psi_l = E\psi_j \tag{4.22}$$

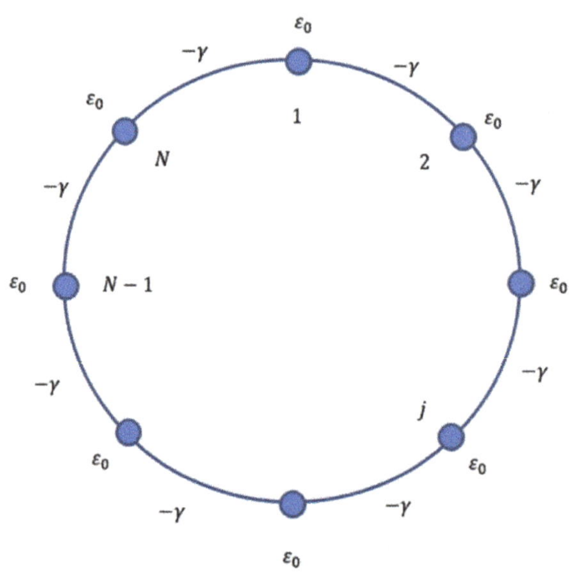

Figure 4.5. A chain of N atoms labelled j, with periodic boundary conditions.

When written out in full, this takes the form

$$\begin{pmatrix} \varepsilon_o & -\gamma & \cdot & \cdot & \cdot & \cdot & \cdot & \cdot & \cdot & \cdot & -\gamma \\ \cdot & \cdot & \cdot & \cdot & \cdot & \cdot & \cdot & \cdot & \cdot & \cdot & \cdot \\ \cdot & \cdot & \varepsilon_o & -\gamma & 0 & 0 & 0 & \cdot & \cdot \\ \cdot & \cdot & -\gamma & \varepsilon_o & -\gamma & 0 & 0 & \cdot & \cdot \\ \cdot & \cdot & 0 & -\gamma & \varepsilon_o & -\gamma & 0 & \cdot & \cdot \\ \cdot & \cdot & 0 & 0 & -\gamma & \varepsilon_o & -\gamma & \cdot & \cdot \\ \cdot & \cdot & 0 & 0 & 0 & -\gamma & \varepsilon_o & \cdot & \cdot \\ \cdot & \cdot & \cdot & \cdot & \cdot & \cdot & \cdot & \cdot & \cdot & \cdot & \cdot \\ -\gamma & \cdot & \cdot & \cdot & \cdot & \cdot & \cdot & \cdot & \cdot & -\gamma & \varepsilon_o \end{pmatrix} \begin{pmatrix} \psi_1 \\ \cdot \\ \cdot \\ \psi_{j-2} \\ \psi_{j-1} \\ \psi_j \\ \psi_{j+1} \\ \psi_{j+2} \\ \cdot \\ \cdot \\ \psi_N \end{pmatrix} = \mathrm{E} \begin{pmatrix} \psi_1 \\ \cdot \\ \cdot \\ \psi_{j-2} \\ \psi_{j-1} \\ \psi_j \\ \psi_{j+1} \\ \psi_{j+2} \\ \cdot \\ \cdot \\ \psi_N \end{pmatrix} \tag{4.23}$$

Note that the presence of the terms $-\gamma$ in the top right and bottom left corners are a consequence of the periodic boundary conditions. When written out row by row, this yields the following N simultaneous equations:

$$\varepsilon_o\psi_1 - \gamma\psi_N - \gamma\psi_2 = E\psi_1 \tag{4.24}$$

$$\varepsilon_o\psi_j - \gamma\psi_{j-1} - \gamma\psi_{j+1} = E\psi_j \text{ for } j = 2, 3, \dots N-1 \tag{4.25}$$

$$\varepsilon_o\psi_N - \gamma\psi_{N-1} - \gamma\psi_1 = E\psi_N \tag{4.26}$$

Clearly the fact that the two equations (4.24) and (4.26) differ from the other $N - 2$ equations is a consequence of the periodic boundary conditions. Equations (4.24) and (4.26) are therefore called 'boundary conditions'.

As for equations (4.8), the solution to the $N - 2$ equations (4.25) is

$$\psi_j = Ae^{ikj} \tag{4.27}$$

where A is an arbitrary constant. This solution can be verified by substituting equation (4.27) into equation (4.25). After cancelling Ae^{ikj} from both sides of the resulting equation, one again obtains the dispersion relation

$$E(k) = \varepsilon_o - 2\gamma \cos k \tag{4.28}$$

On the other hand, the solution to equation (4.23) must also satisfy equations (4.24) and (4.26). Since equation (4.24) can be written $\varepsilon_o\psi_1 - \gamma\psi_0 - \gamma\psi_2 + [\gamma\psi_0 - \gamma\psi_N] = E\psi_1$ and since the function Ae^{ikj} satisfies $\varepsilon_o\psi_1 - \gamma\psi_0 - \gamma\psi_2 = E\psi_1$, the boundary condition (4.24) reduces to

$$[\gamma\psi_0 - \gamma\psi_N] = 0 \tag{4.29}$$

Similarly, since Ae^{ikj} satisfies $\varepsilon_o\psi_N - \gamma\psi_{N+1} - \gamma\psi_{N-1} = E\psi_N$, comparison with equation (4.26) yields a simpler form for the second boundary condition:

$$[\gamma\psi_1 - \gamma\psi_{N+1}] = 0 \tag{4.30}$$

Equations (4.29) and (4.30) are simply the boundary conditions (4.24) and (4.26) written in a shorter form. Substituting equation (4.27) into (4.29) yields $e^{ikN} = 1$, and substituting equation (4.26) into equation (4.30) yields the equivalent equation $e^{ik(N+1)} = e^{ik}$. Both of these equations are solved by requiring that $kN = 2n\pi$, where n is an integer. Hence the allowed values of k are $k = k_n$, where

$$k_n = \frac{2n\pi}{N} \tag{4.31}$$

In contrast with the doubly infinite chain, where k was found to be a continuous variable, k now takes only discrete values. In other words, k has become 'quantized'. The doubly infinite system is an 'open system' with no boundaries, whereas the periodic system with N sites is a 'closed system'. In general, the eigenvalues of closed systems are quantized, whereas (at least some of) the eigenvalues of open systems form a continuum.

Just as in the case of the doubly infinite chain, to avoid including linearly-dependent eigenstates, k must be restricted to an interval of 2π. The precise choice is a matter of taste, provided values of k which differ by 2π are not allowed. Furthermore, as discussed above, because the eigenvalues are degenerate, one could either choose the single set of functions

$$\psi_j^{(n)} = Ae^{ik_nj}, \quad \text{with} \quad -\pi \leqslant k_n < \pi \tag{4.32}$$

or one could choose the pair of functions

$$\psi_j^{(n)} = B\cos k_nj \text{ with } 0 \leqslant k_n \leqslant \pi \text{ and } \psi_j^{(n)} = C\sin k_nj \text{ with } 0 < k_n < \pi \tag{4.33}$$

where A, B, C are arbitrary constant. More generally, one could choose linear combinations of the form

$$\psi_j^{(n)} = Ae^{ik_nj} + Be^{-ik_nj}, \tag{4.34}$$

where A and B are arbitrary constants. Since the arbitrary constants are not determined, their choice is a matter of taste. For example they could be chosen such that $\psi_j^{(n)}$ is normalised and satisfies $\sum_{j=1}^{N}\left|\psi_j^{(n)}\right|^2 = 1$. In the case of equation (4.32) this means that one should choose $A = \frac{e^{i\theta}}{\sqrt{N}}$, where θ is an arbitrary real constant, which does not affect the normalisation, because $|e^{i\theta}| = 1$. In the case of equation (4.33), where the eigenstates are chosen to be real, one should choose $C = \sqrt{2/N}$ and furthermore for $n \neq 0$, $B = \sqrt{2/N}$, while for $n = 0$, $B = \sqrt{1/N}$.

If the solution of equation (4.32) is chosen, then $-\frac{N}{2} \leqslant n < \frac{N}{2}$. For example if $N = 6$, $n = -3, -2, -1, 0, 1, 2$, so as expected, there are 6 linearly-independent eigenstates. The eigenvalues corresponding to $n = -3$ and $n = 0$ are non-degenerate, while those corresponding to $n = \pm 1$ and $n = \pm 2$ are double degenerate. On the other hand, for example if $N = 7$, then $n = -3, -2, -1, 0, 1, 2, 3$, so there are 7 linearly-independent states. The eigenvalue corresponding to $n = 0$ is non-degenerate, while those corresponding to $n = \pm 1$, $n = \pm 2$ and $n = \pm 3$ are double degenerate.

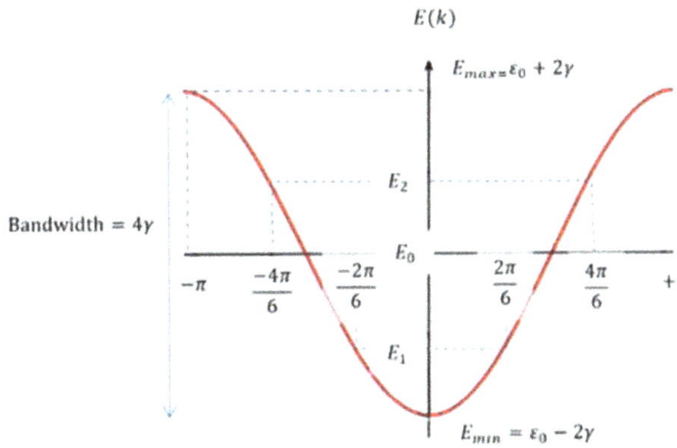

Figure 4.6. The eigenvalues and k values for a $N = 6$ periodic chain. Note that E_1 and E_2 are double degenerate, while E_0 and $E_{-3} = E_{max}$ are non-degenerate.

Figure 4.6 shows the eigenvalues for the case $N = 6$.

Orthogonality. Although eigenstates belonging to different eigenvalues are necessarily orthogonal, there is no guarantee that those belonging to the same degenerate eigenvalue are orthogonal. In fact it turns out that for the choice shown in equation (4.32), all eigenstates are orthogonal. To prove this, note that

$$\langle\psi_m|\psi_n\rangle = \sum_{j=1}^{N}\psi_j^{(m)*}\psi_j^{(n)} = \sum_{j=1}^{N}A^*Ae^{-ik_mj}e^{ik_nj} = |A|^2\sum_{j=1}^{N}e^{i(k_n-k_m)j} = |A|^2S$$

In this expression,

$$S = \sum_{j=1}^{N}e^{i(k_n-k_m)j} = \sum_{j=1}^{N}x^j$$

where $x = e^{i(k_n-k_m)} = e^{i\frac{2\pi(n-m)}{N}}$.

Clearly, if $n = m$, $x = 1$ and $S = N$. Therefore, as noted above, if $\psi_j^{(n)}$ is chosen to be normalised, then $|A|^2 = 1/N$. i.e. $A = \frac{e^{i\theta}}{\sqrt{N}}$, where θ is an arbitrary constant, which will be chosen to be zero. On the other hand if $n \neq m$, so that $x \neq 1$, note that $S = x + x^2 + x^3 + \ldots + x^N = x + x(x + x^2 + x^3 + \ldots + x^{N-1}) = x + x(S - x^N)$. Hence $S - xS = x(1 - x^N)$. That is $(1 - x)S = x(1 - x^N)$. Hence

$$S = \frac{x(1 - x^N)}{1 - x}$$

Since $x^N = e^{i2\pi(n-m)} = e^{i2\pi(\text{integer})} = 1$, one finds $S = 0$. This proves that the following eigenstates are orthonormal:

$$\psi_j^{(n)} = \frac{1}{\sqrt{N}}e^{i\frac{2\pi n}{N}j} \text{ where } n \text{ is an integer in the range } -\frac{N}{2} \leqslant n < \frac{N}{2} \tag{4.35}$$

That is they satisfy the orthonormality condition

$$\langle\psi_m|\psi_n\rangle = \sum_{j=1}^{N} \psi_j^{(m)*}\psi_j^{(n)} = \delta_{nm}, \tag{4.36a}$$

where δ_{nm} is the Kronecker delta (i.e. $\delta_{nm} = 1$ for $n = m$ and $\delta_{nm} = 0$ for $n \neq m$).

Completeness. Note that the range of n is written as $-\frac{N}{2} \leqslant n < \frac{N}{2}$, which is valid, even when N is odd. E.g. if $N = 7$, this yields $n = -3, -2, -1, 0, 1, 2, 3$. It should be emphasised that n and k_n are simply labels and are a matter of choice. Only the eigenvectors and eigenvalues are physically relevant and the same set of eigenvalues and eigenvectors is obtained whatever range of n is chosen, provided only linearly-independent eigenvectors are retained. One could have chosen $n = 1, 2, \ldots N$. This would yield exactly the same set of eigenstates $\psi_j^{(n)}$ and exactly the same set of eigenvalues E_n. Only the labels n would be different. With this choice, it is interesting to note that the above evaluation of the sum over j can be repeated for the sum over n to yield the completeness condition:

$$\sum_{n=1}^{N} |\psi_n\rangle\langle\psi_n| = I \tag{4.36b}$$

i.e.

$$\sum_{n=1}^{N} \langle i|\psi_n\rangle\langle\psi_n|j\rangle = \sum_{n=1}^{N} \psi_j^{(n)*}\psi_i^{(n)} = \delta_{ij} \tag{4.36c}$$

To prove equation (4.36c), note that equation (4.36a) is

$$\delta_{nm} = \frac{1}{N}\sum_{j=1}^{N} e^{i\frac{2\pi(n-m)}{N}j}$$

i.e. if p, q are integers,

$$\delta_{pq} = \frac{1}{N}\sum_{j=1}^{N} e^{i\frac{2\pi(p-q)}{N}j}$$

Since j is a dummy variable, replacing j by the dummy variable n yields

$$\delta_{pq} = \frac{1}{N}\sum_{n=1}^{N} e^{i\frac{2\pi(p-q)}{N}n}$$

Replacing p, q by i, j yields equation (4.36c), i.e.

$$\delta_{ij} = \frac{1}{N}\sum_{n=1}^{N} e^{i\frac{2\pi(j-i)}{N}n} = \sum_{n=1}^{N} \psi_i^{(n)*}\psi_j^{(n)}$$

This is simply the i, jth element of equation (4.36b).

4.8 Group velocity revisited

At first sight, assigning a velocity (4.13) to the 'plane wave' eigenstate $\psi_j = e^{ikj}$ is nonsense, because $|\psi_j|^2 = 1$ for all j, so an electron in such an eigenstate is equally likely to be anywhere. In other words the electron has no definite position. Since velocity is equal to the rate of change of position with time, strictly, such a state cannot be assigned a velocity.

To prove that the group velocity (4.13) is indeed the correct expression for the velocity of an electron, it is necessary to consider the dynamics of a wave packet, formed by superposing several plane waves of the form $\psi_j = e^{ik_n j}$, such that the modulus squared of the wave packet has a maximum at some value of j. Then the rate of change of the position of the maximum with time will be defined to be the group velocity. To this end, consider the general solution to the time-dependent Schrodinger equation, discussed in chapter 3, which takes the form

$$|\varphi(t)\rangle = \exp-\frac{iHt}{\hbar}|\varphi(0)\rangle = \sum_{n=1}^{N}\exp-i\omega_n t|\psi_n\rangle A_n$$

where $|\psi_n\rangle$ is an eigenstate of H satisfying $H|\psi_n\rangle = E_n|\psi_n\rangle$ and $\omega_n = E_n/\hbar$. The amplitude of this time-dependent wave function on site j is

$$\varphi_j(t) = \sum_n A_n\psi_j^{(n)}\exp(-i\omega_n t) \text{ where } -\frac{N}{2} \leqslant n < \frac{N}{2} \tag{4.37a}$$

In the case of a crystalline linear chain with periodic boundary conditions, described by the Hamiltonian (4.23), $\psi_j^{(n)} = \frac{1}{\sqrt{N}}e^{i\frac{2\pi n}{N}j}$ and $\hbar\omega_n = E(k_n) = \varepsilon_o - 2\gamma\cos k_n = \varepsilon_o - 2\gamma\cos\frac{2\pi n}{N}$.

If the wave function $\varphi_j(0)$ is known at time $t = 0$, then equation (4.37a) predicts the wave function $\varphi_j(t)$ at all other times. In equation (4.37a), the initial wave function $\varphi_j(0)$ is

$$\varphi_j(0) = \sum_n A_n\psi_j^{(n)} \tag{4.37b}$$

Therefore the initial wave function is determined by the choice of the amplitudes A_n. In what follows, they will be chosen such that $|\varphi_j(0)|^2$ has a maximum (i.e. a peak) at position $j = 0$. A wave function with a maximum at some position is called a wave packet. Examples of a wave packets with peaks at position $j = j_0$ are shown in figure 4.7, where the green curve shows a fairly narrow wave packet and the red curve shows an example of a broader wave packet. In what follows, our task is to choose the amplitudes A_n such that a wave packet is created with a peak at $j = 0$ at time $t = 0$. Equation (4.37a) will then be used to determine how the wave packet evolves in time.

Normalisation. Note that since the wave packet is chosen to describe a single electron, then it should be normalised and satisfy $\langle\varphi(t)|\varphi(t)\rangle = 1$. From equation (4.37a)

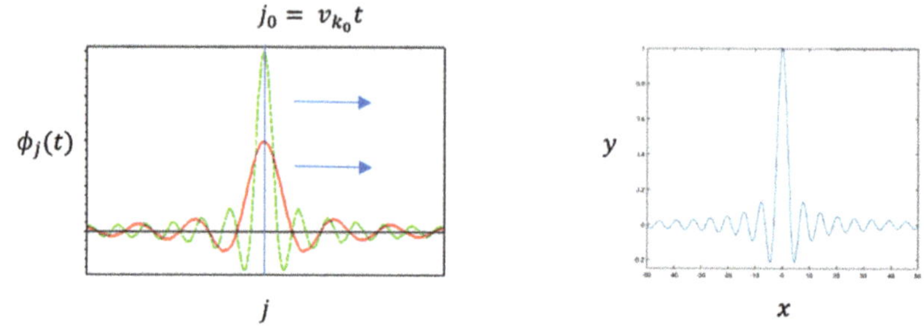

Figure 4.7. (Left panel) A plot of $\phi_j(t) = \frac{\sin(j - v_{k_0}t)\Delta k}{(j - v_{k_0}t)}$ versus j. The green curve corresponds to a large value of Δk, and therefore a narrow wave packet. The red curve corresponds to a smaller value of Δk and therefore a more spread-out wave packet. (Right panel) A plot of $y = [\sin x]/x$.

$$\langle\varphi(t)|\varphi(t)\rangle = \sum_{j=1}^{N}\varphi_j^*(t)\varphi_j(t) = \sum_{nm}A_m^*A_n\exp(-i(\omega_n - \omega_m)t)\left[\sum_{j=1}^{N}\psi_j^{(m)*}\psi_j^{(n)}\right]$$

Since the quantity is the square brackets is $\sum_{j=1}^{N}\psi_j^{(m)*}\psi_j^{(n)} = \delta_{nm}$, one finds

$$\langle\varphi(t)|\varphi(t)\rangle = \sum_{mn}A_m^*A_n\delta_{nm} = \sum_n|A_n|^2 \tag{4.38}$$

Since $\sum_n|A_n|^2$ does not depend on time, this means that the normalisation of the state $\varphi_j(t)$ does not change with time. Therefore in what follows, the amplitudes A_n will be chosen to satisfy

$$\sum_n|A_n|^2 = 1 \tag{4.39}$$

so that the state is normalised at $t = 0$ and remains normalised for all time.

The large N limit. To describe how the wave packet evolves in a large linear system and to avoid effects associated with periodic boundary conditions, it is convenient to take the limit of infinitely large N, where the sum in equation (4.37) becomes an integral:

$$\begin{aligned}\sum_n A_n\psi_j^{(n)}\exp(-i\omega_n t) &= \int_{-\frac{N}{2}}^{\frac{N}{2}} dnA_n\psi_j^{(n)}\exp(-i\omega_n t)\\ &= \frac{N}{2\pi}\int_{-\pi}^{\pi} dkA_k\psi_j^{(k)}\exp(-i\omega_k t)\end{aligned} \tag{4.40}$$

In the last step, the fact that $k = 2\pi n/N$ was used and therefore $dn = \frac{N}{2\pi}dk$.

Similarly the normalisation condition (4.39) becomes

$$\sum_n|A_n|^2 = \int_{-\frac{N}{2}}^{\frac{N}{2}} dn|A_n|^2 = \frac{N}{2\pi}\int_{-\pi}^{\pi} dk|A_k|^2 = 1$$

As an example, if A_k is independent of k, this would mean that to obtain a normalised state, one must choose $|A_k|^2 = \frac{1}{N}$, because $\frac{1}{2\pi}\int_{-\pi}^{\pi} dk = 1$. More generally, one can choose $A_k = a_k/\sqrt{N}$, where a_k is independent of N. Hence equations (4.35), (4.27) and (4.40) combine to yield

$$\varphi_j(t) = \frac{1}{2\pi}\int_{-\pi}^{\pi} dk a_k e^{i(kj-\omega_k t)}, \tag{4.41}$$

where $\varphi_j(t)$ will be normalised, provided

$$\frac{1}{2\pi}\int_{-\pi}^{\pi} dk|a_k|^2 = 1 \tag{4.42}$$

Creating a wave packet at time *t* = *0*. As a simple example of equation (4.41), consider the case where $\varphi_j(t)$ is formed from a group of plane waves, with wave vectors centred on some value k_0. As an example, choose

$$a_k = \left[\frac{2\pi}{2\Delta k}\right]^{1/2} \quad \text{for } k_0 - \Delta k < k < k_0 + \Delta k \tag{4.43}$$

and $a_k = 0$ otherwise. This clearly satisfies the normalisation condition (4.42), because for this choice of a_k, equation (4.42) becomes

$$\frac{1}{2\pi}\int_{-\pi}^{\pi} dk|a_k|^2 = \frac{1}{2\Delta k}\int_{k_0-\Delta k}^{k_0+\Delta k} dk = \frac{1}{2\Delta k}[k]_{k_0-\Delta k}^{k_0+\Delta k} = 1$$

This choice of a_k means that a_k is non-zero only in a range of k of width $2\Delta k$. The quantity $2\Delta k$ is referred to as the width of the wave packet in k-space. Substituting equation (4.43) into equation (4.41) yields for the initial wave function at $t = 0$

$$\varphi_j(0) = \frac{1}{\sqrt{4\pi\Delta k}}\int_{k_0-\Delta k}^{k_0+\Delta k} dk e^{ikj}$$

In what follows, it will be interesting to consider the case where the width $2\Delta k$ is small, so that the values of k in the integral are all very close to k_0. In this case the wave packet is 'almost a plane wave' with k centred on $k = k_0$.

Evolution of the wave packet. Substituting equation (4.43) into equation (4.41) yields for $t \neq 0$,

$$\varphi_j(t) = \frac{1}{\sqrt{4\pi\Delta k}}\int_{k_0-\Delta k}^{k_0+\Delta k} dk e^{i(kj-\omega_k t)}$$

For small Δk, it is convenient to write $k = k_0 + q$ and to expand ω_k about k_0 by writing $\omega_k \approx \omega_{k_0} + \frac{d\omega_{k_0}}{dk_0}q$. That is $\omega_k \approx \omega_{k_0} + v_{k_0}q$, where $v_{k_0} = \frac{d\omega_{k_0}}{dk_0}$.

This yields

$$\varphi_j(t) \approx \frac{e^{i(k_0 j-\omega_{k_0}t)}}{\sqrt{4\pi\Delta k}}\int_{-\Delta k}^{\Delta k} dq e^{i(j-v_{k_0}t)q} = \frac{e^{i(k_0 j-\omega_{k_0}t)}}{\sqrt{\pi\Delta k}}\frac{\sin(j-v_{k_0}t)\Delta k}{(j-v_{k_0}t)} \tag{4.44}$$

The modulus square of this 'group of plane waves' is

$$|\varphi_j(t)|^2 = \frac{1}{\pi\Delta k}\left[\frac{\sin(j - v_{k_0}t)\Delta k}{(j - v_{k_0}t)}\right]^2 \tag{4.45}$$

Since $\sin x \approx x$ for small x, the function $\left[\frac{\sin x}{x}\right]^2$ is equal to 1 at $x = 0$ and decays like $1/x^2$ for large x. Hence $|\varphi_j(t)|^2$ has a peak value of $|\varphi_j(t)|^2 = \Delta k/\pi$ for $j = v_{k_0}t$ and decays like $1/(j - v_{k_0}t)^2$ for values of j far from this peak position. This means that $\varphi_j(t)$ is a wave packet, peaked at $j = v_{k_0}t$. The position $j = v_{k_0}t$ of the peak changes with time, with a velocity $v_{k_0} = \frac{d\omega_{k_0}}{dk_0}$. Hence this 'group velocity' is given by equation (4.13), namely

$$\hbar v_{k_0} = \frac{dE_{k_0}}{dk_0} \tag{4.46}$$

Note that this is a very general result and equation (4.46) is an expression for the group velocity, whatever the k dependence of E_k. It relies on the approximation $\omega_k \approx \omega_{k_0} + v_{k_0}q$, which is valid for small q, which means that Δk must be small. Since $\left[\frac{\sin y}{y}\right]^2$ decays appreciably for $y > 1$, $|\varphi_j(t)|^2$ decays appreciably when j differs from the peak position by $\Delta j = 1/\Delta k$. Hence the spatial width of the wave packet is proportional to $1/\Delta k$. This means that in the limit that Δk tends to zero, the wave packet is almost a plane wave, with a very large spatial extent.

4.9 The eigenvalues and eigenvectors of a linear chain of N identical atoms

Consider the linear chain of N atoms shown below, in which each site or atom possesses a single atomic orbital of energy ε_0, with nearest neighbour couplings $-\gamma$ (figure 4.8):

The aim is to solve the equation

$$H|\psi\rangle = E|\psi\rangle \tag{4.47}$$

or equivalently

$$\sum\nolimits_{l=1}^{N} H_{jl}\psi_l = E\psi_j \tag{4.48}$$

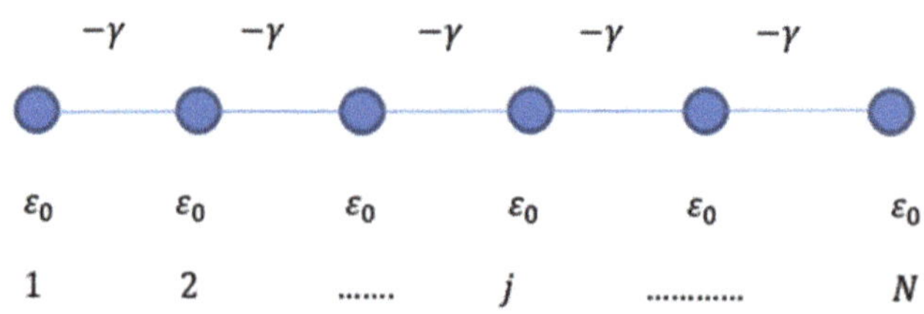

Figure 4.8. A linear chain of N atoms labelled $j = 1, 2, \ldots, N$.

When written out in full, this takes the form

$$\begin{pmatrix} \varepsilon_o & -\gamma & \cdot & \cdot & \cdot & \cdot & \cdot & \cdot & \cdot & 0 \\ \cdot & \cdot & \cdot & \cdot & \cdot & \cdot & \cdot & \cdot & \cdot & \cdot \\ \cdot & \cdot & \varepsilon_o & -\gamma & 0 & 0 & 0 & \cdot & \cdot & \cdot \\ \cdot & \cdot & -\gamma & \varepsilon_o & -\gamma & 0 & 0 & \cdot & \cdot & \cdot \\ \cdot & \cdot & 0 & -\gamma & \varepsilon_o & -\gamma & 0 & \cdot & \cdot & \cdot \\ \cdot & \cdot & 0 & 0 & -\gamma & \varepsilon_o & -\gamma & \cdot & \cdot & \cdot \\ \cdot & \cdot & 0 & 0 & 0 & -\gamma & \varepsilon_o & \cdot & \cdot & \cdot \\ \cdot & \cdot & \cdot & \cdot & \cdot & \cdot & \cdot & \cdot & \cdot & \cdot \\ 0 & \cdot & \cdot & \cdot & \cdot & \cdot & \cdot & \cdot & -\gamma & \varepsilon_o \end{pmatrix} \begin{pmatrix} \psi_1 \\ \cdot \\ \cdot \\ \psi_{j-2} \\ \psi_{j-1} \\ \psi_j \\ \psi_{j+1} \\ \psi_{j+2} \\ \cdot \\ \cdot \\ \psi_N \end{pmatrix} = \mathrm{E} \begin{pmatrix} \psi_1 \\ \cdot \\ \cdot \\ \psi_{j-2} \\ \psi_{j-1} \\ \psi_j \\ \psi_{j+1} \\ \psi_{j+2} \\ \cdot \\ \cdot \\ \psi_N \end{pmatrix} \tag{4.49}$$

Note that the presence of the zeros in the top right and bottom left corners are consequences of the 'free-end boundary conditions', in which atom 1 has no neighbour to the left and atom N has no neighbour to the right. When written out row by row, this yields the following N simultaneous equations:

$$\varepsilon_o\psi_1 - \gamma\psi_2 = E\psi_1 \tag{4.50}$$

$$\varepsilon_o\psi_j - \gamma\psi_{j-1} - \gamma\psi_{j+1} = E\psi_j \text{ for } j = 2, 3, \ldots N - 1 \tag{4.51}$$

$$\varepsilon_o\psi_N - \gamma\psi_{N-1} = E\psi_N \tag{4.52}$$

Clearly the fact that equations (4.50) and (4.52) differ from the other $N - 2$ equations is a consequence of the free-end boundary conditions.

Following equation (4.34), note that the most general solution of equation (4.51) is

$$\psi_j^{(n)} = Ae^{ik_nj} + Be^{-ik_nj} \tag{4.53}$$

where A, B and k_n are to be determined, by requiring that the boundary conditions (4.50) and (4.52) are satisfied. Note that the function $\psi_j^{(n)}$ satisfies equation (4.51) for any value of j, even for $j = 1, 0, -1$ etc and $j = N, N+1, N+2$, etc. Of course there is no requirement that it satisfies equation (4.51) for these values of j. It is simply a matter of fact that it does. This leads to a simplification of equations (4.50) and (4.52), because since $\psi_j^{(n)}$ satisfies equation (4.51) for $j = 1$, it satisfies

$$\varepsilon_o\psi_1 - \gamma\psi_2 - \gamma\psi_0 = E\psi_1 \tag{4.51a}$$

Since it should also satisfy equation (4.50), subtracting equation (4.50) from (4.51a) yields

$$\psi_0 = 0 \tag{4.54}$$

which is a simpler version of the boundary condition (4.50). Similarly since the function $\psi_j^{(n)}$ satisfies equation (4.51) for any value of j, even for $j = N$, it satisfies

$$\varepsilon_o\psi_N - \gamma\psi_{N-1} - \gamma\psi_{N+1} = E\psi_N \tag{4.51b}$$

On the other hand it is also required to satisfy equation (4.52). Subtracting equation (4.52) from (4.51b) yields

$$\psi_{N+1} = 0 \tag{4.55}$$

which is a simpler version of the boundary condition (4.52). Substituting equation (4.53) into (4.55) yields

$$A + B = 0$$

Hence $\psi_j^{(n)} = A(e^{ik_n j} - e^{-ik_n j})$ and therefore

$$\psi_j^{(n)} = C \sin k_n j \tag{4.56}$$

where C is an arbitrary constant. Substituting (4.56) into the second boundary condition (4.55) yields

$$\sin k_n(N + 1) = 0$$

Hence $k_n(N + 1) = n\pi$ or alternatively

$$k_n = \frac{n\pi}{(N + 1)} \tag{4.57}$$

The range of the integers n is obtained by noting that since $\sin\,(k_n + 2\pi)j = \sin k_n j$, so as in section 4.4, to avoid the double counting of linearly-dependent solutions, k_n must be restricted to a range of width 2π, such as $-\pi \leqslant k_n < \pi$. However, $\sin\theta = -\sin(-\theta)$, so that $\psi_j^{(-n)} = C \sin -k_n j = -C \sin k_n j = -\psi_j^{(n)}$. Hence, due to the sinusoidal form of $\psi_j^{(n)}$ in equation (4.56), $\psi_j^{(-n)}$ and $\psi_j^{(n)}$ are linearly dependent. This means that to avoid double counting, k_n must be further restricted to the range $0 \leqslant k_n < \pi$. Also, since j is an integer, $\sin \pi j = 0$ for all j. Hence $k_n = \pi$ is not allowed. Furthermore $\sin 0j = 0$ for all j, so $k_n = 0$ is not allowed. Hence $0 < k_n < \pi$, which means that the allowed values of k_n are

$$k_n = \frac{n\pi}{(N + 1)},\ n = 1, 2, 3, \ldots N \tag{4.58}$$

As expected, exactly N linearly-independent eigenvectors have been found. Note that the solution (4.56) and the allowed values of k_n (4.58) are completely different from those of equation (4.27) with periodic boundary conditions, where $\psi_j = Ae^{ikj}$, $k_n = \frac{2n\pi}{N}$ and $n = p, p + 1, p + 2, p + 3, \ldots p + N - 1$, where p is any integer. For the linear chain, the eigenvalues $E(k_n) = \varepsilon_o - 2\gamma \cos k_n$ are non-degenerate and the spacing between k values is $k_{n+1} - k_n = \frac{\pi}{(N+1)}$, whereas with periodic boundary conditions most eigenvalues are doubly degenerate and the spacing between k values is $k_{n+1} - k_n = \frac{2\pi}{(N+1)}$, which for large N is twice as large as the spacing for the linear chain.

In the solution (4.56), C is an arbitrary constant. As an example, one could choose C such that $\psi_j^{(n)}$ is normalised and real, in which case $C = \sqrt{2/(N+1)}$ and equation (4.56) becomes

$$\psi_j^{(n)} = \sqrt{\frac{2}{N+1}} \sin \frac{n\pi}{(N+1)} j,\ n = 1, 2, 3, \dots N \tag{4.59}$$

4.10 The two-level system revisited

As a check on the above solution, consider the case $N = 2$, where the Hamiltonian for the linear chain reduces to

$$H = \begin{pmatrix} \varepsilon_0 & -\gamma \\ -\gamma & \varepsilon_0 \end{pmatrix} \tag{4.60}$$

and from the analysis below equation (3.22), the two solutions of

$$H|\psi\rangle = E|\psi\rangle \tag{4.61}$$

are

$$|\psi_1\rangle = \begin{pmatrix} \psi_1^{(1)} \\ \psi_2^{(1)} \end{pmatrix} = \sqrt{\frac{1}{2}} \begin{pmatrix} 1 \\ 1 \end{pmatrix} \text{ with eigenvalue } E_1 = \varepsilon_0 - \gamma \tag{4.62}$$

and

$$|\psi_2\rangle = \begin{pmatrix} \psi_1^{(2)} \\ \psi_2^{(2)} \end{pmatrix} = \sqrt{\frac{1}{2}} \begin{pmatrix} 1 \\ -1 \end{pmatrix} \text{ with eigenvalue } E_2 = \varepsilon_0 + \gamma \tag{4.63}$$

For this case equation (4.59) yields

$$\psi_j^{(n)} = \sqrt{\frac{2}{3}} \sin \frac{n\pi}{3} j,\ n = 1, 2 \tag{4.64}$$

Since $\sin \frac{\pi}{3} = \sin \frac{2\pi}{3} = \frac{\sqrt{3}}{2}$ and $\sin \frac{4\pi}{3} = -\frac{\sqrt{3}}{2}$, this yields $\psi_1^{(1)} = \psi_2^{(1)} = \sqrt{\frac{2}{3}} \sin \frac{\pi j}{3}$ $= \sqrt{\frac{1}{2}}$, with $E_1 = \varepsilon_o - 2\gamma \cos \frac{\pi}{3} = \varepsilon_o - \gamma$ in agreement with equation (4.62). Also, equation (4.59) yields $\psi_1^{(2)} = \sqrt{\frac{2}{3}} \sin \frac{2\pi j}{3}$ $= \sqrt{\frac{1}{2}}$ and $\psi_{j2}^{(2)} = -\sqrt{\frac{1}{2}}$, with $E_2 = \varepsilon_o - 2\gamma \cos \frac{2\pi}{3} = \varepsilon_o + \gamma$ in agreement with (4.63), as expected.

On the other hand, equations (4.58) and (4.56) are valid for any value of N and therefore provide the solution of the eigenvalue equation $H|\psi\rangle = E|\psi\rangle$ for an arbitrarily large number of sites.

4.11 A three-level system

Having checked the solution for the two-level system, it is interesting to note that for a chain of three sites, equation (4.59) yields

$$\psi_j^{(n)} = \sqrt{\frac{2}{4}} \sin \frac{n\pi}{4} j \text{ and } E_n = \varepsilon_o - 2\gamma \cos \frac{n\pi}{4}, \text{ where } n = 1, 2, 3 \tag{4.64}$$

These solutions are shown in figure 4.9.

Notice that these are mutually orthonormal, as expected, because the eigenenergies are non-degenerate. In figure 4.9, the blue and red circles are pictorial representations of the eigenstates, using the convention introduced in figures 3.7 and 3.8, where a blue circle represents a positive amplitude, a red circle a negative amplitude and the radius of the circle is proportional to the magnitude of the amplitude on a given site. (A small black dot represents zero.) Notice that in moving from left to right along the circles of equation (4.65), there are no sign changes. Moving from left to right along the circles of equation (4.66) there is one sign change and from left to right along the circles of equation (4.67) there are two sign changes. This increase in the number of sign changes between successive eigenstates is necessary, otherwise the eigenstates could not be mutually orthogonal.

4.12 Normal modes of a vibrating atomic chain

Consider the chain of vibrating masses shown in figure 4.10, where x_j represents the displacement of mass j from equilibrium.

$$E_1 = \varepsilon_o - \gamma\sqrt{2} \quad |\psi_1\rangle = \begin{pmatrix} \psi_1^{(1)} \\ \psi_2^{(1)} \\ \psi_3^{(1)} \end{pmatrix} = \frac{1}{2}\begin{pmatrix} 1 \\ \sqrt{2} \\ 1 \end{pmatrix}, \tag{4.65}$$

$$E_2 = \varepsilon_o \quad |\psi_2\rangle = \begin{pmatrix} \psi_1^{(2)} \\ \psi_2^{(2)} \\ \psi_3^{(2)} \end{pmatrix} = \sqrt{\frac{1}{2}}\begin{pmatrix} 1 \\ 0 \\ -1 \end{pmatrix}, \tag{4.66}$$

$$E_3 = \varepsilon_o + \gamma\sqrt{2} \quad |\psi_3\rangle = \begin{pmatrix} \psi_1^{(3)} \\ \psi_2^{(3)} \\ \psi_3^{(3)} \end{pmatrix} = \frac{1}{2}\begin{pmatrix} 1 \\ -\sqrt{2} \\ 1 \end{pmatrix}, \tag{4.67}$$

Figure 4.9. The eigenstates of a 3-site chain.

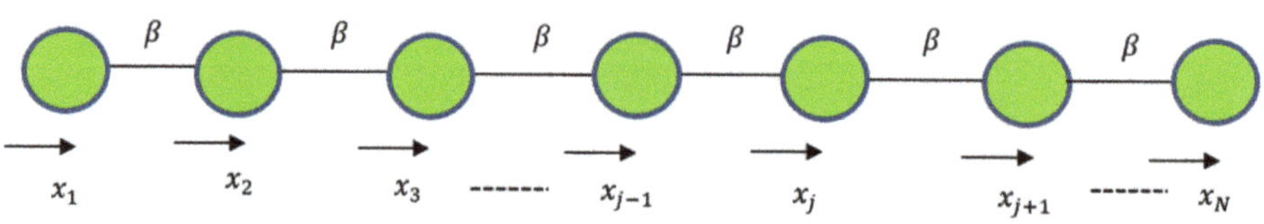

Figure 4.10. A chain of vibrating masses of mass m, connected by Hooke springs of strength β.

Following the discussion leading to equations (3.4)–(3.6) and more generally to equation (3.29), the dynamical matrix describing this chain is

$$H = \begin{pmatrix} \gamma & -\gamma & 0 & 0 & 0 & 0 & . & . \\ -\gamma & 2\gamma & -\gamma & 0 & 0 & 0 & . & . \\ 0 & -\gamma & 2\gamma & -\gamma & 0 & 0 & . & . \\ 0 & 0 & -\gamma & 2\gamma & -\gamma & 0 & . & . \\ 0 & 0 & 0 & -\gamma & 2\gamma & . & . & . \\ . & . & . & . & . & . & . & . \\ 0 & . & . & . & . & . & . & \gamma \end{pmatrix} \tag{4.68}$$

where $\gamma = \beta/m$. The normal modes are obtained by solving

$$H|\psi\rangle = E|\psi\rangle \tag{4.69}$$

$$\sum_{j=1}^{N} H_{ij}\psi_{ij} = E\psi_{ij} \tag{4.70}$$

When written out in full, this becomes

$$\gamma\psi_1 - \gamma\psi_2 = E\psi_1 \tag{4.71}$$

$$2\gamma\psi_j - \gamma\psi_{j-1} - \gamma\psi_{j+1} = E\psi_j \quad 2 \leqslant j \leqslant (N-1) \tag{4.72}$$

$$\gamma\psi_N - \gamma\psi_{N-1} = E\psi_N \tag{4.73}$$

Following equation (4.34), the most general solution of equation (4.73) is

$$\psi_j^{(n)} = Ae^{ik_nj} + Be^{-ik_nj} \tag{4.74}$$

where A, B and k_n are to be determined, by requiring that the boundary conditions (4.71) and (4.73) are satisfied. As in section 4.8, the function $\psi_j^{(n)}$ satisfies equation (4.72) for any value of j, even for $j = 1, 0, -1$ etc and $j = N, N+1, N+2$, etc. This allows equations (4.71) and (4.73) to be simplified, because since $\psi_j^{(n)}$ satisfies equation (4.72) for $j = 1$, it satisfies

$$\varepsilon_o\psi_1 - \gamma\psi_2 - \gamma\psi_0 = E\psi_1 \tag{4.75}$$

On the other hand it should also satisfy equation (4.71). Subtracting equation (4.50) from (4.75) yields

$$\psi_0 = \psi_1 \tag{4.76}$$

which is a simpler version of the boundary condition (4.71). Similarly since the function $\psi_j^{(n)}$ satisfies equation (4.72) for any value of j, even for $j = N$, it satisfies

$$\varepsilon_o\psi_N - \gamma\psi_{N-1} - \gamma\psi_{N+1} = E\psi_N \tag{4.77}$$

Since it should also satisfy equation (4.73), subtracting equation (4.73) from (4.77) yields a simplified form of the second boundary condition:

$$\psi_{N+1} = \psi_N \tag{4.78}$$

which is a simpler version of the boundary condition (4.73). Substituting (4.53) into (4.55) yields $A = De^{-ik_n/2}$ and $B = De^{ik_n/2}$, where D is an arbitrary constant. Hence $\psi_j^{(n)} = C(e^{ik_n(j-1/2)} - e^{-ik_n(j-1/2)})$ and therefore

$$\psi_j^{(n)} = C \cos k_n(j - 1/2) \tag{4.79}$$

where C is an arbitrary constant. Substituting equation (4.79) into the second boundary condition (4.78) yields

$$\cos k_n\left(N + \frac{1}{2}\right) = \cos k_n\left(N - \frac{1}{2}\right)$$

That is $\cos k_n N \cos k_n/2 - \sin k_n N \sin \frac{k_n}{2} = \cos k_n N \cos k_n/2 + \sin k_n N \sin k_n/2$

Hence $\sin k_n N \sin k_n/2 = 0$, which implies $k_n N = n\pi$ or alternatively

$$k_n = \frac{n\pi}{N} \tag{4.80}$$

The range of the integers n is obtained by noting that, since $\cos(k_n + 2\pi)(j - 1/2) = \cos k_n(j - 1/2)$, to avoid double counting linearly-dependent solutions, k_n must be restricted to a range of width 2π. Therefore one can choose $-\pi \leqslant k_n < \pi$. However, $\cos\theta = \cos(-\theta)$, so that $\psi_j^{(-n)} = \psi_j^{(n)}$. Hence, $\psi_j^{(-n)}$ and $\psi_j^{(n)}$ are linearly dependent, so to avoid double counting, k_n must be restricted to the range $0 \leqslant k_n < \pi$. (Note that since j is an integer, $\cos \pi(j - 1/2) = 0$ for all j. Hence $k_n = \pi$ is not allowed. Furthermore $\cos 0(j - 1/2) = 1$ for all j, so $k_n = 0$ is allowed.) This means that the allowed values of k_n are

$$k_n = \frac{n\pi}{N},\ n = 1, 2, 3, \dots N \tag{4.81}$$

As expected, exactly N linearly-independent eigenvectors have been found. Note that the solution (4.79) and the allowed values of k_n (4.81) are completely different from those of equation (4.27) for a linear chain with periodic boundary conditions, where $\psi_j = Ae^{ikj}$, $k_n = \frac{2n\pi}{N}$ and $-\frac{N}{2} \leqslant n < \frac{N}{2}$. They are also completely different from those of equation (4.56) for a linear chain with free-end boundary conditions, where $\psi_j^{(n)} = C \sin k_n j$, $k_n = \frac{n\pi}{(N+1)}$, $n = 1, 2, 3, \dots N$.

In the solution (4.79), C is an arbitrary constant. Choosing C such that $\psi_j^{(n)}$ is normalised, yields for $n \neq 0$, $C = \sqrt{\frac{2}{N}}$, whereas for $n = 0$, $C = \sqrt{\frac{1}{N}}$. Hence for $n \neq 0$ the corresponding normalised eigenstates are

$$\psi_j^{(n)} = \sqrt{\frac{2}{N}} \cos\frac{n\pi}{N}\left(j - \frac{1}{2}\right),\ n = 1, 2, 3, \dots N - 1 \tag{4.82}$$

and for $n = 0$,

$$\psi_j^{(0)} = \sqrt{\frac{1}{N}} \tag{4.83}$$

Note that the N eigenvalues are

$$E_n = 2\gamma - 2\gamma \cos\frac{n\pi}{N}, \quad \text{where } n = 0, 1, 2, 3, \ldots N - 1 \tag{4.84}$$

These are all positive, which is to be expected, since they are related to normal mode frequencies ω_n by $E_n = \omega_n^2$ and the right hand side of this equation is necessarily positive.

4.13 Radicals and Huckel's rule

It should be noted that in sections 4.8 and 4.10, the number of electrons in each system is not specified. Consider the linear chain of figure 4.8 formed from N atoms and described by $N \times N$ Hamiltonian. This has a series of non-degenerate eigenenergies E_1, E_2, E_3, …, E_N, where $E_1 < E_2 < E_3$, etc,. The state of lowest energy of the molecule is obtained by placing electrons into the lowest energy molecular orbitals. According to the Pauli exclusion principle, each orbital can accommodate at most two electrons (one spin 'up' and the other spin 'down'). Therefore if the molecule has two electrons, the state of lowest energy is obtained by placing both electrons in the state E_1. In this case, the highest occupied molecular orbital (HOMO) is E_1 and the lowest unoccupied molecular orbital (LUMO) is E_2. Furthermore the energy gap between the HOMO and LUMO is $E_2 - E_1$. If the molecule has four electrons, the state of lowest energy is obtained by placing two electrons into the state E_1 and two electrons in the state E_2. In this case, the HOMO is E_2, the LUMO is E_3 and the HOMO–LUMO gap is $E_3 - E_2$. Molecules in which the HOMO is completely filled and the LUMO is completely empty are particularly stable. On the other hand, molecules in which the HOMO is only partly filled are called radicals and tend to be highly reactive.

For the tight-binding chain in figure 4.8, consider the case in which each atom donates one electron to the chain. If the number of atoms N is even, then in the state of lowest energy, the first $N/2$ energy levels will be filled (each containing two electrons) and the molecule will be stable. On the other hand, if N is odd, then the first $(N - 1)/2$ energy levels will be filled and the next level will contain only one electron. For this reason, linear chains with odd numbers of monovalent atoms tend to be less stable than chains with even numbers of atoms.

Now consider the tight-binding ring with periodic boundary conditions, shown in figure 4.5, whose energy levels are given by $E_n = \varepsilon_o - 2\gamma \cos 2\pi n/N$, where $n = 0, \pm 1, \pm 2, \ldots$. The lowest level is non-degenerate and can accommodate two electrons. The next level $E_2 = E_{-2}$ is doubly degenerate and can accommodate four electrons. Hence a ring with two electrons, is stable and a ring with six electrons is also stable. Since $E_3 = E_{-3}$ is doubly degenerate and can accommodate four electrons, a ring with ten electrons is stable. In contrast rings with 1, 3, 4, 5, 7, 8,

9 electrons etc are radicals and tend to be reactive, because the HOMO is only partly filled.

This leads us to Huckel's rule which states that rings with $4m + 2$ electrons (where m is an integer) tend to be stable. Such rings are said to be aromatic. On the other hand, rings with $4m$ electrons tend to be unstable and are called anti-aromatic.

4.14 Summary of results for the eigenstates and eigenvalues of one-dimensional chains

Table 4.1 shows the eigenstates and eigenvalues of the three finite systems considered in this chapter. Note that k does not appear in these equations, because k is simply a label. For finite systems, it is preferable to use the label n. Also note that the states are orthonormal and complete, as can be verified explicitly for the examples shown.

4.15 A quantum spider's web

This section is intended as a bit of fun and provides an opportunity for the reader to test their understanding of the material covered so far in this chapter and in chapter 3. Therefore the reader is encouraged to try to solve this problem before reading the solution below. The aim is to find all linearly-independent eigenstates of the quantum spider's web shown in figure 4.11, which contains $NM + 1$ sites. It turns out that $NM - 1$ solutions are easy to find. The last two solutions are the twist in the tale.

In the tight-binding spider's web shown in figure 4.11, each blue site has a site energy ϵ_0, the central red site labelled 0 has a site energy ϵ_1. The nearest neighbour couplings within a branch are $-\gamma$, while those between branches are $-\alpha$. Apart from the central site, all blue sites are labelled by the pair of indices j, L, where L denotes the branch and j denotes a site within a branch. If the wave function on site j, L is denoted $\psi_{j,L}$ and the wave function on site 0 is denoted ϕ then the Schrödinger equation is as follows:
For $L \neq M$ or 1

$$\epsilon_0\psi_{j,L} - \gamma\psi_{j-1,L}(1 - \delta_{j1}) - \gamma\psi_{j+1,L}(1 - \delta_{jN}) - \alpha\psi_{j,L+1} - \alpha\psi_{j,L-1} - \gamma\phi\delta_{j1} = E\psi_{j,L} \tag{4.85}$$

For N, $L = M$

$$\epsilon_0\psi_{j,M} - \gamma\psi_{j-1,M}(1 - \delta_{j1}) - \gamma\psi_{j+1,M}(1 - \delta_{jN}) - \alpha\psi_{j,1} - \alpha\psi_{j,M-1} - \gamma\phi\delta_{j1} = E\psi_{j,M} \tag{4.86}$$

For N, $L = 1$

$$\epsilon_0\psi_{j,1} - \gamma\psi_{j-1,1}(1 - \delta_{j1}) - \gamma\psi_{j+1,1}(1 - \delta_{jN}) - \alpha\psi_{j,2} - \alpha\psi_{j,M} - \gamma\phi\delta_{j1} = E\psi_{j,1} \tag{4.87}$$

Table 4.1. The eigenstates and eigenvalues of the three finite systems considered in this chapter.

	A linear quantum chain with free-end boundary conditions	A 1-d chain with periodic boundary conditions	A 1-d chain of vibrating masses with free-end boundary conditions
Cartoon of the system	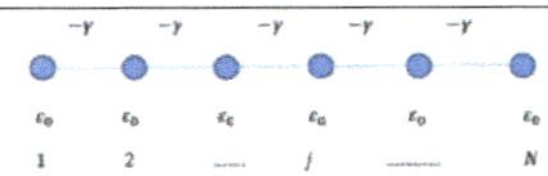	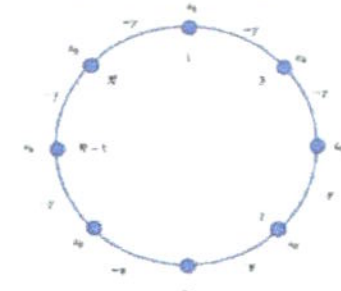	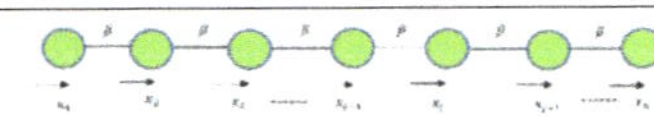
Amplitude of eigenvector $\|\psi_n\rangle$ on site j. This is denoted $\psi_j^{(n)}$ or equivalently $\langle j\|\psi_n\rangle$.	$\psi_j^{(n)} = \sqrt{\frac{2}{N+1}}\sin\frac{n\pi}{N+1}j$	$\psi_j^{(n)} = \frac{1}{\sqrt{N}}e^{i\frac{2\pi n}{N}j}$ or linear combinations of degenerate states $\psi_j^{(n)}$ and $\psi_j^{(-n)}$	$\psi_j^{(n)} = \sqrt{\frac{2}{N}}\cos\frac{n\pi}{N}\left(j-\frac{1}{2}\right)$ for $n \neq 0$ $\psi_j^{(0)} = \sqrt{\frac{1}{N}}$ for $n = 0$
Allowed values of n	$n = 1, 2, \ldots, N$	$-\frac{N}{2} \leqslant n < \frac{N}{2}$ or $n = 0, 1, 2, \ldots, N-1$	$n = 0, 1, 2, \ldots, N-1$
Energy eigenvalues	$E_n = \varepsilon_o - 2\gamma\cos\frac{n\pi}{N+1}$	$E_n = \varepsilon_o - 2\gamma\cos\frac{2\pi n}{N}$	$E_n = 2\gamma - 2\gamma\cos\frac{n\pi}{N} = 4\gamma\sin^2\frac{n\pi}{2N}$ If β= spring constant, $\gamma = \frac{\beta}{m}$ and angular frequency $\omega_n = \sqrt{E_n}$
Degenerate eigenvalues	None	All doubly degenerate, except $n = 0$. Also, if N is even, $n = -N/2$ is non-degenerate.	None
Example: $N = 3$	$E_1 = \varepsilon_o - \gamma\sqrt{2}, \|\psi_1\rangle = \begin{pmatrix}\psi_1^{(1)}\\ \psi_2^{(1)}\\ \psi_3^{(1)}\end{pmatrix} = \frac{1}{2}\begin{pmatrix}1\\ \sqrt{2}\\ 1\end{pmatrix},$ $E_2 = \varepsilon_o, \|\psi_2\rangle = \begin{pmatrix}\psi_1^{(2)}\\ \psi_2^{(2)}\\ \psi_3^{(2)}\end{pmatrix} = \sqrt{\frac{1}{2}}\begin{pmatrix}1\\ 0\\ -1\end{pmatrix},$ $E_3 = \varepsilon_o + \gamma\sqrt{2}, \|\psi_3\rangle = \begin{pmatrix}\psi_1^{(3)}\\ \psi_2^{3}\\ \psi_3^{(3)}\end{pmatrix} = \frac{1}{2}\begin{pmatrix}1\\ -\sqrt{2}\\ 1\end{pmatrix},$	$E_0 = \varepsilon_o - 2\gamma, \|0\rangle = \begin{pmatrix}\psi_1^{(0)}\\ \psi_2^{(0)}\\ \psi_3^{(0)}\end{pmatrix} = \frac{1}{\sqrt{3}}\begin{pmatrix}1\\ 1\\ 1\end{pmatrix},$ $E_1 = \varepsilon_o + \gamma, \|\psi_1\rangle = \begin{pmatrix}\psi_1^{(1)}\\ \psi_2^{(1)}\\ \psi_3^{(1)}\end{pmatrix} = \frac{1}{\sqrt{3}}\begin{pmatrix}e^{i\frac{2\pi}{3}}\\ e^{i\frac{4\pi}{3}}\\ 1\end{pmatrix},$ $E_{-1} = \varepsilon_o + \gamma, \|\psi_{-1}\rangle = \begin{pmatrix}\psi_1^{(-1)}\\ \psi_2^{(-1)}\\ \psi_3^{(-1)}\end{pmatrix} = \frac{1}{\sqrt{3}}\begin{pmatrix}e^{-i\frac{2\pi}{3}}\\ e^{-i\frac{4\pi}{3}}\\ 1\end{pmatrix}$	$E_0 = 0, \|\psi_0\rangle = \begin{pmatrix}\psi_1^{(0)}\\ \psi_2^{(0)}\\ \psi_3^{(0)}\end{pmatrix} = \frac{1}{\sqrt{3}}\begin{pmatrix}1\\ 1\\ 1\end{pmatrix},$ $E_1 = \gamma, \|\psi_1\rangle = \begin{pmatrix}\psi_1^{(1)}\\ \psi_2^{(1)}\\ \psi_3^{(1)}\end{pmatrix} = \sqrt{\frac{2}{3}}\begin{pmatrix}\frac{\sqrt{3}}{2}\\ 0\\ -\frac{\sqrt{3}}{2}\end{pmatrix},$ $E_2 = 3\gamma, \|\psi_2\rangle = \begin{pmatrix}\psi_1^{(2)}\\ \psi_2^{(2)}\\ \psi_3^{(2)}\end{pmatrix} = \sqrt{\frac{2}{3}}\begin{pmatrix}\frac{1}{2}\\ -1\\ \frac{1}{2}\end{pmatrix}$

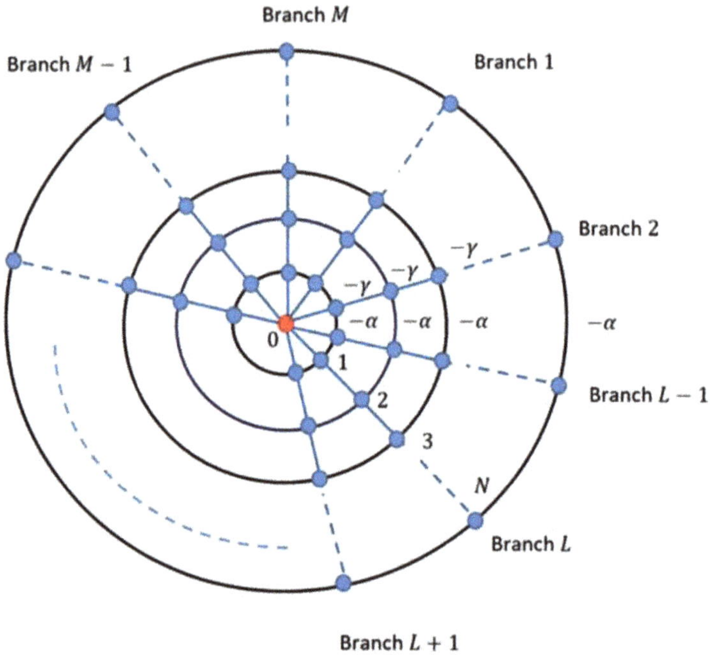

Figure 4.11. A quantum spider's web.

In the above equations, δ_{ij} is a Kronecker delta and the factors $(1 - \delta_{jN})$ and $(1 - \delta_{j1})$ account for the free-end boundary conditions at the outside end of each branch. Similarly the factor $-\gamma\phi\delta_{j1}$ accounts for the fact that sites 1 of each branch are attached to the central (red) site. The only remaining equation is

$$\epsilon_1\phi - \gamma\sum\nolimits_{L=1}^{M}\psi_{1,L} = E\phi \tag{4.88}$$

Since there are $ML + 1$ sites on the web, our task is to find the $ML + 1$ linearly-independent solutions to the above equations. Having defined the problem, this is the point at which the reader is encouraged to try to solve the problem themselves.

From the discussion in section 4.8, the intra-branch dependence on j must be of the form

$$\psi_{j,L} = A_L e^{ikj} + B_L e^{-ikj}$$

For $j = N$, equations (4.85)–(4.87) require that (cf equation (4.55)), $\psi_{N+1,L} = 0$. This is satisfied by choosing A_L and B_L such that

$$\psi_{j,L} = C_L \sin k[j - (N + 1)] \tag{4.89}$$

where k is arbitrary. This j dependence arises because the branches have free-end boundary condition at their outer ends. Similarly, equations (4.85)–(4.87) correspond to periodic boundary conditions for the label L, such that $C_{L+M} = C_L$ and therefore we choose $C_L = De^{\frac{i2n\pi L}{M}}$, where D is an arbitrary constant and $n = 0, 1, 2, \dots M - 1$.

Hence

$$\psi_{j,L} = De^{\frac{i2n\pi L}{M}} \sin k[j - (N + 1)] \tag{4.90}$$

For $j \neq 1$, substituting this into equations (4.85)–(4.87) demonstrates that it is indeed an eigenvector, provided

$$E = \epsilon_0 - 2\gamma \cos k - 2\alpha \cos \frac{2n\pi}{M} \tag{4.91}$$

To determine the allowed values of k, note that for $j = 1$, equations (4.85)–(4.87) yield (by continuity) $\psi_{0,L} = \phi$. That is

$$-De^{\frac{i2n\pi L}{M}} \sin k(N + 1) = \phi \tag{4.92}$$

Since ϕ does not depend on L, this yields for $n \neq 0$, $\sin k(N + 1) = 0$. Hence for $M > 1$ and $n = 1, 2, M - 1$,

$$k = m\pi/(N + 1), m = 1, 2, \dots N \tag{4.93}$$

When substituted into equations (4.91) and (4.92), equation (4.93) yields $N(M - 1)$ linearly-independent eigenstates and their corresponding eigenvalues. Therefore the remaining $N + 1$ solutions must be obtained by choosing $n = 0$ in equation (4.92). For $n = 0$, equation (4.92) yields

$$-D \sin k(N + 1) = \phi \tag{4.94}$$

and equation (4.88) yields

$$\epsilon_1\phi + \gamma MD \sin kN = E\phi \tag{4.95}$$

These combine to yield $\gamma M \sin kN = (\epsilon_1 - E) \sin k(N + 1)$. Since for $n = 0$, equation (4.91) yields $E = \epsilon_0 - 2\gamma \cos k$, this simplifies to

$$M \sin kN = (a + 2 \cos k) \sin k(N + 1) \tag{4.96}$$

where $a = (\epsilon_1 - \epsilon_0)/\gamma$. As an example, for $M = 1$ and $a = 0$, for which the structure of figure 4.9 reduces to a single chain of $N + 1$ sites, this yields

$$\sin kN = (2 \cos k) \sin k(N + 1) \tag{4.97}$$

Since $\sin kN = \sin k(N + 1 - 1) = \cos k \sin k(N + 1) - \sin k \cos k(N + 1)$ equation (4.97) simplifies to $-\sin k \cos k(N + 1) = \cos k \sin k(N + 1)$. Since $\sin k(N + 2) = \cos k \sin k(N + 1) + \sin k \cos k(N + 1)$, this simplifies to

$$\sin k(N + 2) = 0 \tag{4.98}$$

Since the case $M = 1$, $a = 0$ corresponds to a single linear chain with $N + 1$ sites, as expected equation (4.98) yields $k = \frac{m\pi}{N+2}$, $m = 1, 2, \dots N + 1$.

For $M = 2$ and $a = 0$, the structure of figure 4.11 reduces to a linear chain of $2N + 1$ sites and equation (4.96) yields

$$\sin kN = \cos k \sin k(N+1) \tag{4.99}$$

Since $k \neq 0, \pi$ (otherwise $\psi_{j,L}$ would vanish for all j) and since $\sin kN = \cos k \sin k(N+1) - \sin k \cos k(N+1)$, this yields $\cos k(N+1) = 0$ and therefore $k(N+1) = (2m+1)\pi/2$, $m = 0, 1, 2, \ldots N$.

i.e. $k = \frac{(2m+1)\pi}{2(N+1)}$ $m = 0, 1, 2, \ldots N$. Note that equation (4.93) yields $k = 2m\pi/2(N+1)$, $m = 1, 2, \ldots N$. Hence, as expected for a chain of $2N + 1$ sites with free-end boundary conditions, the allowed values of k are $k = p\pi/(2N+2)$, $p = 1, 2, 2N + 1$.

More generally, to solve equation (4.96), one can plot a graph of the following function

$$f(k) = (a + 2\cos k)\sin k(N+1) - M \sin kN \tag{4.100}$$

over the range $0 < k < \pi$. Then the allowed values of k correspond to the zeros of $f(k)$. We are expecting to find $N + 1$ solutions. As an example, figure 4.12 shows plots of $f(k)$ for $N = 4$, $M = 1$ and $a = 1$ or $a = 1.5$. For the former, excluding $k = 0$, $k = \pi$, there are indeed $N + 1 = 5$ zeros. However for the latter, there are only four zeros and therefore one solution is missing. How do we find the missing solution?

In fact there is no reason to suppose that k is real. In equation (4.89), one could choose $k = ib$ or $k = ib + \pi$, for which $\sin k[j - (N+1)]$ is proportional to $\sinh b[j - (N+1)]$ or $(-1)^{(j+(N+1))} \sinh b[j - (N+1)]$ respectively. For the former, equation (4.96) becomes

$$f(ib) = (a + 2\cosh b)\sinh b(N+1) - M \sinh bN = 0 \tag{4.101}$$

For the latter, equation (4.96) becomes

$$g(b) = f(ib + \pi) = (a - 2\cosh b)\sinh b(N+1) + M \sinh bN = 0 \tag{4.102}$$

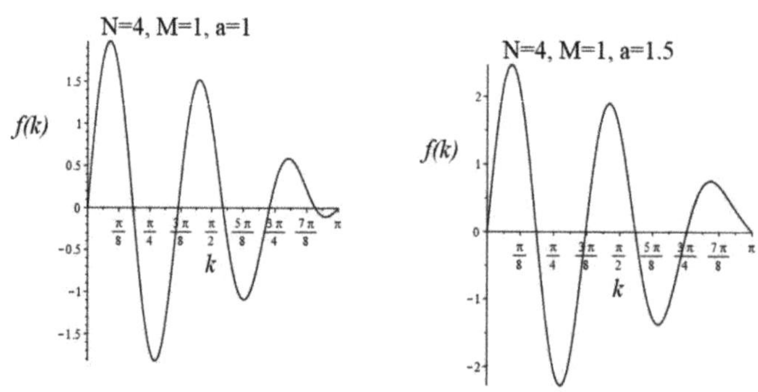

Figure 4.12. Plots of $f(k)$ for $N = 4$, $M = 1$ and $a = 1$ (left panel) or $a = 1.5$ (right panel).

Note that for small b, $g(b) \approx [(a - 2)(N + 1) + MN]b$ and therefore the slope at small b is positive, for $a > a_0$, where $a_0 = 2 - \frac{MN}{N+1}$. On the other hand, for large b, $g(b) \approx [(a - e^b)e^b + M]e^{bN}/2$, which is negative for large enough b. Hence a graph of $g(b)$ versus b will pass through zero, provided $a > a_0$. Similarly a graph of $f(ib)$ versus b will pass through zero, provided $a < -a_0$.

If $M = 1$, the spider's web is a linear chain of $N + 1$ sites with a site energy ϵ_1 at one end and if $M = 2$, the spider's web is a linear chain of $2N + 1$ sites with a site energy ϵ_1 on the middle site. In both cases, $a_0 > 0$ and therefore at most, either $g(b)$ or $f(ib)$ provide a solution, but not both. The appearance of these solutions signals the presence of a bound state, with a large amplitude on site zero, which decays towards the ends of the branches of the web. When $f(ib)$ provides a solution, the corresponding wave function $\sinh b[j - (N + 1)]$ decays monotonically with increasing j. When $g(b)$ provides a solution, the wave function $(-1)^{(j+(N+1))} \sinh b[j - (N + 1)]$ decays in magnitude with increasing j, but oscillates rapidly, switching sign between successive values of j. On the other hand, for $-a_0 < a < a_0$, neither graph passes through zero, but in this range of a, equation (4.9) provides the required $N + 1$ solutions. For $M = 1$, graphs of these functions are shown in figure 4.13 and indeed reveal the missing zeros.

For $M \geqslant 3$ a_0 is negative and therefore at least one of $g(b)$ or $f(ib)$ will provide a solution. Furthermore, in the range $-|a_0| < a < |a_0|$, both will provide a solution and the quantum spider's web possesses two bound states. In this range of a, equation (4.9) provides only $N - 1$ solutions.

Note that as $N \to \infty$ equations (4.101) and (4.102) simplify to

$$ae^{-b} + 1 = (M - 1)e^{-2b} \tag{4.103}$$

and

$$-ae^{-b} + 1 = (M - 1)e^{-2b} \tag{4.104}$$

respectively. For $M = 1$, provided $|\, a \,| \geqslant 1$, these yield for the decay constant b,

$$b = \ln|a| \tag{4.105}$$

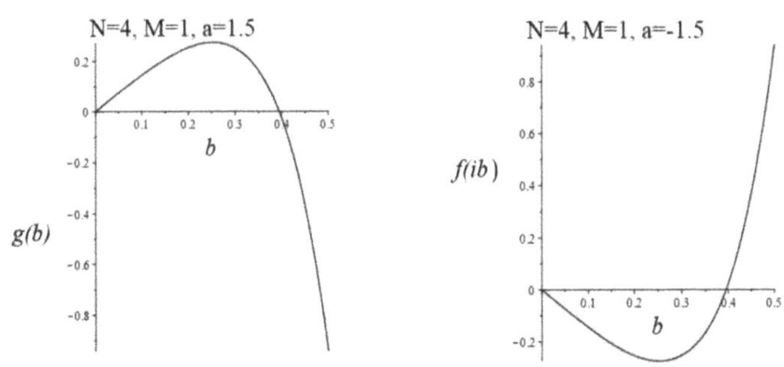

Figure 4.13. Graphs of the functions $g(b)$ and $f(ib)$ corresponding to complex k.

More generally, equation (4.103) has a solution for $a < (M - 2)$, given by

$$b = \ln\left[\frac{\sqrt{a^2 + 4(M - 1)} - a}{2}\right] \tag{4.106}$$

and equation (4.104) has a solution for $a > -(M - 2)$ given by

$$b = \ln\left[\frac{\sqrt{a^2 + 4(M - 1)} + a}{2}\right] \tag{4.107}$$

4.16 Eigenstates of a semi-infinite one-dimensional chain

To gain a little more insight into the appearance of bound states, consider the case of a semi-infinite chain, shown in figure 4.14, in which the end site has a site energy ϵ_1 and all other sites have site energies ϵ_0.

For $j \neq 1$ equation (4.8) is satisfied, leading to equation (4.10) and the most general solution (4.19). Since E is real, k is either real, or complex and of the form $k = ib$ or $k = \pi + ib$, where b is real. In the latter case, if b is positive then in equation (4.19), $B = 0$ otherwise the solution diverges at ∞. If b is negative then in equation (4.19), $A = 0$ otherwise the solution again diverges at ∞. In either case, the same result for ψ_j is obtained, so in what follows, b will be chosen arbitrarily to be positive.

For $j = 1$, $\epsilon_1\psi_1 - \gamma\psi_2 = E\psi_1$, but since ψ_j satisfies equation (4.8) for all j, it also satisfies $\epsilon_0\psi_1 - \gamma\psi_2 - \gamma\psi_0 = E\psi_1$. Comparing these equations yields $(\epsilon_1 - \epsilon_0)\psi_1 + \gamma\psi_0 = 0$. That is

$$a\psi_1 + \psi_0 = 0 \tag{4.108}$$

where $a = (\epsilon_1 - \epsilon_0)/\gamma$. This equation requires that $\frac{\psi_1}{\psi_0} = -\frac{1}{a}$ which is real. Hence for real k, from equation (4.19) $\frac{Ae^{ik} + B^{-ik}}{A + B}$ is real, which implies $B = A^*$. Choosing $A = \frac{C}{2i}e^{i\theta}$, where θ and C are real yields

$$\psi_j^k = C \sin(kj + \theta) \tag{4.109}$$

and

$$a \sin(k + \theta) + \sin\theta = 0 \tag{4.110}$$

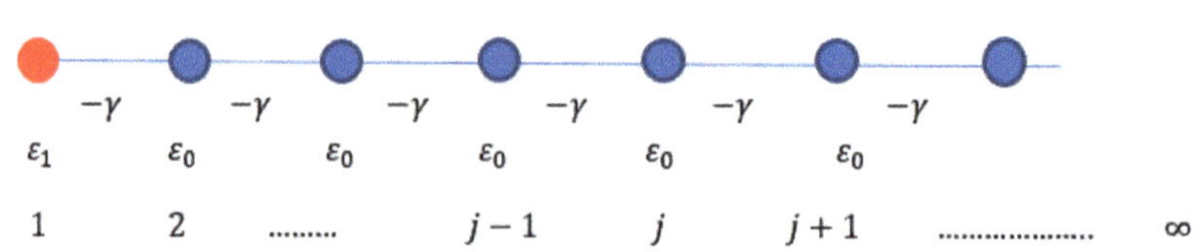

Figure 4.14. A semi-infinite chain with a terminal site energy ϵ_1.

For each choice of k, this is equation should be solved for θ. Note that if k, θ is a solution, then $-k$, $-\theta$ is also a solution. On the other hand,

$$\psi_j^{k,\theta} = -\psi_j^{-k,-\theta}$$

Therefore to avoid double counting linearly dependent solutions real values of k should be restricted to the range $0 < k \leqslant \pi$. The solution is

$$\theta = -\mathrm{a}\tan\left[\frac{a \sin k}{1 + a \cos k}\right] \tag{4.111}$$

On the other hand, if $k = ib$, $\psi_j = Ae^{-bj}$ and equation (4.104) yields

$$ae^{-b} + 1 = 0 \tag{4.112}$$

Since $b \geqslant 0$, this has a solution only if $a \leqslant -1$ and yields for the decay constant b,

$$b = \ln|a| \tag{4.113}$$

Similarly if $k = ib + \pi$, $\psi_j = A(-1)^j e^{-bj}$ and equation (4.104) yields

$$-ae^{-b} + 1 = 0 \tag{4.114}$$

This has a solution only if $a \geqslant 1$ and yields the same result for the decay constant b. Clearly, equation (4.113) coincides equations (4.106) and (4.107) when $M = 1$. Interestingly, for real k, choosing $\theta = -k(N + 2)$ in equation (4.106) reproduces equation (4.98).

4.17 Summary

The aim of this chapter has been to gain a little practice solving eigenvalue problems for tight-binding models involving linear chains. These are models of one-dimensional crystals and can be used as simple models of electrodes, which could act as sources and drains for electrons entering or leaving a molecular or other phase-coherent objects. In the next chapter, models of finite-width crystalline nanoribbons or chain with alternating bonds will be discussed.

Appendix A Relationship to Fourier analysis

As an example of the use of the above eigenvectors in table 4.1, it is of interest to see how they lead to Fourier analysis. Consider, for example a function $f(x)$ and imagine that we are interested in this function over the interval $0 \leqslant x \leqslant L$. Divide the interval into N equally spaced points labelled x_j, where $x_j = (j - 1)a$ and $a = L/(N - 1)$. In other words, $x_1 = 0$, $x_2 = a$, $x_3 = 2a$, … $x_N = L$. Let the value of the function at x_j be denoted $f_j = f(x_j)$ and define the vector $|f\rangle$ by

$$|f\rangle = \sqrt{a}\begin{pmatrix} f_1 \\ \vdots \\ f_{j-2} \\ f_{j-1} \\ f_j \\ f_{j+1} \\ f_{j+2} \\ \vdots \\ f_N \end{pmatrix} \tag{A.1}$$

Introducing the vectors, which have zeros everywhere, except for a 1 in row j

$$|j\rangle = \frac{1}{\sqrt{a}}\begin{pmatrix} 0 \\ \cdot \\ \cdot \\ 0 \\ 0 \\ 1 \\ 0 \\ 0 \\ \cdot \\ \cdot \\ 0 \end{pmatrix} \begin{matrix} \\ \\ \\ \\ \\ \longleftarrow \text{Row } j \\ \\ \\ \\ \\ \\ \end{matrix} \tag{A.2}$$

yields $\langle j|f\rangle = f_j$.

Then for any one of the three examples in table 4.1, since the N vectors $|\psi_n\rangle$ form a complete orthonormal set, one can write

$$|f\rangle = I|f\rangle = \sum\nolimits_{n=1}^{N} |\psi_n\rangle\langle\psi_n|f\rangle = \sum\nolimits_{n=1}^{N} A_n|\psi_n\rangle \tag{A.3}$$

where

$$A_n = \langle\psi_n|f\rangle \tag{A.4}$$

Multiplying both sides of equation (A.3) by $\langle j|$ yields

$$\langle j|f\rangle = \sum\nolimits_{n=1}^{N} A_n\langle j|\psi_n\rangle \tag{A.5}$$

or equivalently

$$f_j = \frac{1}{\sqrt{a}}\sum\nolimits_{n=1}^{N} A_n\psi_j^{(n)} \tag{A.6}$$

where from equation (4.89) and the completeness condition $\frac{1}{a} = \sum_{j=1}^{N}|j\rangle\langle j|$

$$A_n = a\sum_{j=1}^{N} \langle \psi_n | j\rangle \langle j | f\rangle = \sqrt{a}\sum_{j=1}^{N} \psi_j^{(n)} f_j \tag{A.7}$$

So far we have not made a choice for the eigenvectors $|\psi_n\rangle$. As an example, consider choosing the eigenvectors corresponding to the linear quantum chain with free-end boundary conditions:

$$\psi_j^{(n)} = \sqrt{\frac{2}{N+1}} \sin\frac{n\pi}{N+1} j, \; n = 1, 2, \ldots N \tag{A.8}$$

Then equation (A.6) becomes

$$f_j = \sqrt{\frac{2}{a(N+1)}} \sum_{n=1}^{N} A_n \sin\frac{n\pi}{N+1} j \tag{A.9}$$

where

$$A_n = \sqrt{\frac{2}{N+1}} \sum_{j=1}^{N} \sqrt{a} f_j \sin\frac{n\pi}{N+1} j \tag{A.10}$$

Now consider the limit N tends to infinity and the spacing $a = L/(N-1)$ tends to zero. Since the eigenstates are normalised

$$1 = \langle \psi_n | \psi_n \rangle = \sum_{j=1}^{N} \left[\psi_j^{(n)}\right]^2 = \sum_{j=1}^{N} \frac{2}{N+1} \left[\sin\frac{n\pi}{N+1} j\right]^2$$

In the limit, N tends to infinity and the spacing $a = L/(N-1)$ tends to zero, it is convenient to write $a = \Delta x$ this becomes

$$1 = \sum_{j=1}^{N} \frac{2\Delta x}{L} \left[\sin\frac{n\pi x_j}{L}\right]^2 = \frac{2}{L} \int_0^L dx \left[\sin\frac{n\pi x}{L}\right]^2 \tag{A.11}$$

Similarly equation (A.10) becomes

$$A_n = \sqrt{\frac{2}{L}} \int_0^L dx f(x) \sin\frac{n\pi x}{L} \tag{A.12}$$

and equation (A.9) becomes

$$f(x) = \sqrt{\frac{2}{L}} \sum_{n=1}^{N} A_n \sin\frac{n\pi x}{L} \tag{A.13}$$

Writing $k = \frac{n\pi}{L}$ and $dk = dn\frac{\pi}{L}$ yields

$$f(x) = \frac{1}{\pi}\sqrt{2L} \int_0^{\pi} dk A_n \sin kx \tag{A.14}$$

Finally writing $F_n = \sqrt{\frac{L}{2}} A_n$ yields from equation (4.97) and $F(k) = F_n$

$$F(k) = \int_0^L dx f(x) \sin kx \tag{A.15}$$

and from equation (4.99)

$$f(x) = \frac{2}{\pi}\int_0^{\pi} dkF(k)\sin kx \tag{A.16}$$

Equations (A.15) and (A.16) are standard equations of Fourier analysis.

On the other hand, if we choose eigenvectors corresponding to the linear quantum chain with periodic boundary conditions:

$$\psi_j^{(n)} = \sqrt{\frac{1}{N}}e^{i\frac{2n\pi}{N}j}, -\frac{N}{2} \leqslant n < \frac{N}{2} \tag{A.17}$$

Then equation (4.91) becomes

$$f_j = \sqrt{\frac{1}{aN}}\sum\nolimits_{n=1}^{N} A_n e^{i\frac{2n\pi}{N}j} \tag{A.18}$$

where

$$A_n = \sqrt{\frac{1}{N}}\sum\nolimits_{j=1}^{N} \sqrt{a}f_j e^{-i\frac{2n\pi}{N}j} \tag{A.19}$$

In the limit, N tends to infinity and the spacing $a = L/(N-1)$ tends to zero, writing $a = \Delta x$ yields

$$A_n = \sqrt{\frac{1}{L}}\int_0^{L} dxf(x)e^{-i\frac{2n\pi}{N}j} \tag{A.20}$$

and equation (A.18) becomes

$$f(x) = \sqrt{\frac{1}{L}}\sum\nolimits_{n=1}^{N} A_n e^{i\frac{2n\pi}{N}j} \tag{A.21}$$

Writing $k = \frac{2n\pi}{L}$ and $dk = dn\frac{2\pi}{L}$ yields

$$f(x) = \frac{1}{2\pi}\sqrt{L}\int_{-\pi}^{\pi} dkA_n e^{ikx} \tag{A.22}$$

Finally writing $F_n = \sqrt{L}A_n$ yields from equation (A.12) and $F(k) = F_n$

$$F(k) = \int_0^{L} dxf(x)e^{-ikx} \tag{A.23}$$

and from equation (A.23)

$$f(x) = \frac{1}{2\pi}\int_{-\pi}^{\pi} dkF(k)e^{ikx} \tag{A.24}$$

Again, these are standard equations of Fourier analysis.

Finally, a note about why the factor of $\sqrt{a}$ is included in the definition of $|f\rangle$ in equation (4.86).

To see why this is convenient, note that

$$\langle f|f\rangle = a\left(f_1^* f_2^* \ldots f_j^* \ldots f_N^*\right)\begin{pmatrix} f_1 \\ \vdots \\ f_j \\ \vdots \\ f_N \end{pmatrix} = a\sum_{j=1}^{N}|f_j|^2 \tag{A.25}$$

In the limit $a \to \Delta x \to 0$ and $N \to \infty$, the sum on the right hand side is simply the area under the curve of the function $g(x) = |f(x)|^2$. Hence

$$\langle f|f\rangle = \int_0^L dx\,|f(x)|^2 \tag{A.26}$$

The relationship between vectors and functions expressed by equation (A.1) and including the factor of $\sqrt{a}$ therefore allows us to equate the squared length $\langle f|f\rangle$ of the vectors, with the integral of the modulus squared of the function $|f(x)|^2$, in the limit that the spacing a between the discrete points x_i tends to zero.

Appendix B Continuity equations

Consider the equation

$$\varepsilon_o f_j - \gamma f_{j-1} - \gamma f_{j+1} = E f_j \tag{B.1}$$

where

$$E = \varepsilon_o - 2\gamma \cos k \tag{B.2}$$

The general solution to this is

$$f_j = Ae^{ikj} + Be^{-ikj} \tag{B.3}$$

Proof. Since Ae^{ikj} and Be^{-ikj} each satisfy equation (B.1) and since equation (B.1) is a linear equation, then their sum must also satisfy equation (B.1). This can be verified by substituting equation (B.3) into the left-hand side of equation (B.1) to yield

$$\begin{aligned}\varepsilon_o f_j - \gamma f_{j-1} - \gamma f_{j+1} &= A[\varepsilon_o e^{ikj} - \gamma e^{ikj}e^{-ikj} - \gamma e^{ikj}e^{ikj}] \\ &\quad + B[\varepsilon_o e^{-ikj} - \gamma e^{-ikj}e^{ikj} - \gamma e^{-ikj}e^{-ikj}] \\ &= Ae^{ikj}[\varepsilon_o - \gamma e^{-ikj} - \gamma e^{ikj}] + Be^{-ikj}[\varepsilon_o - \gamma e^{ikj} - \gamma e^{-ikj}] \\ &= Ae^{ikj}[\varepsilon_o - 2\gamma \cos k] + Be^{-ikj}[\varepsilon_o - 2\gamma \cos k] \\ &= E[Ae^{ikj} + Be^{-ikj}] = Ef_j\end{aligned}$$

B.1 Forward continuity

Clearly the function f_j satisfies equation (B.1) for all j. Rearranging equation (B.1) yields the recurrence relation

$$f_{j+1} = [(\varepsilon_o - E)/\gamma]f_j - f_{j-1} \tag{B.4}$$

Now consider the equation

$$X = [(\varepsilon_o - E)/\gamma]f_j - f_{j-1} \tag{B.5}$$

where f_j and f_{j-1} are given equation by (B.3). Subtracting equation (B.5) from equation (B.4) yields

$$X = f_{j+1} \tag{B.6}$$

That is

$$X = Ae^{ik(j+1)} + Be^{-ik(j+1)} \tag{B.7}$$

In other words, if f_j and f_{j-1} are given by (B.3) and (B.5) is satisfied, then X is also given by equation (B.3). Equation (B.5) is called a continuity equation, because if f_j and f_{j-1} are given by equations (B.3) and (B.5), then equation (B.5) implies that the same functional form continues onto X. As an example, if f_{-1} and f_{-2} are given by equation (B.3) and the following equation is satisfied

$$\varepsilon_o f_{-1} - \gamma .\ X - \gamma f_{-2} = Ef_{-1} \tag{B.8}$$

Then this implies

$$X = f_0 = A + B \tag{B.9}$$

B.2 Backward continuity

Rearranging equation (B.4) also yields the backward recurrence relation

$$f_{j-1} = [(\varepsilon_o - E)/\gamma]f_j - f_{j+1} \tag{B.10}$$

Now consider the equation

$$X = [(\varepsilon_o - E)/\gamma]f_j - f_{j+1} \tag{B.11}$$

where f_j and f_{j+1} are given by equation (B.3). Subtracting equation (B.10) from equation (B.11) yields

$$X = f_{j-1} \tag{B.12}$$

That is

$$X = Ae^{ik(j-1)} + Be^{-ik(j-1)} \tag{B.13}$$

In other words, if f_j and f_{j+1} are given by equation (B.3) and equation (B.11) is satisfied, then X is also given by equation (B.3). Equation (B.10) is called a backward continuity equation, because if f_j and f_{j+1} are given by equations (B.3) and (B.5), then equation (B.10) implies that the same functional form continues onto X. As an example, if f_{N+1} and f_{N+2} are given by equation (B.3) and the following equation is satisfied

$$\varepsilon_o f_{N+1} - \gamma . X - \gamma f_{N+2} = Ef_{N+1} \tag{B.14}$$

Then this implies

$$X = f_N = Ae^{ikN} + Be^{-ikN} \tag{B.15}$$

B.3 Egalitarian continuity

A more symmetric way of presenting the above argument is as follows:

Consider the equation

$$\varepsilon_o f_j - \gamma f_{j-1} - \gamma f_{j+1} = Ef_j \tag{B.16}$$

where

$$E = \varepsilon_o - 2\gamma \cos k \tag{B.17}$$

The general solution to this is

$$f_j = Ae^{ikj} + Be^{-ikj} \tag{B.18}$$

Consequently, if

$$\varepsilon_o f_j - \gamma X - \gamma f_{j+1} = Ef_j \tag{B.19}$$

then comparing equation (B.16) with equation (B.19) yields

$$X = f_{j-1} \tag{B.20}$$

Similarly, if

$$\varepsilon_o f_j - \gamma f_{j-1} - \gamma X = Ef_j \tag{B.21}$$

then comparing equation (B.16) with equation (B.21) yields

$$X = f_{j+1} \tag{B.22}$$

These more egalitarian continuity equations are useful for simplifying boundary conditions.

Chapter 5

Quantum properties of electrodes in higher dimensions

5.1 Introduction

The aim of the present chapter is to discuss properties of linear chains, nanoribbons and nanotubes, in which each unit cell contains more than one site. These infinitely-long, finite-width structures will be used as electrodes in later chapters, when discussing single-molecule junctions. The band structure of simple 2-dimensional crystals are also discussed.

5.2 Peierl's distortion and oligoynes

Following on from the discussion of Huckel's rule in chapter 2, it should be noted that even when the HOMO is filled and the LUMO is empty, a linear chain can be unstable to dimerization. To illustrate this, consider the infinite linear chain shown in figure 4.3, but where odd-numbered bonds are shorter and even-numbered bonds are longer, such that the nearest neighbour bonds alternate between two values $\gamma_1 = \gamma + \delta$ and $\gamma_2 = \gamma - \delta$, where δ and γ are positive. The resulting chain is shown in figure 5.1.

The chain consists of a series of identical cells, with two atoms in each cell. Let ϕ_J denote the amplitude of an eigenstate on the left-most atom of cell J and ψ_J the amplitude on the right-most atom. Then the Schrödinger equation is

$$\varepsilon_0\phi_J - \gamma_1\psi_{J-1} - \gamma_2\psi_J = E\phi_J \tag{5.1}$$

$$\varepsilon_0\psi_J - \gamma_1\phi_{J+1} - \gamma_2\phi_J = E\psi_J \tag{5.2}$$

Writing

$$\phi_J = ue^{ikJ} \text{ and } \psi_J = ve^{ikJ} \tag{5.3}$$

doi:10.1088/978-0-7503-3639-0ch5

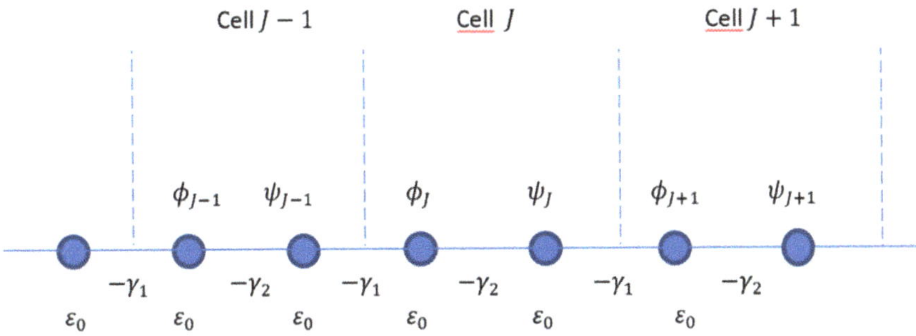

Figure 5.1. A doubly infinite linear chain with bond alternation.

and substituting these into (5.1) and (5.2) yields

$$\varepsilon_o u - v[\gamma_1 e^{-ik} + \gamma_2] = Eu \tag{5.4}$$

$$\varepsilon_o v - u[\gamma_1 e^{ik} + \gamma_2] = Ev \tag{5.5}$$

This can be written

$$\begin{pmatrix} \varepsilon_o & \Delta(k) \\ \Delta^*(k) & \varepsilon_o \end{pmatrix}\begin{pmatrix} u \\ v \end{pmatrix} = E\begin{pmatrix} u \\ v \end{pmatrix} \tag{5.6}$$

where

$$\Delta(k) = -[\gamma_1 e^{-ik} + \gamma_2] \tag{5.7}$$

Hence from the solution to the two-level system shown in section 3.16, if $\varepsilon_o = 0$, one obtains $E = \pm|\Delta(\mathrm{k})|$ and $\frac{u}{v} = \frac{\Delta(k)}{E} = \frac{\Delta(k)}{\pm|\Delta(\mathrm{k})|}$. When $\varepsilon_o \neq 0$, since ε_o is simply an additive constant, $\frac{u}{v}$ remains the same and the eigenvalues are simply shifted by ε_o. Hence the eigenvalues are

$$E_\pm(k) = \varepsilon_o \pm |\Delta(\mathrm{k})| \tag{5.8}$$

where

$$\begin{aligned} |\Delta(\mathrm{k})|^2 &= |\gamma_1 e^{-ik} + \gamma_2|^2 = |(\gamma_1 \cos k + \gamma_2) - i(\gamma_1 \sin k)|^2 \\ &= (\gamma_1 \cos k + \gamma_2)^2 + (\gamma_1 \sin k)^2 \\ &= \gamma_1^2 + \gamma_2^2 + 2\gamma_1\gamma_2 \cos k = (\gamma_1 - \gamma_2)^2 + 4\gamma_1\gamma_2 \cos^2 \frac{k}{2} \end{aligned}$$

Hence

$$|\Delta(\mathrm{k})|^2 = \delta^2 + 4\gamma_1\gamma_2 \cos^2 k/2 \tag{5.9}$$

This shows that $|\Delta(\mathrm{k})|^2$ varies between δ^2 and $(2\gamma)^2$, where $\delta = ||\gamma_1| - |\gamma_2||$ and $2\gamma = |\gamma_1| + |\gamma_2|$. The dispersion relation (5.8) (also called the band structure) is shown in figure 5.2.

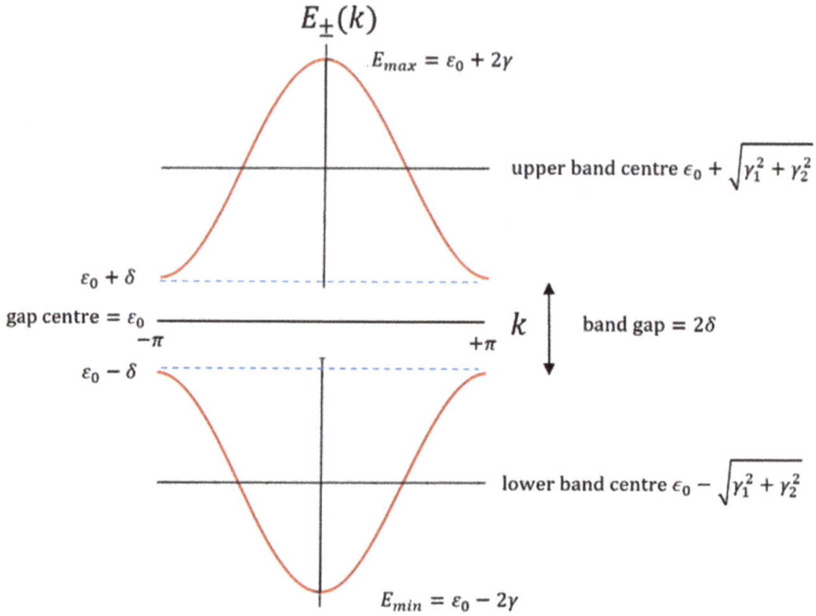

Figure 5.2. The band structure of the chain of figure 5.1, with bond alternation and two atoms per unit cell. γ_1 and γ_2 are assumed to be positive. The width of the two energy bands is $2\gamma - \delta$, which is equal to twice the smaller of $|\gamma_1|$ and $|\gamma_2|$.

This shows that when δ is non-zero, the energies of the lower half of the energy eigenvalues are decreased and the energies of the upper half of the energy eigenvalues are increased, compared to those of the linear chain of figure 4.3. For a chain in which each atom donates one electron, the lower half of the energy eigenvalues will be occupied by electrons, whereas upper half of the energy eigenvalues will be empty. Consequently the total energy of the system is decreased by dimerization. This instability to dimerization is called a Peierl's instability and the lattice is said to undergo a Peierl's distortion. In chemistry, a linear chain of carbon atoms, with equal bonds connecting nearest neighbour atoms is called a cumulene. However long cumulenes are unstable and tend to undergo a Peierl's distortion. Carbon chains with alternating bonds are called oligoynes and are more stable than cumulenes. For a (fictitious) cumulene chain with no bond alternation, the energy gap would vanish. For long oligoynes, the Peierl's distortion creates an energy gap of order 2 eV.

5.3 Conduction bands versus valence bands

As shown in figure 4.4 and equation (4.10), the linear chain of figure 4.3 has a single energy band, because each unit cell in figure 4.3 contains only a single atomic orbital. On the other hand, the unit cell in figure 5.1 contains two atomic orbitals and therefore after using Bloch's theorem (5.3) and solving the resulting 2×2 matrix equation (5.6), one obtains two energy the bands. In general, for a linear periodic structure with M atomic orbitals per unit cell, there will be M energy bands.

As noted above, if each atom of figure 5.1 donates one electron, then in figure 5.2, the lower energy band will be completely filled and because it takes a finite energy to excite an electron from the top of the lower band to the bottom of the upper band, the system is called an insulator or a semiconductor, depending of the size of δ. On the other hand, if more electrons are present, so that the upper band is partially filled up to some level (i.e. up to some Fermi energy), then it takes a negligible energy to excite an electron from the Fermi energy to another state within the upper band. In this case the system is a metal. In solid state physics, the lower band is called the valence band and the upper band is called the conduction band.

5.4 Flexibility in labelling states

In the limit $\delta = 0$, there is no bond alternation, so the band structure of figure 5.2, should reduce to the band structure of an infinite linear chain with no bond alternation, shown in figure 4.4 or equivalently the blue curve in figure 5.3(a). However in this limit, figure 5.2 reduces to the band structure in the left-hand panel of figure 5.3, which at first sight does not equal that of figure 5.3(a).

To show that these are equivalent, it is important to note that only the eigenenergies and eigenstates are of interest. The variable k is simply a label. In equation (5.8) and in the left panel of figure 5.3, two labels are used to label energies. The first label is k and the second label is the variable $= \pm$. From (5.8) and (5.9), when $\delta = 0$,

$$E_\sigma(k) = \epsilon_0 + \sigma 2\gamma \cos k/2 \quad -\pi \leqslant k < \pi \tag{5.10}$$

where $\sigma = 1$ labels the upper band and $\sigma = -1$ labels the lower band. Also when $\delta = 0$, the two eigenstates of equation (5.6) are $\frac{u_\sigma}{v_\sigma} = \frac{\Delta}{\sigma|\Delta|}$, where

$$\Delta(k) = -2\gamma \cos \frac{k}{2} e^{-ik/2}$$

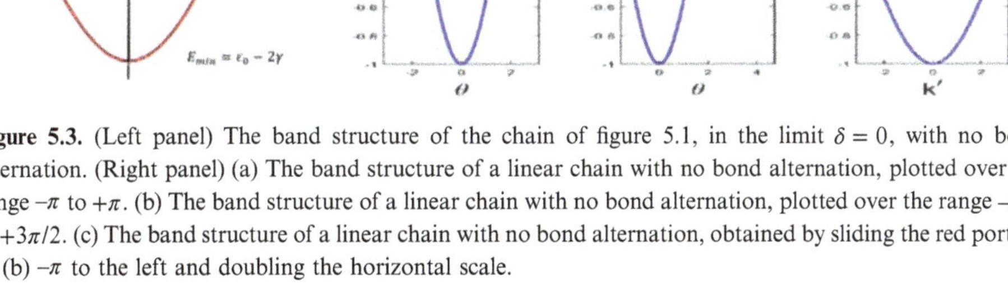

Figure 5.3. (Left panel) The band structure of the chain of figure 5.1, in the limit $\delta = 0$, with no bond alternation. (Right panel) (a) The band structure of a linear chain with no bond alternation, plotted over the range $-\pi$ to $+\pi$. (b) The band structure of a linear chain with no bond alternation, plotted over the range $-\pi/2$ to $+3\pi/2$. (c) The band structure of a linear chain with no bond alternation, obtained by sliding the red portion of (b) $-\pi$ to the left and doubling the horizontal scale.

Hence $\frac{u_\sigma}{v_\sigma} = -\sigma e^{-\frac{ik}{2}}$ and therefore from (equation (5.3))

$$\begin{pmatrix} \phi_J \\ \psi_J \end{pmatrix} = \begin{pmatrix} u_\sigma \\ v_\sigma \end{pmatrix} e^{ikJ} = \begin{pmatrix} 1 \\ -\sigma e^{\frac{ik}{2}} \end{pmatrix} e^{ikJ} \tag{5.11}$$

On the other hand from equation (4.10), the dispersion relation of a linear chain with no bond alternation uses only a single label, k. To distinguish this from the label k in equation (5.10), the former will be denoted θ, so that the dispersion relation (4.10), for a linear chain with no bond alternation, becomes

$$E(\theta) = \varepsilon_o - 2\gamma \cos\theta \tag{5.12}$$

As discussed below equation (4.16), if a is any constant, θ lies in the range $-\pi + a \leqslant \theta < \pi + a$. For the blue curve of figure 5.3(a), a is chosen to be zero.

Furthermore, the eigenstate (4.9) of the linear chain becomes $f_j = e^{i\theta j}$. In this equation, j labels individual sites and f_j is the amplitude of an eigenstate on site j. So for the first site of cell J in figure 5.1, $j = 2J$ and $f_j = f_{2J} = e^{i\theta 2J}$ Similarly for the second site, $j = 2J + 1$ and $f_j = f_{2J+1} = e^{i\theta(2J+1)}$. That is

$$\begin{pmatrix} f_{2J} \\ f_{2J+1} \end{pmatrix} = \begin{pmatrix} 1 \\ e^{i\theta} \end{pmatrix} e^{i\theta 2J} \tag{5.13}$$

The label θ in equation (5.13) is not the same as the label k in equation (5.12). To compare equation (5.11) with equation (5.12), it is convenient to choose the constant $a = \frac{\pi}{2}$, so that $-\pi/2 \leqslant \theta < 3\pi/2$. Figure 5.3(b) shows the resulting dispersion curve when this range is used.

Now divide this range of θ into two sub-ranges. The first is $-\pi/2 \leqslant \theta < \pi/2$ and the second is $\pi/2 \leqslant \theta < 3\pi/2$. In figure 5.3(b), the blue portion of the plot shows $E(\theta)$ for $-\pi/2 \leqslant \theta < \pi/2$ and the red portion of the plot shows $E(\theta)$ for $\pi/2 \leqslant \theta < 3\pi/2$.

For $-\pi/2 \leqslant \theta < \pi/2$, define $k' = 2\theta$, so that $-\pi \leqslant k' < \pi$ and (5.11) and (5.13) become

$$E(k) = \epsilon_0 - 2\gamma \cos\frac{k'}{2} \qquad -\pi \leqslant k < \pi \tag{5.14}$$

$$\begin{pmatrix} f_{2J} \\ f_{2J+1} \end{pmatrix} = \begin{pmatrix} 1 \\ e^{\frac{ik'}{2}} \end{pmatrix} e^{ik'J} \tag{5.15}$$

This reproduces the energies (5.10) and eigenstates (5.11) for the band $\sigma = -1$ and is shown as the blue curve in figure 5.3(c).

For $\pi/2 \leqslant \theta < 3\pi/2$, define $k' = 2\theta - 2\pi$, so that $-\pi \leqslant k' < \pi$.

In this case, since $\theta = \frac{k'}{2} + \pi$, and $\cos\left(\frac{k'}{2} + \pi\right) = -\cos\frac{k'}{2}$, equation (5.11) yields

$$E(k) = \epsilon_0 + 2\gamma \cos k'/2 \quad -\pi \leqslant k' < \pi \tag{5.16}$$

Similarly, since $e^{i\pi 2J} = 1$ and $e^{i\pi} = -1$, equation (5.13) becomes

$$\begin{pmatrix} \psi_{2J} \\ \psi_{2J+1} \end{pmatrix} = \begin{pmatrix} 1 \\ -e^{ik'/2} \end{pmatrix} e^{ik'J} \tag{5.17}$$

This reproduces the energies (5.10) and eigenstates (5.11) for the band $\sigma = +1$ and is shown as the red curve in figure 5.3(c).

In summary equations (5.12) and (5.13) reproduce the two bands of equation (5.10) and the eigenstates (5.15) and (5.17) are identical to (5.13).

5.5 A finite-width electrodes formed from a linear chain of cells

The above example of a linear chain of unit cells, each containing two sites, is a simple example of the more general structure shown in figure 5.4, which consists of a linear chain of cells, each containing N sites, with nearest neighbour interactions only between cells. Figure 5.4 shows an example of a linear chain of identical cells with intra-cell Hamiltonians H_0 and inter-cell Hamiltonians H_1. Since H_0 and H_1 are the same for all cells, this is a model of a crystalline wire, (or equivalently a ribbon or a chain).

Figure 5.5 shows a more general example of a linear chain of cells with different intra-cell Hamiltonians $H_{J,J}$ and different inter-cell Hamiltonians $H_{J,J+1}$. Since the cells are no longer identical this is a model of a non-uniform wire, (or equivalently an inhomogeneous wire or ribbon).

Our aim is to solve the Schrödinger equation $H|\psi\rangle = E|\psi\rangle$ for such structures, where

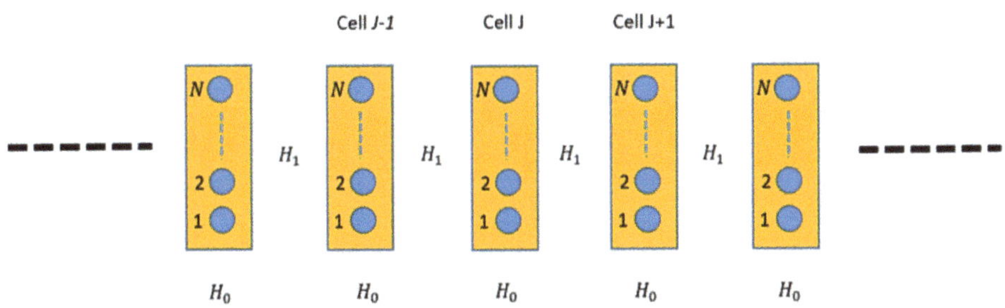

Figure 5.4. A linear chain of identical cells with intra-cell Hamiltonians H_0 and inter-cell Hamiltonians H_1.

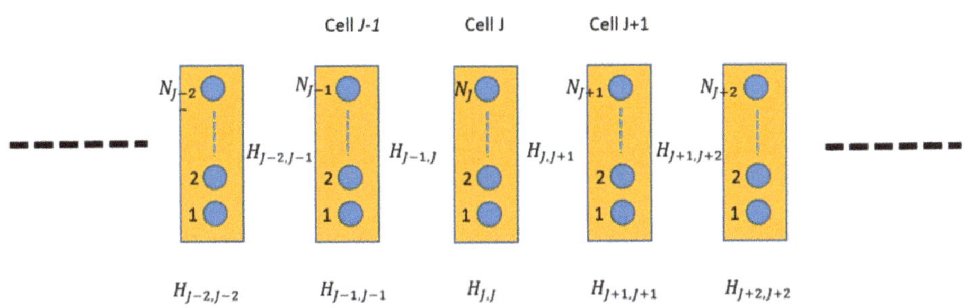

Figure 5.5. A non-uniform chain of different cells with intra-cell Hamiltonians $H_{J,J}$ and inter-cell Hamiltonians $H_{J,J+1}$.

$$|\psi\rangle = \begin{pmatrix} \vdots \\ |\psi_{J-2}\rangle \\ |\psi_{J-1}\rangle \\ |\psi_J\rangle \\ |\psi_{J+1}\rangle \\ |\psi_{J+2}\rangle \\ \vdots \end{pmatrix} \tag{5.18}$$

In this equation, following appendix A of chapter 2, where the idea of dividing matrices into submatrices is discussed, $|\psi_J\rangle$ is a vector containing the amplitudes of the eigenstate on the sites within cell J. If the N_J sites of cell J are labelled J, α where $\alpha = 1, 2, \ldots.N_J$, then

$$|\psi_J\rangle = \begin{pmatrix} \psi_{J,1} \\ \psi_{J,2} \\ \vdots \\ \psi_{J,N_J} \end{pmatrix} \tag{5.19}$$

With this notation the equation $H|\psi\rangle = E|\psi\rangle$ takes the form

$$\begin{pmatrix} \ddots & \ddots & 0 & 0 & 0 & 0 & 0 \\ \ddots & \ddots & \ddots & 0 & 0 & 0 & 0 \\ 0 & H_{J-1,J-2} & H_{J-1,J-1} & H_{J-1,J} & 0 & 0 & 0 \\ 0 & 0 & H_{J,J-1} & H_{JJ} & H_{J,J+1} & 0 & 0 \\ 0 & 0 & 0 & H_{J+1,J} & H_{J+1,J+1} & H_{J+1,J+2} & 0 \\ 0 & 0 & 0 & 0 & \ddots & \ddots & \ddots \\ 0 & 0 & 0 & 0 & 0 & \ddots & \ddots \end{pmatrix} \begin{pmatrix} \vdots \\ |\psi_{J-1}\rangle \\ |\psi_J\rangle \\ |\psi_{J+1}\rangle \\ \vdots \end{pmatrix} = \mathrm{E} \begin{pmatrix} \vdots \\ |\psi_{J-1}\rangle \\ |\psi_J\rangle \\ |\psi_{J+1}\rangle \\ \vdots \end{pmatrix} \tag{5.20}$$

Since $H = H^\dagger$, the sub-matrices satisfy $H_{JJ} = H^\dagger_{JJ}$ and $H_{JJ+1} = H^\dagger_{J+1J}$. The Jth row of this equation is

$$H_{JJ}|\psi_J\rangle + H_{JJ+1}|\psi_{J+1}\rangle + H_{JJ-1}|\psi_{J-1}\rangle = E|\psi_J\rangle \tag{5.21}$$

For the special case of a crystalline chain, this reduces to

$$\begin{pmatrix} \ddots & \ddots & 0 & 0 & 0 & 0 & 0 \\ \ddots & \ddots & \ddots & 0 & 0 & 0 & 0 \\ 0 & H_{-1} & H_0 & H_1 & 0 & 0 & 0 \\ 0 & 0 & H_{-1} & H_0 & H_1 & 0 & 0 \\ 0 & 0 & 0 & H_{-1} & H_0 & H_1 & 0 \\ 0 & 0 & 0 & 0 & \ddots & \ddots & \ddots \\ 0 & 0 & 0 & 0 & 0 & \ddots & \ddots \end{pmatrix} \begin{pmatrix} \vdots \\ |\psi_{J-1}\rangle \\ |\psi_J\rangle \\ |\psi_{J+1}\rangle \\ \vdots \end{pmatrix} = \mathrm{E} \begin{pmatrix} \vdots \\ |\psi_{J-1}\rangle \\ |\psi_J\rangle \\ |\psi_{J+1}\rangle \\ \vdots \end{pmatrix} \tag{5.22}$$

Since $H = H^{\dagger}$, the sub-matrices H_0 and H_1 satisfy $H_0 = H_0^{\dagger}$ and $H_{-1} = H_1^{\dagger}$. The J th row of this equation is

$$H_0|\psi_J\rangle + H_1|\psi_{J+1}\rangle + H_{-1}|\psi_{J-1}\rangle = E|\psi_J\rangle \tag{5.23}$$

In this case, equation (5.23) is solved by writing

$$|\psi_J\rangle = |\chi\rangle e^{ikJ} \tag{5.24}$$

Since $e^{i(k+2\pi)J} = e^{ikJ}$, k must be restricted to an interval of width 2π, to avoid introducing linearly-dependent states. For example one could choose $-\pi \leqslant k < \pi$ or $0 \leqslant k < 2\pi$.
Substituting this into equation (5.23) yields

$$H_0|\chi\rangle e^{ikJ} + H_1|\chi\rangle e^{ik(J+1)} + H_{-1}|\chi\rangle e^{ik(J-1)} = E|\chi\rangle e^{ikJ} \tag{5.25}$$

After dividing by e^{ikJ}, this simplifies to

$$H(k)|\chi\rangle = E(k)|\chi\rangle \tag{5.26}$$

where

$$H(k) = H_0 + H_1 e^{ik} + H_{-1}e^{-ik} \tag{5.27}$$

Hence equation (5.24) is a solution to equation (5.23) provided $|\chi\rangle$ is chosen to be an eigenstate of $H(k)$. Since $H_0 = H_0^{\dagger}$ and $H_{-1} = H_1^{\dagger}$, $H(k)$ is Hermitian, so if $H(k)$ is a $N \times N$ matrix, there will be N linearly independent solutions to equation (5.26), which can be labelled $n = 1, 2, \ldots N$. If the eigenvectors are denoted $|\chi_n\rangle$ and the corresponding eigenvalues denoted $E_n(k)$, then there will be N energy bands, each with an individual energy dispersion curve $E_n(k)$.

5.6 A linear chain of cells, each containing two sites

As a simple example, consider the doubly infinite linear chain shown in figure 5.6, which is an alternative representation of the chain in figure 5.1.

In this example, if the sites have site energies ϵ_0, and the couplings $-\gamma_1$ and $-\gamma_2$ are real,

$$H_0 = \begin{pmatrix} \varepsilon_0 & -\gamma_2 \\ -\gamma_2 & \varepsilon_0 \end{pmatrix},\ H_1 = \begin{pmatrix} 0 & 0 \\ -\gamma_1 & 0 \end{pmatrix},\ H_{-1} = \begin{pmatrix} 0 & -\gamma_1 \\ 0 & 0 \end{pmatrix},\ |\chi\rangle = \begin{pmatrix} \chi_1 \\ \chi_2 \end{pmatrix}$$

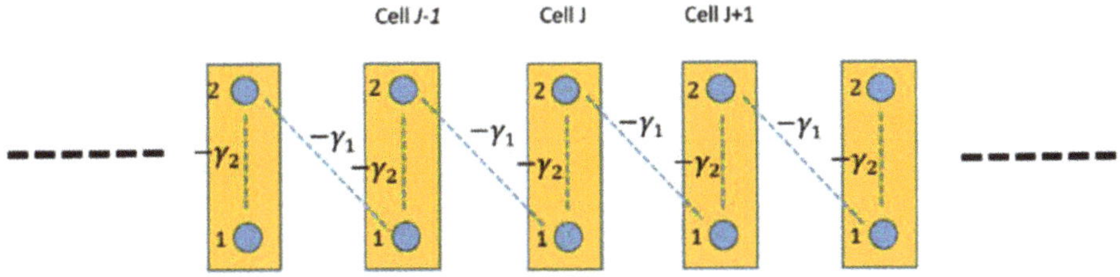

Figure 5.6. A crystalline linear chain of identical cells.

and

$$H(k) = H_0 + H_1 e^{ik} + H_{-1} e^{-ik} = \begin{pmatrix} \varepsilon_0 & \Delta(\mathrm{k}) \\ \Delta^*(\mathrm{k}) & \varepsilon_0 \end{pmatrix} \tag{5.28}$$

where $\Delta(\mathrm{k}) = -\gamma_2 - \gamma_1 e^{-ik}$. Hence equation (5.28) becomes

$$\begin{pmatrix} \varepsilon_0 & \Delta(\mathrm{k}) \\ \Delta^*(\mathrm{k}) & \varepsilon_0 \end{pmatrix} \begin{pmatrix} \chi_1 \\ \chi_2 \end{pmatrix} = E \begin{pmatrix} \chi_1 \\ \chi_2 \end{pmatrix} \tag{5.29}$$

As expected, this is identical to equations (5.6) and (5.7) and there are two energy bands, with dispersion curves given by equation (5.8).

5.7 A linear chain of cells, each containing N sites, with an intra-cell Hamiltonian H_l proportional to the unit matrix

As a second example, consider the doubly infinite crystalline ribbon shown in figure 5.7, in which each site of a cell is connected to the corresponding site of its neighbours by a bond $-\gamma$.

In this case,

$$H_1 = -\gamma I \tag{5.30}$$

where I is the $N \times N$ unit matrix, and equation (5.27) simplifies to

$$H(k) = H_0 - 2\gamma I \cos k \tag{5.31}$$

Since $H(k)$ differs from H_0 by a constant x I, they share the same eigenvectors. Hence if the eigenvalues and eigenvectors of H_0 are ϵ_n and $|\chi_n\rangle$ respectively, where $n = 1, 2, \ldots .N$, satisfying

$$H_0|\chi_n\rangle = \epsilon_n|\chi_n\rangle \tag{5.32}$$

then $|\chi_n\rangle$ is also an eigenvector of $H(k)$. From equation (5.26), the corresponding eigenvalue of $H(k)$ is

$$E_n(k) = \epsilon_n - 2\gamma \cos k \tag{5.33}$$

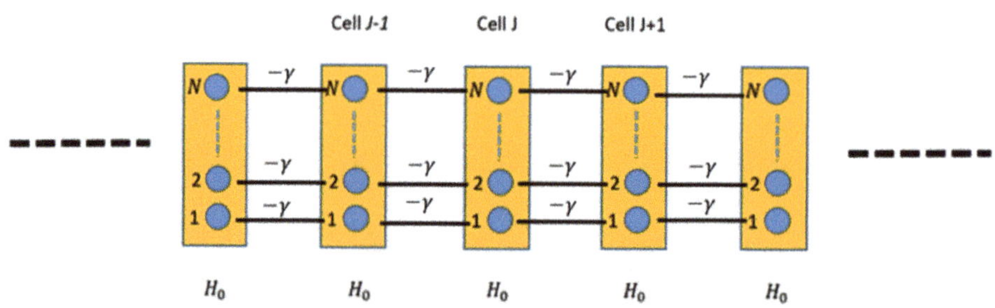

Figure 5.7. A crystalline linear chain of identical cells, in which each site of a cell is connected to the corresponding site of its neighbours by a bond $-\gamma$.

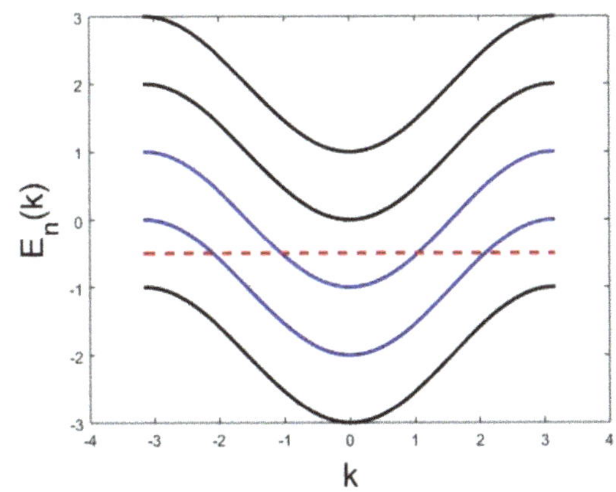

Figure 5.8. Plots of mini-bands. At an energy corresponding to the dashed red line, the mini-bands coloured blue are 'open', because they intersect the red line, while the others are 'closed,' because they do not intersect the red line.

This shows that there are N energy bands with energy dispersion curves given by equation (5.33), and plotted in figure 5.8, where to avoid introducing linearly-dependent solutions, k is restricted to an interval of width 2π.

5.8 A linear chain of cells, each containing N sites, with an intra-cell Hamiltonian H_l proportional to the unit matrix and an intra-cell Hamiltonian H_0 describing a linear chain with free ends

In most of the discussion below, H_0 will not be specified, so the discussion will be rather general. Nevertheless, for illustrative purposes, it will be helpful to consider a couple of specific examples. As an example, consider the case where each unit cell consists of a linear chain of N sites, with nearest neighbour couplings $-\gamma_1$ and site energies ϵ_0, as shown in figure 5.9.

In this example, H_0 takes the form

$$H_0 = \begin{pmatrix} \varepsilon_o & -\gamma_1 & . & . & . & . & . & . & 0 \\ . & . & . & . & . & . & . & . & . \\ . & . & \varepsilon_o & -\gamma_1 & 0 & 0 & 0 & . & . \\ . & . & -\gamma_1 & \varepsilon_o & -\gamma_1 & 0 & 0 & . & . \\ . & . & 0 & -\gamma_1 & \varepsilon_o & -\gamma_1 & 0 & . & . \\ . & . & 0 & 0 & -\gamma_1 & \varepsilon_o & -\gamma_1 & . & . \\ . & . & 0 & 0 & 0 & -\gamma_1 & \varepsilon_o & . & . \\ . & . & . & . & . & . & . & . & . \\ 0 & . & . & . & . & . & . & -\gamma_1 & \varepsilon_o \end{pmatrix} \tag{5.34}$$

So from the analysis in section 4.8 of chapter 4, the amplitude of $|\chi_n\rangle$ on site α is (cf equation (4.59))

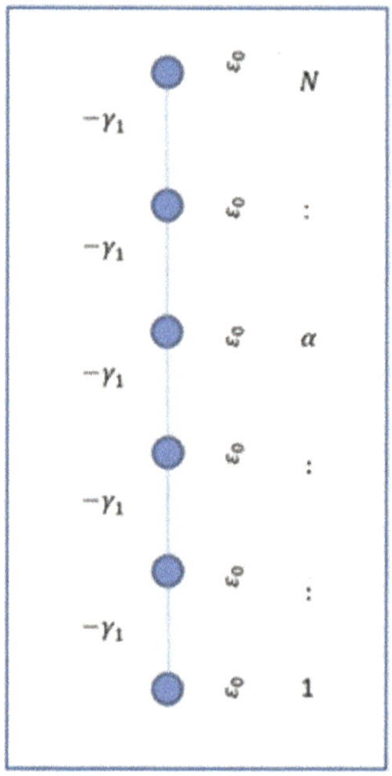

Figure 5.9. Example of a unit cell of the crystalline chain of figure 5.7, comprising a linear chain of sites with free ends.

$$\chi_\alpha^{(n)} = \sqrt{\frac{2}{N+1}}\sin\frac{n\pi}{(N+1)}\alpha, \quad n = 1, 2, 3, \; .... N \tag{5.35}$$

and

$$\epsilon_n = \epsilon_0 - 2\gamma_1 \cos\frac{n\pi}{N+1}, \quad n = 1, \; 2, \; ... \; N \tag{5.36}$$

Hence equation (5.33) becomes

$$E_n(k) = \epsilon_n - 2\gamma \cos k = \epsilon_0 - 2\gamma_1 \cos k_n - 2\gamma \cos k$$

where

$$k_n = \frac{n\pi}{N+1}$$

5.9 A linear chain of cells, each containing N sites, with an intra-cell Hamiltonian H_1 proportional to the unit matrix and an intra-cell Hamiltonian H_0 describing a linear chain with periodic boundary conditions

As a second example, consider the case where each unit cell consists of a linear chain of N sites, with nearest neighbour couplings $-\gamma_1$ and site energies ϵ_0, as shown in figure 5.10.

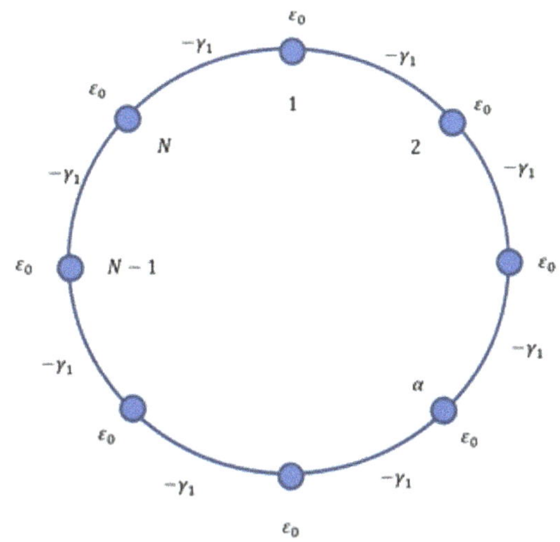

Figure 5.10. Example of a unit cell of the crystalline chain of figure 5.7, comprising a linear chain of sites with periodic boundary conditions.

In this example, H_0 takes the form

$$H_0 = \begin{pmatrix} \varepsilon_o & -\gamma_1 & \cdot & \cdot & \cdot & \cdot & \cdot & \cdot & -\gamma_1 \\ \cdot & \cdot & \cdot & \cdot & \cdot & \cdot & \cdot & \cdot & \cdot \\ \cdot & \cdot & \varepsilon_o & -\gamma_1 & 0 & 0 & 0 & \cdot & \cdot \\ \cdot & \cdot & -\gamma_1 & \varepsilon_o & -\gamma_1 & 0 & 0 & \cdot & \cdot \\ \cdot & \cdot & 0 & -\gamma_1 & \varepsilon_o & -\gamma_1 & 0 & \cdot & \cdot \\ \cdot & \cdot & 0 & 0 & -\gamma_1 & \varepsilon_o & -\gamma_1 & \cdot & \cdot \\ \cdot & \cdot & 0 & 0 & 0 & -\gamma_1 & \varepsilon_o & \cdot & \cdot \\ \cdot & \cdot & \cdot & \cdot & \cdot & \cdot & \cdot & \cdot & \cdot \\ -\gamma_1 & \cdot & \cdot & \cdot & \cdot & \cdot & \cdot & -\gamma_1 & \varepsilon_o \end{pmatrix} \tag{5.37}$$

which differs from equation (5.34) through the inclusion of bonds $-\gamma_1$ in the top right and bottom left corners. In contrast to the structure obtained by combining figures 5.9 and 5.7, which describes a rectangular ribbon (called a nanoribbon when the width is of order a few nanometres), the structure obtained by combining figures 5.10 and 5.7 is a cylinder and is a simple example of a nanotube.

From the analysis in section 4.8 of chapter 4, the amplitude of $|\chi_n\rangle$ on site α is (cf equation (4.59))

$$\chi_\alpha^{(n)} = \sqrt{\frac{1}{N}} e^{ik_n\alpha} \tag{5.38}$$

and

$$\epsilon_n = \epsilon_0 - 2\gamma_1 \cos k_n, \tag{5.39}$$

where $k_n = \frac{2\pi n}{N}$. In this expression, the integers n take N values, such that $-\pi \leqslant k_n < \pi$, so if N is odd, $-\frac{N-1}{2} \leqslant n \leqslant \frac{N-1}{2}$, whereas if N is even, $-\frac{N}{2} \leqslant n \leqslant \frac{N-2}{2}$.

Hence equation (5.33) becomes

$$E_n(k) = \epsilon_n - 2\gamma \cos k = \epsilon_0 - 2\gamma_1 \cos k_n - 2\gamma \cos k$$

5.10 A two-dimensional crystal on a square lattice

It is interesting to see how band structures of 2-dimensional crystals appear from the above analysis, by considering the chain of equations (5.38) and (5.39) in the limit N tends to infinity. The resulting structure is a crystalline square lattice of sites, extending to infinity in all directions. In this case, k_n in equation (5.39) becomes a continuous variable, which will be denoted k_y It is also useful to change notation and replace the variable k in equation (5.33) by k_x and $E_n(k)$ by $E(k_x, k_y)$. Then from equations (5.39) and (5.33)

$$E(k_x, k_y) = \epsilon_0 - 2\gamma \cos k_x - 2\gamma_1 \cos k_y \tag{5.40}$$

To avoid introducing linearly-dependent eigenstates,

$$-\pi \leqslant k_x < \pi \text{ and } -\pi \leqslant k_y < \pi \tag{5.41}$$

This restricted range of k_x and k_y is called the first Brillouin zone. Figure 5.11 shows a plot of $E(k_x, k_y)$, when $\epsilon_0 = 0$ and $\gamma = \gamma_1 = 1$.

Another way of representing this band structure is to plot contours of fixed energy E, as shown in figure 5.12.

Figure 5.12 shows contour plots obtained by taking horizontal cuts through this surface at fixed energies. The contour corresponding to $E = E_F$ (where E_F is the Fermi energy) is called the Fermi surface. Furthermore, from equation (5.40), for an electron with wave vector k_x, k_y the x-component of the group velocity is given by

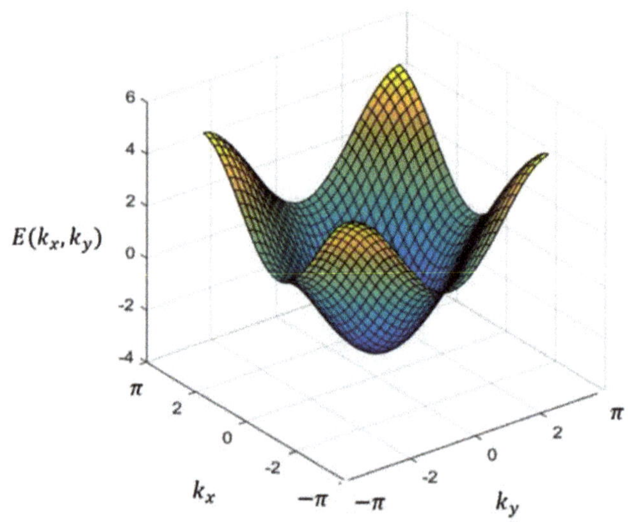

Figure 5.11. The band structure of a square lattice ($\epsilon_0 = 0$, $\gamma = \gamma_1 = 1$).

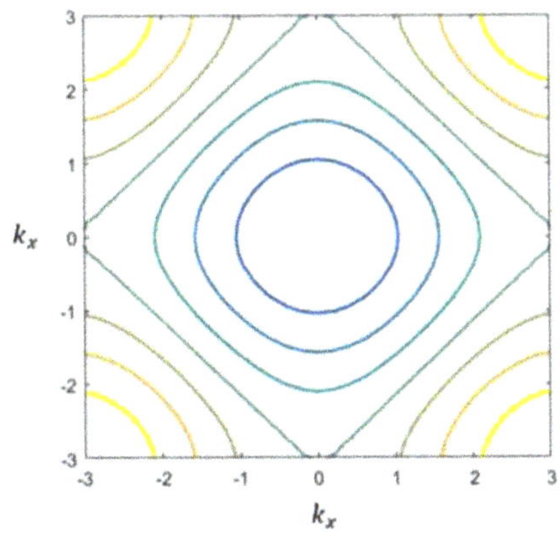

Figure 5.12. Contours of constant energy obtained from the band structure of figure 5.10.

$$\hbar v_x(k_x,\ k_y) = \frac{\partial E}{\partial k_x} = 2\gamma \sin k_x \tag{5.42}$$

and the y-component of the group velocity is given by

$$\hbar v_y(k_x,\ k_y) = \frac{\partial E}{\partial k_y} = 2\gamma_1 \sin k_y \tag{5.43}$$

In vector form, this becomes

$$\hbar \underline{v}(\underline{k}) = \underline{\nabla}_k E(\underline{k}) \tag{5.44}$$

where $\underline{v} = (v_x,\ v_y)$, $\underline{k} = (k_x,\ k_y)$ and $\underline{\nabla}_k = (\frac{\partial}{\partial k_x},\ \frac{\partial}{\partial k_y})$. Since the gradient of a scalar is a vector at 90° to contour lines, the group velocity $\underline{v}(\underline{k})$ of an electron of wave vector $\underline{k}$, energy $E = E(\underline{k})$ is at 90° to the contour of energy E passing through the point $\underline{k}$.

5.11 The most general solution to the Schrödinger equation for the periodic structure of figure 5.7, for which H_l is proportional to the unit matrix

Let us return to equation (5.23) namely

$$H_0|\psi_J\rangle + H_1|\psi_{J+1}\rangle + H_{-1}|\psi_{J-1}\rangle = E|\psi_J\rangle \tag{5.45}$$

and consider for simplicity the case, where H_1 is proportional to the unit matrix; i.e.

$$H_1 = -\gamma I \tag{5.46}$$

In this case,

$$H_0|\psi_J\rangle - \gamma|\psi_{J+1}\rangle - \gamma|\psi_{J-1}\rangle = E|\psi_J\rangle \tag{5.47}$$

Writing

$$|\psi_J\rangle = \psi_J^n|\chi_n\rangle \tag{5.48}$$

where $|\chi_n\rangle$ is one of the N eigenstates of H_0 satisfying $H_0|\chi_n\rangle = \epsilon_n|\chi_n\rangle$, yields

$$\epsilon_n\psi_J^n - \gamma\psi_{J-1}^n - \gamma\psi_{J+1}^n = E\psi_J^n \tag{5.49}$$

This is analogous to the Schrödinger equation for a one-dimensional lead, with site energies ϵ_n. These fictitious, one-dimensional leads are often referred to as channels. Each channel corresponds to the wave function $|\psi_J\rangle$ possessing a distinct transverse spatial dependence $|\chi_n\rangle$, such that the amplitude on site α of each cell is proportional to $\chi_\alpha^{(n)}$. Examples of the latter are given in equations (5.38) and (5.35). Overall, this means that (in the notation of equation 5.19) the amplitude on site α of cell J is simply a product $\psi_{J,\alpha} = \chi_\alpha^{(n)}\psi_J^n$.

The most general solution to equation (5.49) is

$$\psi_J^n = Ae^{ikJ} \tag{5.50}$$

where the plane wave amplitude A is arbitrary. Substituting equation (5.50) into equation (5.49) yields

$$E = \epsilon_n - 2\gamma\cos k \tag{5.51}$$

This equation gives the energy E of an electron in channel n, for each choice of k. The most general solution to the Schrödinger equation (5.47) is obtained by creating a linear superposition of all possible solutions corresponding to a given energy E. If the wave number in channel n corresponding to energy E is denoted $k_n(E)$, then

$$k_n(E) = \text{a}\cos\left(\frac{\epsilon_n - E}{2\gamma}\right) \tag{5.52}$$

and the most general solution to the Schrödinger equation is therefore (cf equation (5.24))

$$|\psi_J\rangle = \sum_{n=1}^{N}\psi_J^n(E)|\chi_n\rangle \tag{5.53}$$

where

$$\psi_J^n(E) = A_n e^{ik_n(E)J} + B_n e^{-ik_n(E)J} \tag{5.54}$$

In this equation, A_n and B_n are arbitrary constants.

Note that if $k_n(E)$ is a solution to equation (5.52), then $-k_n(E)$ is also a solution. Furthermore, $k_n(E)$ could be either real or complex. If $\epsilon_0 - 2\gamma \leqslant E \leqslant \epsilon_0 + 2\gamma$, then $k_n(E)$ is real, and it is useful to adopt the convention of choosing $k_n(E)$ such that the group velocity $v_n(E)$ given by $\hbar v_n(E) = 2\gamma\sin k_n(E)$ is positive. This means that if γ is positive, then $k_n(E)$ is positive, but if γ is negative, then $k_n(E)$ would be chosen to be negative. Similarly if $E < \epsilon_0 - 2\gamma$, then $k_n(E)$ is imaginary and $k_n(E)$ will be chosen to have a positive imaginary part. Finally if $E > \epsilon_0 + 2\gamma$, then $k_n(E) = \pi +$ imaginary part and again the imaginary part will be chosen to be positive.

With this convention, the constants A_n in equation (5.39) are amplitudes of plane waves with real $k_n(E)$ and positive group velocities or amplitudes of states with $\mathrm{Im}k_n(E) > 0$, which decay exponentially with increasing J. Similarly, the constants B_n are amplitudes of plane waves with real $k_n(E)$ and negative group velocities or amplitudes of states positive imaginary $k_n(E)$, which decay exponentially with decreasing J. If a wave vector k_n is real, then the mini-band or channel n is said to be 'open', whereas if k_n is complex, the channel is 'closed'.

5.12 The equivalence between a finite-width lead and many one-dimensional leads

In later chapters, finite-dimensional leads will be attached to a scattering region, as shown in figure 5.13. It will now be demonstrated that such a finite-width lead connected to a scattering region (shown in blue) is mathematically equivalent to N one-dimensional leads connected to the scattering region.

For such a structure, the Schrödinger equation for cells $J \leqslant -1$ is

$$H_0|\psi_J\rangle - \gamma|\psi_{J+1}\rangle - \gamma|\psi_{J-1}\rangle = E|\psi_J\rangle \tag{5.55}$$

On the other hand, for the last cell of the lead,

$$H_0|\psi_0\rangle - \gamma|\psi_{-1}\rangle + V|f\rangle = E|\psi_0\rangle \tag{5.56}$$

where V is a $N \times P$ matrix, coupling the N sites in cell $J = 0$ to the P sites of the scatterer and $|f\rangle$ is a P-component vector containing the amplitudes of the wave function on sites of the scatterer.

$$i.e.|f\rangle = \begin{pmatrix} f_1 \\ \vdots \\ f_j \\ \vdots \\ f_P \end{pmatrix} \tag{5.57}$$

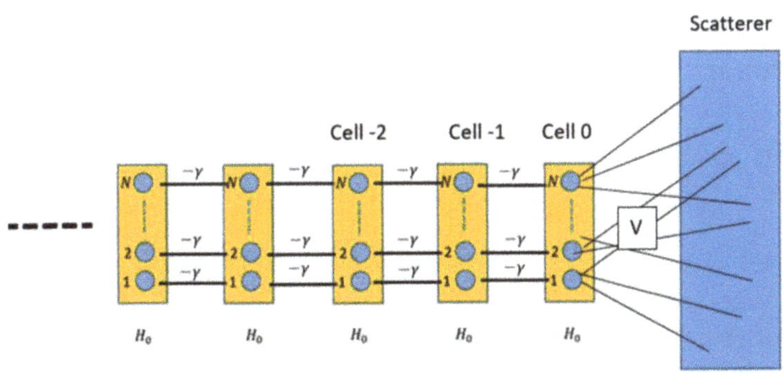

Figure 5.13. A finite-width semi-infinite lead connected to a scatterer. The matrix V contains couplings between the sites of cell $J = 0$ and sites of the scatterer.

Substituting equation (5.48) into (5.54) yields equation (5.49) for cells $J \leqslant -1$. On the other hand, substituting equation (5.48) into (5.55) yields

$$\sum_{n=1}^{N}[\epsilon_n\psi_0^n - \gamma\psi_{-1}^n]|\chi_n\rangle + V|f\rangle = E\sum_{n=1}^{N}\psi_0^n(E)|\chi_n\rangle$$

Multiplying both side by $\langle\chi_m|$ and taking advantage of the orthogonality condition $\langle\chi_m|\chi_n\rangle = \delta_{nm}$ yields the following equation for a one-dimensional chain

$$\epsilon_m\psi_0^m - \gamma\psi_{-1}^m + \langle\chi_m|V|f\rangle = E\psi_0^m(E) \tag{5.58}$$

In this equation,

$$\langle\chi_m|V|f\rangle = \sum_{\substack{\alpha=1,N \\ j=1,P}} \chi_\alpha^{(m)} V_{\alpha j} f_j = \sum_{j=1,P} V_{mj} f_j$$

where

$$V_{mj} = \sum_{\alpha=1,N} \chi_\alpha^{(m)} V_{\alpha j} \tag{5.59}$$

Hence equation (5.57) becomes for $J = 0$

$$\epsilon_m\psi_0^m - \gamma\psi_{-1}^m + \sum_{j=1,P} V_{mj} f_j = E\psi_0^m(E) \tag{5.60}$$

Equation (5.49) for cells $J \leqslant -1$, combined with equation (5.60) for $J = 0$ are the Schrödinger for a semi-infinite linear chain with site energies ϵ_m connected to a scatterer at one end, in which the end site ($J = 0$) is coupled to each site of the scatterer by couplings V_{mj}. In other words, from a mathematical viewpoint, the structure in figure 5.13, in which sites of the last cell are coupled to sites of the scatterer by the matrix V, is equivalent to the structure in figure 5.14, comprising a set of N one-dimensional leads, in which the last site of lead m is connected to sites of the scatterer by matrix elements V_{mj}. This will be helpful in later chapters, when conceptualising scattering in the presence of multi-channel, finite-width leads.

5.13 Summary

In this chapter, the Schrödinger equation has been solved for chains with more than one site per unit cell. For the structure of figure 5.1, the presence of alternating coupling led to the emergence of two energy bands, because there are two sites per unit cell. When the couplings were made to be equal, the band structure of a linear chain of sites was reproduced, but in a folded form, as shown in figure 5.3. In general, a finite-width ribbon with N sites per unit cell possesses N energy bands, which are often referred to as mini-bands or channels. Such ribbons will form the electrical leads used in later chapters to feed electrons into single-molecule junctions. As shown in section 5.11, from a conceptual viewpoint, a finite-width lead with

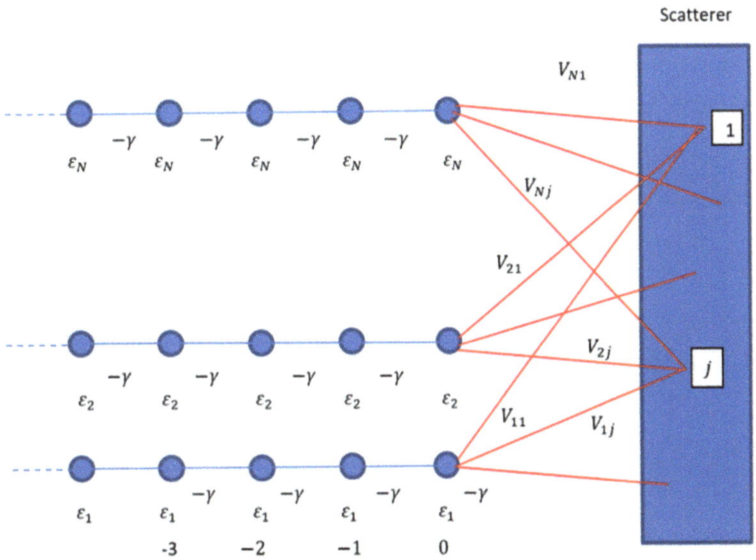

Figure 5.14. A set of N one-dimensional semi-infinite leads (representing scattering channels) connected to a scatterer, which are mathematically equivalent to the finite-width, semi-infinite lead connected to a scatterer shown in figure 5.12. The matrix element V_{nj} is the coupling between channel n and site j of the scatterer.

N channels can be regarded as being equivalent to N one-dimensional leads, each with a different site energy. Therefore from a mathematical viewpoint, solving the Schrödinger equation for a molecule connected to finite-width leads is equivalent to solving the Schrödinger equation for a molecule connected to many one-dimensional leads.

Chapter 6

Scattering theory of electrical conductance and thermopower

6.1 Introduction

When a nanoscale device is created by placing a molecule or some other phase-coherent structure between electrodes, and the electrodes are allowed to carry a current into or out of the device, the combined structure is an open system. The region containing the molecule and nearby surfaces of the electrodes is called a scattering region and the electrodes far from the scattering region are modelled as semi-infinite crystalline leads. The latter ultimately connect to reservoirs, which act as sources and sinks of electrons. The aim of the present chapter is to discuss how to solve the Schrödinger equation for such a doubly infinite system and how to describe currents entering and leaving such a device. This will provide an introduction to the idea of a scattering matrix and a method of predicting the electrical conductance and thermoelectric properties of a material via the Landauer formalism.

6.2 A single impurity in one dimension

Consider modifying the doubly infinite one-dimensional crystal of figure 4.3 by replacing the atom at site $j = 0$ with a different atom, whose orbital energy is ϵ_1, to yield the structure shown in figure 6.1.

The aim is to solve the equation

$$H|\psi\rangle = E|\psi\rangle \tag{6.1}$$

doi:10.1088/978-0-7503-3639-0ch6

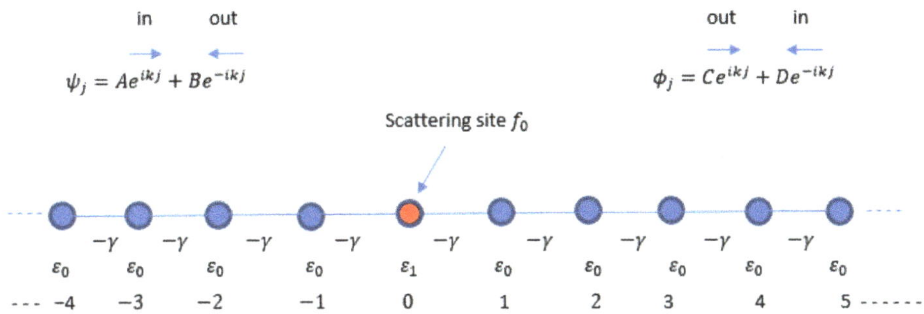

Figure 6.1. A doubly infinite chain containing a single impurity at site $j = 0$.

When written out in full, this takes the form

$$\begin{pmatrix} \cdot & \cdot & \cdot & \cdot & \cdot & \cdot & \cdot & \cdot & \cdot \\ \cdot & \cdot & \cdot & \cdot & \cdot & \cdot & \cdot & \cdot & \cdot \\ \cdot & \cdot & \varepsilon_o & -\gamma & 0 & 0 & 0 & \cdot & \cdot \\ \cdot & \cdot & -\gamma & \varepsilon_o & -\gamma & 0 & 0 & \cdot & \cdot \\ \cdot & \cdot & 0 & -\gamma & \varepsilon_1 & -\gamma & 0 & \cdot & \cdot \\ \cdot & \cdot & 0 & 0 & -\gamma & \varepsilon_o & -\gamma & \cdot & \cdot \\ \cdot & \cdot & 0 & 0 & 0 & -\gamma & \varepsilon_o & \cdot & \cdot \\ \cdot & \cdot & \cdot & \cdot & \cdot & \cdot & \cdot & \cdot & \cdot \\ \cdot & \cdot & \cdot & \cdot & \cdot & \cdot & \cdot & \cdot & \cdot \end{pmatrix} \begin{pmatrix} \cdot \\ \cdot \\ \psi_{-2} \\ \psi_{-1} \\ \psi_0 \\ \psi_1 \\ \psi_2 \\ \cdot \\ \cdot \end{pmatrix} = \mathrm{E} \begin{pmatrix} \cdot \\ \cdot \\ \psi_{-2} \\ \psi_{-1} \\ \psi_0 \\ \psi_1 \\ \psi_2 \\ \cdot \\ \cdot \end{pmatrix} \quad (6.2)$$

Therefore the equation for row j is

$$\sum_{l=-\infty}^{\infty} H_{jl}\psi_l = E\psi_j \quad (6.3)$$

To solve this set of equations, it is useful to denote the eigenvector amplitudes to the left of the impurity by ψ_j, the amplitudes to the right of the impurity by ϕ_j and the amplitude on the impurity by f_0. Then equation (6.2) or equation (6.3) become

$$\varepsilon_o\psi_j - \gamma\psi_{j-1} - \gamma\psi_{j+1} = E\psi_j \qquad \text{for } j < -1 \quad (6.4)$$

$$\varepsilon_o\psi_{-1} - \gamma\psi_{-2} - \gamma f_0 = E\psi_{-1} \qquad \text{for } j = -1 \quad (6.5)$$

$$\varepsilon_1 f_0 - \gamma\psi_{-1} - \gamma\phi_1 = Ef_0 \qquad \text{for } j = 0 \quad (6.6)$$

$$\varepsilon_o\phi_1 - \gamma f_0 - \gamma\phi_2 = E\phi_1 \qquad \text{for } j = 1 \quad (6.7)$$

$$\varepsilon_o\phi_j - \gamma\phi_{j-1} - \gamma\phi_{j+1} = E\phi_j \qquad \text{for } j > 1 \quad (6.8)$$

To the left of the impurity, where the system is a semi-infinite crystalline lead, (i.e. for $j < -1$), the solution to equation (6.5) is

$$\psi_j = Ae^{ikj} + Be^{-ikj} \qquad (j < -1) \quad (6.9)$$

and to the right of the impurity for $j > 0$, the solution to equation (6.8) is

$$\phi_j = Ce^{ikj} + De^{-ikj} \qquad (j > 1) \qquad (6.10)$$

where A, B, C, D are constants. Substituting these into equations (6.4) and (6.8) demonstrates that they are indeed solutions, provided

$$E = \varepsilon_o - 2\gamma \cos k \qquad (6.11)$$

As discussed in appendix B in chapter 4, the function ψ_j has the mathematical property that it satisfies equation (6.4) for all j even though it is only requires to satisfy equation (6.4) for $j < -1$. For example for $j = -1$ it satisfies

$$\varepsilon_o\psi_{-1} - \gamma\psi_{-2} - \gamma\psi_0 = E\psi_{-1} \qquad (6.12)$$

Comparing equation (6.12) with equation (6.5) yields

$$f_0 = \psi_0 = A + B \qquad (6.13)$$

Similarly, the function ϕ_j has the mathematical property that it satisfies equation (6.8) for all j. For example for $j = 1$ it satisfies

$$\varepsilon_o\phi_1 - \gamma\phi_0 - \gamma\phi_2 = E\phi_1 \qquad (6.14)$$

Comparing equation (6.14) with equation (6.7) yields

$$f_0 = \phi_0 = C + D \qquad (6.15)$$

Hence equation (6.13) and equation (6.15) yield

$$A + B = C + D \qquad (6.16)$$

Note that equation (6.13) allows equation (6.6) to be rewritten as

$$\varepsilon_1\psi_0 - \gamma\psi_{-1} - \gamma\phi_1 = E\psi_0 \qquad (6.17)$$

Since function ψ_j has the mathematical property that it satisfies equation (6.4) for all j, setting $j = 0$ yields for equation (6.4),

$$\varepsilon_0\psi_0 - \gamma\psi_{-1} - \gamma\psi_1 = E\psi_0 \qquad (6.18)$$

Comparing equation (6.18) with equation (6.17) yields

$$(\varepsilon_1 - \epsilon_0)\psi_0 + \gamma(\psi_1 - \phi_1) = 0 \qquad (6.19)$$

i.e.

$$\phi_1 = \frac{(\varepsilon_1 - \epsilon_0)}{\gamma}\psi_0 + \psi_1 \qquad (6.20)$$

which is simpler form of equation (6.6). From equation (6.9) and equation (6.10), this becomes

Hence

$$Ce^{ik} + De^{-ik} = A\left[\frac{(\varepsilon_1 - \epsilon_0)}{\gamma} + e^{ik}\right] + B\left[\frac{(\varepsilon_1 - \epsilon_0)}{\gamma} + e^{-ik}\right] \tag{6.21}$$

This can be simplified by writing it as

$$C(e^{ik} - e^{-ik}) + (C + D)e^{-ik} = (A + B)\left[\frac{(\varepsilon_1 - \epsilon_0)}{\gamma} + e^{ik}\right] - B(e^{ik} - e^{-ik}) \tag{6.22}$$

Replacing $C + D$ on the left-hand side by $A + B$, yields

$$(2i \sin k)C = (A + B)\left[\frac{(\varepsilon_1 - \epsilon_0)}{\gamma} + 2i \sin k\right] - B2i \sin k$$

That is

$$C = (\alpha + 1)A + \alpha B \tag{6.23}$$

where

$$\alpha = \frac{(\varepsilon_1 - \epsilon_0)}{\gamma 2i \sin k} \tag{6.24}$$

Equations (6.16) and (6.23) are a pair of equations relating the constants A, B, C, and D.

Before solving these equations it is useful to introduce the idea of a scattering matrix.

6.3 The scattering matrix

As discussed in section 4.5 and appendix B in chapter 4, because the equations (6.4) and (6.5) are two-point recurrence relations, if the value of ψ_j is given for two values of j, then ψ_j is determined for all j. In other words, the most general solution to equation (6.6) and equation (6.7) can contain at most two arbitrary constants. Hence if any two of the constants A, B, C, D are known, then the other two constants are determined. For example if A, D are given, then B, C are determined. This means that there exists a relationship between them of the form

$$\begin{pmatrix} B \\ C \end{pmatrix} = \begin{pmatrix} S_{11} & S_{12} \\ S_{21} & S_{22} \end{pmatrix} \begin{pmatrix} A \\ D \end{pmatrix} \tag{6.25}$$

As indicated in figure 6.1, the plane waves of amplitudes B and C are travelling outwards from the impurity, whereas the plane waves of amplitudes A and D are travelling inwards, towards the impurity. Therefore it is reasonable to write equation (6.25) in the form

$$|\text{out}\rangle = S|\text{in}\rangle, \tag{6.26}$$

where

$$|\mathrm{out}\rangle = \begin{pmatrix} B \\ C \end{pmatrix}, \quad |\mathrm{in}\rangle = \begin{pmatrix} A \\ D \end{pmatrix} \quad \text{and} \quad S = \begin{pmatrix} S_{11} & S_{12} \\ S_{21} & S_{22} \end{pmatrix} \tag{6.27}$$

The matrix S is known as the scattering matrix or simply s-matrix and relates outgoing amplitudes to incoming amplitudes. The problem of solving equation (6.3) is equivalent to that of determining how the scattering matrix elements depend on the energy E and on the impurity site energy ϵ_1.

Equation (6.11) shows that if k is real, E can be any energy in the band between $\varepsilon_o - 2\gamma$ and $\varepsilon_o + 2\gamma$. In what follows, the aim is to compute the scattering matrix for a given choice of E. In other words, since the system is open (i.e. it extends to $\pm\infty$), apart from the constraint that E is lies within the energy band of the crystalline leads, E is not determined and the task is to solve for all properties as functions of E. For a given E, equation (6.11) reveals that the wave vector k is a function of E, given by

$$k(E) = \mathrm{acos}\left[\frac{\epsilon_0 - E}{2\gamma}\right] \tag{6.28}$$

Obviously, if $\epsilon_1 = \epsilon_0$, then the linear chain in figure 6.1 is a perfect, doubly infinite crystal. In this case, $A = C$ and $B = D$, so that in the absence of an impurity, one finds

$$S = \begin{pmatrix} 0 & 1 \\ 1 & 0 \end{pmatrix} \tag{6.29}$$

More generally, the scattering matrix is obtained by noting that equation (6.16) is

$$(1 \quad -1)\begin{pmatrix} B \\ C \end{pmatrix} = (-1 \quad 1)\begin{pmatrix} A \\ D \end{pmatrix} \tag{6.30}$$

and equation (6.23) is

$$(-\alpha \quad 1)\begin{pmatrix} B \\ C \end{pmatrix} = (\alpha + 1 \quad 0)\begin{pmatrix} A \\ D \end{pmatrix} \tag{6.31}$$

These combine to yield

$$\begin{pmatrix} 1 & -1 \\ -\alpha & 1 \end{pmatrix}\begin{pmatrix} B \\ C \end{pmatrix} = \begin{pmatrix} -1 & 1 \\ \alpha + 1 & 0 \end{pmatrix}\begin{pmatrix} A \\ D \end{pmatrix}$$

Multiplying both sides by the inverse of $\begin{pmatrix} 1 & -1 \\ -\alpha & 1 \end{pmatrix}$, namely $\frac{1}{1-\alpha}\begin{pmatrix} 1 & 1 \\ \alpha & 1 \end{pmatrix}$, yields equation (6.26), with

$$S = \frac{1}{1 - \alpha}\begin{pmatrix} \alpha & 1 \\ 1 & \alpha \end{pmatrix} \tag{6.32}$$

As expected, when $\epsilon_1 = \epsilon_0$, so that $\alpha = 0$, equation (6.29) is obtained.

It is interesting to note that for the above single impurity problem, since α is imaginary, $\alpha^* = -\alpha$ and therefore the Hermitian conjugate of S is $S^\dagger = \frac{1}{1+\alpha}\begin{pmatrix} -\alpha & 1 \\ 1 & -\alpha \end{pmatrix}$. Consequently,

$$S^\dagger S = \frac{1}{1-\alpha^2}\begin{pmatrix} -\alpha & 1 \\ 1 & -\alpha \end{pmatrix}\begin{pmatrix} \alpha & 1 \\ 1 & \alpha \end{pmatrix} = \begin{pmatrix} 1 & 0 \\ 0 & 1 \end{pmatrix} \tag{6.33}$$

Hence $S^\dagger$ is the inverse of S:

$$S^\dagger = S^{-1} \tag{6.34}$$

Any matrix whose inverse is equal to its Hermitian conjugate is called a unitary matrix. Hence the scattering matrix (6.32) is unitary.

It will now be shown that all scattering matrices of structures attached to one-dimensional crystalline leads are unitary. To do this, we first need an expression for the current passing through the bond between two atomic orbitals.

6.4 Bond currents and conservation of probability

Consider for example the one-dimensional chain shown in figure 6.2, in which all site energies ϵ_i and all bonds $-\gamma_i$ are arbitrary. In terms of Hamiltonian matrix elements, $H_{jj} = \epsilon_j$ and $H_{j,j+1} = -\gamma_j$.

The time dependent Schrödinger equation for this system is

$$i\hbar\frac{d}{dt}|\varphi(t)\rangle = H|\varphi(t)\rangle,$$

or equivalently

$$i\hbar\frac{d}{dt}\varphi_j(t) = \varepsilon_j\phi_j(t) + H_{j,j-1}\phi_{j-1}(t) + H_{j,j+1}\phi_{j+1}(t) \tag{6.35}$$

Note that the probability for finding an electron on site j at time t is proportional to

$$P_j(t) = |\varphi_j(t)|^2 = \phi_j(t)\phi_j^*(t) \tag{6.36}$$

To obtain an expression for the change in occupation probability on site j with time, note that

$$i\hbar\frac{d}{dt}P_j(t) = i\hbar\left[\phi_j(t)\frac{d}{dt}\phi_j^*(t) + \phi_j^*(t)\frac{d}{dt}\varphi_j(t)\right] \tag{6.37}$$

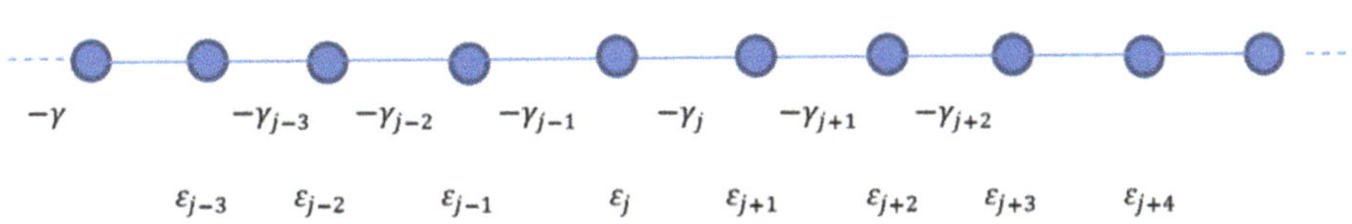

Figure 6.2. A one-dimensional chain in which all site energies ϵ_i and all bonds $-\gamma_i$ are arbitrary.

To evaluate this expression, note that the complex conjugate of equation (6.35) is

$$i\hbar\frac{d}{dt}\varphi_j^*(t) = \left[-\varepsilon_j\phi_j^*(t) + H_{j,j-1}^*\phi_{j-1}^*(t) + H_{j,j+1}^*\phi_{j+1}^*(t)\right]$$

Hence

$$\begin{aligned} i\hbar\frac{d}{dt}P_j(t) = &+ H_{j,j-1}\phi_j^*(t)\phi_{j-1}(t) + H_{j,j+1}\phi_j^*(t)\phi_{j+1}(t) \\ &- H_{j,j-1}^*\varphi_j(t)\phi_{j-1}^*(t) - H_{j,j+1}^*\varphi_j(t)\phi_{j+1}^*(t) \end{aligned} \tag{6.38}$$

Bond currents are defined such that the rate of change with time of the probability P_j is equal to the sum of all currents entering site j minus the sum of all currents leaving site j; i.e.

$$\frac{d}{dt}P_j(t) = J_{j-1\to j} - J_{j\to j+1} \tag{6.39}$$

In this expression, $J_{j-1\to j}$ is the bond current from site $j-1$ to site j, and $J_{j\to j+1}$ is the bond current from site j to site $j+1$. Comparing this with equation (6.38) yields

$$J_{j-1\to j} = \frac{i}{\hbar}\left[H_{j,j-1}^*\varphi_j(t)\phi_{j-1}^*(t) - H_{j,j-1}\phi_j^*(t)\phi_{j-1}(t)\right] \tag{6.40}$$

$$J_{j\to j+1} = \frac{i}{\hbar}\left[H_{j,j+1}\phi_j^*(t)\phi_{j+1}(t) - H_{j,j+1}^*\varphi_j(t)\phi_{j+1}^*(t)\right] \tag{6.41}$$

To check that equations (6.40) and (6.41) are mutually consistent, note that since $H_{j,j-1}^* = H_{j-1,j}$, equation (6.40) can be written

$$J_{j-1\to j} = \frac{i}{\hbar}\left[+H_{j-1,j}\varphi_j(t)\phi_{j-1}^*(t) - H_{j-1,j}^*\phi_j^*(t)\phi_{j-1}(t)\right] \tag{6.42}$$

This shows that the expression (6.40) for $J_{j-1\to j}$ is obtained from equation (6.41) by replacing j by $j-1$ in equation (6.41) and therefore equations (6.41) and (6.42) are indeed mutually consistent. Writing $H_{j,j+1} = -\gamma_j$ yields from equation (6.41),

$$J_{j\to j+1} = \frac{i}{\hbar}\left[-\gamma_j\phi_j^*(t)\phi_{j+1}(t) + \gamma_j^*\varphi_j(t)\phi_{j+1}^*(t)\right] \tag{6.43}$$

If $z = x + iy$ is a complex number, then $-i(z - z^*) = 2y = 2\mathrm{Im}\, z$, where Im z = imaginary part of z. Hence

$$J_{j\to j+1} = \frac{2}{\hbar}\mathrm{Im}\left[\gamma_j\phi_j^*(t)\phi_{j+1}(t)\right] \tag{6.44}$$

A more general expression for bond currents in an arbitrary system is given in equation (A.4) of appendix A.

Equation (6.39) is a law of conservation of probability, because it reveals that the rate of change of the site occupation probability $P_j(t)$ is equal to the probability current entering site j from site $j-1$ minus the probability current leaving site j to site $j+1$ (i.e. current in minus current out). In other words, probability cannot be destroyed; it can only flow from a site to one of its neighbours.

6.5 Current carried by counter-propagating plane waves

As an example, consider the case where $\varphi_j(t)$ is an eigenstate of the Hamiltonian of the form $\varphi_j(t) = \psi_j e^{-iEt/\hbar}$, where ψ_j satisfies equation (6.3). In this case, $P_j(t) = |\varphi_j(t)|^2 = |\psi_j|^2$. Since this is independent of time, $\frac{d}{dt}P_j(t) = 0$, which means that all occupation probabilities are stationary, and therefore from equation (6.39) $J_{j-1\to j} = J_{j\to j+1}$, for all j. This means that all bond currents are equal and the current entering any site is equal to the current leaving the site.

As an example, consider the case of a one-dimensional crystal, where for all j, $\varepsilon_j = \epsilon_0$ and $\gamma_j = \gamma$ and $\psi_j = Ae^{ikj}$. In this case, $\psi_j^* \psi_{j+1} = |A|^2 e^{ik}$ and $\mathrm{Im}\, \psi_j^* \psi_{j+1} = |A|^2 \sin k$. Hence from equation (6.44), the current carried by a plane wave of amplitude A is

$$J_{j\to j+1} = \frac{2\gamma \sin k}{\hbar} |A|^2 = v_k |A|^2 \tag{6.45}$$

where from equation (4.13), v_k is the group velocity of the plane wave.

As a second example, consider the case of a one-dimensional crystal, where for all j, $\varepsilon_j = \epsilon_0$ and $\gamma_j = \gamma$ and $\psi_j = Ae^{ikj} + Be^{-ikj}$. In this case, the eigenstate is a sum of two plane waves and

$$\psi_j^* \psi_{j+1} = [A^* e^{-ikj} + B^* e^{ikj}][Ae^{ik(j+1)} + Be^{-ik(j+1)}]$$

i.e.

$$\psi_j \psi_{j+1}^* = |A|^2 e^{ik} + |B|^2 e^{-ik} + [BA^* e^{-ik(2j+1)} + B^* A e^{ik(2j+1)}]$$

Hence

$$\begin{aligned}\psi_j \psi_{j+1}^* = & \, i(|A|^2 - |B|^2) \sin k + (|A|^2 - |B|^2) \cos k \\ & + [BA^* e^{-ik(2j+1)} + B^* A e^{ik(2j+1)}]\end{aligned}$$

If z is a complex number, $z + z^*$ is real, so the term in the square brackets is real. Hence

$$\mathrm{Im}\, \psi_j \psi_{j+1}^* = (|A|^2 - |B|^2) \sin k$$

and equation (6.44) yields

$$J_{j\to j+1} = v_k[|A|^2 - |B|^2] \tag{6.46}$$

This shows that the bond current is equal to the current flowing to the right carried by the plane wave Ae^{ikj}, minus the current flowing to the left carried by the plane wave Be^{-ikj}.

In view of equations (6.45) and (6.46), the following superposition of plane waves $\psi_j = A\frac{e^{ikj}}{\sqrt{v_k}} + B\frac{e^{-ikj}}{\sqrt{v_k}}$ carries a current $|A|^2 - |B|^2$ and the following plane wave

$$\psi_j = \frac{e^{ikj}}{\sqrt{v_k}} \tag{6.47}$$

carries a unit current. For this reason, the plane wave in equation (6.47) is said to be 'normalised to a unit current'.

6.6 Unitarity of the scattering matrix

We are now in a position to prove unitarity of the scattering matrix, by analysing the more general structure shown in figure 6.3, which consists of two different semi-infinite crystalline leads connected to a more complicated 'scattering region'. The lead on the left has site energies and hopping elements ϵ_L and $-\gamma_L$ and a dispersion relation $E = \epsilon_L - 2\gamma_L \cos k_L$. More precisely, since the aim is to solve the Schrödinger equation for eigenstates of a given E, this means that for a given E, k_L is given by $k_L(E) = \text{acos}[(\epsilon_L - E)/2\gamma_L]$ and the group velocity of the plane waves on the left is $v_L = 2\gamma_L \sin k_L$. Similarly the lead on the right has site energies and hopping elements ϵ_R and $-\gamma_R$ and for a given E, k_R is given by $k_R(E) = \text{acos}[(\epsilon_R - E)/2\gamma_R]$ and the group velocity of the plane waves on the right is $v_R = 2\gamma_R \sin k_R$. In other words, since there is only one E in equation (6.3), the wave vectors and groups velocities in the leads are different.

Since the plane waves in the left and right leads are normalised to unit current, the current in the left lead is $|A|^2 - |B|^2$ and the current in the right lead is $|C|^2 - |D|^2$. Conservation of probability requires that for such an eigenstate $|A|^2 - |B|^2 = |C|^2 - |D|^2$.

Equivalently, the incoming current from both leads is $|A|^2 + |D|^2$ and the outgoing current into both leads is $|B|^2 + |C|^2$. Conservation of probability requires $|B|^2 + |C|^2 = |A|^2 + |D|^2$, which can be written

$$(B^* \ \ C^*)\begin{pmatrix} B \\ C \end{pmatrix} = (A^* \ \ D^*)\begin{pmatrix} A \\ D \end{pmatrix} \tag{6.48}$$

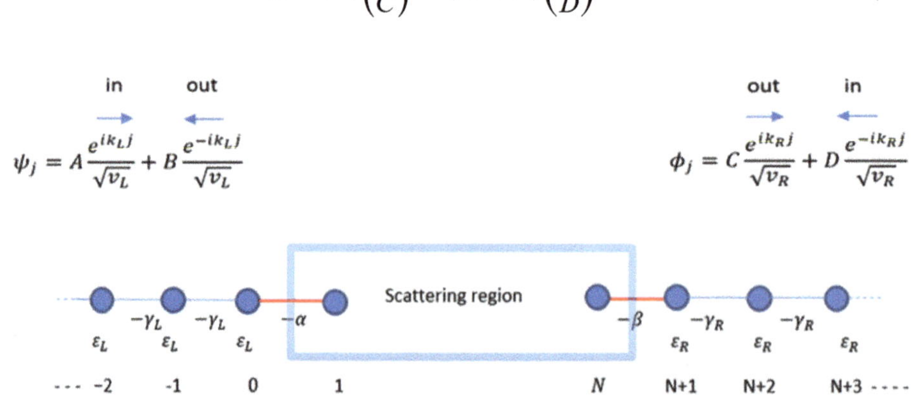

Figure 6.3. An arbitrary scatterer with one-dimensional leads.

Since the scattering matrix is defined to satisfy $\begin{pmatrix} B \\ C \end{pmatrix} = S\begin{pmatrix} A \\ D \end{pmatrix}$, taking the Hermitian conjugate yields $(B^* \ \ C^*) = (A^* \ \ D^*)S^\dagger$. Hence, substituting these into equation (6.48) yields

$$(A^* \ \ D^*)S^\dagger S\begin{pmatrix} A \\ D \end{pmatrix} = (A^* \ \ D^*)\begin{pmatrix} A \\ D \end{pmatrix}$$

Since this must be satisfied for all choices of the incoming amplitudes, the s-matrix must satisfy

$$S^\dagger S = I \tag{6.49}$$

where $I = \begin{pmatrix} 1 & 0 \\ 0 & 1 \end{pmatrix}$ is the unit matrix. In other words, S is unitary. Note that this proof requires that A, B, C, D are the amplitudes of plane waves normalised to unit current. This is important when the leads are different. If the leads are the same, as in figure 6.1, then the square roots of group velocities would cancel everywhere and simple plane waves, without the square roots could be used in the definition of S.

6.7 Physical meaning of scattering matrix elements

For the structure of figure 6.4, with one-dimensional leads, the scattering matrix is defined to satisfy

$$\begin{pmatrix} B \\ C \end{pmatrix} = \begin{pmatrix} S_{11} & S_{12} \\ S_{21} & S_{22} \end{pmatrix}\begin{pmatrix} A \\ D \end{pmatrix} \tag{6.50}$$

To understand the physical meaning of the scattering matrix elements, consider the case where a plane wave of unit current enters only from the left, so that $A = 1$ and $D = 0$. In this case, B is the amplitude of the reflected wave on the left and C is the amplitude of the transmitted wave on the right. To draw attention to this fact, it is useful to write $B = r$, where r is the amplitude of the reflected wave due to an incoming wave from the left and t is the amplitude of the transmitted wave. In this case, equation (6.50) becomes

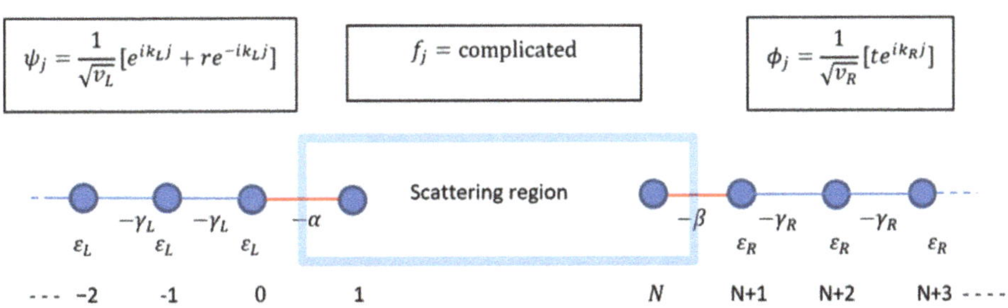

Figure 6.4. An arbitrary scattering region connected to one-dimensional leads, with an incident wave of unit current from the left.

$$\begin{pmatrix} r \\ t \end{pmatrix} = \begin{pmatrix} S_{11} & S_{12} \\ S_{21} & S_{22} \end{pmatrix} \begin{pmatrix} 1 \\ 0 \end{pmatrix} = \begin{pmatrix} S_{11} \\ S_{21} \end{pmatrix} \tag{6.51}$$

Hence the physical meaning of $S_{11} = r$ is that it is the reflection amplitude associated with an incoming plane wave from the left and similarly $S_{21} = t$ is the corresponding transmission amplitude from the left to the right lead. This situation is depicted in figure 6.4.

Now consider the case where a plane wave of unit current enters only from the right, so that $A = 0$ and $D = 1$. In this case, B is the amplitude of the transmitted wave on the left and C is the amplitude of the reflected wave on the right. To draw attention to this fact, it is useful to write $B = r'$, where r' is the amplitude of the reflected wave due to an incoming wave from the right and t' is the amplitude of the transmitted wave from right to left. In this case, (6.50) becomes

$$\begin{pmatrix} t' \\ r' \end{pmatrix} = \begin{pmatrix} S_{11} & S_{12} \\ S_{21} & S_{22} \end{pmatrix} \begin{pmatrix} 0 \\ 1 \end{pmatrix} = \begin{pmatrix} S_{12} \\ S_{22} \end{pmatrix} \tag{6.52}$$

Hence the physical meaning of $S_{12} = t'$ is that it is the transmission amplitude associated with an incoming plane wave from the right and similarly $S_{21} = r'$ is the corresponding reflection amplitude. Collecting these results together yields

$$S = \begin{pmatrix} S_{11} & S_{12} \\ S_{21} & S_{22} \end{pmatrix} = \begin{pmatrix} r & t' \\ t & r' \end{pmatrix} \tag{6.53}$$

where all quantities r, r', t', t' are functions of the eigenstate energy E.

6.8 Consequences of unitarity of the scattering matrix

The aim of this section is to show that unitarity of S leads to general relationships between reflection and transmission amplitudes. From equation (6.49)

$$\begin{pmatrix} 1 & 0 \\ 0 & 1 \end{pmatrix} = S^{\dagger}S = \begin{pmatrix} r^* & t^* \\ t'^* & r'^* \end{pmatrix} \begin{pmatrix} r & t' \\ t & r' \end{pmatrix} = \begin{pmatrix} r^*r + t^*t & r^*t' + t^*r' \\ r'^*t + t'^*r & r'^*r' + t'^*t' \end{pmatrix} \tag{6.54}$$

If $\begin{pmatrix} 1 & 0 \\ 0 & 1 \end{pmatrix} = S^{-1}S$, then $\begin{pmatrix} 1 & 0 \\ 0 & 1 \end{pmatrix} = SS^{-1}$. Hence

$$\begin{pmatrix} 1 & 0 \\ 0 & 1 \end{pmatrix} = SS^{\dagger} = \begin{pmatrix} r & t' \\ t & r' \end{pmatrix} \begin{pmatrix} r^* & t^* \\ t'^* & r'^* \end{pmatrix} = \begin{pmatrix} r^*r + t'^*t' & r'^*t' + t^*r \\ r^*t + t'^*r' & r'^*r' + t^*t \end{pmatrix} \tag{6.55}$$

The diagonal elements of these equations yield

$$r^*r + t^*t = r'^*r' + t'^*t' = r^*r + t'^*t' = r'^*r' + t^*t = 1,$$

from which one obtains

$$r^*r = r'^*r' \quad \text{and} \quad t^*t = t'^*t' \tag{6.56}$$

In summary, if reflection and transmission coefficients[1] due to a plane wave from the right are defined by $R(E) = r^*r$ and $T(E) = t^*t$, and reflection and transmission coefficients due to an incoming plane wave from the left by $R'(E) = r'^*r'$ and $T'(E) = t'^*t'$ then

$$R(E) = R'(E),\ T(E) = T'(E) \tag{6.57}$$

and

$$R(E) + T(E) = 1 \tag{6.58}$$

Finally the off-diagonal terms yield $r'^*t' + t^*r = r^*t' + t^*r' = 0$

That is

$$\frac{r'}{r^*} = -\frac{t'}{t^*} \tag{6.59}$$

As an example, for the single impurity problem

$$S = \frac{1}{1 - \alpha}\begin{pmatrix} \alpha & 1 \\ 1 & \alpha \end{pmatrix}$$

Consequently

$$R = |\alpha|^2/(1 + |\alpha|^2) \quad \text{and} \quad T = \frac{1}{1 + |\alpha|^2} = \frac{1}{1 + \left|\frac{(\varepsilon_1 - \epsilon_0)}{\gamma 2 \sin k}\right|^2} \tag{6.60}$$

Furthermore, since $\alpha = -\alpha^*$, the relationship $\frac{r'}{r^*} = -\frac{t'}{t^*}$ is satisfied, as expected.

6.9 Bound states

Before leaving this discussion of the single impurity problem, it is interesting to note that in addition to the continuum of energies within the band, there is also a zero-current solution to equation (6.3), obtained by setting $A = D = 0$ in equations (6.30) and (6.31). From equation (6.30), this yields $B = C$ and from equation (6.31) $C = \alpha B$. Hence a zero-current state is possible if $\alpha = 1$. From equation (6.24), thus means that

$$(\varepsilon_1 - \epsilon_0) = \gamma 2i \sin k$$

Since the left-hand side is real, the right-hand side must also be real. Hence, since $i \sin i\theta = -\sinh \theta$ and $i \sin i(\theta + \pi) = \sinh \theta$, one finds that if θ is real, one possible solution is $k = i\theta$, which yields

$$(\varepsilon_1 - \epsilon_0) = -\gamma 2 \sinh \theta$$

Since θ must be positive, to avoid the wave function diverging at $\pm\infty$, this means that if γ is positive, this is the solution when $(\varepsilon_1 - \epsilon_0) \leqslant 0$

[1] In what follows, the complex numbers r, r', t, t' are referred to as reflection and transmission amplitudes, whereas the positive real numbers R, T are referred to as reflection and transmission coefficients.

On the other hand, another possible solution is $k = i\theta + \pi$, in which case

$$(\varepsilon_1 - \epsilon_0) = 2\gamma \sinh \theta$$

If γ is positive, this is the solution when $(\varepsilon_1 - \epsilon_0) \geqslant 0$.

For the case $k = i\theta$, equation (6.9) becomes

$$\psi_j = Be^{\theta j} \qquad (j < -1)$$

(which tends to zero as $j \to -\infty$) and to the right of the impurity for $j > 0$, equation (6.10) becomes

$$\phi_j = Be^{-\theta j} \qquad (j > 1)$$

(which tends to zero as $j \to +\infty$).

For the case $=i\theta + \pi$, equation (6.9) becomes

$$\psi_j = (-1)^j Be^{\theta j} \qquad (j < -1)$$

(which again tends to zero as $j \to -\infty$) and to the right of the impurity for $j > 0$, equation (6.10) becomes

$$\phi_j = (-1)^j Be^{-\theta j} \qquad (j > 1)$$

(which tends to zero as $j \to +\infty$).

Hence zero-current states are bound states, which have a large amplitude on the impurity and decay exponentially at values of j remote from the impurity.

6.10 The *S* matrix in higher dimensions

Section 5.10 of chapter 5 discussed the properties of an infinite finite-width ribbon composed of a series of identical unit cells, shown in figure 5.4. Now consider the scattering properties of a finite dimensional scatter (labelled 'B' below) connected to two such finite-width leads, as shown in figure 6.5. The semi-infinite left lead is composed of a series of identical unit cells, with intra-cell Hamiltonians H_0

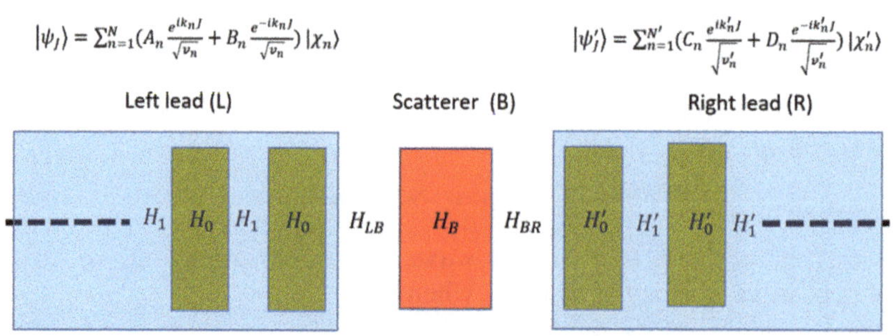

Figure 6.5. A scatterer 'B' connected to left and right semi-infinite leads. H_{LB} and H_{BR} are sub-matrices of the Hamiltonian connecting the scatterer to the left and right leads respectively.

and nearest neighbour inter-cell Hamiltonians H_1, as described in equations (5.21)–(5.23). As described in section (5.10), when H_1 is proportional to the unit matrix, the most general solution to the Schrödinger equation in the left lead is

$$|\psi_J\rangle = \sum_{n=1}^{N}\left(A_n\frac{e^{ik_nJ}}{\sqrt{v_n}} + B_n\frac{e^{-ik_nJ}}{\sqrt{v_n}}\right)|\chi_n\rangle \tag{6.61}$$

where (in contrast with equation (5.54)) all plane waves have now been normalised to unit current. This restriction is helpful, because it allows us to simplify the notation and focus on key concepts. As discussed in chapter 5, each wave vector k_n belongs to a distinct mini-band. When attaching such a lead to a scatterer, these mini-band can carry electrons to and from the scatter and are therefore called 'scattering channels'. If k_n is real, then the channel is said to be 'open' and can carry a current, otherwise it is 'closed' and cannot carry a current.

Similarly, the most general solution in the right lead is of the form shown in equation (16.62), where the 'primes' allow for the possibility that for a given energy E, the wave vectors and group velocities of the right lead could differ from those of the left lead.

$$|\psi_J'\rangle = \sum_{n=1}^{N'}\left(C_n\frac{e^{ik_n'J}}{\sqrt{v_n'}} + D_n\frac{e^{-ik_n'J}}{\sqrt{v_n'}}\right)|\chi_n'\rangle \tag{6.62}$$

If the amplitudes A_n are used to define a vector $|A\rangle$ such that

$$|A\rangle = \begin{pmatrix} A_1 \\ \vdots \\ A_N \end{pmatrix} \tag{6.63}$$

and similarly for $|B\rangle$, $|C\rangle$, $|D\rangle$, then (cf equation (6.50)) the corresponding scattering matrix $\hat{S}$ can be defined to satisfy

$$\begin{pmatrix} |B\rangle \\ |C\rangle \end{pmatrix} = \hat{S}\begin{pmatrix} |A\rangle \\ |D\rangle \end{pmatrix} \tag{6.64}$$

However in general, $\hat{S}$ will not be unitary, because all wave vectors are not necessarily real and therefore not all channels are necessarily open. Adopting the labelling convention described after equation (5.47) of chapter 5, to avoid closed channels diverging at $-\infty$, the amplitudes A_n associated with closed channels must be set to zero. Similarly to avoid closed channels diverging at $+\infty$, the amplitudes D_n associated with closed channels must be set to zero. The remaining closed channels on the left will then decay to zero as $J \to -\infty$ and the closed channels on the right will decay to zero as $J \to +\infty$. If the amplitudes of open channels are denoted A_n^{open}, B_n^{open} etc and used to define vectors such as $|A^{\text{open}}\rangle$, $|B^{\text{open}}\rangle$ etc and similarly the amplitudes of closed channels are denoted B_n^{closed}, C_n^{closed}, with associated vectors $|B^{\text{closed}}\rangle$, $|C^{\text{closed}}\rangle$, then equation (6.64) can be rewritten in the form

$$\begin{pmatrix} |\text{out}^{\text{open}}\rangle \\ |\text{closed}\rangle \end{pmatrix} = \begin{pmatrix} S \\ \bar{S} \end{pmatrix}|\text{in}^{\text{open}}\rangle \tag{6.65}$$

where $|\text{out}^{\text{open}}\rangle = \begin{pmatrix} |B^{\text{open}}\rangle \\ |C^{\text{open}}\rangle \end{pmatrix}$, $|\text{in}^{\text{open}}\rangle = \begin{pmatrix} |A^{\text{open}}\rangle \\ |D^{\text{open}}\rangle \end{pmatrix}$ and $|\text{closed}\rangle = \begin{pmatrix} |B^{\text{closed}}\rangle \\ |C^{\text{closed}}\rangle \end{pmatrix}$

Hence the amplitudes associated with open channels satisfy

$$|\text{out}^{\text{open}}\rangle = S|\text{in}^{\text{open}}\rangle \tag{6.66}$$

In the notation of equation (6.53), this can be written

$$\begin{pmatrix} |B^{\text{open}}\rangle \\ |C^{\text{open}}\rangle \end{pmatrix} = \begin{pmatrix} r & t' \\ t & r' \end{pmatrix} \begin{pmatrix} |A^{\text{open}}\rangle \\ |D^{\text{open}}\rangle \end{pmatrix} \tag{6.67}$$

where r, t, r', t' are now reflection and transmission matrices. The resulting scattering matrix is

$$S = \begin{pmatrix} r & t' \\ t & r' \end{pmatrix} \tag{6.68}$$

As shown in appendix B, since the incident and outgoing currents must be equal for any choice of incoming amplitudes, S is unitary. In appendix B, properties of S arising from time-reversal symmetry are also discussed, where it is shown that if the Hamiltonian is real t is the transpose of t' and both r and r' are symmetric. If the number of open channels on the left is M and the number of open channels on the right is M', then r is a $M \times M$ square matrix and t is a rectangular matrix with M columns and M' rows, etc.

Similarly from equation (6.65), the amplitudes of the decaying plane wave components are related to the amplitudes of the incoming waves by

$$|\text{closed}\rangle = \bar{S}|\text{in}^{\text{open}}\rangle \tag{6.69}$$

That is

$$\begin{pmatrix} |B^{\text{closed}}\rangle \\ |C^{\text{closed}}\rangle \end{pmatrix} = \bar{S} \begin{pmatrix} |A^{\text{open}}\rangle \\ |D^{\text{open}}\rangle \end{pmatrix} \tag{6.70}$$

where the matrix $\bar{S}$ is not generally unitary. This demonstrates that incoming waves can generate either current-carrying outgoing waves and/or evanescent waves bound to the scatterer, which decay exponentially within the leads. Methods for computing S and $\bar{S}$ will be discussed in chapter 11.

6.11 The Landauer formula

Reprinted in part from Lambert C J 2015 *Chem. Soc. Rev.* **44** 875–8, with permission of RSC (https://doi.org/10.1039/C4CS00203B)

Knowledge of the scattering matrix allows the electrical conductance of a scatterer connected to semi-infinite leads to be predicted using the Landauer formula. To illustrate this standard approach to computing the electrical conductance of single molecules and other nanostructures, consider the abstract structure shown in figure 6.5, which could be used, for example, to represent the structure of figure 6.6, comprising a molecule located between two gold electrodes.

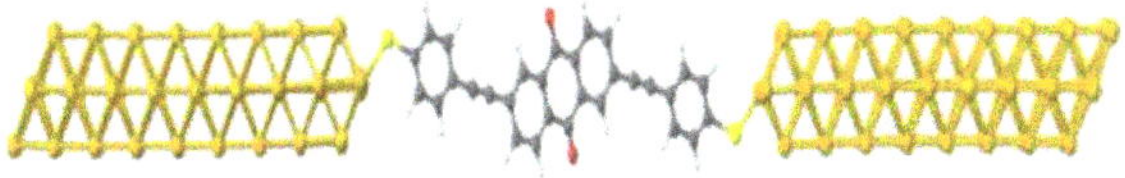

Figure 6.6. A molecule in contact with two gold electrodes, represented by semi-infinite crystalline leads.

The electrodes are connected to external reservoirs (not shown), which feed electrons of energy E into the scattering channels belonging to the gold electrodes. The energy distribution of electrons entering the left (right) gold lead from the left (right) reservoir is $f_{\text{left}}(\text{E})$ ($f_{\text{right}}(\text{E})$) and according to the Landauer formula, the current passing from left to right is

$$I = \left(\frac{2e}{h}\right) \int_{-\infty}^{\infty} dE\ T(E)[f_{\text{left}}(\text{E}) - f_{\text{right}}(\text{E})] \tag{6.71}$$

where $T(E)$ is the transmission coefficient for electrons passing from one gold lead to the other via the molecule. Clearly $I = 0$ when $f_{\text{left}}(\text{E}) = f_{\text{right}}(\text{E})$, because only differences in the distributions contribute to the net current. A derivation of the Landauer formula for a structure connected to one-dimensional leads and multi-dimensional leads is presented in the next section, along with a derivation of corresponding formulae for thermoelectric properties of a scattering region connected to external leads.

Close to equilibrium the energies of electrons in the left and right reservoirs are distributed according to their respective Fermi distributions, $f_{\text{left}}(\text{E}) = [e^{\beta(E-E_{\text{F}}^{\text{left}})} + 1]^{-1}$ and $f_{\text{right}}(\text{E}) = [e^{\beta(E-E_{\text{F}}^{\text{right}})} + 1]^{-1}$, where $E_{\text{F}}^{\text{left}}(E_{\text{F}}^{\text{right}})$ is the Fermi energy of the left (right) reservoir and $\beta = \frac{1}{k_{\text{B}}T}$, where T is the temperature. If V is the voltage difference between the left and right reservoirs, then $E_{\text{F}}^{\text{left}} = E_{\text{F}} + \frac{eV}{2}$ and $E_{\text{F}}^{\text{right}} = E_{\text{F}} - \frac{eV}{2}$. This means that at zero temperature, but finite voltage

$$I = \left(\frac{2e}{h}\right) \int_{E_{\text{F}}-eV/2}^{E_{\text{F}}+eV/2} dE\ T(E) \tag{6.72}$$

Consequently, the electrical conductance I/V is obtained by averaging $T(E)$ over an energy window of width eV centred on the Fermi energy. On the other hand, if $T(E)$ does not vary significantly over an energy range of order eV, then as shown in section 6.12, the Fermi functions can be Taylor expanded to yield the electrical conductance in the zero voltage, finite temperature limit:

$$I/V = G_0 \int_{-\infty}^{\infty} dE\ T(E)\left(-\frac{df(E)}{dE}\right) \tag{6.73}$$

where G_0 is the quantum of conductance,

$$G_0 = \left(\frac{2e^2}{h}\right) = 77\ \mu\text{S} \tag{6.74}$$

Since the quantity $-df(E)/dE)$ is a normalised probability distribution of width approximately k_BT, (see figure 15.9) the above integral represents a thermal average of the transmission function $T(E)$ over an energy window of width k_BT.

Finally, in the limit of zero voltage and zero temperature, one obtains

$$I/V = G_0T(E_F) \tag{6.75}$$

The transmission coefficient $T(E)$ is a property of the whole system comprising the leads, the molecule and the contact between the leads and the molecule. Nevertheless, if the contact to the electrodes is weak, then as discussed in chapter 1, a graph of $T(E)$ versus E will reflect the energy-level structure of the isolated molecule. In particular if the isolated molecule has energy levels E_n, (where n represents the quantum numbers labelling the energy levels), then $T(E)$ will possess a series of peaks (i.e. resonances) located at energies in the vicinity of the levels E_n. These 'Breit–Wigner' resonances were introduced in chapter 1 and will be discussed in more detail in chapter 11.

The above expressions assume that during the time it takes for an electron to pass from one lead to the other, the system of leads plus molecule can be described by a time-independent mean-field Hamiltonian H. This means that an electron remains phase coherent as it passes from one lead to the next and does not undergo inelastic scattering (i.e. the energy E of the electron does not change). For short molecules, there is evidence that this is a reasonable assumption. However as the length of a molecule increases, the probability of inelastic scattering (e.g. from phonons or other electrons) becomes non-negligible. In what follows, the discussion is confined to systems where the above expressions are valid and the energy of an electron remains constant as it passes through a scatterer.

On the other hand, just because a Hamiltonian H is static on the timescale required to pass through the molecule, does not mean that the system is completely stationary, because on a longer timescale, the molecule may change its shape, the electrodes may distort or the environment surrounding the molecule may change. In this case, in the spirit of the Born–Oppenheimer approximation, one should construct an ensemble of Hamiltonians describing each realisation of the system (i.e. electrodes, molecule and environment) and compute $T(E)$ for each case. Since the environment and molecular conformation can fluctuate many times on the timescale of a typical electrical measurement, an ensemble average of such transmission functions is required. This ensemble average (i.e. average over snapshots) does not include the even slower changes in H associated with the pulling apart of electrodes in a break-junction experiment, or with the changes associated with the switching on of a dc bias or gate voltage. In the presence of such slow changes, histograms of conductances should be measured and analysed, as discussed in section 2.8 of chapter 2.

6.12 Derivation of the Landauer formula and formulae for thermoelectric properties of a quantum object connected to external electrodes

The derivation of the Landauer formula, is based on the idea of a perfect reservoir, which can be used to deliver electrons to a scattering region. To define a 'perfect reservoir' consider the one-dimensional ring of N sites with periodic boundary conditions shown in figure 4.5, whose normalised eigenstates are

$$\psi_j^{(n)} = Ae^{ik_n j}$$

In this equation, $A = \sqrt{1/N}$ and $k_n = \frac{2n\pi}{N}$. If N is even, the integer n lies in the range $-\frac{N}{2} \leqslant n < \frac{N}{2}$, whereas if N is odd, the integer n lies in the range $-\frac{N-1}{2} \leqslant n \leqslant \frac{N-1}{2}$. The energy levels of such a system are given by $E_n = E(k_n)$, where $E(k) = \varepsilon_o - 2\gamma \cos k$. Note that the level $E_0 = 0$ is singly degenerate. For N odd, all levels E_n for non-zero n are doubly degenerate. Similarly for N even, all levels E_n for non-zero n are doubly degenerate, except the level $n = -\frac{N}{2}$, which is singly degenerate.

Consider the case where the system contains $4m + 2$ electrons, where m is an integer. As discussed in section 5.1, since the level $E_0 = 0$ is singly degenerate and levels E_n for non-zero n are doubly degenerate, and since each level can accommodate two electrons, the lowest energy state of the system (called the ground state) is obtained by filling the lowest energy $2m + 1$ levels E_n, where $-m \leqslant n \leqslant m$. In such a system, the HOMO energy is E_m, the LUMO energy is E_{m+1} and the HOMO–LUMO gap is $E_{m+1} - E_m$ and the Fermi energy E_F lies between the HOMO and LUMO energies; i.e. $E_m < E_F < E_{m+1}$. For a finite molecule, the Fermi energy is not well defined, because the HOMO–LUMO gap is finite. On the other hand for an infinitely large molecule, where the HOMO–LUMO gap tends to zero, E_F is well defined and is trapped between the HOMO and LUMO. A fundamental result of statistical physics is that for a system at equilibrium, the probability that an energy level is occupied by an electron is given by the Fermi distribution $f(E) = \left[\exp\left(\frac{E - E_F}{k_B T}\right) + 1\right]^{-1}$, where T is the temperature and k_B is Boltzmann's constant. In the low temperature limit $T \to 0$, $f(E) = 1$ for $E < E_F$ and $f(E) = 0$ for $E > E_F$. This means that at $T = 0$, the ground state of the system is obtained by filling all energy levels below the Fermi energy and consequently all levels with wave vectors in the range $-k_F < k < k_F$ are filled. Since plane waves with positive k carry a current which is equal and opposite to those with negative k, the ground state carries no net current.

For a linear chain in the limit of large N, consider adding electrons with positive group velocities to the energy interval E to $E + \Delta E$, as shown below (figure 6.7).

In this interval, the number of states of the form

$$\psi_j^{(n)} = \frac{e^{ik_n j}}{\sqrt{N}}$$

is $\Delta n = \frac{dn}{dk}\Delta k = \frac{dn}{dk}\frac{dk}{dE}\Delta E$. Since $k_n = \frac{2n\pi}{N}$, $n = \frac{N}{2\pi}k_n$ and therefore $\frac{dn}{dk} = \frac{N}{2\pi}$. Also, $\hbar v_k = \frac{dE}{dk}$

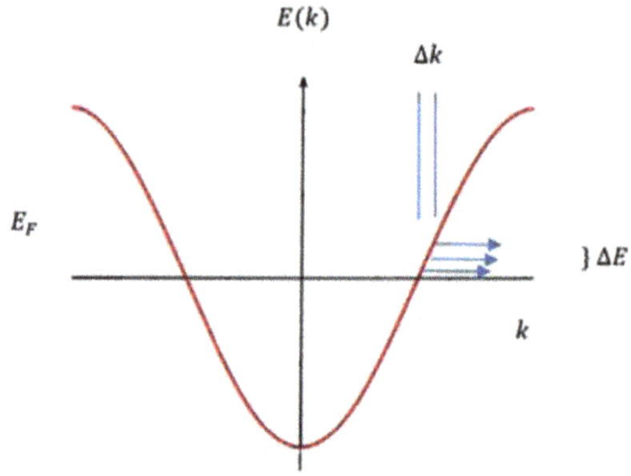

Figure 6.7. The band structure of a 1-dimensional chain of sites, with arrows indicating the right-moving states in an energy interval ΔE.

Hence

$$\Delta n = \frac{N}{hv_k}\Delta E \tag{6.76}$$

This shows that the number of states with positive velocity per unit energy is

$$\frac{\Delta n}{\Delta E} = \frac{N}{hv_k} \tag{6.77}$$

This quantity is sometime called the density of states (for positive velocity electrons) and is discussed in more detail in chapter 15. Since probability current carried by $\psi_j^{(n)}$ is

$$P_k = \frac{v_k}{N}, \tag{6.78}$$

the number of electrons per unit time passing through an arbitrary bond in the ring is

$$\Delta P = \frac{1}{h}\Delta E \qquad \text{per spin} \tag{6.79}$$

This is a fundamental relation, arising from the fact that the group velocity in the denominator of the density of states, is exactly cancelled by the group velocity appearing in the expression for the current. As a consequence, the current per unit energy $\frac{\Delta P}{\Delta E}$ is simply $\frac{1}{h}$, which is independent of k.

Allowing for the fact that electrons have spin, so that each state contains two electrons (one spin up and the other spin down), yields

$$\Delta P = \frac{2}{h}\Delta E \qquad \text{for both spins} \tag{6.80}$$

An ideal reservoir located on the left of a scatterer is defined to be a 'demon' that fills all right-moving states with an occupancy determined by the Fermi function $f_L(E)$

of the left lead and similarly an ideal reservoir located on the right of a scatterer is defined to be a 'demon' that fills all left-moving states with an occupancy determined by the Fermi function $f_R(E)$ of the right lead.

Then the number of electrons per unit time incident on the scatterer from the left reservoir, between E and $E + \Delta E$ is

$$\Delta P_{\text{left}} = \frac{2}{h} f_L(E) \Delta E \tag{6.81}$$

and the number of electrons per unit time incident on the scatterer from the right reservoir, between E and $E + \Delta E$ is

$$\Delta P_{\text{right}} = \frac{2}{h} f_R(E) \Delta E \tag{6.82}$$

Since electrons from the left reservoir can transmit into the right lead and electrons from the right reservoir can reflect into the right lead, the net electron current in the right lead is

$$\Delta P = T(E) \Delta P_{\text{left}} - (1 - R(E)) \Delta P_{\text{right}}$$

Noting that $T(E) + R(E) = 1$ yields

$$\Delta P = \frac{2}{h} [f_L(E) - f_R(E)] T(E) \Delta E \tag{6.83}$$

In this equation, the transmission coefficient $T(E)$ is the transmitted current in the right lead, per unit incident current in the left lead and $R(E)$, is the reflected current in the right lead, per unit incident current in the right lead.

As shown in appendix B, when many open channels are connected to a scatterer, it is useful to define R_{mn} to be the reflected current in channel m of the left lead, due to a unit current incident along channel n of the left lead and $T_{n'n}$ to be the transmitted current in channel n' of the right lead, due to a unit current incident along channel n of the left lead. If the distribution feeding all channels labelled n of the left lead is $f_L(E)$, and the distribution feeding all channels n' of the right lead is $f_R(E)$ then the net current in channel m' of the right lead between E and $E + \Delta E$ is

$$\Delta P_{m'} = \frac{2}{h} \sum_n f_L(E) T_{m'n}(\mathrm{E}) \Delta E - \frac{2}{h} \sum_{n'} [\delta_{m'n'} - R'_{m'n'}(E)] f_R(E) \Delta E$$

Since (see equation (B.14) of appendix B) $\sum_{n'} R'_{m'n'} + \sum_n T_{m'n} = 1$, one obtains

$$\Delta P_{m'} = \frac{2}{h} \sum_n [f_L(E) - f_R(E)] T_{m'n}(\mathrm{E}) \Delta E \tag{6.84}$$

i.e.

$$\Delta P_{m'} = \frac{2}{h} [f_L(E) - f_R(E)] T_{m'}(\mathrm{E}) \Delta E \tag{6.85}$$

where

$$T_{m'}(E) = \sum_{n} T_{m'n}(\mathrm{E}) \tag{6.86}$$

Hence the total number of electrons per unit time between E and $E + \Delta E$ entering the left reservoir (which is the negative of the number per unit time entering the right reservoir) is

$$\Delta P_L = \frac{2}{h}[f_R(E) - f_L(E)]T(\mathrm{E})\Delta E \tag{6.87}$$

where

$$T(E) = \sum_{m',n} T_{m'n}(\mathrm{E}) \tag{6.88}$$

In this equation, n sums over all open channels in the left lead and m' sums over all open channels in the right lead.

Since the charge on an electron is $-|e|$, the total electrical current entering the left reservoir is

$$I = \frac{-2|e|}{h}\int_{-\infty}^{\infty} dE[f_R(E) - f_L(E)]T(\mathrm{E}) \tag{6.89}$$

Similarly, the 'heat energy' (i.e. the energy in excess of the Fermi energy) per unit time entering the left reservoir is

$$Q = \frac{2}{h}\int_{-\infty}^{\infty} dE(E - E_{\mathrm{F}}^{L})[f_R(E) - f_L(E)]T(\mathrm{E}) \tag{6.90}$$

where E_{F}^{L} is the Fermi energy of the left reservoir. If T_L is the temperature of the left reservoir, then

$$f_L(E) = f(x_L)$$

where $f(x_L) = \frac{1}{e^{x_L}+1}$ and $x_L = \frac{E - E_{\mathrm{F}}^{L}}{k_{\mathrm{B}}T_L}$,. Similarly

$$f_R(E) = f(x_R)$$

where $x_R = \frac{E - E_{\mathrm{F}}^{R}}{k_{\mathrm{B}}T_R}$.

An analytic function of a variable x can be Taylor expanded about some point a as follows

$$f(x) = f(a) + \frac{df(a)}{da}(x - a) + \cdots$$

and if $(x - a)$ is sufficiently small, a reasonable approximation involves retaining only the first two terms. Therefore if the left and right Fermi distributions do not differ by too much, so that $(x_R - x_L)$ is small, then Taylor expanding yields

$$f(x_R) \approx f(x_L) + \frac{df(x_L)}{dx_L}(x_R - x_L)$$

In the above expression, $\frac{df(x_L)}{dx_L} = \frac{df_L(E)}{dE}\frac{dE}{dx_L} = k_B T_L \frac{df_L(E)}{dE}$. Furthermore,

$$x_R - x_L \approx \frac{\partial x_L}{\partial T_L}(T_R - T_L) + \frac{\partial x_L}{\partial E_F^L}(E_F^R - E_F^L)$$
$$= -\frac{E - E_F^L}{k_B T_L^2}(T_R - T_L) - \frac{1}{k_B T_L}(E_F^R - E_F^L)$$

and consequently

$$f_R(E) - f_L(E) = \left[-\frac{df_L(E)}{dE}\right]\left[\frac{(E - E_F^L)}{T_L}(T_R - T_L) + (E_F^R - E_F^L)\right]$$

Hence if the left and right Fermi distributions are only slightly different,

$$I = \frac{-2|e|}{h}\int_{-\infty}^{\infty} dE\left[-\frac{df_L(E)}{dE}\right]T(E)\left[\frac{(E - E_F^L)}{T_L}(T_R - T_L) + (E_F^R - E_F^L)\right] \quad (6.91)$$

Similarly, the 'heat energy' per unit time entering the left reservoir is

$$Q = \frac{2}{h}\int_{-\infty}^{\infty} dE\left[-\frac{df_L(E)}{dE}\right]T(\mathrm{E}) \times\left[\frac{(E - E_F^L)^2}{T_L}(T_R - T_L) + (E - E_F^L)(E_F^R - E_F^L)\right] \quad (6.92)$$

If the difference in Fermi levels is written

$$(E_F^R - E_F^L) = -|e|(V_R - V_L)$$

where $V_R - V_L$ is the voltage difference between the left and right reservoirs, then to linear order (i.e. first order) in the small differences $V_R - V_L$ and $T_R - T_L$, equations (6.91) and (6.92) yield

$$\begin{pmatrix} I \\ Q \end{pmatrix} = \frac{2}{h}\begin{pmatrix} e^2 L_0 & -\frac{|e|}{T_L}L_1 \\ -|e|L_1 & L_2 \end{pmatrix}\begin{pmatrix} V_R - V_L \\ T_R - T_L \end{pmatrix} \quad (6.93)$$

where

$$L_i = \int dE(E - E_F)^i\left[-T(E)\frac{df(E)}{dE}\right] \quad (6.94)$$

In equation (6.94), E_F is the average Fermi energy of the two reservoirs, and $f(E) = \left[e^{\frac{E-E_F}{k_B T}} + 1\right]^{-1}$, where T is the average temperature of the two reservoirs.

The electrical conductance G is defined to be the ratio $G = I/(V_R - V_L)$, when $T_R - T_L = 0$, whereas the Seebeck coefficient S is defined to be the ratio

$S = -(V_R - V_L)/(T_R - T_L)$, when $I = 0$. (Note the minus sign in the definition of S.) Hence

$$I/(V_R - V_L) = \frac{2e^2}{h}L_0 \tag{6.95}$$

and

$$S = \frac{-1}{|e|T}\frac{L_1}{L_0} \tag{6.96}$$

6.13 Summary

Starting from the simplest problem of a single impurity placed between two semi-infinite, 1-dimensional crystalline leads, the discussion in sections 6.2–6.8 demonstrates that open systems possesses a continuum of energy levels due to waves extending to infinity within the electrodes, accompanied by bound states, which are localised on the scattering region. For finite-width leads, this is reflected in the appearance of open and closed scattering channels. By separating the current-carrying states from the bound states, a unitary scattering matrix can be defined, which yields the transmission coefficient appearing in the Landauer formula and in the expression (6.96) for the Seebeck coefficient of a nanoscale junction. These formulae will be further analysed in chapter 7.

Appendix A A general expression for bond currents

In general, the equation $i\hbar\frac{d}{dt}|\varphi(t)\rangle = H|\varphi(t)\rangle$, is equivalent to

$$i\frac{d}{dt}\varphi_j(t) = \frac{1}{\hbar}\sum_l H_{j,l}\phi_l(t) \tag{A.1}$$

Also, if $z = x + iy$, then $\mathrm{Re}[z] = \mathrm{Im}[iz]$, so equation (6.37) yields

$$\begin{aligned}\frac{d}{dt}P_j(t) &= \left[\phi_j(t)\frac{d}{dt}\phi_j^*(t) + \phi_j^*(t)\frac{d}{dt}\varphi_j(t)\right]\\ &= 2\mathrm{Re}\left[\phi_j^*(t)\frac{d}{dt}\varphi_j(t)\right] = 2\mathrm{Im}\left[\phi_j^*(t)i\frac{d}{dt}\varphi_j(t)\right],\end{aligned}$$

Hence

$$\frac{d}{dt}P_j(t) = \frac{2}{\hbar}\mathrm{Im}\left[\sum_{l\neq j}\phi_j^*(t)H_{j,l}\phi_l(t)\right] \tag{A.2}$$

where the term $l = j$ is excluded from the sum, because $\phi_j^*(t)H_{j,j}\phi_j(t)$ is real.

Hence

$$\frac{d}{dt}P_j(t) = \sum_{l \neq j} J_{l \to j}, \tag{A.3}$$

where

$$J_{l \to j} = \frac{2}{\hbar}\text{Im}\left[\phi_j^*(t)H_{j,l}\phi_l(t)\right] = -\frac{2}{\hbar}\text{Im}\left[\phi_l^*(t)H_{l,j}\phi_j(t)\right] \tag{A.4}$$

If $\phi_l(t)$ is an eigenstate of energy E, with a time dependence $\phi_l(t) = \phi_l e^{-iEt/\hbar}$, then the bond currents are time independent and of the form

$$J_{l \to j} = \frac{2}{\hbar}\text{Im}\left[\phi_j^* H_{j,l}\phi_l\right] = -\frac{2}{\hbar}\text{Im}\left[\phi_l^* H_{l,j}\phi_j\right] \tag{A.5}$$

Appendix B Unitarity of the multi-channel *S* matrix

Equation (A.5) will now be used to prove unitarity of the multi-channel scattering matrix for a scatterer connected to leads of the type shown in figure 6.5. To simplify the discussion, consider the case where H_1 and H_1' are proportional to unit matrices.

From equation (6.61), the amplitude of a wave function on site j of cell J of the left lead is

$$\phi_{j,J} = \langle j|\psi_J\rangle = \sum_{n=1}^{N}\left(A_n\frac{e^{ik_nJ}}{\sqrt{v_n}} + B_n\frac{e^{-ik_nJ}}{\sqrt{v_n}}\right)\langle j|\chi_n\rangle \tag{B.1}$$

Hence the current from site l of cell J to site j of cell $J + 1$ is

$$J_{l,J \to j,J+1} = -\frac{2}{\hbar}\text{Im}\left[\phi_{l,J}^* H_{l,j}\phi_{j,J+1}\right]$$

Therefore the sum of the currents from all sites in cell J to all sites in cell $J + 1$ is

$$\begin{aligned} J_{J \to J+1} &= -\frac{2}{\hbar}\sum_{l,j=1}^{N}\text{Im}\left[\phi_{l,J}^* H_{l,j}\phi_{j,J+1}\right] \\ &= -\frac{2}{\hbar}\text{Im}\sum_{n,m=1}^{N}\left(A_n^*\frac{e^{-ik_n^*J}}{\sqrt{v_n}} + B_n^*\frac{e^{ik_n^*J}}{\sqrt{v_n}}\right) \\ &\times\left(A_m\frac{e^{ik_m(J+1)}}{\sqrt{v_m}} + B_m\frac{e^{-ik_m(J+1)}}{\sqrt{v_m}}\right)\sum_{l,j=1}^{N}\langle\chi_n|l\rangle H_{l,j}\langle j|\chi_m\rangle \end{aligned}$$

Since H_1 is proportional to the unit matrix, write $H_1 = -\gamma I$, where I is the $N \times N$ unit matrix. That is $H_{l,j} = -\gamma\delta_{lj}$, where δ_{lj} is a Kronecker delta.

Hence $\sum_{l,j=1}^{N} \langle \chi_n | l \rangle H_{l,j} \langle j | \chi_m \rangle = -\gamma \langle \chi_n | \chi_m \rangle = -\gamma \delta_{nm}$ and therefore

$$J_{J\to J+1} = \frac{2\gamma}{\hbar}\mathrm{Im}\sum_{n=1}^{N}\frac{1}{v_n}\left(A_n^* e^{-ik_n^* J} + B_n^* e^{ik_n^* J}\right)\left(A_n e^{ik_n(J+1)} + B_n e^{-ik_n(J+1)}\right)$$

Note that for open channels, k_n is real. For closed channels, either $k_n = i\kappa_n$ or $k_n = \pi + i\kappa_n$, where κ_n is a positive real number. For closed channels $A_n = 0$, therefore

$$\left(A_n^* e^{-ik_n^* J} + B_n^* e^{ik_n^* J}\right)\left(A_n e^{ik_n(J+1)} + B_n e^{-ik_n(J+1)}\right) = \pm\left(B_n^* e^{ik_n^* J}\right)\left(B_n e^{-ik_n(J+1)}\right)$$

Since this is real, closed channels do not carry a current, as expected. For open channels

$$\mathrm{Im}\left(A_n^* e^{-ik_n^* J} + B_n^* e^{ik_n^* J}\right)\left(A_n e^{ik_n(J+1)} + B_n e^{-ik_n(J+1)}\right) = (|A_n|^2 - |B_n|^2)\sin k_n$$

Hence, since $\hbar v_n = 2\gamma \sin k_n$,

$$J_{J\to J+1} = \sum_{n=\mathrm{open}} (|A_n|^2 - |B_n|^2)$$

where the sum is over open channels only. Similarly the inter-cell current in the right lead is

$$J_{J\to J+1} = \sum_{n'\mathrm{open}} (|C_{n'}|^2 - |D_{n'}|^2)$$

The incident current is $J_{\mathrm{in}} = \sum_{n=\mathrm{open}} |A_n|^2 + \sum_{n'\mathrm{open}} |D_{n'}|^2$, which in the notation of equation (6.66) becomes

$$J_{\mathrm{in}} = \langle \mathrm{in}^{\mathrm{open}} | \mathrm{in}^{\mathrm{open}} \rangle$$

Similarly, the outgoing current is $J_{\mathrm{out}} = \langle \mathrm{out}^{\mathrm{open}} | \mathrm{out}^{\mathrm{open}} \rangle = \langle \mathrm{in}^{\mathrm{open}} | S^\dagger S | \mathrm{in}^{\mathrm{open}} \rangle$.

Since $J_{\mathrm{in}} = J_{\mathrm{out}}$ for any choice of incident amplitudes, this proves that $S^\dagger S = I$, where I is the unit matrix.

From equation (6.68), this implies that

$$\begin{pmatrix} I & 0 \\ 0 & I \end{pmatrix} = S^\dagger S = \begin{pmatrix} r^\dagger & t^\dagger \\ t'^\dagger & r'^\dagger \end{pmatrix}\begin{pmatrix} r & t' \\ t & r' \end{pmatrix} = \begin{pmatrix} r^\dagger r + t^\dagger t & r^\dagger t' + t^\dagger r' \\ r'^\dagger t + t'^\dagger r & r'^\dagger r' + t'^\dagger t' \end{pmatrix} \tag{B.2}$$

and

$$\begin{pmatrix} I & 0 \\ 0 & I \end{pmatrix} = SS^\dagger = \begin{pmatrix} r & t' \\ t & r' \end{pmatrix}\begin{pmatrix} r^\dagger & t^\dagger \\ t'^\dagger & r'^\dagger \end{pmatrix} = \begin{pmatrix} rr^\dagger + t't'^\dagger & rt^\dagger + t'r'^\dagger \\ tr^\dagger + r't'^\dagger & tt^\dagger + r'r'^\dagger \end{pmatrix} \tag{B.3}$$

Hence

$$r^\dagger r + t^\dagger t = rr^\dagger + t't'^\dagger = I \tag{B.4}$$

and

$$r'^{\dagger}r' + t'^{\dagger}t' = tt^{\dagger} + r'r'^{\dagger} = I' \tag{B.5}$$

Note that r is a square matrix, whereas t and t' are rectangular matrices. Therefore $r^{\dagger}r$, $t^{\dagger}t$, $t't'^{\dagger}$ and I are all $M \times M$ square matrices, where M is the number of open channels in the left lead. Similarly $r'^{\dagger}r'$, $t'^{\dagger}t'$, $tt^{\dagger}$, $r'r'^{\dagger}$ and I' are $M' \times M'$ square matrices, where M' is the number of open channels in the right lead. The diagonal elements of the above equations yield

$$\sum_{n} |r_{nm}|^2 + \sum_{n'} |t_{n'm}|^2 = 1 \tag{B.6}$$

where the indices n and n' sum over the open channels of the left and right leads respectively.

Similarly

$$\sum_{n} |r_{mn}|^2 + \sum_{n'} |t'_{mn'}|^2 = 1 \tag{B.7}$$

$$\sum_{n'} |r'_{n'm'}|^2 + \sum_{n} |t'_{nm'}|^2 = 1 \tag{B.8}$$

$$\sum_{n'} |r'_{m'n'}|^2 + \sum_{n} |t_{m'n}|^2 = 1 \tag{B.9}$$

Noting that all amplitudes refer to open channels only, so that equation (6.67) becomes

$$\begin{pmatrix} |B\rangle \\ |C\rangle \end{pmatrix} = \begin{pmatrix} r & t' \\ t & r' \end{pmatrix} \begin{pmatrix} |A\rangle \\ |D\rangle \end{pmatrix} \tag{B.10}$$

Equation (B.6) arises when considering the case of a single incident channel m from the left, so that $A_m = 1$ and all other incoming amplitudes A_n and $D_{n'}$ are zero. Then $B_n = r_{nm}$ and $C_{n'} = t_{n'm}$ and since there is unit incident current, they must sum to unity. i.e.

$$\sum_{n} |r_{nm}|^2 + \sum_{n'} |t_{n'm}|^2 = 1$$

Similarly equation (B.8) shows the sum of all outgoing currents when a unit current is incident from the right along channel m'.

Reflection and transmission coefficients between channels are defined by $|r_{nm}|^2 = R_{nm}$, $|t_{n'm}|^2 = T_{n'm}$, $|r'_{m'n'}|^2 = R'_{n'm'}$ and $|t'_{nm'}|^2 = T'_{nm'}$. In these expressions, R_{nm} is the reflected current in channel n of the left lead, due to a unit current incident along channel n of the left lead and $T_{n'm}$ is the transmitted current in channel n' of the right lead, due to a unit current incident along channel n of the left lead etc. These satisfy

$$\sum_{n} R_{nm} + \sum_{n'} T_{n'm} = 1 \tag{B.11}$$

$$\sum_{n'} R'_{n'm'} + \sum_{n} T'_{nm'} = 1 \tag{B.12}$$

$$\sum_{n} R_{mn} + \sum_{n'} T'_{mn'} = 1 \tag{B.13}$$

$$\sum_{n'} R'_{m'n'} + \sum_{n} T_{m'n} = 1 \tag{B.14}$$

The reflection and transmission coefficients of the structure in figure 6.5 are defined by

$$R = \sum_{n,m} R_{nm} = Tr[rr^{\dagger}], \tag{B.15}$$

$$R' = \sum_{m'n'} R'_{n'm'} = Tr[r'r'^{\dagger}], \tag{B.16}$$

$$T = \sum_{mn'} T_{n'm} = Tr[tt^{\dagger}] \tag{B.17}$$

and

$$T' = \sum_{nm'} T'_{nm'} = Tr[t't'^{\dagger}], \tag{B.18}$$

From equations (B.11)–(B.14) these satisfy

$$R + T = M \tag{B.19}$$

$$R' + T' = M' \tag{B.20}$$

$$R + T' = M \tag{B.21}$$

and

$$R' + T = M' \tag{B.22}$$

Hence

$$T = T' \text{ and } R = R' \tag{B.23}$$

Note that from a conceptual viewpoint, the division of channels into those belonging to different leads is not needed and is a little clumsy. If all incoming channels are labelled i with amplitudes A_i and outgoing channels are labelled o, with amplitudes B_o then $|B\rangle = S|A\rangle$ or equivalently

$$B_o = \sum_{i} S_{oi} A_i \tag{B.24}$$

Equations (B.11) and (B.12) then reduce to a single equation

$$\sum_o |S_{oi}|^2 = 1 \tag{B.25}$$

Similarly equations (B.13) and (B.14) become

$$\sum_i |S_{oi}|^2 = 1 \tag{B.26}$$

Furthermore, unitarity of S yields

$$(SS^\dagger)_{oo'} = \sum_i S_{oi}(S^\dagger)_{io'} = \sum_i S_{oi}S_{o'i} = \delta_{oo'} \tag{B.27}$$

and

$$(S^\dagger S)_{ii'} = \sum_o (S^\dagger)_{io}S_{oi'} = \sum_o S_{oi}S_{oi'} = \delta_{ii'} \tag{B.28}$$

If one incoming amplitude $A_l = 1$ and all others are zero, then channel l is a source. In this case $B_o = S_{ol}$ and equation (B.25) shows that the outgoing currents sum to unity; i.e. $\sum_o |B_o|^2 = 1$. On the other hand if the incoming amplitudes are chosen to be $A_i = S^*_{io'}$, then equation (B.24) shows that $B_o = \sum_i S_{oi}S^*_{io'}$. Hence from equation (B.27) this choice of incoming amplitudes yields a unit outgoing current along channel o' and no current along the other channels. In other words, for this choice of incoming amplitudes, channel o' becomes a sink. The fact that a sink along channel o' can be produced from this choice of incoming amplitudes is a consequence of time-reversal symmetry. More precisely if expressions (6.61) and (6.62) are eigenstates of a Hamiltonian H, then taking the complex conjugate of the Schrödinger equation shows that the complex conjugates of these are solutions of the Schrödinger equation with a Hamiltonian H^*. Complex conjugation changes incoming channels into outgoing channels, and coverts a source into a sink.

Since

$$|B\rangle = S(H)|A\rangle, \tag{B.29}$$

where $S(H)$ is the scattering matrix of a Hamiltoian H, time reversal symmetry implies

$$|A^*\rangle = S(H^*)|B^*\rangle \tag{B.30}$$

Hence $|B^*\rangle = S^\dagger(H^*)|A^*\rangle = S^{*t}(H^*)|A^*\rangle$. Taking the complex conjugate of this equation yields

$$|B\rangle = S^t(H^*)|A\rangle$$

Comparing this with equation (B.29) yields

$$S(H) = S^t(H^*) \tag{B.31}$$

This means that the S-matrix of a system possessing a Hamiltonian H is the transpose of the S-matrix of a system possessing a Hamiltonian H^*. Consequently, if H is real, $S = S^t$; i.e. S is symmetric. Returning to the notation of equation (6.68), this means that

$$t' = t^t,\ r = r^t \text{ and } r' = r'^t \tag{B.32}$$

IOP Publishing

Chapter 7

Thermoelectricity in nanostructures and molecules

7.1 Introduction

The aim of this chapter is to address some of the issues associated with predicting and improving the thermoelectric performance of molecular-scale junctions and nanoscale devices. Low-grade thermal energy is abundant on Earth, much of it in the form of waste heat from transport, manufacturing and consumer electronics. For example, waste heat below 130 °C from factories accounts for about half of the energy demand in the United States. The need for technologies that can harvest this waste heat has led to a world-wide race to develop high-efficiency thermoelectric materials that can convert heat into electricity via the Seebeck effect. Potential waste heat from car exhausts, data farms and the human body could be used to generate electricity economically, provided materials with a high thermoelectric efficiency (characterised by a high dimensionless thermoelectric figure of merit ZT) could be identified. Conversely, Peltier cooling using such materials would have applications to on-chip cooling of semiconductor-based devices.

In section 6.12, a quantitative expression (6.93) for the flow of electricity and heat between two reservoirs was derived, which will be used in this chapter to explore molecular-scale and nanoscale thermoelectricity. Before discussing technical details, section 7.2 begins with a qualitative discussion, which shows how a temperature difference can lead to a voltage difference in an arbitrary junction. After discussing analytic formulae for thermoelectric coefficients, section 7.4 discusses the relationship between ZT and thermoelectric efficiency.

7.2 A qualitative description of thermoelectricity

Figure 7.1 shows qualitatively how a voltage difference is created when high-energy electrons can transmit more easily than low-energy electrons and a temperature difference is imposed between two reservoirs.

doi:10.1088/978-0-7503-3639-0ch7

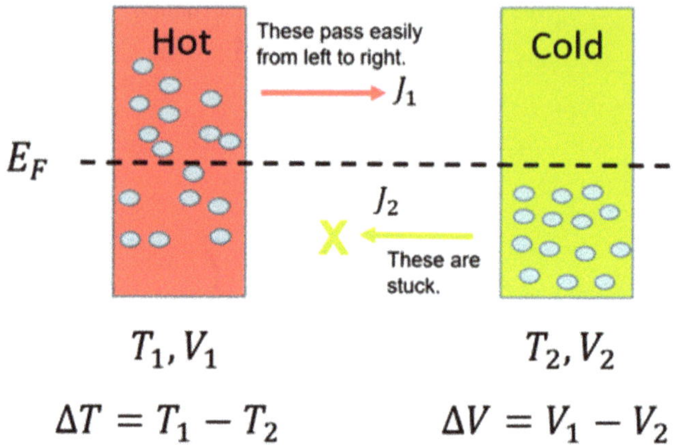

Figure 7.1. A cartoon revealing origin of the Seebeck effect. If high-energy electrons can transmit from left to right more easily than low-energy electrons can transmit from right to left, then negative charge will accumulate on the left reservoir and consequently V_2 is negative and V_1 is positive. That is $\Delta V > 0$. Hence $S = -\Delta V/\Delta T$ is negative.

The two reservoirs (i.e. electrodes) in figure 7.1 are labelled 1 and 2, with voltages and temperatures V_1, T_1 and V_2, T_2. Define $\Delta V = V_1 - V_2$ and $\Delta T = T_1 - T_2$. Then the reservoir chemical potentials are $\mu_1 = E_F + eV_1$, $\mu_2 = E_F - eV_2$, where $e = -|e|$ and their difference is $\Delta\mu = \mu_1 - \mu_2 = -|e|\Delta V$. Define the steady-state particle current J_1 to be the number of electrons per unit time leaving reservoir 1 and entering reservoir 2 and the steady-state particle current J_2 to be the number of electrons per unit time leaving reservoir 2 and entering reservoir 1. The net particle current leaving reservoir 1 and entering reservoir 2 is $J = J_1 - J_2$ and the electrical current is $I = -|e|J$. The Seebeck coefficient S is defined to be $S = -\left(\frac{\Delta V}{\Delta T}\right)_{I=0}$, where ΔV is the open circuit (i.e. zero-current) voltage difference produced by a temperature difference ΔT.

To obtain the sign of S, start from the equilibrium condition $\Delta V = 0, \Delta T = 0$ and consider what happens when the temperature T_1 of the left reservoir is increased, such that the population of higher-energy electrons in reservoir 1 is higher than that of reservoir 2. If these higher-energy electrons find it easier to pass from reservoir 1 to reservoir 2 than lower-energy electrons, then J_1 will increase. To achieve the steady-state condition $J = 0$, μ_2 must increase, so that J_2 increases. This demonstrates that if $T_1 > T_2$, $\mu_2 > \mu_1$. Hence if $\Delta T > 0$, $\Delta\mu < 0$, $\Delta V > 0$ and $S < 0$. This above situation occurs when the Fermi energy is located within the HOMO–LUMO gap of a single molecule and is located close to the LUMO, because in this case the electron transmission coefficient is an increasing function of electron energy. It also occurs when the Fermi energy is in the band gap of a semiconductor and lies closer to the conduction band.

Conversely, if the Fermi energy is located closer to the HOMO as shown for example in figure 1.3, or to the valence band, higher-energy electrons find it more difficult to pass from reservoir 1 to reservoir 2 than lower-energy electrons, then if T_1 is increased relative to T_2, J_1 will decrease and $S > 0$.

7.3 Linear response formulae for thermoelectric coefficients

As derived in the section 6.12 of chapter 6, in the linear response regime, the electric current I and heat flux Q through a junction is related to the voltage difference ΔV and temperature difference ΔT by

$$\begin{pmatrix} I \\ Q \end{pmatrix} = \frac{2}{h}\begin{pmatrix} e^2 L_0 & -\frac{|e|}{T}L_1 \\ -|e|L_1 & L_2 \end{pmatrix}\begin{pmatrix} \Delta V \\ \Delta T \end{pmatrix} \tag{7.1}$$

where

$$L_n = \int_{-\infty}^{\infty} (E - E_F)^n T(E)\left(-\frac{\partial f(E, T)}{\partial E}\right)dE \tag{7.2}$$

In equation (7.2), E_F is the average Fermi energy of the two reservoirs, $f(E) = \left[e^{\frac{E-E_F}{k_B T}} + 1\right]^{-1}$ and T is the average temperature of the two reservoirs.

The electrical conductance G is defined to be the ratio $G = I/(\Delta V)$, when $\Delta T = 0$, whereas the Seebeck coefficient S is defined to be the ratio $S = -\Delta V/\Delta T$, when $I = 0$. (Note the minus sign in the definition of S.)

Hence

$$G = \frac{2e^2}{h}L_0 \tag{7.3}$$

and

$$S = \frac{-1}{|e|T}\frac{L_1}{L_0} \tag{7.4}$$

Substituting these back into equation (7.1) yields

$$\begin{pmatrix} I \\ Q \end{pmatrix} = \begin{pmatrix} G & GS \\ GST & K_e \end{pmatrix}\begin{pmatrix} \Delta V \\ \Delta T \end{pmatrix} \tag{7.5}$$

where $K_e = 2L_2/h$. Alternatively, this equation can be rearranged into the following form

$$\begin{pmatrix} \Delta V \\ Q \end{pmatrix} = \begin{pmatrix} 1/G & -S \\ \Pi & \kappa_e \end{pmatrix}\begin{pmatrix} I \\ \Delta T \end{pmatrix} \tag{7.6}$$

where κ_e is the thermal conductance due to electrons, given by

$$\kappa_e = K_e - GS^2T \tag{7.7}$$

In equation (7.6), Π is the Peltier coefficient, which is define to be $\left(\frac{Q}{I}\right)_{\Delta T=0}$. This describes the heat flux created by an electrical current, which can be used to cool at the nanoscale. Setting $\Delta T = 0$ in equation (7.5) yields

$$\Pi = ST \tag{7.8}$$

Consequently,

$$\kappa_e = \frac{2}{h}\left(L_2 - \frac{(L_1)^2}{TL_0}\right) \tag{7.9}$$

and

$$\Pi = \frac{-1}{|e|}\frac{L_1}{L_0} \tag{7.10}$$

7.4 An expressions for thermoelectric efficiency

In terms of these quantities, the following discussion shows that the thermoelectric efficiency is determined by the dimensionless figure of merit

$$ZT = S^2GT/\kappa \tag{7.11}$$

where κ is the thermal conductance given by

$$\kappa = \kappa_e + \kappa_p \tag{7.12}$$

where κ_p is the thermal conductance due to phonons.

Note that if the thermal conductance due to phonons is included, then equation (7.6) becomes

$$\begin{pmatrix} \Delta V \\ Q \end{pmatrix} = \begin{pmatrix} 1/G & -S \\ \Pi & \kappa \end{pmatrix}\begin{pmatrix} I \\ \Delta T \end{pmatrix} \tag{7.13}$$

and equation (7.5) is replaced by

$$\begin{pmatrix} I \\ Q \end{pmatrix} = \begin{pmatrix} G & GS \\ GST & K \end{pmatrix}\begin{pmatrix} \Delta V \\ \Delta T \end{pmatrix} \tag{7.14}$$

where the parameter K replaces K_e in equation (7.7) and is given by

$$K = \kappa + GS^2T \tag{7.15}$$

To obtain an expression for thermoelectric efficiency, note that $\Delta\mu = |e|S\Delta T$. If $S < 0$ and $\Delta T > 0$ then $\Delta\mu < 0$, which means that to achieve equilibrium under open circuit conditions, electrons have been transported from a lower chemical potential in 1 to a higher chemical potential in 2 (figure 7.2). If in steady state a current is allowed to flow through an external circuit, the device works as an engine in which a temperature difference drives electrons against a chemical potential difference with efficiency

$$\eta = \frac{\text{work done per unit time against the chemical potential difference}}{\text{heat extracted from the hot reservoir per unit time}}$$

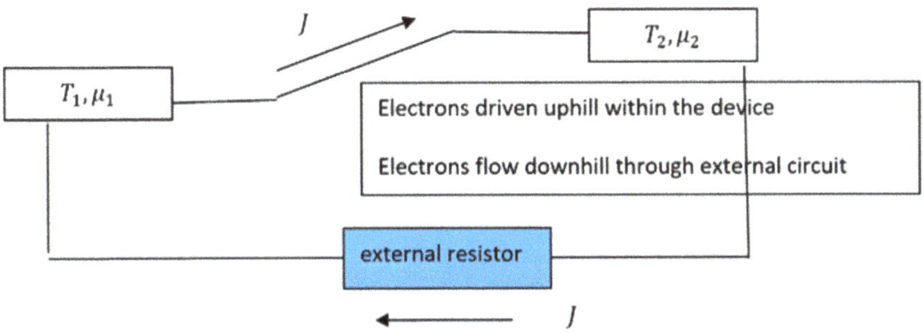

Figure 7.2. A sketch showing the energetics of a thermoelectric generator.

The work done per unit time against the chemical potential difference = number of electrons transferred per unit time (i.e. J) × $(\mu_2 - \mu_1) = J(\mu_2 - \mu_1) = -|e|J(V_2 - V_1) = I(V_2 - V_1) = -I\Delta V$. Hence

$$\eta = \frac{-I\Delta V}{\dot{Q}} = \frac{-\Delta V(G\Delta V + GS\Delta T)}{GST\Delta V + K\Delta T} \tag{7.16}$$

7.5 Relationship between thermoelectric efficiency and ZT

The relevant efficiency depends on whether the heat is free or expensive. If it is free (e.g. waste heat) and the goal is to maximise the power to an external load, then the efficiency at maximum power is relevant. On the other hand if the heat is a valuable resource, as in the case of a deep-space satellite, then the maximum efficiency is relevant. Expressions for both efficiencies are provided below.

7.5.1 Efficiency at maximum power

The output power is a maximum when the work done per unit time against the voltage (i.e. $W = -I\Delta V$) is a maximum. That is when $W = -\Delta V(G\Delta V + GS\Delta T)$ is a maximum. Differentiating this expression with respect to ΔV and setting the result to zero, shows that W is a maximum when $\Delta V = -\frac{S}{2}\Delta T$.

With this choice of ΔV, the current at maximum power is

$$I_{(\text{max power})} = G\Delta V + GS\Delta T = GS\Delta T/2 \tag{7.17}$$

and the a maximum value of W is

$$W_{\max} = GS^2\left(\frac{\Delta T}{4}\right)^2 \tag{7.18}$$

Similarly (from the bottom row of equation (7.5)) the heat flux at maximum power is

$$Q_{(\text{max power})} = STI + \kappa\Delta T = \left(\frac{GS^2T}{2} + \kappa\right)\Delta T$$

Hence

$$\eta_{(\text{max power})} = \frac{W_{\text{max}}}{\dot{Q}_{(\text{max power})}} = \frac{\Delta T}{2T}\frac{ZT}{ZT + 2} \tag{7.19}$$

7.5.2 Maximum efficiency

The above efficiency at maximum power is not the maximum efficiency. The maximum efficiency $\eta_{(\text{max})}$ is obtained by differentiating equation (7.16) with respect to ΔV and setting the result to zero. This yields (see below)

$$\eta_{(\text{max})} = \frac{\Delta T}{T}\frac{\sqrt{ZT + 1} - 1}{\sqrt{ZT + 1} + 1} \tag{7.20}$$

Both $\eta_{(\text{max power})}$ and $\eta_{(\text{max})}$ achieve their maximum values when ZT tends to infinity. The derivation of equation (7.20) is as follows.

To simplify the notation, write $x = \Delta V$, $a = S\Delta T$ and $b = \frac{K\Delta T}{GST}$

From equation (7.16), this yields

$$\eta = \frac{-1}{ST}\frac{x^2 + ax}{x + b} = \frac{-\Delta T}{T}\frac{x/a + 1}{b/x + 1} \tag{7.21}$$

The maximum of η is obtained by computing $\frac{d\eta}{dx}$ and equating the result to zero. Since $\frac{d\eta}{dx} = \frac{-1}{ST}\left[\frac{2x + a}{x + b} - \frac{x^2 + ax}{(x + b)^2}\right]$, this yields $(2x + a)(x + b) - x^2 - ax = 0$ Solving this quadratic equation yields

$$\frac{x}{b} = -1 + \sqrt{1 - a/b}$$

Since $K = \kappa + GS^2T$

$$\frac{b}{a} = \frac{K}{GS^2T} = 1 + \frac{1}{ZT} = \frac{1 + ZT}{ZT}$$

Hence $1 - \frac{a}{b} = 1 - \frac{ZT}{1 + ZT} = \frac{1}{1 + ZT}$ and therefore the maximum efficiency occurs when

$$\frac{x}{b} = -1 + \frac{1}{\sqrt{1 + ZT}} = \frac{1 - \sqrt{1 + ZT}}{\sqrt{1 + ZT}}$$

To substitute this back into equation (7.21), note that

$$\frac{x}{a} + 1 = \frac{x}{b}\frac{b}{a} + 1 = \frac{1 - \sqrt{1 + ZT}}{\sqrt{1 + ZT}}\frac{1 + ZT}{ZT} + 1 = \frac{\sqrt{1 + ZT} - 1}{ZT}$$

and

$$\frac{b}{x} + 1 = \frac{\sqrt{1 + ZT}}{1 - \sqrt{1 + ZT}} + 1 = \frac{1}{1 - \sqrt{1 + ZT}}$$

Hence from equation (7.21), the maximum efficiency is

$$\eta_{(\max)} = \frac{\Delta T}{T}\frac{(\sqrt{1 + ZT} - 1)^2}{ZT} = \frac{\Delta T}{T}\frac{\sqrt{1 + ZT} - 1}{\sqrt{1 + ZT} + 1}$$

7.6 Expressions for thermoelectric coefficients

The above discussion shows that a major challenge in designing an efficient thermoelectric material is to maximise ZT. To discuss this challenge, it is convenient to write

$$ZT = \frac{ZT_e}{1 + \kappa_p/\kappa_e)} \tag{7.22}$$

where ZT_e is the electronic thermoelectric figure $ZT_e = S^2GT/\kappa_e$ is given by

$$ZT_e = \frac{(L_1)^2}{L_0L_2 - (L_1)^2} \tag{7.23}$$

ZT_e is a property of electrons only and is determined by the electron transmission coefficient $T(E)$. To discuss the main features of ZT and other transport coefficients, it is convenient to consider the low-temperature limit. If the temperature is low enough, so that $T(E)$ varies approximately linearly with E on the scale of k_BT then $L_0 \approx T(E_F)$, $L_1 \approx (\mathrm{eT})^2\alpha\left(\frac{dT(E)}{dE}\right)_{E=E_F}$ and $L_2 \approx (\mathrm{eT})^2\alpha T(E_F)$, where α is the Lorentz number given by $\alpha = \left(\frac{1}{\mathrm{eT}}\right)^2 \int_{-\infty}^{\infty} (E - E_F)^2\left(-\frac{\partial f(E, T)}{\partial E}\right)dE = \left(\frac{k_B}{e}\right)^2\frac{\pi^2}{3} = 2.44 \times 10^{-8}$ $\mathrm{W\Omega K^{-2}}$. Hence in this limit, the electrical conductance, Seebeck coefficient and thermal conductance due to electrons take the form:

$$G(E_F) \approx \left(\frac{2e^2}{h}\right)T(E_F) \tag{7.24}$$

$$S(E_F) \approx -\alpha|e|T\left(\frac{d\,lnT(E)}{dE}\right)_{E=E_F}, \tag{7.25}$$

$$\kappa_e \approx \alpha TG \tag{7.26}$$

Equation (7.26) is known as the Wiedemann–Franz law and shows that when $T(E_F)$ varies slowly with E_F on the scale of k_BT, then κ_e is proportional to the electrical conductance.

As expected, if $T(E)$ is an increasing function of E, (so that high-energy electrons can transmit more easily than low-energy electrons, as in figure 7.1) then

$\left(\frac{d\, lnT(E)}{dE}\right)_{E=E_F} > 0$ and $S < 0$. Equation (7.25) demonstrates that S is enhanced by increasing the slope of $lnT(E)$ and hence quantum-interference-induced resonances or other features in $T(E)$ with steep slopes close to E_F are desirable. To estimate what constitutes a 'steep slope', note that the Wiedemann–Franz Law, (equation (7.26)) yields

$$ZT_e = \frac{S^2}{\alpha} \tag{7.27}$$

Therefore in the low-temperature limit, to achieve a value of $ZT_e > 1$, the Seebeck coefficient should satisfy $S^2 > \alpha$, i.e. $S > 150\ \mu V\ K^{-1}$. It is also useful to note that if energy E is measured in electron volts rather than SI units, equation (7.25) becomes

$$S(E_F) \approx -S_0\left(\frac{d\, lnT(E)}{dE}\right)_{E=E_F} \tag{7.28}$$

where $S_0 = \alpha T$, which at $T = 300$ K, takes the value $S_0 = 7.3\ \mu V\ K^{-1}$. This value sets the scale for the Seebeck coefficients encountered in single-molecule thermoelectric structures.

7.7 Proof that ZT_e is positive

Note that because of the minus sign in equation (7.23), it is of interest to prove that ZT_e is always positive. This can be achieved by taking a 'statistical' view of equation (7.2), which demonstrates why the denominator of equation (7.23) must be positive from a mathematical viewpoint. If the Fermi energy of the reservoirs is E_F, then it is convenient to introduce the non-normalised distribution $P(E) = -T(E)\frac{\partial f(E)}{\partial E}$ and the corresponding normalised distribution $\rho(E) = \frac{P(E)}{L_0}$. Then the mean of $(E - E_F)$ is $\langle E - E_F\rangle = \int dE\ \rho(E)(E - E_F) = L_1/L_0$ and the variance is $\sigma^2 = \langle(E - E_F - \langle E - E_F\rangle)^2\rangle = \langle(E - E_F)^2\rangle - \langle E - E_F\rangle^2 = \frac{L_2}{L_0} - \left(\frac{L_1}{L_0}\right)^2$. This yields for equation (7.4)

$$S = \frac{-1}{|e|T}\langle E - E_F\rangle \tag{7.29}$$

Furthermore equation (7.23) becomes

$$ZT_e = \frac{\langle E - E_F\rangle^2}{\sigma^2}, \tag{7.30}$$

which is clearly positive.

7.8 Strategies for maximising thermoelectric performance

The mean $\langle E - E_F\rangle$ and standard deviation σ capture essential features regarding the shape of $\rho(E)$ and $P(E)$. Equations (7.29) and (7.30) reveal that S and ZT_e depend only on these shape parameters and are independent of L_0. Only the electrical and electronic thermal conductances depend on L_0. This feature can be traced to the fact that $G = \frac{2e^2}{h}L_0$ and $k_e = \frac{2L_0\sigma^2}{hT}$ describe the magnitudes of currents and therefore depend on the normalisation L_0 of $P(E)$, whereas S and ZT_e involve only ratios. Clearly $\langle E - E_F \rangle$ will be non-zero only if $P(E) = -T(E)\frac{\partial f(E)}{\partial E}$ is an asymmetric function of $(E - E_F)$ and since $-\frac{\partial f(E)}{\partial E}$ is a symmetric function of $(E - E_F)$, $T(E)$ should be asymmetric with respect to E_F. Typically, the Seebeck coefficient is maximised when E_F lies within approximately k_BT of a region of steep slope in $T(E)$. In the presence of a very narrow transmission resonance, where σ is small, equation (7.30) shows that ZT_e could be very large. However in practice, only the full ZT is relevant from the point of view of thermoelectric efficiency. Clearly $ZT_e = S^2GT/\kappa_e$ is large when κ_e is small, which means that that the thermal conductance due to phonons becomes the dominant contribution to thermal conductance.

The thermal conductance due to phonons is obtained by solving for the phonon transmission coefficient $T_{ph}(\omega)$ from one electrode to the other via the molecule. Since phonons of frequency ω carry an energy $\hbar\omega$ and since the equilibrium distribution of phonons is given by the Bose–Einstein distribution $f_{BE}(\omega, T) = (e^{\hbar\omega/k_BT} - 1)^{-1}$, the Landauer-type equation for the thermal conductance due to phonons takes the form

$$\kappa_{ph}(T) = \frac{1}{2\pi}\int_0^\infty \hbar\omega T_{ph}(\omega)\frac{\partial f_{BE}(\omega, T)}{\partial T}d\omega \tag{7.31}$$

Since

$$\frac{\partial f_{BE}(\omega, T)}{\partial T} = [\hbar\omega/k_BT^2]e^{\hbar\omega/k_BT}(e^{\hbar\omega/k_BT} - 1)^{-2} \tag{7.32}$$

this yields

$$\kappa_{ph}(T) = \frac{k_B}{2\pi}\int_0^\infty T_{ph}(\omega)[\hbar\omega/k_BT]^2e^{\hbar\omega/k_BT}(e^{\hbar\omega/k_BT} - 1)^{-2}d\omega \tag{7.33}$$

Introducing the dimensionless variable $y = \hbar\omega/k_BT$, yields

$$\kappa_{ph}(T) = \frac{3g_0}{\pi^2}\int_0^\infty T_{ph}(y)y^2e^y(e^y - 1)^{-2}dy \tag{7.34}$$

where g_0 is the quantum of thermal conductance, given by $g_0 = \frac{(\frac{\pi}{6})(k_B)^2}{\hbar}T = (9.456 \times 10^{-13}\ \text{W K}^{-2})T$.

For a junction with N open phonon channels and no reflection $T_{ph}(y) = N$, and since $\int_0^\infty y^2e^y(e^y - 1)^{-2}dy = \pi^2/3$, one would obtain $\kappa_{ph}(T) = Ng_0$. This means that the phonon thermal conductance per open channel is g_0. As an example, at room temperature, g_0 is 284 pW K^{-1}, which is the maximum room-temperature phonon thermal conductance per channel and occurs when the upper cut off of the phonon channel is greater than approximately 5 k_BT. In contrast, for electrons, the quantum of thermal conductance (associated with a single electronic channel, whose transmission coefficient is unity) is $\frac{2(k_B)^2T}{h} = \frac{(k_B)^2T}{\pi\hbar} = 173$ pW K^{-1} at room temperature.

As noted in chapters 3 and 4 (see for example, section 4.11), the equations describing phonons are almost identical to those describing electrons and therefore the techniques used to compute $T(E)$ can also be used to compute $T_{ph}(\omega)$. Typically the thermal conductance due to phonons in a single-molecule junction is of the order of tens of picowatts per Kelvin and therefore to minimise the ratio κ_p/κ_e, it is desirable to increase κ_e beyond that value. In view of the Wiedemann–Franz law, this means that high-electrical conductance junctions are needed to achieve high ZT.

7.9 Summary

In this chapter, the key concepts controlling the thermoelectric performance of molecular-scale and nanoscale junctions have been addressed, starting with a qualitative discussion of the origin of the Seebeck effect. From equation (7.25), it is clear that high slopes in the transmission coefficient are desirable. Figure 7.3 shows why molecules exhibiting DQI are helpful in this respect. Both molecules have the same molecular core, but different connectivities to the electrodes. In both cases, the HOMO–LUMO gap is $2b$. The top connectivity yields DQI near the middle of the HOMO–LUMO gap, and a low mid-gap transmission coefficient whose logarithm is $\ln T(0) = -a'$, whereas the lower connectivity yields CQI and a higher mid-gap transmission coefficient $\ln T(0) = -a$. In the absence of information about the Fermi energy, a crude estimate of S can be obtained from the slopes of the triangles, which satisfy $\frac{a'}{b} \gg \frac{a}{b}$. In other words, from equation (7.25), one expects the Seebeck coefficient to be on the scale of $-\alpha|e|T$ multiplied by the minimum value of $\ln T(E)$ divided by half of the HOMO–LUMO gap. This shows that DQI tends to create higher values of S than CQI. On the other hand the mid-gap conductance is lower and

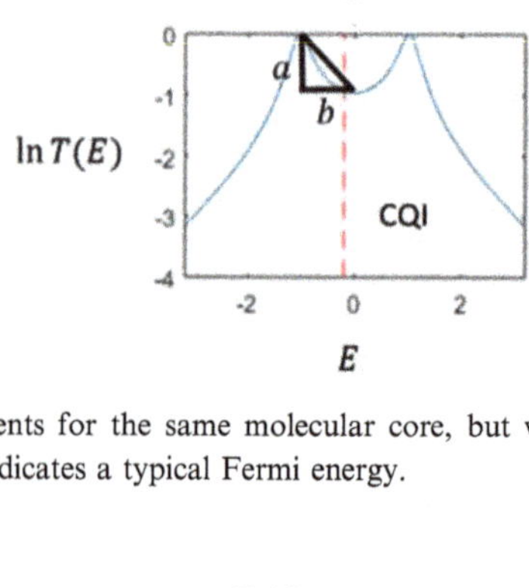

Figure 7.3. Transmission coefficients for the same molecular core, but with different connectivities to the electrodes. The red dashed line indicates a typical Fermi energy.

therefore careful tuning of the Fermi level towards either the HOMO or LUMO is needed to maintain a high electrical conductance.

In this chapter, the relationship between the thermoelectric figure of merit ZT and thermoelectric efficiency was also discussed. Since the denominator of ZT contains the thermal conductances of both electrons and phonons, and since the phonons have no useful thermoelectric function, it is desirable that the thermal conductance due to electrons significantly exceeds that of phonons, so that the latter can be neglected.

IOP Publishing

Quantum Transport in Nanostructures and Molecules
An introduction to molecular electronics
Colin John Lambert

Chapter 8

A very useful formula (VUF) for the transmission coefficient of an arbitrary scatterer connected to one-dimensional leads and a magic ratio theory for intra-molecular currents

8.1 Introduction

In this chapter, a very useful formula (VUF) for the transmission coefficient of a quantum system connected to one-dimensional leads is presented and the currents arising within the core of a quantum structure are discussed. This leads to an extension of magic ratio theory for intra-molecular currents, which expresses ratios of bond currents within the core of a molecule to magic numbers or equivalently core Green's functions, via equation (8.59).

8.2 The Schrödinger equation for a scatterer connected to two one-dimensional leads

Consider the structure shown below, in which a one-dimensional left lead has sites labelled $j = -1, -2, \ldots -\infty$, with sites energies ϵ_L and hopping elements $-\gamma_L$ and a one-dimensional right lead has sites labelled $j = N+1, N+2, \ldots\infty$, with sites energies ϵ_R and hopping elements $-\gamma_R$. The leads are connected to a scattering region containing N sites labelled $j = 1, 2, \ldots N$ by matrix elements $-\alpha$ and $-\beta$ respectively (figure 8.1).

Our aim is to solve the Schrödinger equation

$$H\,|\psi\rangle = E\,|\psi\rangle \tag{8.1}$$

or equivalently

$$\sum_{l=-\infty}^{\infty} H_{jl}\psi_l = E\psi_j \tag{8.2}$$

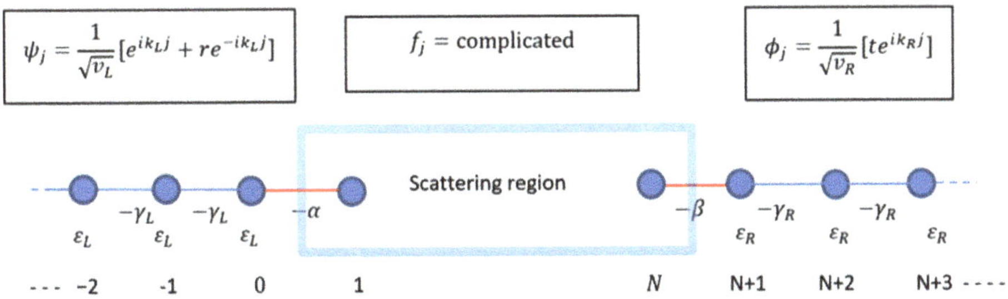

Figure 8.1. An arbitrary scattering region connected to one-dimensional leads.

For $j = -1, -2, \ldots -\infty$, the solution to the Schrödinger equation will be denoted ψ_j and for $j = N + 1, N + 2, \ldots\infty$, it will be denoted ϕ_j. Within the scattering region, where $j = 1, 2, \ldots N$, the solution will be denoted f_j. With this notation, the Schrödinger equation for this structure takes the form

$$\varepsilon_L\psi_j - \gamma_L\psi_{j-1} - \gamma_L\psi_{j+1} = E\psi_j \text{ for } j = -1, -2, \ldots -\infty \tag{8.3}$$

$$\varepsilon_L\psi_0 - \gamma_L\psi_{-1} - \alpha f_1 = E\psi_0 \text{ for } j = 0 \tag{8.4}$$

$$\sum_{l=1}^{N} H_{1l} f_l - \alpha\psi_0 = Ef_1 \text{ for } j = 1 \tag{8.5}$$

$$\sum_{l=1}^{N} H_{jl} f_l = Ef_j \text{ for } j = 2, \ldots N - 1 \tag{8.6}$$

$$\sum_{l=1}^{N} H_{Nl} f_l - \beta\phi_{N+1} = Ef_N \text{ for } j = N \tag{8.7}$$

$$\varepsilon_L\phi_{N+1} - \beta f_N - \gamma_R\phi_{N+2} = E\phi_{N+1} \text{ for } j = N + 1 \tag{8.8}$$

$$\varepsilon_L\phi_j - \gamma_R\phi_{j-1} - \gamma_R\phi_{j+1} = E\phi_j \text{ for } j = N + 2, N + 3, \ldots\infty \tag{8.9}$$

In equation (8.6) H_{jl} is a matrix element of the Hamiltonian describing the scattering region.

To obtain the reflection and transmission amplitudes r and t, the aim is to compute the eigenstates of this system when ψ_j is of the form

$$\psi_j = \frac{1}{\sqrt{v_L}}[e^{ik_Lj} + re^{-ik_Lj}] \tag{8.10}$$

and ϕ_j is of the form

$$\phi_j = \frac{1}{\sqrt{v_R}}[te^{ik_Rj}] \tag{8.11}$$

The solution to this problem will also yield the amplitudes f_j of the wave function within the scattering region.

8.3 Note about the choice of wave vectors

Since $e^{ik_L j}$ is chosen to be a plane wave carrying a positive incident bond current, such that the number of electrons per unit time passing along any bond from site j to site $j+1$ is positive, v_L is required to be positive. This means that k_L must be chosen such that $\hbar v_L = 2\gamma_L \sin k_L$ is positive. In other words, k_L must be chosen such that $\sin k_L$ and γ_L have the same sign. Similarly, k_R must be chosen such that $\sin k_R$ and γ_R have the same sign.

8.4 Solution to the Schrödinger equation

Within the scattering region, where $j = 1, 2, \ldots N$, the solution f_j is the wave function created by the incoming wave from the left and may be rather complicated. To compute f_j and the reflection and transmission amplitudes r and t for an arbitrary scattering Hamiltonian matrix H_{jl}, substitute equation (8.10) into (8.3). This yields

$$E = \epsilon_L - 2\gamma_L \cos k_L \tag{8.12}$$

Substituting equation (8.10) into (8.4) yields, after taking advantage of continuity (see appendix B of chapter 4),

$$\gamma_L \psi_1 = \alpha f_1 \tag{8.13}$$

Similarly, substituting equation (8.9) into (8.6) and (8.7) yields

$$E = \epsilon_R - 2\gamma_R \cos k_R \tag{8.14}$$

and

$$\gamma_R \phi_N = \beta f_N \tag{8.15}$$

Equations (8.12) and (8.14) reveal that in general, the wave vectors in the left and right leads are not equal. Equations (8.13) and (8.15) relate the wave function on the inside of the scatterer to the wave function on the outside. To obtain a second relation between the wave function on the inside and the outside, note that equations (8.5)–(8.7) can be written

$$\begin{pmatrix} H_{11} & H_{12} & H_{1N} \\ H_{21} & H_{22} & H_{2N} \\ & & \\ H_{N1} & H_{N2} & H_{NN} \end{pmatrix} \begin{pmatrix} f_1 \\ \vdots \\ f_j \\ \vdots \\ f_N \end{pmatrix} + \begin{pmatrix} -\alpha\psi_0 \\ 0 \\ \vdots \\ 0 \\ -\beta\phi_{N+1} \end{pmatrix} = \mathrm{E} \begin{pmatrix} f_1 \\ \vdots \\ f_j \\ \vdots \\ f_N \end{pmatrix} \tag{8.16}$$

This can be written more compactly in the form,

$$(E - h)\,|f\rangle = |s\rangle \tag{8.17}$$

where

$$|f\rangle = \begin{pmatrix} f_1 \\ \vdots \\ f_j \\ \vdots \\ f_N \end{pmatrix}, \; |s\rangle = \begin{pmatrix} -\alpha\psi_0 \\ 0 \\ \vdots \\ 0 \\ -\beta\phi_{N+1} \end{pmatrix} \tag{8.18}$$

and the Hamiltonian of the isolated scatterer is

$$h = \begin{pmatrix} H_{11} & H_{12} & \cdots & H_{1N} \\ H_{21} & H_{22} & \cdots & H_{2N} \\ \vdots & & & \vdots \\ \vdots & & & \vdots \\ H_{N1} & H_{N2} & \cdots & H_{NN} \end{pmatrix} \tag{8.19}$$

Multiplying both sides of equation (8.17) by the inverse matrix $(E - h)^{-1}$ yields

$$|f\rangle = g\,|s\rangle \tag{8.20}$$

where

$$g = (E - h)^{-1} \tag{8.21}$$

Equation (8.20) is similar to equation (2.6) and relates the wave function inside the scatterer to an external source. The $N \times N$ matrix g is the Green's function of the isolated scatterer and has the form

$$g = \begin{pmatrix} g_{11} & g_{12} & \cdots & g_{1N} \\ g_{21} & g_{22} & \cdots & g_{2N} \\ \vdots & & & \vdots \\ \vdots & & & \vdots \\ g_{N1} & g_{N2} & \cdots & g_{NN} \end{pmatrix} \tag{8.22}$$

Consequently equation (8.20) has the form

$$\begin{pmatrix} f_1 \\ \vdots \\ f_j \\ \vdots \\ f_N \end{pmatrix} = \begin{pmatrix} g_{11} & g_{12} & \cdots & g_{1N} \\ g_{21} & g_{22} & \cdots & g_{2N} \\ \vdots & & & \vdots \\ \vdots & & & \vdots \\ g_{N1} & g_{N2} & \cdots & g_{NN} \end{pmatrix} \begin{pmatrix} -\alpha\psi_0 \\ 0 \\ \vdots \\ 0 \\ -\beta\phi_{N+1} \end{pmatrix} \tag{8.23}$$

The top and bottom rows of this equation yield the following equation for f_1 and f_N

$$\begin{pmatrix} f_1 \\ f_N \end{pmatrix} = \begin{pmatrix} g_{11} & g_{1N} \\ g_{N1} & g_{NN} \end{pmatrix} \begin{pmatrix} -\alpha\psi_0 \\ -\beta\phi_{N+1} \end{pmatrix} \tag{8.24}$$

or alternatively,

$$\begin{pmatrix} f_1 \\ f_N \end{pmatrix} = \tilde{g} \begin{pmatrix} -\alpha\psi_0 \\ -\beta\phi_{N+1} \end{pmatrix} \tag{8.25}$$

where $\tilde{g}$ is the 2×2 sub-matrix of g given by

$$\tilde{g} = \begin{pmatrix} g_{11} & g_{1N} \\ g_{N1} & g_{NN} \end{pmatrix} \tag{8.26}$$

This yields

$$[\tilde{g}]^{-1} \begin{pmatrix} f_1 \\ f_N \end{pmatrix} = \begin{pmatrix} -\alpha\psi_0 \\ -\beta\phi_{N+1} \end{pmatrix} \tag{8.27}$$

where the inverse of the 2×2 sub-matrix of the Green's functions is

$$[\tilde{g}]^{-1} = \begin{pmatrix} g_{11} & g_{1N} \\ g_{N1} & g_{NN} \end{pmatrix}^{-1} = \frac{1}{d} \begin{pmatrix} g_{NN} & -g_{1N} \\ -g_{N1} & g_{11} \end{pmatrix} \tag{8.28}$$

and

$$d = \det\tilde{g} = g_{11}g_{NN} - g_{1N}g_{N1} \tag{8.29}$$

Equation (8.27) relates f_1 and f_N to ψ_0 and ϕ_{N+1}. In contrast, equations (8.13) and (8.15) relate f_1 and f_N to ψ_1 and ϕ_N and take the form

$$\psi_1 = \frac{\alpha}{\gamma_L} f_1 \text{ and } \phi_N = \frac{\beta}{\gamma_R} f_N \tag{8.30}$$

To solve these for $\begin{pmatrix} f_1 \\ f_N \end{pmatrix}$ and $\begin{pmatrix} \psi_0 \\ \phi_{N+1} \end{pmatrix}$, one needs to relate ψ_1 and ϕ_N to ψ_0 and ϕ_{N+1}, by noting that equations (8.10) and (8.11) yield

$$\psi_1 = \frac{1}{\sqrt{v_L}}[e^{ik_L} + re^{-ik_L}] = \frac{1}{\sqrt{v_L}}[e^{ik_L} - e^{-ik_L} + (1 + r)e^{-ik_L}] = \frac{1}{\sqrt{v_L}}[2i \sin k_L] + \psi_0 e^{-ik_L}$$

and

$$\phi_N = \phi_{N+1} e^{-ik_R}$$

Hence

$$\begin{pmatrix} -\alpha\psi_0 \\ -\beta\phi_{N+1} \end{pmatrix} = \begin{pmatrix} -\alpha e^{ik_L}\psi_1 \\ -\beta e^{ik_R}\phi_N \end{pmatrix} + \begin{pmatrix} \frac{\alpha e^{ik_L}}{\sqrt{v_L}}[2i \sin k_L] \\ 0 \end{pmatrix} \tag{8.31}$$

Utilising equation (8.29) yields

$$\begin{pmatrix} -\alpha e^{ik_L}\psi_1 \\ -\beta e^{ik_R}\phi_N \end{pmatrix} = \begin{pmatrix} \dfrac{-\alpha^2 e^{ik_L}}{\gamma_L} f_1 \\ \dfrac{-\beta^2 e^{ik_R}}{\gamma_R} f_N \end{pmatrix}$$

Hence equation (8.31) becomes

$$\begin{pmatrix} -\alpha\psi_0 \\ -\beta\phi_{N+1} \end{pmatrix} = \begin{pmatrix} \Sigma_L & 0 \\ 0 & \Sigma_R \end{pmatrix}\begin{pmatrix} f_1 \\ f_N \end{pmatrix} + \begin{pmatrix} \dfrac{\alpha e^{ik_L}}{\sqrt{v_L}}[2i \sin k_L] \\ 0 \end{pmatrix} \tag{8.32}$$

where Σ_L and Σ_R are self energies, given by

$$\Sigma_L = \frac{-\alpha^2 e^{ik_L}}{\gamma_L} \quad \text{and} \quad \Sigma_R = \frac{-\beta^2 e^{ik_R}}{\gamma_R} \tag{8.33}$$

Substituting equation (8.32) into equation (8.26) yields

$$[\tilde{g}]^{-1}\begin{pmatrix} f_1 \\ f_N \end{pmatrix} = \Sigma\begin{pmatrix} f_1 \\ f_N \end{pmatrix} + \begin{pmatrix} \dfrac{\alpha e^{ik_L}}{\sqrt{v_L}}[2i \sin k_L] \\ 0 \end{pmatrix} \tag{8.34}$$

where

$$\Sigma = \begin{pmatrix} \Sigma_L & 0 \\ 0 & \Sigma_R \end{pmatrix} \tag{8.35}$$

Equation (8.34) yields

$$([\tilde{g}]^{-1} - \Sigma)\begin{pmatrix} f_1 \\ f_N \end{pmatrix} = \begin{pmatrix} \dfrac{\alpha e^{ik_L}}{\sqrt{v_L}}[2i \sin k_L] \\ 0 \end{pmatrix} \tag{8.36}$$

Hence

$$\begin{pmatrix} f_1 \\ f_N \end{pmatrix} = G\begin{pmatrix} \dfrac{\alpha e^{ik_L}}{\sqrt{v_L}}[2i \sin k_L] \\ 0 \end{pmatrix} \tag{8.37}$$

where

$$G = ([\tilde{g}]^{-1} - \Sigma)^{-1} \tag{8.38}$$

Since G has the form

$$G = \begin{pmatrix} G_{11} & G_{1N} \\ G_{N1} & G_{NN} \end{pmatrix} \tag{8.39}$$

Equation (8.36) yields

$$f_N = G_{N1}\frac{\alpha e^{ik_L}}{\sqrt{v_L}}[2i \sin k_L] \tag{8.40}$$

Hence from equation (8.30),

$$\phi_N = \frac{\beta}{\gamma_R} f_N = G_{N1}\frac{\alpha\beta e^{ik_L}}{\gamma_R\sqrt{v_L}}[2i \sin k_L]$$

8.5 Expression for the transmission amplitude in terms of the full Green's function *G*

Since $\phi_N = \frac{1}{\sqrt{v_R}}[te^{ik_R N}]$, one obtains

$$t = G_{N1}\frac{\alpha\beta e^{ik_L}}{\gamma_R\sqrt{v_L}}[2i \sin k_L]\sqrt{v_R}e^{-ik_R N}$$

Since the group velocities are given by $\hbar v_R = 2\gamma_R \sin k_R$ and $\hbar v_L = 2\gamma_L \sin k_L$, this yields

$$\boxed{t = iG_{N1}\,\alpha\beta e^{ik_L}\sqrt{\frac{2 \sin k_L}{\gamma_L}}\sqrt{\frac{2 \sin k_R}{\gamma_R}}e^{-ik_R N}} \tag{8.41}$$

and finally, one obtains the following very useful formula (VUF):

$$T = |t|^2 = 4\left[\frac{\alpha^2 \sin k_L}{\gamma_L}\right]\left[\frac{\beta^2 \sin k_R}{\gamma_R}\right]|G_{N1}|^2 \tag{8.42a}$$

or equivalently

$$T = |t|^2 = 4\Gamma_L G_{N1}\Gamma_R G^*_{N1} \tag{8.42b}$$

where $\Gamma_L = -\text{Im}\Sigma_L = \frac{\alpha^2 \sin k_L}{\gamma_L}$ and $\Gamma_R = -\text{Im}\Sigma_R = \frac{\beta^2 \sin k_R}{\gamma_R}$.

An early version of this formula appears in an analysis of metal–insulator–metal junctions by Caroli *et al* [1].

8.6 Expression for the transmission amplitude in terms of the Green's function of the isolated scatterer

Note that equations (8.28), (8.35) and (8.38) combine to yield,

$$G = \left[\frac{1}{d}\begin{pmatrix} g_{NN} & -g_{1N} \\ -g_{N1} & g_{11} \end{pmatrix} - \begin{pmatrix} \Sigma_L & 0 \\ 0 & \Sigma_R \end{pmatrix}\right]^{-1}$$

That is

$$G = \left[\frac{1}{d}\right]^{-1}\left[\begin{pmatrix} g_{NN} & -g_{1N} \\ -g_{N1} & g_{11} \end{pmatrix} - \begin{pmatrix} \mathrm{d}\Sigma_L & 0 \\ 0 & \mathrm{d}\Sigma_R \end{pmatrix}\right]^{-1}$$

$$G = d\begin{pmatrix} g_{NN} - \mathrm{d}\Sigma_L & -g_{1N} \\ -g_{N1} & g_{11} - \mathrm{d}\Sigma_R \end{pmatrix}^{-1}$$

That is

$$G = \frac{d}{D}\begin{pmatrix} g_{11} - \mathrm{d}\Sigma_R & g_{1N} \\ g_{N1} & g_{NN} - \mathrm{d}\Sigma_L \end{pmatrix} \tag{8.43}$$

where $D = (g_{11} - \mathrm{d}\Sigma_R)(g_{NN} - \mathrm{d}\Sigma_L) - g_{N1}g_{1N}$. Hence $\frac{d}{D} = \frac{1}{\Delta}$, where

$$\Delta = 1 - g_{11}\Sigma_L - g_{NN}\Sigma_R + \Sigma_L\Sigma_R(g_{11}g_{22} - g_{12}g_{21}) \tag{8.44}$$

From this expression, equation (8.41) becomes

$$\boxed{t = \frac{ig_{N1}}{\Delta}\alpha\beta e^{ik_L}\sqrt{\frac{2\sin k_L}{\gamma_L}}\sqrt{\frac{2\sin k_R}{\gamma_R}}e^{-ik_R N}} \tag{8.45}$$

and (VUF) equation (8.42) becomes

$$\boxed{T = |t|^2 = 4\left[\frac{\alpha^2 \sin k_L}{\gamma_L}\right]\left[\frac{\beta^2 \sin k_R}{\gamma_R}\right]\left|\frac{g_{N1}}{\Delta}\right|^2} \tag{8.46}$$

8.7 Expression for the reflection amplitude

Since $\psi_1 = \frac{1}{\sqrt{v_L}}[e^{ik_L} + re^{-ik_L}]$, from equation (8.30) one obtains

$$e^{ik_L} + re^{-ik_L} = \frac{\alpha\sqrt{v_L}}{\gamma_L}f_1$$

From equation (8.37), this becomes

$$e^{ik_L} + re^{-ik_L} = G_{11}\frac{\alpha^2 e^{ik_L}}{\gamma_L}[2i\sin K_L]$$

$$\boxed{r = e^{i2k_L}\left[G_{11}\frac{\alpha^2}{\gamma_L}[2i\sin k_L] - 1\right]} \tag{8.47}$$

where, from equation (8.43),

$$G_{11} = (g_{11} - \mathrm{d}\Sigma_R)/\Delta \tag{8.48}$$

Note that if $\alpha = 0$, $r = -e^{i2k_L}$ and therefore $\psi_1 = 0$ as expected, since the wave function of a semi-infinite one-dimensional lead must vanish at the first missing site. Also note that the labelling of sites within the scatterer is arbitrary, so that site '1' could be any chosen site attached to the left lead and site 'N' could be chosen to be any site connected to the right lead.

8.8 Expression for the wave function inside the scatterer

From equation (8.23), the wave function inside the scatterer is given by

$$f_j = g_{j1}(-\alpha\psi_0) + g_{jN}(-\beta\phi_{N+1})$$

That is

$$f_j = -\alpha g_{j1}\left(\frac{1+r}{\sqrt{v_L}}\right) - \beta g_{jN}\left(\frac{te^{ik_R(N+1)}}{\sqrt{v_R}}\right) \tag{8.49}$$

8.9 Expressions for bond currents

From equation (5) in appendix B of chapter 6, the bond current between sites l and j within the scatterer is

$$J_{l\to j} = -\frac{2}{\hbar}\mathrm{Im}\left[f_l^* H_{l,j} f_j\right] \tag{8.50}$$

Similarly, the current from the left lead to site 1 of the scatterer is

$$J_{0\to 1} = \frac{2\alpha}{\hbar}\mathrm{Im}\left[\psi_0^* f_1\right] = \frac{2\gamma_L}{\hbar}\mathrm{Im}[\psi_0^*\psi_1] \tag{8.51}$$

where the last step follows from equation (8.13). Consequently $J_{0\to 1} = (1 - R) = T$.

Similarly the current from site N of the scatter to the right lead is

$$J_{N\to N+1} = \frac{2\beta}{\hbar}\mathrm{Im}\left[f_N^*\phi_{N+1}\right] = \frac{2\gamma_R}{\hbar}\mathrm{Im}\left[\phi_N^*\phi_{N+1}\right] = T \tag{8.52}$$

where equation (8.15) was employed.

As an example of these formulae, figure 8.2 shows various properties of the pyrene lattice, obtained using the above equations. All site energies ϵ_0 are chosen to be zero and all nearest neighbour couplings γ are chosen to be unity. The resulting

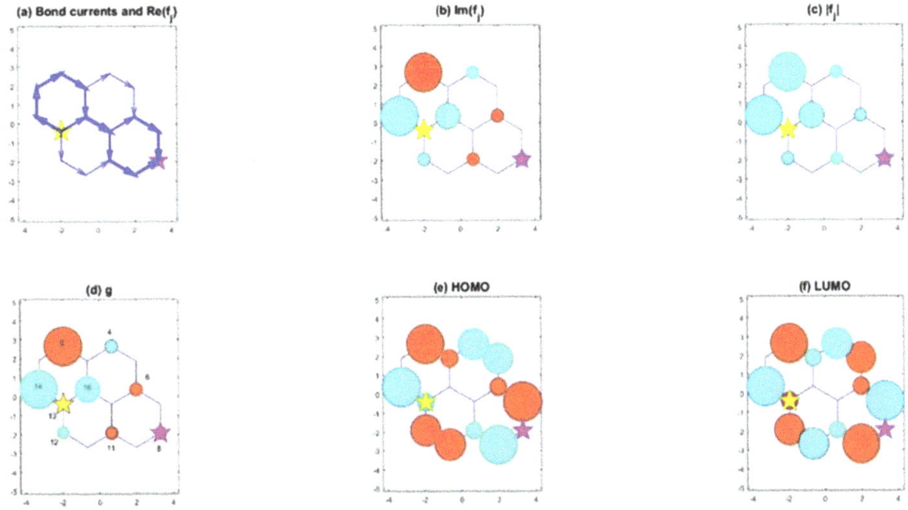

Figure 8.2. (a)–(d) Bond currents, wave function and core Green's function at $E = 0$ for a pyrene tight-binding lattice. (e) and (f) The frontier orbitals of a pyrene tight-binding lattice.

HOMO and LUMO energies of the isolated pyrene core are −0.445 and +0.445 respectively (in units of γ).

In figure 8.2, site 13 (indicated by a yellow star) is connected to the left lead and site N=8 (indicated by a pink star) is connected to the right lead. The couplings to the leads are $\alpha = 0.6$ and $\beta = 0.8$. The energy of the injected electron is chosen to be $E = 0$, which corresponds to the centre of the HOMO–LUMO gap.

Figure 8.2(a) shows the bond currents and also the real part of the wave function f_j within the scatterer. The magnitudes of the bond currents are proportional to the widths of the arrows. Figures 8.2(b) and (c) show the imaginary part and modulus of f_j respectively. In these plots, blue represents a positive amplitude and red a negative amplitude. The radius of each circle is proportional to the magnitude of the function on each site. These show that for the chosen injection energy $E = 0$, the wave function within the scatterer is almost imaginary.

Figure 8.2(d) shows the Greens function $g_{j,14}$ of the isolated core, which is of course real. Clearly this resembles the wave function f_j shown in figure 8.2(b), which is the core assertion of the magic ratio theory discussed in chapter 2. Indeed the radii of the circles in figure 8.2(d) are proportional to the numbers appearing in column 13 of the magic number table, shown in figure 8.3.

Finally figures 8.2(e) and (f) show the HOMO and LUMO. Clearly these do not resemble g_{j1} or f_j, because wave functions of electrons injected into the HOMO–LUMO gap do not coincide with any particular molecular orbital (MO).

Injection into the HOMO–LUMO gap is typical of a single-molecule junction at room temperature under ambient conditions, in which transport is almost always off resonance. On the other hand at low temperatures, if a MO energies are tuned using a gate electrode, then it is possible to inject electrons into a chosen MO, to realise on-resonance transport. To illustrate the difference between off-resonance and

$\bar{M} =$

M	1	3	5	7	9	11	13	15
2	3	3	-3	3	-3	3	-3	0
4	-1	1	5	-3	3	-1	1	-2
6	1	-1	1	3	-3	1	-1	2
8	-1	1	-1	3	3	-1	1	-2
10	1	-1	1	-3	3	1	-1	2
12	-1	1	-1	3	-3	5	1	-2
14	3	-3	3	-3	3	-3	3	0
16	-2	2	-2	0	0	-2	2	2

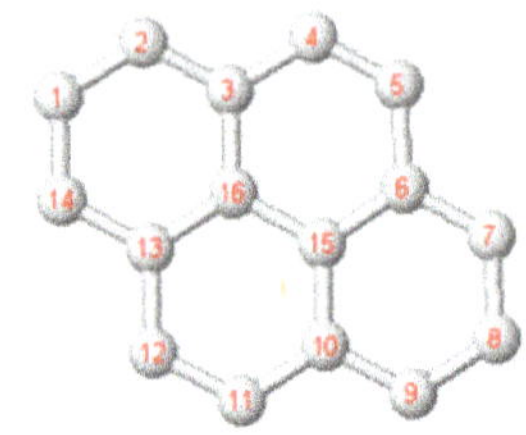

Figure 8.3. The M-table and a site numbering system for a tight-binding pyrene lattice.

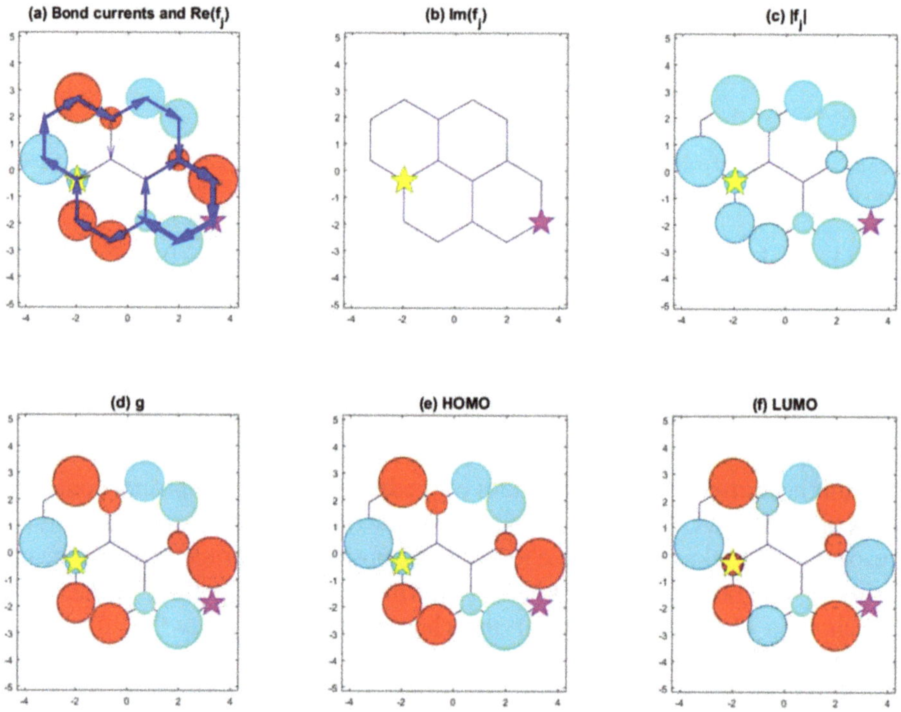

Figure 8.4. (a)—(d) Bond currents, wave function and core Green's function at E is equal to the energy of the HOMO for a pyrene tight-binding lattice. (e) and (f) The frontier orbitals of a pyrene tight-binding lattice.

on-resonance transport, figure 8.4 shows the same plots as figure 8.2, except the injection energy is now chosen to coincide with the HOMO energy. In this case, the Greens function $g_{j,14}$ of the isolated core, and the injected wave function f_j closely resemble the HOMO, as expected.

8.10 Magic ratio theory for intra-molecular currents

In this section, using the above expressions for bond currents, a magic ratio theory for intra-molecular currents is developed.

From equation (8.23), the wave function on site l within the scatterer is

$$f_l = g_{l1}s_1 + g_{lN}s_N \tag{8.53}$$

where $s_1 = -\alpha\psi_0$ and $s_N = -\beta\phi_{N+1}$

Since g and h are real, the bond current from site l to site m is

$$J_{l\to m} = -\frac{2}{\hbar}\mathrm{Im}\left[(g_{l1}s_1^* + g_{lN}s_N^*)h_{lm}(g_{m1}s_1 + g_{mN}s_N)\right]$$

That is

$$J_{l\to m} = I_{1,\,N}^{l\to m}\ \mathrm{Im}(s_1^* s_N) \tag{8.54}$$

where

$$I_{1,\,N}^{l\to m} = -\frac{2}{\hbar}(g_{l1}g_{mN} - g_{lN}g_{m1})h_{lm} \tag{8.55}$$

More generally,

$$J_{l\to m} = -\frac{2}{\hbar}\sum_{i,j=1}^{N}\mathrm{Im}\left[(g_{li}s_i^*)h_{lm}(g_{mj}s_j)\right] = -\frac{2}{\hbar}h_{lm}\sum_{i>j}(g_{li}g_{mj} - g_{lj}g_{mi})\ \mathrm{Im}(s_i^* s_j)$$

That is

$$J_{l\to m} = \sum_{i>j} I_{i,\,j}^{l\to m}\ \mathrm{Im}(s_i^* s_j) \tag{8.56}$$

where

$$I_{ij}^{l\to m} = -\frac{2}{\hbar}(g_{li}g_{mj} - g_{lj}g_{mi})h_{lm} \tag{8.57}$$

The current $I_{ij}^{l\to m}$ is a property of the core of the molecule only and therefore determined by the core Green's function.

Since the sources are

$$s_1 = -\alpha\psi_0 = -\frac{\alpha}{\sqrt{v_L}}(1 + r) = -\frac{\alpha e^{ik_L}2i\sin k_L}{\sqrt{v_L}}[G_{11}\frac{\alpha^2}{\gamma_L}e^{ik_L} - 1]$$

and

$$s_N = -\beta\phi_{N+1} = -\frac{\beta}{\sqrt{v_R}}te^{ik_R(N+1)} = -iG_{N1}\frac{\alpha\beta^2}{\gamma_R}e^{i(k_L+k_R)}\sqrt{\frac{2\sin k_L}{\gamma_L}}$$

one obtains, from equations (8.33)

$$\begin{pmatrix} s_1 \\ s_N \end{pmatrix} = \begin{pmatrix} \Sigma_L G_{11} + 1 \\ \Sigma_R G_{N1} \end{pmatrix}\frac{\alpha e^{ik_L}}{\sqrt{v_L}}[2i\sin k_L]$$

Therefore $s_1^* s_N = \frac{\hbar 2\alpha^2 \sin k_L}{\gamma_L}(\Sigma_L G_{11} + 1)^* \Sigma_R G_{N1}$, where $G_{N1} = \frac{g_{N1}}{\Delta}$. Also from equations (8.48), (8.43) and (8.29), $\Sigma_L G_{11} + 1 = \frac{1 - g_{NN}\Sigma_R}{\Delta}$. Hence

$$\mathrm{Im}[s_1^* s_N] = \frac{2\hbar\alpha^2 \sin k_L}{\gamma_L \, |\Delta|^2} g_{N1} \mathrm{Im}\, \Sigma_R = \frac{-2\hbar\alpha^2\beta^2 \sin k_L \sin k_R}{\gamma_L \, |\Delta|^2 \, \gamma_R} g_{N1}$$

That is

$$\mathrm{Im}[s_1^* s_N] = \frac{-\hbar T_{N1}}{2 g_{N1}} \tag{8.58}$$

where T_{N1} is the transmission coefficient corresponding to a connectivity N, 1 to the electrodes.

In summary,

$$I_{1,\,N}^{l \to m} = T_{N1} \frac{(g_{l1} g_{mN} - g_{lN} g_{m1})}{g_{N1}} h_{lm} \tag{8.59}$$

If the core Hamiltonian is proportional to a connectivity matrix, such that $h_{lm} = -\gamma$ for connected sites, then

$$\frac{J_{l \to m}}{J_{l' \to m'}} = \frac{I_{1,\,N}^{l \to m}}{I_{1,\,N}^{l' \to m'}} = \frac{(g_{l1} g_{mN} - g_{lN} g_{m1})}{g_{l'1} g_{m'N} - g_{l'N} g_{m'1}} \tag{8.60}$$

This demonstrates that from a knowledge of the core Green's function alone, ratios of bond currents can be predicted. Interestingly, this does not require that the core is weakly coupled to the electrodes; it is valid for any choice of α and β.

A similar argument shows that the current from the left lead to site 1 is equal to the current from site N to the right lead, which in turn is equal to T_{N1}, as expected.

8.11 Properties of the very useful formula for the transmission coefficient

Note that in figure 8.1, the labelling of sites within the scatterer is arbitrary, so that site '1' could be any chosen site labelled a, attached to the left lead and site 'N' could be chosen to be any site labelled b,connected to the right lead. With this change of labels, equation (8.46) becomes the very useful formula (VUF) shown below

$$\boxed{T_{ba}(E) = A \left| \frac{g_{ba}(E)}{1 + x_{ab}} \right|^2 \text{(The VUF)}} \tag{8.61}$$

In this expression,

$$A = 4\left[\frac{\alpha^2 \sin k_L}{\gamma_L}\right]\left[\frac{\beta^2 \sin k_R}{\gamma_R}\right] \tag{8.62}$$

and from equation (8.44)

$$x_{ab} = g_{aa}\Sigma_L - g_{bb}\Sigma_R + \Sigma_L\Sigma_R(g_{aa}g_{bb} - g_{ab}g_{ba}) \tag{8.63}$$

If transport takes place within the HOMO–LUMO gap and the molecular core is weakly coupled to (compound) electrodes, then the term x_{ab} can be neglected. On the other hand, if E coincides with the HOMO or LUMO, then x_{ab} becomes very large, even in the weak coupling limit, because g_{aa}, g_{bb}, g_{ab}, g_{ba} diverge (see section 10.2). This change in the significance of x_{ab} means that a description of transport near the HOMO or LUMO differs qualitatively from a description of transport within the HOMO–LUMO gap, even though they are described by the same VUF. Magic ratio theory is valid when x_{ab} is negligible. When x_{ab} becomes large, resonant transport takes place and as discussed below, the VUF reproduces the Breit–Wigner formula.

In the VUF (8.61), g_{ba} is the a, b th matrix element of the Green's function matrix

$$g(E) = (E - H)^{-1} \tag{8.64}$$

where H is the Hamiltonian of the molecule. The VUF looks a little complicated, because it describes many key phenomena relevant to molecular electronics. Nevertheless many of its properties are intuitive. Three immediately obvious properties of the VUF are:

1. If either of the couplings α or β are zero, then $T_{ab}(E) = 0$. This is reasonable, because for a current to flow, the molecule must be connected to both leads.
2. If $g_{ba}(E) = 0$, then $T_{ab}(E) = 0$. Again this is reasonable, because as discussed in chapter 1, g_{ba} is the wave function amplitude at atom b, created by a source at atom a. If a source at a creates no amplitude at b, then an electronic de Broglie wave cannot propagate from a to b and there can be no transmission.
3. If $\sin k = 0$, then $T_{ab}(E) = 0$. This is reasonable, because it would mean that the velocity of electrons in the leads is zero, in which case electrons of energy E would carry no current. The VUF is useful, because analytic formulae can be derived for the Green's functions of many model structures. For later reference, examples of such formulae, which are derived in later chapters are shown in table 8.1.

The remainder of this chapter is aimed at discussing the various aspects of VUF and demonstrating how it leads to the Breit–Wigner formula and to a naïve form of magic ratio theory.

8.12 The relationship between magic ratio theory and the VUF

If x_{ab} is neglected, then equation (8.61) reduces to $T_{ba}(E_F) = A|g_{ba}(E_F)|^2$. Since the corresponding electrical conductance is $\sigma_{ba} = \left(\frac{2e^2}{h}\right)T_{ba}(E_F)$, this immediately leads to the following expression for conductance ratios of structures such as figure 8.1

$$\frac{\sigma_{ba}}{\sigma_{b'a'}} = \frac{[g_{ba}(E_F)]^2}{[g_{b'a'}(E_F)]^2} \tag{8.65}$$

Table 8.1. Formulae for retarded Green's functions of model structures.

Structure	Equation	Retarded Green's function
Doubly infinite chain	(10.40)	$G_{jp}(E) = \frac{e^{ik(E)\lvert j-p\rvert}}{i\hbar v(E)}$
Surface Green's function of a semi-infinite chain	(10.46)	$G_{ll}(E) = -\frac{e^{ik(E)}}{\gamma}$
Aromatic ring	(10.59)	$G_{jp}(E) = A\cos[k(E)(\lvert j-p\rvert - N/2)]$, $A = \frac{1}{2\gamma\sin k(E)\sin k(E)N/2}$
Linear chain	(10.70)	$G_{jp}(E) = C\sin k[p-(N+1)\sin kj \quad \text{for} \quad j \leqslant p,$ and $G_{jp}(E) = C\sin kp\sin k[j-(N+1)] \quad \text{for} \quad j \geqslant p,$
Single impurity	(11.35)	$G_{ij} = g_{ij} + \frac{g_{il}\epsilon_l g_{lj}}{1-g_{ll}\epsilon_l},$ where g_{ij} is the Green's function of the unperturbed structure
Pendant site		Replace ϵ_l in equation (11.35) by $\epsilon_1 = \epsilon_p + \alpha^2/(E-\epsilon_B)$
Oligoyne chain	(12.35)	$G_{1N} = \frac{g_{1N}}{\mathrm{D}}, \quad \text{where} \quad g_{1N} = \frac{\sin k}{-\gamma\sin k(N+1)}$
Fluorene core	(13.41)	$G_{ab} = \hat{g}_{ab} + \frac{\alpha}{D}(\hat{g}_{a1} + \hat{g}_{a2})(\hat{g}_{1b} + \hat{g}_{2b})$
Multi-path molecule	(13.64)	$\hat{g}_{ab} \approx \sum_{i=3}^{M} A_i, \quad \text{where} \quad A_i = g_{a1'} v_i g_{ii'} v_i g_{2'b}$

This is analogous to equation (2.44), except that (2.44) is an expression for ratios of most-probable conductances, after making the assumption that the statistics of the contacts to electrodes is independent of connectivity. Furthermore the right-hand side of equation (2.44) involves the core Green's function and the electrodes could be compound electrodes, not simple linear chains as in the present discussion. These additional steps make magic ratio theory applicable to real molecular junctions.

8.13 The relationship between the Breit–Wigner formula, on-resonance transport and the VUF

As will be discussed in section 10.2, when the energy E of an electron passing through a scatterer approaches one of the energy levels of the isolated scatterer, the Green's function of the isolated scatterer simplifies. As an example, consider the case where E approaches the HOMO energy E_{H} of an isolated molecular core. In this case, the Green's function element $g_{ab}(E)$ simplifies to (see equation (10.10))

$$g_{ab}(E) \approx \frac{\psi_a^{(E_{\mathrm{H}})}\psi_b^{(E_{\mathrm{H}})}}{E - E_{\mathrm{H}}} \tag{8.66}$$

where $\psi_a^{(E_{\mathrm{H}})}$ and $\psi_b^{(E_{\mathrm{H}})}$ are the amplitudes of the HOMO on sites a and b respectively.

Clearly, when $E = E_{\mathrm{H}}$, this becomes infinite. This means that although the self energies Σ_L and Σ_R become small in the limit of very weak coupling, the products $g_{aa}\Sigma_L$ and $g_{bb}\Sigma_R$ cannot be neglected. Therefore the term x_{ab} cannot be neglected.

From equation (8.65) $g_{aa}g_{bb} - g_{ab}g_{ba} = 0$ and therefore equation (8.62) simplifies to

$$x_{ab} = -g_{aa}\Sigma_L - g_{bb}\Sigma_R = -\frac{\psi_a^{(E_{\mathrm{H}})}\psi_a^{(E_{\mathrm{H}})}\Sigma_L + \psi_b^{(E_{\mathrm{H}})}\psi_b^{(E_{\mathrm{H}})}\Sigma_R}{E - E_{\mathrm{H}}} \tag{8.67}$$

After substituting equations (8.33) into equation (8.66), one obtains

$$x_{ab} = \frac{-(\sigma_L + \sigma_R) + i(\Gamma_L + \Gamma_R)}{E - E_{\mathrm{H}}}$$

where

$$\sigma_L = \frac{-\psi_a^{(E_{\mathrm{H}})}\psi_a^{(E_{\mathrm{H}})}\alpha^2 \cos k_L}{\gamma_L} \quad \text{and} \quad \sigma_R = \frac{-\psi_b^{(E_{\mathrm{H}})}\psi_b^{(E_{\mathrm{H}})}\beta^2 \cos k_R}{\gamma_R} \tag{8.68}$$

$$\Gamma_L = \frac{\psi_a^{(E_{\mathrm{H}})}\psi_a^{(E_{\mathrm{H}})}\alpha^2 \sin k_L}{\gamma_L} \quad \text{and} \quad \Gamma_R = \frac{\psi_b^{(E_{\mathrm{H}})}\psi_b^{(E_{\mathrm{H}})}\beta^2 \sin k_R}{\gamma_R} \tag{8.69}$$

Writing $\sigma = \sigma_L + \sigma_R$ and $\Gamma = \Gamma_L + \Gamma_R$ yields for the VUF (8.61),

$$T_{ba}(E) = A\left|\frac{g_{ba}(E)}{1 + x_{ab}}\right|^2 = A\left|\frac{\psi_a^{(E_H)}\psi_b^{(E_H)}}{(E - E_H + \sigma) + i\Gamma}\right|^2$$
$$= A\frac{[\psi_a^{(E_H)}\psi_b^{(E_H)}]^2}{(E - E_H + \sigma)^2 + \Gamma^2} \tag{8.70}$$

From substituting equation (8.62) into equation yields the Breit–Wigner formula [2].

$$\boxed{T_{ba}(E) = \frac{4\Gamma_L\Gamma_R}{(E - E_H + \sigma)^2 + \Gamma^2}} \tag{8.71}$$

Note that although the core Green's function becomes infinite at $E = E_H$, this infinity cancels in the VUF, so that the resulting transmission coefficient is well behaved.

Figure 8.5 shows plots of equation (8.70), when $\Gamma_L = \Gamma_R$ (left panel) and when $\Gamma_L = 10.\ \Gamma_R$ (right panel). In both case, $\Gamma = \Gamma_L = \Gamma_R = 0.1$ and $E_H - \sigma = 0$.

When $E = E_H - \sigma$, the transmission coefficient reaches a maximum value T_{max} and transport is said to be 'on resonance'. Note that the energetic location of the resonance is not equal to the HOMO energy E_H. Instead it is shifted slightly by the real part of the self energy, σ. For the case $\Gamma_L = \Gamma_R$, the transmission coefficient reaches a maximum value of $T_{max} = 1$. For the case $\Gamma_L \neq \Gamma_R$, the maximum value of the transmission coefficient is less than unity and given by

$$T_{max} = \frac{4\Gamma_L\Gamma_R}{(\Gamma_L + \Gamma_R)^2} \approx \frac{4\Gamma_R}{\Gamma_L} \ll 1 \tag{8.72}$$

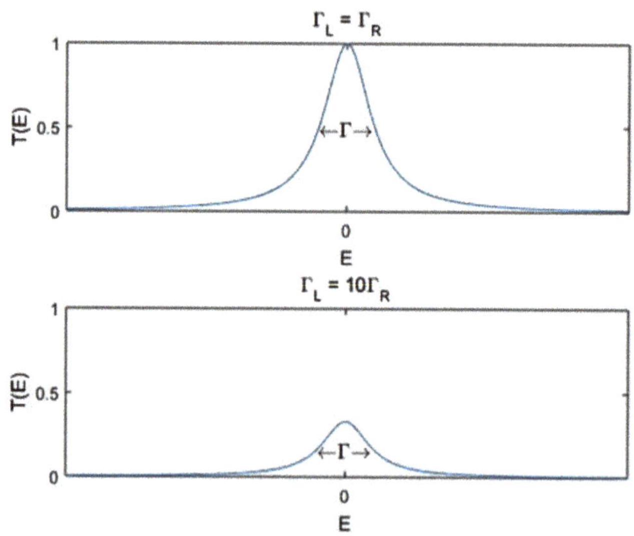

Figure 8.5. Plots of the Breit–Wigner formula, equation (2.32).

In other words, the on-resonance transmission coefficient of asymmetric junctions is lower than that of symmetric junctions. Finally, one notes that the width of the resonance is Γ, which is small in the weak coupling limit (where the coupling s to the electrodes α and β are small) and therefore narrow resonances arise in molecules which are weakly coupled to electrodes.

In the following chapters, all of the above formulae and their generalisations to finite-width electrodes and more realistic structures will be derived, and various viewpoints for interpreting and approximating transport properties will be discussed.

8.14 Summary

In this chapter, formulae for reflection and transmission coefficients of an arbitrary scatterer connected to one-dimensional leads were obtained, by matching the wave functions of leads to the wave function inside the scatterer. During the calculation, the Green's function of the scatterer emerged as a natural tool for solving the problem. By utilising the expression for the wave function inside the scatterer, a new magic ratio theory of intra-molecular currents was obtained, which shows that ratios of these current are a property of the isolated molecule alone, even in the limit that the scatterer is strongly coupled to the electrodes. A VUF for the transmission coefficient was also discussed, which shows how magic ratio theory and the Breit–Wigner formula are contained within the same very useful formula.

References

[1] Caroli C *et al* 1971 Direct calculation of the tunneling current *J. Phys. C: Solid State Phys.* **4** 916
[2] Breit G and Wigner E 1936 Capture of slow neutrons *Phys. Rev.* **49** 519

IOP Publishing

Quantum Transport in Nanostructures and Molecules
An introduction to molecular electronics
Colin John Lambert

Chapter 9

A quantum system connected to many scattering channels

9.1 Introduction

In this chapter, the analysis of chapter 8 is generalised to a scatterer connected to many one-dimensional leads. It serves as an introduction to Dyson's equation in the presence of many channels and shows how the multi-channel S-matrix can be obtained by wave-function matching. In chapter 11, all formulae will be re-derived using Green's functions, to demonstrate the equivalence between the two approaches and to obtain a formula for the transmission coefficient in the presence of finite-width multi-channel leads.

9.2 Solving the Schrödinger equation

To generalise the analysis of chapter 8, consider the scattering region shown below, consisting of N sites, connected to M one-dimensional leads. The n th lead has site energies ϵ_n and nearest neighbour couplings $-\gamma_n$. The wave vector in lead n is k_n and the group velocity is v_n, where

$$k_n = \text{acos}\,(\epsilon_n - E)/2\gamma_n \tag{9.1}$$

and

$$\hbar v_n = 2\gamma_n \sin k_n \tag{9.2}$$

The sites in each lead are labelled 0, 1, 2, 3, … starting from the end of the lead nearest to the scatterer, so that plane waves of the form $\frac{1}{\sqrt{v_n}}e^{ik_n j}$ carry a unit current in the outgoing direction.

To ensure that transmitted waves are outgoing or decaying, the wave vectors satisfy two conditions: if k_n is real, then the sign of k_n must chosen such that $\gamma_n \sin k_n$ is positive, whereas if k_n is complex, the imaginary part of k_n must be chosen to be

doi:10.1088/978-0-7503-3639-0ch9

positive. Our aim is to solve the Schrödinger equation for the case where lead n carries a unit incident current. The transmission amplitude from lead n to lead m is denoted t_{mn}.

With this notation, notice that the reflected wave in lead n is treated on the same footing as the transmitted waves. Furthermore, in contrast with chapter 8, the numbering system for the sites in lead n follows that of other leads; i.e. sites are labelled by integers j, starting from $j = 0$ nearest to the scatterer.

In what follows, the end site of each lead is allowed to interact with any site within the scattering region and the Hamiltonian matrix element between end site of lead l and site j of the scatterer is denoted V_{lj}, as indicated by figure 9.1. In what follows, if the sites of lead l are labelled $j_l = 0_l, 1_l, 2_l, \ldots$, so that the end site of lead l is given a unique label 0_l and a site in the scatterer is labelled j_s, then the Hamiltonian matrix element $H_{0_l j_s}$ will be denoted V_{lj}. Similarly $H_{j_s 0_l}$ will be written $H_{j_s 0_l} = W_{jl}$. Since the Hamiltonian is Hermitian, $W_{jl} = V_{lj}^*$

The wave function of site j of lead m will be denoted ψ_j^m and the wave function on site j of the scatterer will be denoted f_j. The $N \times N$ Hamiltonian of the isolated scatterer is denoted h. By analogy with equations (8.3)–(8.9), the Schrödinger equation for sites in lead m is

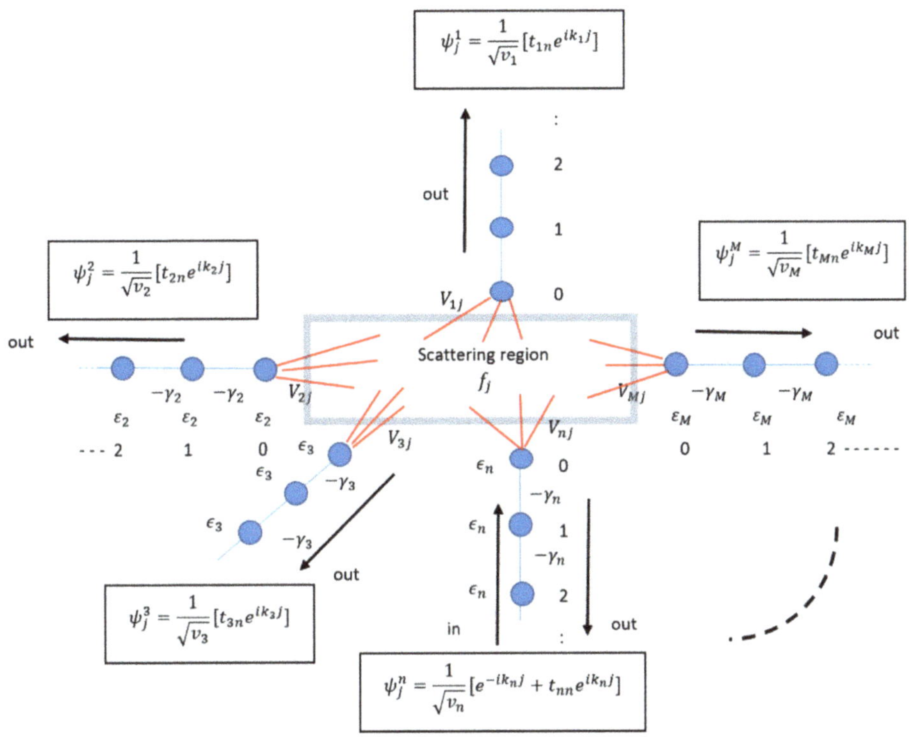

Figure 9.1. M leads connected to an arbitrary scatterer, with an incoming wave in lead n and outgoing waves of amplitude $t_{mn}/\sqrt{v_n}$ in all leads $m = 1, 2, \ldots M$.

$$\varepsilon_m\psi_j^m - \gamma_m\psi_{j-1}^m - \gamma_m\psi_{j+1}^m = E\psi_j^m \text{ for } j = 1, 2, \ldots.\infty \tag{9.3}$$

$$\varepsilon_m\psi_0^m - \gamma_m\psi_1^m + \sum_{j=1}^{N} V_{mj}f_j = E\psi_0^m \text{ for } j = 0 \tag{9.4}$$

Similarly for site i of the scatterer,

$$\sum_{j=1}^{N} h_{ij}f_j + \sum_{m=1}^{M} W_{im}\psi_0^m = Ef_i \text{ for } i = 1\ldots N, \tag{9.5}$$

where in the last equation, the indices i, j label sites within the scatterer. By continuity (cf equations (8.13) and (8.15)), equations (9.3) and (9.4) combine to yield

$$-\gamma_m\psi_{-1}^m = \sum_{l=1}^{N} V_{ml}f_l \tag{9.6}$$

Since $\psi_j^m = \frac{1}{\sqrt{v_n}}[e^{-ik_nj}\delta_{mn} + t_{mn}e^{ik_mj}]$, where δ_{mn} is a Kronecker delta (i. e. $\delta_{mn} = 1$ for $n = m$ and $\delta_{mn} = 0$ for $n \neq m$), one obtains

$$\psi_0^m = \frac{1}{\sqrt{v_n}}[\delta_{mn} + t_{mn}] \tag{9.7}$$

and

$$\psi_{-1}^m = \frac{1}{\sqrt{v_n}}[e^{ik_n}\delta_{mn} + t_{mn}e^{-ik_m}] \tag{9.8}$$

To solve these equations, it is useful to use the following definitions:

γ is a $M \times M$ diagonal matrix with elements $\gamma_{ml} = \gamma_l\delta_{ml}$

$\sin k$ is a $M \times M$ diagonal matrix with elements $(\sin k)_{ml} = \sin k_l\delta_{ml}$

e^{ik} is a $M \times M$ diagonal matrix with elements $(e^{ik})_{ml} = e^{ik_l}\delta_{ml}$

$\sqrt{v}$ is a $M \times M$ diagonal matrix with elements $(\sqrt{v})_{ml} = \sqrt{v_l}\delta_{ml}$

Similarly, $v^{-1/2} = 1/\sqrt{v}$ is a $M \times M$ diagonal matrix with elements $\sqrt{1/v_l}\delta_{ml}$

V is a $M \times N$ rectangular matrix (i.e. M rows and N columns) with elements V_{mj}.

W is a $N \times M$ rectangular matrix (i.e. N rows and M columns) with elements $W_{jm} = V_{mj}^*$.

For example if there are only two leads (i.e. $M = 2$),

$$\sin k = \begin{pmatrix} \sin k_1 & 0 \\ 0 & \sin k_2 \end{pmatrix},\ e^{ik} = \begin{pmatrix} e^{ik_1} & 0 \\ 0 & e^{ik_2} \end{pmatrix},\ \gamma = \begin{pmatrix} \gamma_1 & 0 \\ 0 & \gamma_2 \end{pmatrix} \text{ and } \sqrt{v} = \begin{pmatrix} \sqrt{v_1} & 0 \\ 0 & \sqrt{v_2} \end{pmatrix}.$$

$$1/\sqrt{v} = \begin{pmatrix} \sqrt{1/v_1} & 0 \\ 0 & \sqrt{1/v_2} \end{pmatrix}$$

Similarly, if there are M leads, and N sites in the scatterer,

$$V = \begin{pmatrix} V_{11} & V_{12} & \dots & V_{1N} \\ \vdots & \vdots & \dots & \vdots \\ V_{M1} & V_{M2} & \dots & V_{MN} \end{pmatrix} \text{ and } W = V^{\dagger} = \begin{pmatrix} W_{11} \dots & W_{1M} \\ W_{21} \dots & W_{2M} \\ \vdots \dots & \vdots \\ W_{N1} \dots & W_{MN} \end{pmatrix} = \begin{pmatrix} V_{11}^{*} \dots & V_{M1}^{*} \\ V_{12}^{*} \dots & V_{M2}^{*} \\ \vdots \dots & \vdots \\ V_{1N}^{*} \dots & V_{MN}^{*} \end{pmatrix}$$

The following definitions are also useful, where the symbol $|\rangle$ is used to denote M-component vectors and $|)$ is used to denote N-component vectors. In particular, $|l\rangle$ is a M-component, unit column vector, such that l th element of $|l\rangle$ is 1 and all other elements are zero, while $|j)$ is a N-component, unit column vector, such that the jth element of $|j)$ is 1 and all other elements are zero. That is

$$|l> = \begin{pmatrix} 0 \\ \vdots \\ 1 \\ \vdots \\ 0 \end{pmatrix} \leftarrow \text{row } l \quad \text{length } M$$

$$|j) = \begin{pmatrix} 0 \\ \vdots \\ 1 \\ \vdots \\ 0 \end{pmatrix} \leftarrow \text{row } j \quad \text{length } N$$

These unit vectors satisfy $\langle m|n\rangle = \delta_{nm}$, $(i|j) = \delta_{ij}$ and $V_{lj} = (l\,|V|\,j\rangle$. They can be used to define the following vectors

$$(V_l| = \langle l|\, V = \begin{pmatrix} V_{l1} & V_{l2} & \dots & V_{lN} \end{pmatrix} \tag{9.9}$$

$$|V_j\rangle = V\,|\,j) = \begin{pmatrix} V_{1j} \\ V_{2j} \\ \vdots \\ \vdots \\ V_{Mj} \end{pmatrix} \tag{9.10}$$

With this notation,

$$V = \sum_{l=1}^{M} |l\rangle(V_l) = \sum_{j=1}^{N} |V_j\rangle(j) = \sum_{j=1}^{N}\sum_{j=1}^{N} |l\rangle V_{lj}(j) \tag{9.11}$$

and

$$W = V^{\dagger} = \sum_{l=1}^{M} (V_l)\langle l| = \sum_{j=1}^{N} |j)\langle V_j| = \sum_{j=1}^{N}\sum_{j=1}^{N} (j) W_{jl}\langle l| = \sum_{j=1}^{N}\sum_{j=1}^{N} (j) V_{lj}^{*}\langle l| \tag{9.12}$$

Furthermore, $|f\rangle$ is defined to be a N component vector containing the wave function amplitudes within the scattering region. i.e.

$$|f\rangle = \begin{pmatrix} f_1 \\ \vdots \\ f_j \\ \vdots \\ f_N \end{pmatrix}$$

Similarly $|\psi_j\rangle$ is defined to be a M component vector containing the amplitudes of the wave functions on site j of each lead. E.g.

$$|\psi_0\rangle = \begin{pmatrix} \psi_0^1 \\ \vdots \\ \psi_0^m \\ \vdots \\ \psi_0^M \end{pmatrix} \text{ and } |\psi_{-1}\rangle = \begin{pmatrix} \psi_{-1}^1 \\ \vdots \\ \psi_{-1}^m \\ \vdots \\ \psi_{-1}^M \end{pmatrix}$$

Finally, $|t\rangle$ is defined to be a M component vector containing the transmission amplitudes. That is

$$|t\rangle = \begin{pmatrix} t_{1n} \\ \vdots \\ t_{ln} \\ \vdots \\ t_{Mn} \end{pmatrix}$$

With this notation, equations (9.7) and (9.8) become

$$|\psi_0\rangle = \frac{1}{\sqrt{v}}\,|t\rangle + \frac{1}{\sqrt{v}}\,|n\rangle \tag{9.13}$$

and

$$|\psi_{-1}\rangle = \frac{e^{-ik}}{\sqrt{v}}\,|t\rangle + \frac{e^{ik}}{\sqrt{v}}\,|n\rangle \tag{9.14}$$

Similarly equation (9.6) takes the form

$$-\gamma|\psi_{-1}\rangle = V\,|f\rangle \tag{9.15}$$

and equation (9.5) becomes $h\,|f\rangle + W\,|\psi_0\rangle = E\,|f\rangle$. i.e. $(E - h)|f\rangle = W\,|\,\psi_0\rangle$ or equivalently

$$|f\rangle = gW\,|\psi_0\rangle \tag{9.16}$$

where

$$g = (E - h)^{-1} \tag{9.17}$$

To solve these for $|t\rangle$, substitute equation (9.13) into (9.16) to obtain

$$|f\rangle = gW|\psi_0\rangle = gW\frac{1}{\sqrt{v}}[|t\rangle + |n\rangle] \tag{9.18}$$

while equation (9.15) becomes

$$V|f\rangle = -\gamma|\psi_{-1}\rangle = -\gamma\left[\frac{e^{-ik}}{\sqrt{v}}|t\rangle + \frac{e^{ik}}{\sqrt{v}}|n\rangle\right] \tag{9.19}$$

These can now be solved directly for $|t\rangle$, by noting that equation (9.19) yields

$$|t\rangle = -\frac{\sqrt{v}}{\gamma}e^{ik}V|f\rangle - e^{2ik}|n\rangle$$

Substituting this into equation (9.18) yields

$$|f\rangle = gW|\psi_0\rangle = gW\frac{1}{\sqrt{v}}[-\sqrt{v}\gamma^{-1}e^{ik}V|f\rangle - e^{2ik}|n\rangle + |n\rangle]$$

That is

$$(g^{-1} - \Sigma_0)|f\rangle = W\frac{1}{\sqrt{v}}[-e^{ik}2i\sin k\,|n\rangle]$$

where

$$\Sigma_0 = -W\gamma^{-1}e^{ik}V \tag{9.20}$$

Hence

$$|f\rangle = -GW\frac{e^{ik}}{\sqrt{v}}2i\sin k\,|n\rangle \tag{9.21}$$

where

$$G = (g^{-1} - \Sigma_0)^{-1} \tag{9.22}$$

Finally, from equation (9.20)

$$|t\rangle = \frac{\sqrt{v}}{\gamma}e^{ik}VGW\frac{e^{ik}}{\sqrt{v}}2i\sin k\,|n\rangle - e^{2ik}|n\rangle$$

That is

$$|t\rangle = i\frac{\sqrt{v}}{\gamma}e^{ik}VGWe^{ik}\frac{\sqrt{v}}{\gamma}|n\rangle - e^{2ik}|n\rangle \tag{9.23}$$

Hence

$$t_{mn} = \langle m|t\rangle = i\langle m|\frac{\sqrt{v}}{\gamma}e^{ik}VGWe^{ik}\frac{\sqrt{v}}{\gamma}|n\rangle - e^{2ik}\langle m|n\rangle \quad (9.24a)$$

That is

$$t_{mn} = i\frac{\sqrt{v_m}}{\gamma_m}e^{ik_m}\langle m|VGW|n\rangle e^{ik_n}\frac{\sqrt{v_n}}{\gamma_n} - e^{2ik_n}\delta_{mn} \quad (9.24b)$$

9.3 Unitarity of the many-channel scattering matrix

In this section, unitarity of S is demonstrated explicitly by noting that equation (9.23) is of the form

$$|t\rangle = S|n\rangle \quad (9.25)$$

where S is the S-matrix, given by

$$S = i\frac{\sqrt{v}}{\gamma}e^{ik}VGWe^{ik}\frac{\sqrt{v}}{\gamma} - e^{2ik}I \quad (9.26)$$

Note that unitarity of the S matrix requires $SS^\dagger = I$, where

$$S^\dagger = -i\frac{\sqrt{v}}{\gamma}e^{-ik}VG^\dagger We^{-ik}\frac{\sqrt{v}}{\gamma} - e^{-2ik}I \quad (9.27)$$

$$SS^\dagger = \frac{\sqrt{v}}{\gamma}e^{ik}VGW\frac{v}{\gamma^2}VG^\dagger We^{-ik}\frac{\sqrt{v}}{\gamma} + 1 - i\frac{\sqrt{v}}{\gamma}e^{ik}VGWe^{-ik}\frac{\sqrt{v}}{\gamma} + i\frac{\sqrt{v}}{\gamma}e^{ik}VG^\dagger We^{-ik}\frac{\sqrt{v}}{\gamma}$$

$$SS^\dagger = 1 + \frac{\sqrt{v}}{\gamma}e^{ik}[VGW\frac{v}{\gamma^2}VG^\dagger W - iVGW + iVG^\dagger W]e^{-ik}\frac{\sqrt{v}}{\gamma}$$

Hence unitarity requires

$$VGW\frac{v}{\gamma^2}VG^\dagger W - iVGW + iVG^\dagger W = 0$$

That is

$$GW\frac{v}{\gamma^2}VG^\dagger - iG + iG^\dagger = 0$$

Multiply from the left by G^{-1} and from the right by G^{-1} to yield the requirement

$$W\frac{v}{\gamma^2}V - iG^{-1\dagger} + iG^{-1} = 0$$

Writing $\Sigma_0 = \sigma - i\Gamma$, where $\sigma = -W\frac{\cos k}{\gamma}V$ and $\Gamma = W\frac{\sin k}{\gamma}V$ yields
$G^{-1} = g^{-1} - \sigma + i\Gamma$ and $G^{-1\dagger} = g^{-1} - \sigma - i\Gamma$.

Hence unitarity requires

$$W\frac{v}{\gamma^2}V - 2\Gamma = 0$$

Since $\Gamma = W\frac{\sin k}{\gamma}V = W\frac{v}{2\gamma^2}V$, this is indeed correct and completes the proof of unitarity.

Note that the S-matrix has the form

$$S = \begin{pmatrix} t_{11} & \cdots & t_{1M} \\ \vdots & \ddots & \vdots \\ t_{M1} & \cdots & t_{MM} \end{pmatrix} \text{ and } S^\dagger = \begin{pmatrix} t_{11}^* & \cdots & t_{M1}^* \\ \vdots & \ddots & \vdots \\ t_{1M}^* & \cdots & t_{MM}^* \end{pmatrix} \tag{9.28}$$

Hence the unitarity conditions $S^\dagger S = I$ and $SS^\dagger = I$ imply for any row or column l,

$$\sum_{n=1}^{M} |t_{nl}|^2 = 1 \tag{9.29}$$

and

$$\sum_{n=1}^{M} |t_{ln}|^2 = 1 \tag{9.30}$$

9.4 Expressions for the transmission coefficients

For $m \neq n$, equation (9.18) yields

$$t_{mn} = ie^{i(k_m+k_n)}\sqrt{\frac{4\sin k_m \sin k_n}{\gamma_m \gamma_n}}\langle m|VGV^\dagger|n\rangle \quad (m \neq n) \tag{9.31}$$

Hence the transmission coefficient from open channel n to open m is $T_{mn} = |t_{mn}|^2$, where

$$T_{mn} = \frac{4\sin k_m \sin k_n}{\gamma_m \gamma_n}|\langle m|VGW|n\rangle|^2 \tag{9.32}$$

This is a generalisation of equation (8.41a). To obtain an expression which resembles equation (8.41b), note that since $\langle m|VGW|n\rangle^* = \langle n|W^\dagger G^\dagger V^\dagger|m\rangle = \langle n|VG^\dagger W|m\rangle$, equation (9.32) can be written

$$T_{mn} = \frac{4 \sin k_m \sin k_n}{\gamma_m \gamma_n} \langle m|VGW|n\rangle\langle n|VG^\dagger W|m\rangle \tag{9.33}$$

It is useful to note that for any two vectors $|\psi\rangle$ and $|\phi\rangle$

$$\langle\psi|\phi\rangle = \sum_j \psi_j^* \phi_j \tag{9.34}$$

and

$$|\phi\rangle\langle\psi| = \begin{pmatrix} \phi_1 \\ \phi_2 \\ \vdots \\ \phi_M \end{pmatrix} \begin{pmatrix} \psi_1^* & \psi_2^* & \cdots & \psi_M^* \end{pmatrix} = \begin{pmatrix} \phi_1\psi_1^* & \phi_1\psi_2^* \cdots \phi_1\psi_M^* \\ \phi_2\psi_1^* & \phi_2\psi_2^* \cdots \phi_2\psi_M^* \\ \vdots & \vdots \quad\quad \vdots \\ \phi_M\psi_1^* & \phi_M\psi_2^* \cdots \phi_M\psi_M^* \end{pmatrix} \tag{9.35}$$

Comparing equation (9.34) with the sum of the diagonal elements (i.e. the trace) of the matrix $|\phi\rangle\langle\psi|$ yields

$$\langle\psi|\phi\rangle = \mathrm{Tr}\, |\phi\rangle\langle\psi| \tag{9.36}$$

where Tr indicates the trace of the matrix i.e the sum of the diagonal elements. This is a particular example of cyclic invariance under the trace, (i.e. for any two matrices A and B, $\mathrm{Tr}AB = \mathrm{Tr}BA$), because since $\langle\psi|\phi\rangle$ is simply a number, $\langle\psi|\phi\rangle = \mathrm{Tr}\langle\psi|\phi\rangle$.

This means that equation (9.33) can be written

$$T_{mn} = \frac{4 \sin k_m \sin k_n}{\gamma_m \gamma_n} \mathrm{Tr}\left\{ |m\rangle\langle m|VGW|n\rangle\langle n|VG^\dagger W \right\}$$

That is

$$T_{mn} = \frac{4 \sin k_m \sin k_n}{\gamma_m \gamma_n} \mathrm{Tr}\left\{ [W|m\rangle\langle m|V]G[W|n\rangle\langle n|V]G^\dagger \right\}$$

To simplify this expression, it is convenient to define the $N \times N$ matrix

$$\Gamma(n) = W|n\rangle \frac{\sin k_n}{\gamma_n} \langle n|V, \tag{9.37}$$

with elements

$$\Gamma_{ij}(n) = V_{ni}^* \frac{\sin k_n}{\gamma_n} V_{nj}, \tag{9.38}$$

This leads to the following equation, which is analogous to (8.41b).

$$T_{mn} = 4Tr\{\Gamma(m)G\Gamma(n)G^{\dagger}\} \quad (m \neq n) \tag{9.39}$$

In contrast with (8.41b), which only involves products of numbers, equation (9.39) involves products of non-commuting matrices and therefore it is important to preserve their order.

9.5 Expressions for the reflection coefficients

For $m = n$, equation (9.24) yields

$$t_{nn} = e^{2ik_n}\left[i2 \sin k_n \frac{\langle n|VGW|n\rangle}{\gamma_n} - 1\right] \tag{9.40}$$

As a check, note that when $V = 0$, this yields

$$t_{nn} = -e^{2ik_n} \tag{9.41}$$

On the other hand, when $V = 0$, the following boundary condition should be satisfied

$$0 = \psi_{-1}^{n} = t_{nn}e^{-ik_n} + e^{ik_n} = e^{-ik_n}[t_{nn} + e^{2ik_n}]$$

Hence equation (9.19) is equivalent to the boundary condition $\psi_{-1}^{n} = 0$

9.6 The self-energy matrix

Note that equation (9.20) is

$$\Sigma_0 = -We^{ik}\gamma^{-1}V = -\sum_{n=1}^{M} We^{ik}\gamma^{-1}\,|n\rangle\langle n|V$$

Hence

$$\Sigma_0 = -\sum_{n=1}^{M} W\,|n\rangle\frac{e^{ik_n}}{\gamma_n}\langle n|V \tag{9.42}$$

and therefore the matrix elements of Σ_0 are

$$\Sigma_{\mathrm{ij}} = \sum_{n=1}^{M}\Sigma_{\mathrm{ij}}(n) \tag{9.43}$$

where

$$\Sigma_{ij}(n) = -V_{ni}^{*}\frac{e^{ik_n}}{\gamma_n}V_{nj} \tag{9.44}$$

Note that Σ_0 is not Hermitian, because $\Sigma_0 = -V^{\dagger}e^{ik}\gamma^{-1}V$, whereas $\Sigma_0^{\dagger} = -V^{\dagger}e^{-ik^{*}}\gamma^{-1}V$.

Therefore it is convenient to write

$$\Sigma_0 = \frac{\Sigma_0 + \Sigma_0^{\dagger}}{2} + \frac{\Sigma_0 - \Sigma_0^{\dagger}}{2}$$

where $\frac{\Sigma_0 + \Sigma_0^{\dagger}}{2}$ is Hermitian and $\frac{\Sigma_0 - \Sigma_0^{\dagger}}{2}$ is anti-Hermitian. In other words,

$$\left(\frac{\Sigma_0 - \Sigma_0^{\dagger}}{2}\right)^{\dagger} = -\left(\frac{\Sigma_0 - \Sigma_0^{\dagger}}{2}\right), \text{ where as } \left(\frac{\Sigma_0 + \Sigma_0^{\dagger}}{2}\right)^{\dagger} = \left(\frac{\Sigma_0 + \Sigma_0^{\dagger}}{2}\right)$$

Note that only real wave vectors contribute to the anti-Hermitian term, because if k is imaginary, $ik = -ik^{*}$. Therefore it is convenient to define

$$\begin{aligned} \Gamma(n) &= W|n\rangle\frac{\sin k_n}{\gamma_n}\langle n|V \quad (\text{for } k_n \text{ real}) \quad \text{and} \\ \Gamma(n) &= 0 \quad (\text{for } k_n \text{ imaginary}) \end{aligned} \tag{9.45}$$

Therefore if the Hermitian part of the self-energy is defined to be

$$\sigma = \frac{\Sigma_0 + \Sigma_0^{\dagger}}{2} \tag{9.46}$$

then the self-energy can be written

$$\Sigma_0 = \sigma - i\Gamma \tag{9.47}$$

In this expression, both σ and Γ are Hermitian matrices. From equation (9.43)

$$\sigma = \sum_{n=1}^{M}\sigma(n), \quad \text{with } \sigma_{ij}(n) = -V_{ni}^{*}\frac{\cos k_n}{\gamma_n}V_{nj} \tag{9.48}$$

and since real wave vectors correspond to open channels,

$$\Gamma = \sum_{n=1\ (\text{open channels})}^{M}\Gamma(n), \quad \text{where, for real } k_n,\ \Gamma_{ij}(n) = +V_{ni}^{*}\frac{\sin k_n}{\gamma_n}V_{nj} \tag{9.49}$$

With this notation, the expression $G = (g^{-1} - \Sigma_0)^{-1}$ becomes

$$G = (g^{-1} - \sigma + i\Gamma)^{-1} \tag{9.50}$$

In summary, with equation (9.39), the following expressions have been obtained:

$$T_{mn} = 4\mathrm{Tr}\{\Gamma(m)G\Gamma(n)G^{\dagger}\} \tag{9.51}$$

where

$$G = (g^{-1} - \sigma + \mathrm{i}\Gamma)^{-1} \tag{9.52}$$

$$\sigma = \sum_{n=1}^{M}\sigma(n), \quad \text{with } \sigma_{ij}(n) = -V_{ni}^{*}\frac{\cos k_n}{\gamma_n}V_{nj} \tag{9.53}$$

$$\Gamma = \sum_{n=1\ (\text{open channels})}^{M}\Gamma(n), \quad \text{with } \Gamma_{ij}(n) = +V_{ni}^{*}\frac{\sin k_n}{\gamma_n}V_{nj} \tag{9.54}$$

9.7 Dyson's equation

As noted in chapter 2, the equation

$$G = (g^{-1} - \Sigma_0)^{-1} \tag{9.55}$$

is known as Dyson's equation. It is equivalent to the equation

$$G = g + g\Sigma_0 G \tag{9.56}$$

To see this, write equation (9.56) as $(1 - g\Sigma_0)G = g$, then multiply from the left by g^{-1} to yield $(g^{-1} - \Sigma_0)G = 1$ and multiply from the left by $(g^{-1} - \Sigma_0)^{-1}$ to yield equation (9.55).

Equation (9.56) is useful, because if Σ_{ij} is non-zero only for a subset of N' sites, one can define a $N' \times N'$ submatrix of Σ_0, denoted $\tilde{\Sigma}$ containing only these non-zero elements If $\tilde{g}$ and $\tilde{G}$ are the corresponding sub-matrices of g and G, then equation (9.56) yields

$$\tilde{G} = \tilde{g} + \tilde{g}\tilde{\Sigma}\tilde{G} \tag{9.57}$$

To see this, write equation (9.56) in component form, to yield

$$G_{i'j'} = g_{i'j'} + \sum_{k'l'} g_{ik'}\Sigma_{k'l'}G_{l'j}$$

where k',l' sum over the N' non-zero elements of Σ. Hence if i, j are chosen to belong to this set, equation (9.57) is obtained, which can be solved to yield

$$\tilde{G} = (\tilde{g}^{-1} - \tilde{\Sigma})^{-1} \tag{9.58}$$

This involves smaller matrices than equation (9.56), although to obtain $\tilde{g}$, one must first compute the larger matrix $g = (E - h)^{-1}$ and then extract the submatrix $\tilde{g}$. Similarly, in equation (9.40), only sub-matrices $\tilde{V}\tilde{G}\tilde{W}$ contribute. Using these sub-matrices, equations (9.51)–(9.54) become

$$T_{mn} = 4\mathrm{Tr}\{\tilde{\Gamma}(m)\tilde{G}\tilde{\Gamma}(n)\tilde{G}^{\dagger}\} \tag{9.59}$$

where

$$\tilde{G} = (\tilde{g}^{-1} - \tilde{\sigma} + \mathrm{i}\tilde{\Gamma})^{-1} \tag{9.60}$$

$$\tilde{\sigma} = \sum_{n=1}^{M} \tilde{\sigma}(n), \quad \text{with } \tilde{\sigma}_{i'j'}(n) = -V^{*}_{ni'}\frac{\cos k_n}{\gamma_n}V_{nj'} \tag{9.61}$$

$$\tilde{\Gamma} = \sum_{n=1\text{ (open channels)}}^{M} \tilde{\Gamma}(n), \quad \text{with } \tilde{\Gamma}_{i'j'}(n) = +V^{*}_{ni'}\frac{\sin k_n}{\gamma_n}V_{nj'} \tag{9.62}$$

Dyson's equation for structures connected to finite-width, multi-channel leads will be discussed in more detail in chapter 11.

9.8 Example: a scatterer connected to two 1-dimensional leads

As a check, it is interesting to show that equation (9.65) reproduces the expressions for transmission coefficient in chapter 8. For two leads, labelled $n = L$ and $n = R$, which connect to site 1 and N of the scatterer (where $N \neq 1$) $V_{L1} = -\alpha$ and $V_{Li} = 0$ for $i \neq 1$. Also $V_{RN} = -\beta$ and $V_{Li} = 0$ for $i \neq N$. Hence, assuming both wave vectors are real,

$\tilde{\sigma}_{i'j'}(L) = -\alpha^2\frac{\cos k_L}{\gamma_L}$ for $i' = j' = 1$ and $\tilde{\sigma}_{i'j'}(L) = 0$ otherwise

$\tilde{\Gamma}_{i'j'}(L) = \alpha^2\frac{\sin k_L}{\gamma_L}$ for $i' = j' = 1$ and $\tilde{\Gamma}_{i'j'}(L) = 0$ otherwise

$\tilde{\sigma}_{i'j'}(R) = -\beta^2\frac{\cos k_R}{\gamma_R}$ for $i' = j' = N$ and $\tilde{\sigma}_{i'j'}(LR) = 0$ otherwise

$\tilde{\Gamma}_{i'j'}(R) = \beta^2\frac{\sin k_R}{\gamma_R}$ for $i' = j' = 1$ and $\tilde{\Gamma}_{i'j'}(R) = 0$ otherwise

This means that if one defines $\sigma_L = -\alpha^2\frac{\cos k_L}{\gamma_L}$, $\Gamma_L = \alpha^2\frac{\sin k_L}{\gamma_L}$, $\sigma_R = -\beta^2\frac{\cos k_R}{\gamma_R}$ and $\Gamma_R = \beta^2\frac{\sin k_R}{\gamma_R}$, then

$$\tilde{\sigma}(L) = \begin{pmatrix} \sigma_L & 0 \\ 0 & 0 \end{pmatrix}, \tilde{\sigma}(R) = \begin{pmatrix} 0 & 0 \\ 0 & \sigma_R \end{pmatrix}, \tilde{\Gamma}(L) = \begin{pmatrix} \Gamma_L & 0 \\ 0 & 0 \end{pmatrix}, \tilde{\Gamma}(R) = \begin{pmatrix} 0 & 0 \\ 0 & \Gamma_R \end{pmatrix}$$

Hence

$$T_{RL} = 4\mathrm{Tr}\left\{\tilde{\Gamma}(R)\tilde{G}\tilde{\Gamma}(L)\tilde{G}^{\dagger}\right\} = 4\mathrm{Tr}\begin{pmatrix} 0 & 0 \\ 0 & \Gamma_R \end{pmatrix}\begin{pmatrix} G_{11} & G_{1N} \\ G_{N1} & G_{NN} \end{pmatrix}\begin{pmatrix} \Gamma_L & 0 \\ 0 & 0 \end{pmatrix}\begin{pmatrix} G_{11}^{*} & G_{N1}^{*} \\ G_{1N}^{*} & G_{NN}^{*} \end{pmatrix}$$

That is

$$T_{RL} = 4\mathrm{Tr}\begin{pmatrix} 0 & 0 \\ \Gamma_R G_{N1} & \Gamma_R G_{NN} \end{pmatrix}\begin{pmatrix} \Gamma_L G_{11}^{*} & \Gamma_L G_{N1}^{*} \\ 0 & 0 \end{pmatrix} = 4\mathrm{Tr}\begin{pmatrix} 0 & 0 \\ \Gamma_R G_{N1}\Gamma_L G_{11}^{*} & \Gamma_R G_{N1}\Gamma_L G_{N1}^{*} \end{pmatrix}$$

Hence $T_{RL} = 4\Gamma_R G_{N1}\Gamma_L G_{N1}^{*}$, in agreement with equation (8.41).

9.9 A scatterer attached to M leads, in which each lead connects to only one site of the scatterer

As an example of the above general formulae, consider the case in which lead n couples to only a single site i_n of the scatterer. Then

$\sigma_{ij}(n) = \sigma_n$ for $i = j = i_n$ and $\sigma_{ij}(L) = 0$ otherwise, where $\sigma_n = -V_{ni_n}^{*}\frac{\cos k_n}{\gamma_n}V_{ni_n}$. Similarly

$\Gamma_{ij}(n) = \Gamma_n$ for $i = j = i_n$ and $\Gamma_{ij}(L) = 0$, where $\Gamma_n = V_{ni_n}^{*}\frac{\sin k_n}{\gamma_n}V_{ni_n}$. With this notation,

$$\sigma(1) = \begin{pmatrix} \sigma_1 & \cdots & 0 \\ \vdots & \ddots & \vdots \\ 0 & \cdots & 0 \end{pmatrix}, \ldots \sigma(M) = \begin{pmatrix} 0 & \cdots & 0 \\ \vdots & \ddots & \vdots \\ 0 & \cdots & \sigma_M \end{pmatrix}$$

$$\Gamma(1) = \begin{pmatrix} \Gamma_1 & \cdots & 0 \\ \vdots & \ddots & \vdots \\ 0 & \cdots & 0 \end{pmatrix}, \ldots \Gamma(M) = \begin{pmatrix} 0 & \cdots & 0 \\ \vdots & \ddots & \vdots \\ 0 & \cdots & \Gamma_M \end{pmatrix}$$

$$\sigma = \begin{pmatrix} \sigma_1 & \cdots & 0 \\ \vdots & \ddots & \vdots \\ 0 & \cdots & \sigma_M \end{pmatrix} \text{ and } \Gamma = \begin{pmatrix} \Gamma_1 & \cdots & 0 \\ \vdots & \ddots & \vdots \\ 0 & \cdots & \Gamma_M \end{pmatrix}$$

Also if the notation is simplified by writing $G_{i_n j_m} = G_{nm}$, then

$$\tilde{G} = \begin{pmatrix} G_{11} & \cdots & G_{1M} \\ \vdots & \ddots & \vdots \\ G_{M1} & \cdots & G_{MM} \end{pmatrix} \text{ and } \tilde{G}^{\dagger} = \begin{pmatrix} G_{11}^{*} & \cdots & G_{M1}^{*} \\ \vdots & \ddots & \vdots \\ G_{1M}^{*} & \cdots & G_{MM}^{*} \end{pmatrix}$$

With the above notation, the following product has non-zero elements in row m only

$$\Gamma(m)\tilde{G} = \begin{pmatrix} 0 & 0 & 0 & \ldots & 0 \\ \vdots & \vdots & \vdots & & \vdots \\ \Gamma_m G_{m1} & \Gamma_m G_{m2} & \ldots. & \ldots & \Gamma_m G_{mM} \\ \vdots & \vdots & \vdots & & 0 \end{pmatrix}$$

and the product below has non-zero elements in row n only.

$$\Gamma(n)\tilde{G}^{\dagger} = \begin{pmatrix} 0 & 0 & 0 & \dots & 0 \\ \vdots & \vdots & \vdots & & \vdots \\ \Gamma_n G_{1n}^* & \Gamma_n G_{2n}^* & \dots & \dots & \Gamma_n G_{Mn}^* \\ \vdots & \vdots & \vdots & & 0 \end{pmatrix}$$

Hence the following product has non-zero elements in row m only

$$\Gamma(m)\tilde{G}\Gamma(n)\tilde{G}^{\dagger} = \begin{pmatrix} 0 & 0 & 0 & \dots & 0 \\ \vdots & \vdots & \vdots & & \vdots \\ \Gamma_m G_{mn}\Gamma_n G_{1n}^* & \Gamma_m G_{mn}\Gamma_n G_{2n}^* & \dots & \dots & \Gamma_m G_{mn}\Gamma_n G_{Mn}^* \\ \vdots & \vdots & \vdots & & 0 \end{pmatrix}$$

and finally

$$\mathrm{T_{mn}} = 4\mathrm{Tr}\Gamma(m)\tilde{G}\Gamma(n)\tilde{G}^{\dagger} = 4\Gamma_m G_{mn}\Gamma_n G_{mn}^* \tag{9.63}$$

where $\tilde{G}$ is the $M \times M$ matrix $\tilde{G} = (\tilde{g}^{-1} - \sigma + \mathrm{i}\Gamma)^{-1}$

This is the same as the two channel case, except it is now necessary to invert a $M \times M$ matrix. The size of the matrix is equal to the number of leads, not the number of sites in the scatterer, so for a large scatterer this is a significant reduction in size.

9.10 Scattering theory in the presence of an open scatterer

Note that since self-energies are additive, leads can be added to a multi-lead structure iteratively. To illustrate this feature, consider the case where $\Sigma_0 = \Sigma_1 + \Sigma_2$. Then

$$G = (g^{-1} - \Sigma_0)^{-1} = \left(g_1^{-1} - \Sigma_2\right)^{-1}, \tag{9.64}$$

where

$$g_1 = (g^{-1} - \Sigma_1)^{-1} \tag{9.65}$$

If Σ_1 is the self-energy representing $M - 2$ leads, and one would like the transmission coefficient between the remaining two leads, labelled n and m, then this shows that the two-lead formula (8.37) can be used, provided $\tilde{g}$ in equation (8.37) is replaced by g_1. Note that g_1 is non-Hermitian, because the scatterer is now an open system, even before attaching the final two leads. Naturally, this means that the sum of the reflection coefficients between leads n and m is no longer unity, because current entering the open system can be lost to the other $M - 2$ leads.

9.11 The effect of closed channels

If, for example, channel n is closed, then it carries no current and e^{ik_n} will be real. That is k_n is either imaginary, or equal to $\pi+$ an imaginary number. To ensure that the wave e^{ik_nj} does not diverge as j tends to infinity, the imaginary part of k_n should be chosen to be positive. In this case, as noted above, the imaginary part of $\Sigma(n)$ is zero (i.e. $\Gamma(n) = 0$) and only $\sigma(n)$ will be non-zero. In other words, only open channels contribute to the level broadening Γ.

9.12 A scatterer connected to finite-dimensional leads

As shown in section 5.11, when a finite-dimensional lead comprising unit cells (each containing N sites) is connected to scatterer, the resulting structure is mathematically equivalent to N one-dimensional leads connected to the scatter. Each one-dimensional lead represents a scattering channel of the finite-width lead. Therefore the above analysis also describes the problem of finite-width leads connected to a scatterer. One simply needs to assign an appropriate number of one-dimensional leads in figure 9.1 to each of the finite-dimensional leads connected to the scatterer.

9.13 Summary

In this chapter, expressions for reflection and transmission coefficients of an arbitrary scatterer connected to many one-dimensional leads were obtained. These are a generalisation of the formulae obtained in chapter 8 and reveal how multiple channels contribute to the self-energy in an iterative manner. Since one-dimensional leads are mathematically equivalent to scattering channels in finite-width leads, comparison with the Green's function approach in later chapters will show that scattering theory based on wave function matching is entirely equivalent to a theory based on Green's functions alone.

IOP Publishing

Chapter 10

Relationship between Green's functions, wave functions and scattering amplitudes

10.1 Introduction

In previous chapters, the Schrödinger equation was solved for the wave function associated with a scatterer connected to scattering channels and formulae for reflection and transmission amplitudes were obtained. It was found that Green's functions arose naturally when trying to find a solution to the Schrödinger equation of such open systems. In this chapter, a more intimate connection between Green's functions and wave functions will be discussed. By examining how reflection and transmission coefficients can be obtained directly from Green's functions, the multi-channel formulae for reflection and transmission amplitudes will be re-derived using Green's functions alone, without explicitly solving the Schrödinger equation. This opens the way to using efficient methods of solving scattering problems, based on Dyson's equation.

The general definition of a Green's function $G(E)$ belonging to a Hamiltonian H is

$$(EI - H)G(E) = I \tag{10.1}$$

where I is the unit matrix. Often equation (10.1) is simply written $(E - H)G = I$ or $(E - H)G = 1$, in which case the presence of the unit matrix multiplying E and on the right-hand side of the equation is implied.

In what follows, this equation will first be discussed when H is finite, corresponding to a closed system and then later, when H is an infinite matrix, corresponding to an open system. From a mathematical viewpoint, E is an arbitrary parameter, which could be chosen to be real or complex. For most of the discussion below, E will be chosen to be real.

doi:10.1088/978-0-7503-3639-0ch10

10.2 Green's functions of closed systems

If H is a $N \times N$ matrix, then I and G are $N \times N$ matrices. For example, if $N = 2$ and $H = \begin{pmatrix} \epsilon & -\gamma \\ -\gamma & -\epsilon \end{pmatrix}$, then $I = \begin{pmatrix} 1 & 0 \\ 0 & 1 \end{pmatrix}$ and $G = \begin{pmatrix} G_{11} & G_{12} \\ G_{21} & G_{22} \end{pmatrix}$. In this case, equation (10.1) becomes

$$\begin{pmatrix} E - \epsilon & \gamma \\ \gamma & E + \epsilon \end{pmatrix}\begin{pmatrix} G_{11} & G_{12} \\ G_{21} & G_{22} \end{pmatrix} = I \tag{10.2}$$

Since all matrices are assumed to be finite, equation can be solved by multiplying both sides by the inverse of $(EI - H)$ to yield

$$G(E) = (EI - H)^{-1} = \frac{1}{\det(E - H)} C^t \tag{10.3}$$

where C is the cofactor matrix of $E - H$. This inversion can be obtained from the following line of MATLAB code:

$$G = \text{inv}\,(E \times \text{eye}(N) - H)$$

Consequently, if H is small enough to be contained within the available computer memory, the problem of computing Green's functions of closed systems is solved. Our only task is to elucidate the general properties of such finite-size Green's functions.

For the 2×2 example of equation (10.2), this yields

$$\begin{pmatrix} G_{11} & G_{12} \\ G_{21} & G_{22} \end{pmatrix} = \frac{1}{d(E)}\begin{pmatrix} E + \epsilon & -\gamma \\ -\gamma & E - \epsilon \end{pmatrix} \tag{10.4}$$

In this expression $\text{d}(E) = \det(E - H) = (E + \epsilon)(E + \epsilon) - \gamma^2$, which can be written in the form

$$d(E) = (E - E_1)(E - E_2) \tag{10.5}$$

where $E_1 = \sqrt{\epsilon^2 + \gamma^2}$ and $E_2 = -\sqrt{\epsilon^2 + \gamma^2}$ From the discussion of a two-level system in chapter 3, E_1 and E_2 are recognised as the two eigenvalues of H. Consequently the Green's function diverges when $E = E_1$ and $E = E_2$. To prove that this divergence is a general property of Green's functions of closed systems, note that if $|\psi_n\rangle = E_n|\psi_n\rangle$, then as discussed in chapter 3, the eigenstates $|\psi_n\rangle$ satisfy the completeness condition

$$\sum_{n=1}^{N} |\psi_n\rangle\langle\psi_n| = I$$

So equation (10.3) becomes

$$G(E) = (EI - H)^{-1}I = \sum_{n=1}^{N}(EI - H)^{-1}|\psi_n\rangle\langle\psi_n| = \sum_{n=1}^{N}(EI - E_n)^{-1}|\psi_n\rangle\langle\psi_n|$$

where the last step follows from the fact (see chapter 3 appendix) that a function of H acting on an eigenvector yields the same function of the corresponding eigenvalue.

Hence

$$G(E) = \sum_{n=1}^{N} \frac{|\psi_n\rangle\langle\psi_n|}{E - E_n} \tag{10.6}$$

and the components of G satisfy

$$G_{ij}(E) = \sum_{n=1}^{N} \frac{\psi_n^*(i)\psi_n(j)}{E - E_n} \tag{10.7}$$

Clearly, if E is real, $G_{ij}(E) = G_{ji}^*(E)$ and therefore one concludes that the Green's function of any closed system is Hermitian. Furthermore, if H is real, then the eigenstates can be chosen to be real, which proves that in this case, $G(E)$ is a real symmetric matrix.

Note that from equation (10.4), the mid-gap magic number table (obtained by choosing $\gamma = 1$ and writing $M = d(0)G(0)$) is

$$\begin{pmatrix} M_{11} & M_{12} \\ M_{21} & M_{22} \end{pmatrix} = \begin{pmatrix} \epsilon & -1 \\ -1 & -\epsilon \end{pmatrix} \tag{10.8}$$

For the case $\epsilon = 0$, where the site on the odd lattice ($i = 1$) has the same site energy as the site on the even lattice ($i = 2$), this becomes off diagonal, as expected. That is

$$\begin{pmatrix} M_{11} & M_{12} \\ M_{21} & M_{22} \end{pmatrix} = \begin{pmatrix} 0 & -1 \\ -1 & 0 \end{pmatrix} \tag{10.9}$$

10.3 Approximations to Green's functions of closed systems

Two approximations are commonly used when discussing Green's functions of closed systems, one of which is reasonable and the other is usually unreasonable.

Approximation a. The first arises when E is very close to an eigenvalue E_L of the closed system, in which case one of the terms in equation (10.7) is much larger than the others. Keeping only this term leads to the approximation

$$G_{ij}(E) \approx \frac{\psi_L^*(i)\psi_L(j)}{E - E_L} \tag{10.10}$$

As discussed in chapter 2, this leads to the Breit–Wigner formula and when computing the transmission coefficient, is a reasonable approximation, provided the level broadening due to the coupling to electrodes is smaller than the energy level spacing of the isolated molecule or scatterer.

Approximation b. The second (generally poor) approximation arises when E lies in between two eigenvalues. As an example, consider the common case, where E lies between the HOMO energy $E_{\rm H}$ and a LUMO energy $E_{\rm L}$ of a molecule. Since E is close to $E_{\rm H}$ and $E_{\rm L}$, these correspond to the largest terms in equation (10.7) and therefore one might consider the following approximation

$$G_{ij}(E) \approx \frac{\psi_{\rm H}^*(i)\psi_{\rm H}(j)}{E - E_{\rm H}} + \frac{\psi_{\rm L}^*(i)\psi_{\rm L}(j)}{E - E_{\rm L}} \tag{10.11}$$

To see why this is not a generally a reasonable approximation, consider the mid-gap Green's function evaluated at $E = E_{\rm HL}$, where $E_{\rm HL}$ is the energy of the middle of the HOMO–LUMO gap, i.e. $E_{\rm HL} = (E_{\rm H} + E_{\rm L})/2$. This yields for the mid-gap Green's function

$$G_{ij}(E_{\rm HL}) \approx \frac{\psi_{\rm H}^*(i)\psi_{\rm H}(j) - \psi_{\rm L}^*(i)\psi_{\rm L}(j)}{\delta} \tag{10.12}$$

where δ is one half of the HOMO–LUMO gap, i.e. $\delta = \frac{E_{\rm L} - E_{\rm H}}{2}$.

To show that equation (10.12) is not a very precise approximation, and to estimate the contribution from the other MOs, consider the case where the level spacing is approximately constant. Then $E_n = (2n - 1)\delta$ and the exact equation (10.7) becomes

$$G_{ij}(E_{\rm HL}) = \frac{1}{\delta}\sum\nolimits_n \frac{a_n}{(1 - 2n)} \tag{10.13}$$

where

$$a_n = \psi_n^*(i)\psi_n(j) \tag{10.14}$$

As an example, if the HOMO, HOMO−1, LUMO and LUMO+1 are retained, this becomes

$$G_{ij}(E_{\rm F}) \approx \frac{1}{\delta}\left[\frac{a_{\rm H}}{(+1)} + \frac{a_{\rm L}}{(-1)} + \frac{a_{\rm H-1}}{(+3)} + \frac{a_{\rm L+1}}{(-3)}\right] \tag{10.15}$$

This shows that the contributions from the LUMO+1 and HOMO−1 are on the scale of one third of the contributions from the HOMO and LUMO and in general should not be neglected. In particular, if the HOMO and LUMO interfere destructively, so that

$$\frac{a_{\rm H}}{(+1)} + \frac{a_{\rm L}}{(-1)} = 0, \tag{10.16}$$

then the dominant contribution comes from the other orbitals, not simply the HOMO and LUMO.

10.4 Green's functions of a doubly infinite linear chain

To gain a little more insight into the Green's functions of closed systems, it is useful to obtain analytic formula wherever possible. In chapter 4, it was found that as a stepping stone to computing the eigenstates of a chain with periodic boundary conditions and a linear chain with free-end boundary conditions, an analytic formula for the eigenstates of an infinite linear chain was particularly useful. Here, a similar route leading to the Green's functions of these structures will be followed, starting with the Green's function G of a doubly infinite linear chain.

Since G is a square matrix composed of columns of numbers, the numbers $G_{j,p}$ belonging to the pth column of G can be written in the form of a vector:

$$|\psi^{(p)}\rangle = \begin{pmatrix} \vdots \\ G_{j-2,p} \\ G_{j-1,p} \\ G_{j,p} \\ G_{j+1,p} \\ G_{j+2,p} \\ \vdots \end{pmatrix} \tag{10.17}$$

In other words, G can be presented in the form

$$G = (|\psi^{(1)}\rangle|\psi^{(2)}\rangle \; \ldots\ldots |\psi^{(N)}\rangle) \tag{10.18}$$

To highlight the relationship between Green's functions and wave functions, discussed in detail below, it is useful to introduce the notation

$$G_{jp} = \psi_j^{(p)} \tag{10.19}$$

So that equation (10.17) becomes

$$\begin{pmatrix} \vdots \\ G_{j-2,p} \\ G_{j-1,p} \\ G_{j,p} \\ G_{j+1,p} \\ G_{j+2,p} \\ \vdots \end{pmatrix} = \begin{pmatrix} \vdots \\ \psi_{j-2}^{(p)} \\ \psi_{j-1}^{(p)} \\ \psi_j^{(p)} \\ \psi_{j+1}^{(p)} \\ \psi_{j+2}^{(p)} \\ \vdots \end{pmatrix} = |\psi^{(p)}\rangle \tag{10.20}$$

To obtain an equation for $G_{j,p}$, take the j, p th element of equation (10.1), which yields

$$\sum_{l=1}^{N} (EI - H)_{jl} G_{lp}(E) = \delta_{jp} \tag{10.21}$$

where δ_{jp} is a Kronecker delta, satisfying $\delta_{jj} = 1$ and $\delta_{jp} = 0$ for $j \neq p$. Clearly δ_{jp} is the j, pth element of the unit matrix; i.e. $\delta_{jp} = I_{jp}$. Equation (10.21) is

$\sum_{l=1}^{N}(EI_{jl} - H_{jl})G_{lp}(E) = \delta_{jp}$ or equivalently $EG_{jp}(E) - \sum_{l=1}^{N}H_{jl}G_{lp}(E) = \delta_{jp}$, which can be written as

$$\sum_{l=1}^{N} H_{jl}G_{lp}(E) = EG_{jp}(E) - \delta_{jp} \tag{10.22}$$

Using the notation of equation (10.19), this becomes

$$\sum_{l=1}^{N} H_{jl}\psi_l^{(p)} = E\psi_j^{(p)} - \delta_{jp} \tag{10.23}$$

This is almost identical to equation (4.7), except for the presence of the Kronecker delta on the right-hand side. Therefore with a slight modification, the discussions in chapter 4 can be used to obtain solutions to equation (10.23).

First consider the infinite chain described by the Schrödinger equation (4.8), for which the Green's function equation (10.23) becomes

$$\varepsilon_o\psi_j^{(p)} - \gamma\psi_{j-1}^{(p)} - \gamma\psi_{j+1}^{(p)} = E\psi_j^{(p)} - \delta_{jp} \tag{10.24}$$

Following the discussion below equation (4.8), one might be tempted to guess that the solution to equation (10.24) is $\psi_j^{(p)} = \phi_j$, where $\phi_j = Ae^{ikj}$, or $\psi_j^{(p)} = f_j$, where $f_j = Be^{-ikj}$. Unfortunately, for all j, including $j = p$, these functions satisfy

$$\varepsilon_o\phi_j - \gamma\phi_{j-1} - \gamma\phi_{j+1} = E\phi_j - 0 \tag{10.25}$$

and

$$\varepsilon_o f_j - \gamma f_{j-1} - \gamma f_{j+1} = Ef_j - 0 \tag{10.26}$$

where $E = \epsilon_0 - 2\gamma \cos k$. Since the aim is to solve for the Green's function as a function of E, the latter equation implies that k must be chosen to satisfy

$$k(E) = \text{acos}\left[\frac{\epsilon_0 - E}{2\gamma}\right]. \tag{10.27}$$

From the discussion in chapter 4, the corresponding group velocity $v(E)$ is given by

$$v(E) = \frac{2\gamma \sin k(E)}{\hbar} \tag{10.28}$$

Since equation (10.27) does not determine the sign of k, henceforth the sign of k will be chosen such that when k is real, $v(E)$ is positive. Furthermore, if E lies outside the energy band of the chain, k will be complex and k will be chosen such that $\text{Im}\, k(E) > 0$.

In equations (10.25) and (10.26), the Kronecker delta is missing from the right-hand side. This shows that both Ae^{ikj} and Be^{-ikj} satisfy the Green's function equation (10.24) for all j, except $j = p$. Therefore neither of them is a solution to equation (10.24).

As an alternative consider the guess

$$\psi_j^{(p)} = \phi_j \text{ for } j > p \tag{10.29}$$

and

$$\psi_j^{(p)} = f_j \quad \text{for } j < p \tag{10.30}$$

For $j = p + 1$, equations (10.24) and (10.29) yield

$$\varepsilon_0\phi_{p+1} - \gamma\psi_p^{(p)} - \gamma\phi_{p+2} = E\phi_{p+1} \tag{10.31}$$

On the other hand, for $j = p + 1$, equation (10.25) is

$$\varepsilon_0\phi_{p+1} - \gamma\phi_p - \gamma\phi_{p+2} = E\phi_{p+1} \tag{10.32}$$

Comparing equations (10.32) and (10.31) yields

$$\psi_p^{(p)} = \phi_p \tag{10.33}$$

In other words, the fact that $\psi_j^{(p)} = \phi_j$ for $j > p$ implies, via equation (10.29), that $\psi_j^{(p)} = \phi_j$ for $j \geqslant p$. This is because equation (10.25) can be viewed as a recurrence relation, as discussed in appendix B of chapter 4 and therefore the condition $j > p$ implies $j \geqslant p$ by continuity. Similarly, by continuity, equation (10.30) implies $\psi_j^{(p)} = f_j$ for $j \leqslant p$. Hence choosing $j = p$ yields

$$\psi_p^{(p)} = \phi_p = f_p \tag{10.34}$$

That is $Ae^{ikp} = Be^{-ikp}$, which implies $A = Ce^{-ikp}$ and $B = Ce^{ikp}$. Hence

$$\phi_j = Ce^{ik(j-p)} \quad \text{and} \quad f_j = Ce^{-ik(j-p)}, \tag{10.35}$$

where C is a constant. Combining these with equations (10.29) and (10.30) yields

$$\psi_j^{(p)} = Ce^{ik|j-p|} \tag{10.36}$$

The only remaining equation to be satisfied is equation (10.24) for the case $j = p$, which becomes

$$\varepsilon_0\psi_p^{(p)} - \gamma\psi_{p-1}^{(p)} - \gamma\psi_{p+1}^{(p)} = E\psi_p^{(p)} - 1 \tag{10.37}$$

That is $\varepsilon_0\phi_p - \gamma f_{p-1} - \gamma\phi_{p+1} = E\phi_p - 1$. Since $\varepsilon_0\phi_p - \gamma\phi_{p-1} - \gamma\phi_{p+1} = E\phi_p$, this yields

$$\gamma\phi_{p-1} - \gamma f_{p-1} = -1$$

That is $\gamma C(e^{-ik} - e^{ik}) = -1$, or equivalently

$$C = \frac{1}{2i\gamma \sin k} = \frac{1}{i\hbar v(E)} \tag{10.38}$$

where $v(E)$ is the group velocity. Hence the Green's function of a doubly infinite chain is

$$\boxed{G_{jp}(E) = \psi_j^{(p)} = \frac{e^{ik(E)|j-p|}}{i\hbar v(E)}} \tag{10.39}$$

Note that if E lies outside the energy band k is chosen such that $\text{Im}\, k(E) > 0$, $G_{jp}(E)$ becomes an evanescent state, which decays exponentially with distance $|j - p|$ from the injection point.

It should be noted that equation (10.39) is not the most general solution, because any solution of the Schrödinger equation could be added to $\psi_j^{(p)}$ and it would still satisfy (10.23).

Therefore the most general solution is

$$G_{jp}(E) = \psi_j^{(p)} = \frac{e^{ik(E)|j-p|}}{i\hbar v(E)} + Ae^{ik(E)j} + Be^{-ik(E)j} \tag{10.40}$$

where A and B are arbitrary constants. If A and B are chosen to be zero, then the resulting Green's function in equation (10.39) is called a 'retarded Green's function'. On the other hand, if one chooses $A = -\frac{e^{ik(E)(j-p)}}{i\hbar v(E)}$ and $B = -\frac{e^{-ik(E)(j-p)}}{i\hbar v(E)}$, then equation (10.40) becomes

$$G_{jp}(E) = \psi_j^{(p)} = \frac{-e^{-ik(E)|j-p|}}{i\hbar v(E)} \tag{10.41}$$

This Green's function is called an 'advanced Green's function'.

The presence of the modulus in the exponent of $\psi_j^{(p)}$ in the above equations is the crucial difference between a Green's function and a wave function. Without the modulus, $\psi_j^{(p)}$ would satisfy the Schrödinger equation, $\sum_{l=1}^{N} H_{jl}\psi_l^{(p)} = E\psi_j^{(p)}$ and not equation (10.23).

10.5 Relationship between a Green's function of a linear chain and the wave function of a T-shaped junction

In the above discussion, $\psi_j^{(p)}$ is a Green's function of a doubly infinite chain and not a wave function of the chain. To elucidate the physical meaning of the retarded Green's function (10.39), it is interesting to note that $\psi_j^{(p)}$ is a wave function of the T-shaped junction shown in figure 10.1, formed from a vertical semi-infinite chain coupled to site p of a doubly infinite horizontal chain, by a coupling $-\alpha$. The sites in the vertical arm are labelled $l = 0, -1, -2, \ldots$ and the sites in the horizontal chain are labelled j, where $-\infty < j < \infty$. Consider the case when a plane wave Ae^{ikl} is incident along the vertical arm. This 'source' will produce a reflected wave Are^{ikl} in the vertical arm and transmitted waves in the horizontal arms. If S_j is the wave function in the vertical arm, then the Schrödinger equation for this structure is

$$\varepsilon_0\psi_j^{(p)} - \gamma\psi_{j-1}^{(p)} - \gamma\psi_{j+1}^{(p)} - \alpha S_0\delta_{jp} = E\psi_j^{(p)} \tag{10.42}$$

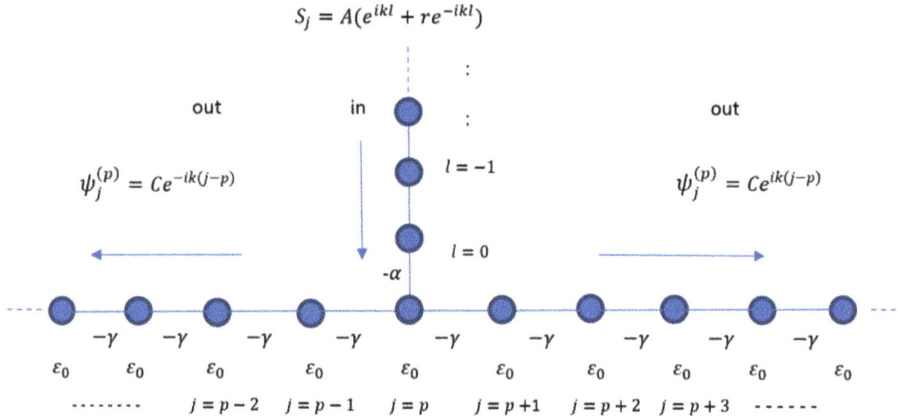

Figure 10.1. A T-shaped junction formed from a vertical semi-infinite chain coupled to site p of a doubly infinite horizontal chain, by a coupling $-\alpha$.

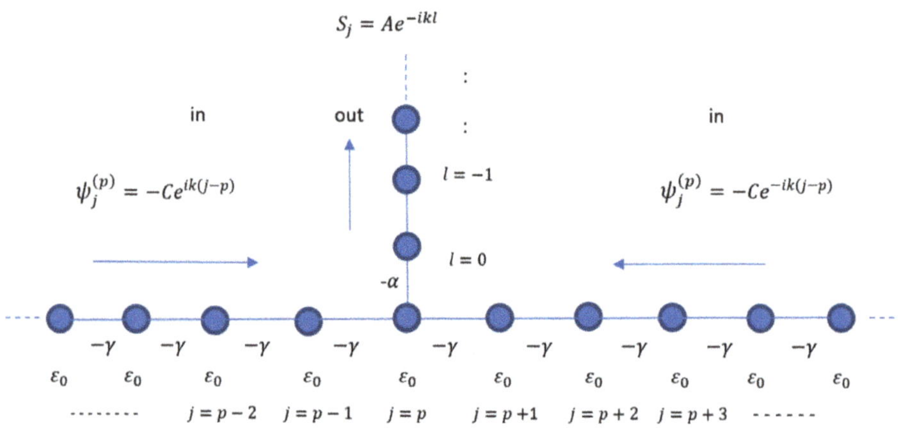

Figure 10.2. A T-shaped junction formed from a vertical semi-infinite chain coupled to site p of a doubly infinite horizontal chain, by a coupling $-\alpha$.

where the term $\alpha S_0 \delta_{jp}$ arises from the coupling between site p and the vertical arm. If the amplitude A of the source S_j is chosen such that $\alpha S_0 = -1$, then equation (10.42) becomes identical to equation (10.37). This demonstrates that the retarded Green's function of the doubly infinite linear chain is a wave function of a T-shaped junction, arising from a source at site p.

Similarly, the advanced Green's function (10.41) is a wave function of a T-shaped junction in the presence of a sink at site p. This is illustrated in figure 10.2, which shows incoming waves from the left and right meeting at site p and then exiting along the vertical arm. Since the advanced Green's function of equation (10.41) satisfies equation (10.42), it indeed corresponds to the wave function shown in figure 10.2.

10.6 Green's functions of a semi-infinite linear chain

Consider the semi-infinite linear chain shown in figure 10.3, which terminates at site $j = l$, where $l \geqslant p$. As in the discussion leading to equation (4.55), the fact that site $j = l + 1$ is missing leads to the boundary condition

$$\psi_{l+1}^{(p)} = 0 \tag{10.43}$$

This can be satisfied by noting that as in equation (10.40) any eigenstate of energy E can be added to a Green's function. Since a plane wave e^{ikj} impinging on the end of the chain will create a reflected wave proportional to e^{-ikj}, a reflected wave $R_j = Ae^{-ikj}$ can be added to the retarded Green's function, as shown in figure 10.3 and R_j can be chosen such that equation (10.43) is satisfied. That is one can choose

$$e^{ik(l+1-p)} + Ae^{-ik(l+1)} = 0$$

This yields $A = -e^{ik(2l+2-p)}$. Hence the retarded Green's function of a semi-infinite linear chain is

$$G_{jp}(E) = \psi_j^{(p)} = \frac{e^{ik(E)|j-p|} - e^{-ik(E)(j+p-2l-2)}}{i\hbar v(E)} \tag{10.44}$$

This satisfies $\psi_{l+1}^{(p)} = 0$ and also $\psi_j^{(l+1)} = 0$. Similarly, if the chain terminates at site $l \leqslant p$, the first missing site is $j = l - 1$ and the boundary condition becomes $\psi_{l-1}^{(p)} = 0$. This leads to

$$G_{jp}(E) = \psi_j^{(p)} = \frac{e^{ik(E)|j-p|} - e^{ik(E)(j+p-2l+2)}}{i\hbar v(E)} \tag{10.45}$$

Note that in both cases, the Green's function on the terminal site l is

$$\boxed{G_{ll}(E) = -\frac{e^{ik(E)}}{\gamma}} \tag{10.46}$$

$G_{ll}(E)$ is called the 'surface Green's function' of a semi-infinite lead and will be used extensively in later calculations.

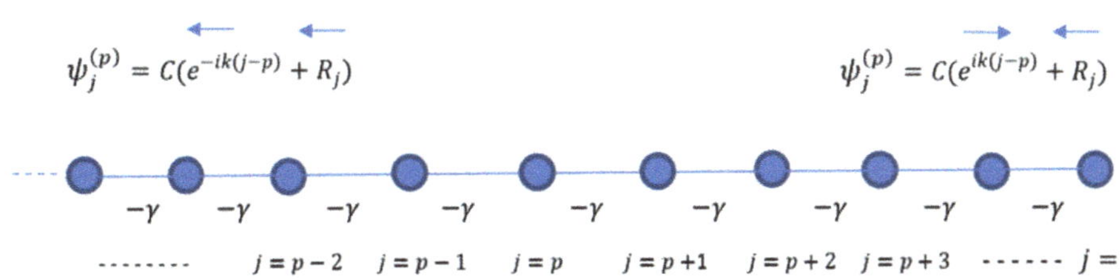

Figure 10.3. A semi-infinite linear chain, which terminates at site $j = l$.

10.7 The relationship between Green's functions and wave functions in the presence of a scatterer

The above argument can be extended to illustrate the relationship between Green's functions and wave functions in the presence of an arbitrary scatterer. To establish this relationship, consider the structure shown in figure 8.1 of chapter 8, whose wave function consists of an incoming wave of unit current from the left, along with the resulting reflected and transmitted waves. Now multiply the wave function by an arbitrary constant A, to yield the solution shown in figure 10.4.

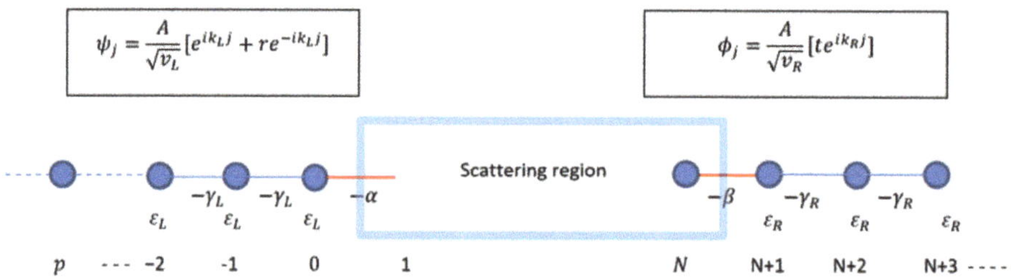

Figure 10.4. An arbitrary scatterer connected to 1-dimensional leads. The incoming current from the left is $|A|^2$.

As discussed in chapter 4, since the Schrödinger equation can be regarded as a recurrence relation, the form of the wave function on the left of the scatterer, uniquely determines the form of the wave function on the right. Similarly, by operating the recurrence backwards, the form of the wave function on the right uniquely determines the form of the wave function on the left. In other words, if the transmitted wave on the right is $\phi_j = \frac{A}{\sqrt{v_R}} e^{ik_Rj}$, then the form of ψ_j is determined.

Now choose $A = \frac{e^{-ik_Lp}}{i\hbar\sqrt{v_L}}$, where k_L is the wave vector in the left lead and p labels some site in the left lead. This yields the wave function shown in figure 10.5, where the form of ψ_j on the left is determined by ϕ_j on the right.

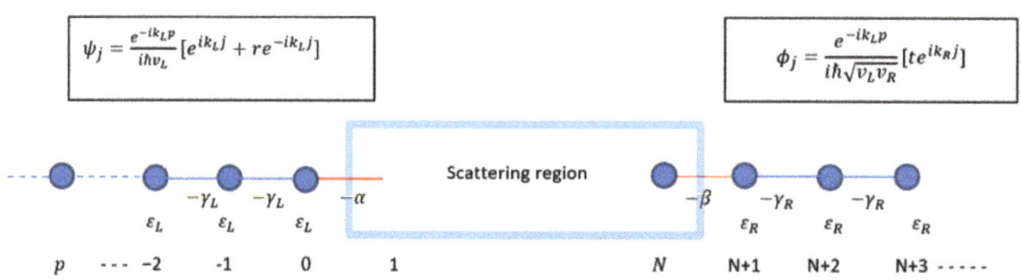

Figure 10.5. The wave function of an arbitrary scatterer connected to 1-dimensional leads.

Now consider the function shown in figure 10.6, which is identical to that of figure 10.5 for $j \geqslant p$, but for $j \leqslant p$ has a slightly different form. This is not a wave function, because for $j \leqslant p$, it does not correspond to the wave function ψ_j of figure 10.5.

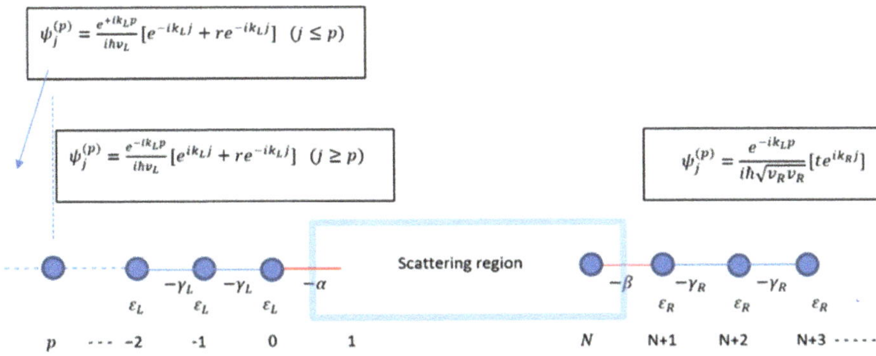

Figure 10.6. A function which is identical to that of figure 10.5 for $j \geqslant p$, but for $j \leqslant p$ has a slightly different form.

Clearly the function of figure 10.6 can be written as shown in figure 10.7, where the presence of the modulus in the exponent means that this function is not a wave function, but instead satisfies equation 10.23. In other words $\psi_j^{(p)}$ is the retarded Green's function $G_{jp}(E)$ of the system, subject to the condition that site p belongs to the left lead;. i.e. $p \leqslant 0$.

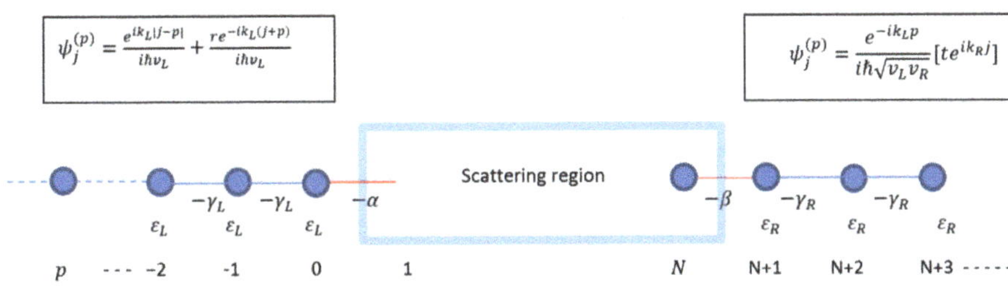

Figure 10.7. The function of figure 10.6, rewritten using a modulus in the exponent.

Since $G_{jp}(E) = \psi_j^{(p)}$, this means that if $G_{jp}(E)$ can be calcuated, then the transmission and reflection amplitudes can be obtained. Indeed from figure 10.7, for $p \leqslant 0$ and $j \leqslant 0$, $G_{jp}(E)$ has the form

$$G_{jp}(E) = \frac{e^{ik_L|j-p|}}{i\hbar v_L} + \frac{re^{-ik_L(j+p)}}{i\hbar v_L} \quad (p \leqslant 0, j \leqslant 0) \tag{10.47}$$

Hence

$$r = i\hbar v_L e^{ik_L(j+p)}\left[G_{jp}(E) - \frac{e^{ik_L|j-p|}}{i\hbar v_L}\right] \quad (p \leqslant 0, j \leqslant 0) \tag{10.48}$$

As an example, choosing $j = p = 0$ yields

$$\boxed{r = i\hbar v_L G_{00}(E) - 1} \tag{10.49}$$

Similarly, for $p \leqslant 0$ and $j \geqslant N + 1$,

$$G_{jp}(E)\frac{e^{-ik_Lp}}{i\hbar\sqrt{v_Lv_R}}[te^{ik_Rj}] \quad (p \leqslant 0, j \geqslant N + 1) \tag{10.50}$$

Hence choosing $p = 0$ and $j = N + 1$ yields

$$\boxed{t = i\hbar\sqrt{v_Lv_R}\,e^{-ik_R(N+1)}G_{N+1,0}(E)} \tag{10.51}$$

Similarly for an incoming wave from the right, choosing $p \geqslant N + 1$ and repreating the above argument yields for reflection and transmission amplitudes from the right,

$$\boxed{r' = e^{-2ik_R(N+1)}[i\hbar v_R G_{N+1,N+1}(E) - 1]} \tag{10.52}$$

and

$$\boxed{t' = i\hbar\sqrt{v_Lv_R}\,e^{-ik_R(N+1)}G_{0,N+1}(E)} \tag{10.53}$$

Taking the modulus squared of these equations yields the reflection coefficient $R = |r|^2 = |r'|^2$ and the transmission coefficient $T = |t|^2 = |t'|^2$:

$$\boxed{T = \hbar v_L \hbar v_R\,|G_{0,N+1}(E)|^2 = \hbar v_L \hbar v_R\,|G_{N+1,0}(E)|^2} \tag{10.54}$$

and

$$\boxed{R = |i\hbar v_L G_{00}(E) - 1|^2 = |i\hbar v_R G_{N+1,N+1}(E) - 1|^2} \tag{10.55}$$

The above expressions for r, t, r' and t' are extremely useful, since as shown in the following chapter, there are powerful methods for computing Green's functions of arbitrary scatterers connected to semi-infinite leads (figure 10.8). Once the Green's function is known, equations (10.49)–(10.53) yield the associated scattering amplitudes, from which quantities such as the electrical conductance can be calculated. In the notation of figure 8.1, the wave function inside the scatterer, due to an incident wave of unit current is denoted f_j. From figures 10.4–10.7, it is clear that if $p \leqslant 0$,

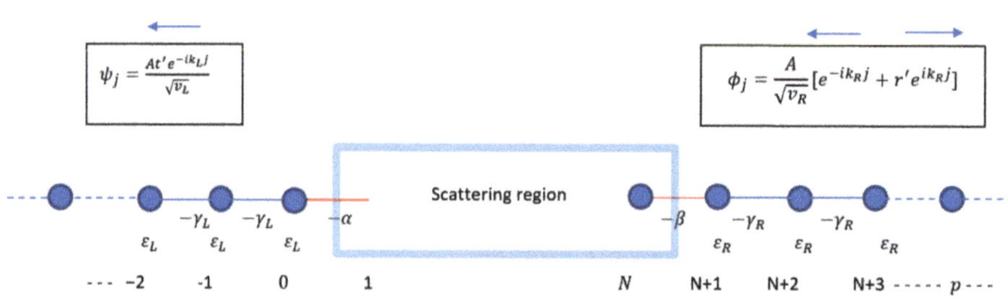

Figure 10.8. An incoming wave from the right, with reflection and transmission amplitudes r' and t' respectively.

$$f_j = i\hbar\sqrt{v_L}\, e^{ik_L p} G_{jp}(E) \quad (p \leqslant 0) \tag{10.56}$$

Similarly the wave function inside the scatterer due to an incoming wave of unit current from the right is

$$f_j = i\hbar\sqrt{v_R}\, e^{-ik_R p} G_{jp}(E) \quad (p \geqslant N + 1) \tag{10.57}$$

The bond current $J_{l \to j}$ between sites l and j within the scatterer is given by equation (8.49). For example if the wave is incident from the left,

$$J_{l \to j} = -\frac{2}{\hbar}\mathrm{Im}\left[f_l^* H_{l,j} f_j\right] = -2\hbar v_L \mathrm{Im}\left[G_{lp}^* H_{l,j} G_{jp}\right] \quad (p \leqslant 0) \tag{10.58}$$

10.8 Green's function of a linear chain with periodic boundary conditions

So far analytic expressions for the Green's functions of open systems have been obtained. Since any wave function can be added to such a Green's function, the Green's functions of open systems are not uniquely defined by the equation $(EI - H)G = I$. The retarded and advanced Green's functions are merely two examples corresponding to particular choices of A and B in equation (10.4). In contrast, for a finite system of N sites, equation (10.3) tells us that there is a unique solution corresponding to the inverse matrix $G = (E - H)^{-1}$. Consequently if an analytic solution for $G_{jp}(E)$ is found, then it is unique. Furthermore, if E is real, G will be Hermitian; i.e. $G_{jp}(E) = G_{pj}^*(E)$ and if H is real, then $G_{jp}(E)$ will be real. This is in contrast with the retarded and advanced Green's functions of infinite chains. In fact the Hermitian conjugate of the retarded Green's function (10.39) is equal to the advanced Green's function (10.41).

In this section, an analytic formula for the Green's function of a finite linear chain with periodic boundary conditions is obtained. Since the Green's function of the structure of figure 4.5, with periodic boundary conditions, must be real, consider a solution of the form

$$G_{jp}(E) = \psi_j^{(p)} = \frac{Ae^{ik(E)|j-p|}}{i\hbar v(E)} - \frac{Be^{-ik(E)|j-p|}}{i\hbar v(E)}$$

Since the retarded Green's function and the advanced Green's function each satisfy $(E - H)G = 1$, the above Green's function will satisfy $(E - H)G = A + B$, the latter must be chosen to satisfy $A + B = 1$. Also since $G_{jp}(E)$ is real, $B = A^*$. Writing $A = iDe^{i\theta}$, where D is real then yields

$$G_{jp}(E) = \frac{2D}{\hbar v(E)} \cos\left[k(E)|j - p| + \theta\right]$$

Periodic boundary conditions require $G_{N+1p}(E) = G_{1p}(E)$. That is

$$\cos\left[k(E)|N + 1 - p| + \theta\right] = \cos\left[k(E)|1 - p| + \theta\right]$$

Since for any choice of p, $|N + 1 - p| = N + 1 - p$ and $|1 - p| = p - 1$, this yields

$$\cos[k(E)(N + 1 - p) + \theta] = \cos[k(E)(p - 1) + \theta]$$

Since for any x, $\cos x = \cos(-x)$, this yields $\theta = -\frac{kN}{2}$. Also, since $A + B = 1$,

$$iDe^{i\theta} - iDe^{-i\theta} = 1$$

That is $D = \frac{-1}{2\sin\theta} = \frac{1}{2\sin kN/2}$.

Hence one finds that the Green's function of a chain of N sites with periodic boundary conditions is

$$\boxed{G_{jp}(E) = A\cos[k(E)(|j - p| - N/2)]} \tag{10.59}$$

where

$$A = \frac{1}{2\gamma\sin k(E)\sin k(E)N/2} \tag{10.60}$$

Again the presence of the modulus ensures that this is a Green's function, not an eigenstate. Note that this diverges when $\sin k(E)N/2 = 0$; i.e. when $k(E) = 2\pi n/N$, where n is an integer. From equation (4.31), this occurs when E coincides with an eigenvalue of the chain. Note that the above result reproduces equation (10.4), when $\varepsilon = 0$, provided that one notes that for $N = 2$, equation (10.59) describes a two-site structure with a nearest neighbour coupling of -2γ, instead of simply $-\gamma$.

The mid-gap magic number table for this chain can be constructed by choosing $E = 0$, which corresponds to $k = \frac{\pi}{2}$. In this case, it is convenient to define $M_{jp} = \frac{G_{jp}(0)}{A} = \cos\left[\frac{\pi}{2}(|j - p| - N/2)\right]$. For a molecule obeying for a molecule obeying Hückel's rule, $N = 4n + 2$,where n is an integer, this yields

$$M_{jp} = (-1)^n \sin\frac{\pi}{2}|j - p|$$

That is $M_{jp} = 0$ if j, p are both odd or both even and $\left|M_{jp}\right| = 1$ otherwise. For example for a phenyl ring, where $N = 6$, if j, p are both odd or both even, this corresponds to a meta connectivity with destructive quantum interference (DQI) at the gap centre, otherwise it corresponds to ortho- or para- connectivity with constructive quantum interference (CQI) at the gap centre. For molecules not obeying Huckel's rule, such that $N = 4n$, the amplitude A diverges at $E = 0$. As discussed in chapter 2, this signals the fact that such molecules are unstable and may undergo a Peierl's distortion.

10.9 Green's functions of a linear chain with free-end boundary conditions

In this section, the task is to derive the Green's function of the linear chain of N tight-binding sites, with free-end boundary conditions, shown in figure 4.8 of chapter 4. In view of the boundary conditions (4.54) and (4.55), the Green's function $\psi_j^{(p)}$ must vanish at the first missing sites at each end of the chain, and therefore

$$\psi_0^{(p)} = \psi_{N+1}^{(p)} = 0 \tag{10.61}$$

When solving for the wave function of a linear chain in chapter 4, these boundary condition led to quantisation of the energy E. When computing a Green's function, E is a continuous variable, which can take any value. Therefore to satisfy the boundary conditions, $\psi_j^{(p)}$ must be chosen to have a functional form, such that for $j \leqslant p$, it vanishes at $j = 0$ and for $j \geqslant p$ it vanishes at $j = N + 1$. Since it must also satisfy equation (10.24), one obtains

$$\psi_j^{(p)} = \chi_j \text{ for } j \leqslant p, \text{ where } \chi_j = A \sin kj \tag{10.62}$$

and

$$\psi_j^{(p)} = \phi_j \text{ for } j \geqslant p, \text{ where } \phi_j = B \sin k\,[j - (N + 1)] \tag{10.63}$$

In these equations, $k = k(E)$. Having satisfied the boundary conditions, note that continuity at $j = p$ yields

$$\chi_p = \phi_p \tag{10.64}$$

That is

$$A \sin kp = B \sin k[p - (N + 1)]$$

Hence

$$A = C \sin k[p - (N + 1)] \text{ and } B = C \sin kp \tag{10.65}$$

where C is a constant.

Finally from equation (10.24), choosing $j = p$ yields

$$\varepsilon_o\psi_p^{(p)} - \gamma\psi_{p-1}^{(p)} - \gamma\psi_{p+1}^{(p)} = E\psi_p^{(p)} - 1 \tag{10.66}$$

i.e.

$$\varepsilon_o\chi_p - \gamma\chi_{p-1} - \gamma\phi_{p+1} = E\chi_p - 1 \tag{10.67}$$

Since $\varepsilon_o\chi_p - \gamma\chi_{p-1} - \gamma\chi_{p+1} = E\chi_p$, this simplifies to

$$\gamma\chi_{p+1} - \gamma\phi_{p+1} = -1 \tag{10.68}$$

Note that

$$\chi_{p+1} = A \sin k(p + 1) = \chi_p \cos k + A \cos kp \sin k$$

and

$$\phi_{p+1} = B \sin k[(p + 1) - (N + 1)] = \phi_p \cos k + B \cos k[p - (N + 1)] \sin k$$

Hence, in view of equation (10.64), this yields

$$\chi_{p+1} - \phi_{p+1} = A \cos kp \sin k - B \cos k[p - (N + 1)] \sin k.$$

Hence

$$\frac{\chi_{p+1} - \phi_{p+1}}{C} = \{\sin k[p - (N + 1)] \cos kp - \sin kp \cos k[p - (N + 1)]\} \sin k$$

The term in the curly brackets is of the form $\sin x \cos y - \sin y \cos x = -\sin(y - x)$. Therefore

$$\frac{\chi_{p+1} - \phi_{p+1}}{C} = -\sin k(N + 1) \sin k$$

From equation (10.68), this yields

$$C = \frac{1}{\gamma \sin k(N + 1) \sin k} \tag{10.69}$$

From equations (10.62), (10.63) and (10.65), one obtains

$$G_{jp}(E) = C \sin k[p - (N + 1) \sin kj \quad \text{for } j \leqslant p, \tag{10.70}$$

and

$$G_{jp}(E) = C \sin kp \sin k[j - (N + 1)] \quad \text{for } j \geqslant p, \tag{10.71}$$

The change in functional form when $j = p$ ensures that this is a Green's function, not an eigenstate. Note that this diverges when $\sin k(E)(N + 1) = 0$; i.e. when $k(E) = n\pi/(N + 1)$, where n is an integer. From equation (4.57), this occurs when E coincides with an eigenvalue of the linear chain, as expected from equation (10.6). At the gap centre $E = 0$,where $k(E) = \frac{\pi}{2}$, C diverges when N is odd, signalling the fact that such chains are unstable. For even N, writing $N = 2n$ yields for the magic number table $M_{jp} = \frac{G_{jp}(0)}{C}$,

$$\begin{aligned} M_{jp} &= (-1)^{n+1} \sin \frac{\pi j}{2} \cos \frac{\pi p}{2} \quad (\text{for } j \leqslant p) \\ M_{jp} &= (-1)^{n+1} \sin \frac{\pi p}{2} \cos \frac{\pi j}{2} \quad (\text{for } j \geqslant p) \end{aligned} \tag{10.72}$$

As expected for such a bipartite lattice, this shows that $M_{jp} = 0$ if j, p are both odd or both even. Furthermore, for $j \leqslant p$, $M_{jp} = 0$ if j is even or p is odd and for $j \geqslant p$, $M_{jp} = 0$ if p is even or j is odd. Otherwise, $|M_{jp}| = 1$.

As a check on the above expressions, it is interesting to note that that for the special case $N = 2$, a ring with periodic boundary conditions and nearest neighbour couplings $-\gamma$ is identical to a linear chain with free-end boundary conditions and nearest neighbour couplings $-\alpha = -2\gamma$. Therefore they should give the same result for e.g. G_{12}. To demonstrate that this is indeed the case, note that for the $N = 2$ ring, equation (10.59) yields

$$G_{12}(E) = \frac{1}{2\gamma \sin^2 k(E)} = \frac{1}{2\gamma[1 - \cos^2 k(E)]} = \frac{1}{2\gamma\left[1 - \left[\frac{\epsilon_0 - E}{2\gamma}\right]^2\right]}$$

where it was noted that $\cos^2 k(E) = \left[\frac{\epsilon_0 - E}{2\gamma}\right]^2$. For the $N = 2$ chain, equation (10.70) yields

$$G_{12}(E) = \frac{-\sin k(E)}{\alpha \sin 3k(E)} = \frac{1}{\alpha(1 - 4\cos^2 k(E))}$$

where it was noted that $\sin 3\theta = (4\cos^2\theta - 1)\sin\theta$. Since $E = \epsilon_0 - 2\alpha\cos k(E)$, $\cos^2 k(E) = \left[\frac{\epsilon_0 - E}{4\gamma}\right]^2$ and therefore the two expressions agree.

10.10 Green's functions of a finite-width nanoribbon

Consider the finite-width ribbon shown in figure 5.7, comprising a crystalline linear chain of identical cells, in which each site of a cell is connected to the corresponding site of its neighbours by a bond $-\gamma$. The most general eigenstate of such a structure in unit cell J is (see equations (5.53) and (5.54)) is

$$|\psi_J\rangle = \sum_{n=1}^{N}(A_n e^{ik_n J} + B_n e^{-ik_n J})|\chi_n\rangle \tag{10.73}$$

where $|\chi_n\rangle$ is a solution of

$$H_0|\chi_n\rangle = \epsilon_n|\chi_n\rangle \tag{10.74}$$

and

$$k_n(E) = \text{acos}\left(\frac{\epsilon_n - E}{2\gamma}\right) \tag{10.75}$$

This satisfies the Schrödinger equation for the structure (see equation (5.25)) namely

$$H_0|\psi_J\rangle + H_1|\psi_{J+1}\rangle + H_{-1}|\psi_{J-1}\rangle = E|\psi_J\rangle \tag{10.76}$$

From this solution, one can construct the retarded Green's function by setting all $B_n = 0$ in equation (10.73) and noting that the following quantity also satisfies equation (10.76), because each individual term in the sum satisfies (10.76)

$$g_J = \sum_{n=1}^{N}(A_n e^{ik_n(E)J})|\chi_n\rangle\langle\chi_n|$$

Choosing $A_n = \frac{e^{-ik_n P}}{i\hbar v_n}$, where P is the label of cell P yields

$$g_J = \sum_{n=1}^{N}\frac{e^{ik_n(E)(J-P)}}{i\hbar v_n(E)}|\chi_n\rangle\langle\chi_n|$$

which still satisfies equation (10.76). To obtain a Green's function $G_{JP}(E)$ satisfying

$$H_0 G_{JP}(E) + H_1 G_{J+1,P}(E) + H_{-1}G_{J-1,P}(E) = EG_{JP}(E) - I\delta_{JP} \tag{10.77}$$

simply introduce a modulus into the above expression to yield the following equation for the retarded Green's function of the nanoribbon

$$\boxed{G_{JP}(E) = \sum_{n=1}^{N} \frac{e^{ik_n(E)|J-P|}}{i\hbar v_n(E)} |\chi_n\rangle\langle\chi_n|} \tag{10.78}$$

For $J \neq P$, this clearly satisfies equations (10.76) and (10.77). To demonstrate that it satisfies equation (10.77) for $J = P$, note that for $J = P$ equation (10.77) becomes

$$H_0 G_{PP}(E) + H_1 G_{P+1,P}(E) + H_{-1} G_{P-1,P}(E) = EG_{PP}(E) - I \tag{10.79}$$

Since $H_0|\chi_n\rangle = \epsilon_n|\chi_n\rangle$ and $H_1 = H_{-1} = -\gamma I$,

$$\begin{aligned} H_0 G_{PP}(E) + H_1 G_{P+1,P}(E) + H_{-1} G_{P-1,P}(E) &= \sum_{n=1}^{N} \frac{(\epsilon_n - \gamma e^{ik_n} - \gamma e^{ik_n})}{i\hbar v_n} |\chi_n\rangle\langle\chi_n| \\ &= \sum_{n=1}^{N} \frac{(\epsilon_n - 2\gamma\cos k_n - 2i\gamma\sin k_n)}{i\hbar v_n} |\chi_n\rangle\langle\chi_n| \\ &= \sum_{n=1}^{N} \frac{(E - 2i\gamma\sin k_n)}{i\hbar v_n} \rangle|\chi_n|\langle\chi_n| \\ &= EG_{PP}(E) - \sum_{n=1}^{N} |\chi_n\rangle\langle\chi_n| \end{aligned}$$

Since the eigenstates $|\chi_n\rangle$ form a complete set, $\sum_{n=1}^{N} |\chi_n\rangle\langle\chi_n| = I$ and hence equation (10.79) is satisfied.

If α labels a site in cell J and β labels a site in cell P, then equation (10.78) yields

$$\boxed{G_{J\alpha,P\beta}(E) = \sum_{n=1}^{N} \frac{e^{ik_n(E)|J-P|}}{i\hbar v_n(E)} \langle\alpha|\chi_n\rangle\langle\chi_n|\beta\rangle} \tag{10.80}$$

As an example from equation (5.35), for a finite-width ribbon, which is N sites wide, with free-end boundary condition in the transverse direction,

$$\langle\alpha|\chi_n\rangle = \chi_\alpha^{(n)} = \sqrt{\frac{2}{N+1}} \sin\frac{n\pi}{(N+1)}\alpha \tag{10.81}$$

By analogy with the discussion leading to equations (10.44)–(10.46), the Green's function of a semi-infinite lead, which terminates at cell $L \leqslant P$ is is

$$\boxed{G_{JP}(E) = \sum_{n=1}^{N} \frac{e^{ik_n(E)|J-P|} - e^{ik_n(E)(J+P-2L+2)}}{i\hbar v_n(E)} |\chi_n\rangle\langle\chi_n|} \tag{10.82}$$

Hence the surface Green's function on the terminal cell L of a semi-infinite lead is

$$\boxed{G_{LL}(E) = \sum_{n=1}^{N} \frac{-e^{ik_n(E)}}{\gamma} |\chi_n\rangle\langle\chi_n|} \tag{10.83}$$

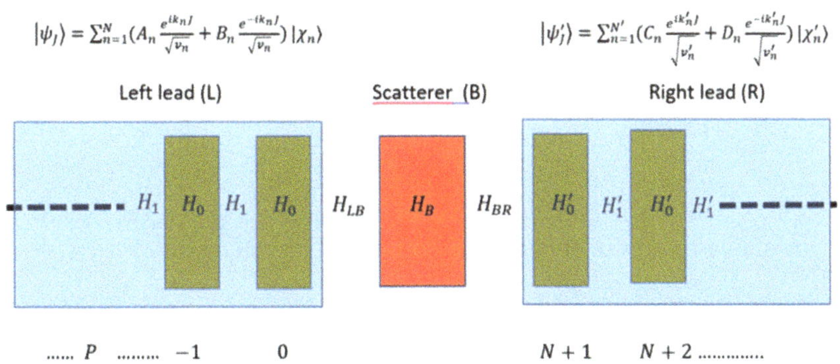

Figure 10.9. A scatterer 'B' connected to left and right semi-infinite leads. H_{LB} and H_{BR} are sub-matrices of the Hamiltonian connecting the scatterer to the left and right leads respectively.

10.11 Relationships between Green's functions and transmission amplitudes

To write down the Green's function in the presence of a scatter, consider the structure shown in figure 6.5, which is reproduced in figure 10.9 for convenience. Choosing the incoming amplitudes on the right to be zero (i.e. $D_n = 0$ for all n) allows the wave function on the left to be written (cf equation (6.60)).

$$|\psi_J\rangle = \sum_{n=1}^{M} A_n \left[\frac{e^{ik_nJ}}{\sqrt{v_n}} |\chi_n\rangle + \sum_m r_{mn} \frac{e^{-ik_mJ}}{\sqrt{v_m}} |\chi_m\rangle \right] \tag{10.84}$$

In this equation, n sums over all open channels in the left lead and m sums over all channels, both open and closed, because evanescent states which decay as $J \to -\infty$ are allowed.

Similarly the wave function on the right is (cf equation (6.61))

$$|\psi'_J\rangle = \sum_n A_n \sum_{m=1}^{N'} \left(t_{mn} \frac{e^{ik'_mJ}}{\sqrt{v'_m}} \right) |\chi'_m\rangle \tag{10.85}$$

where m sums over all channels in the right lead. These equations recognise that an incident wave $\frac{e^{ik_nJ}}{\sqrt{v_n}}|\chi_n\rangle$ produces a sum of reflected waves on the left of the form $\sum_m r_{mn} \frac{e^{-ik_mJ}}{\sqrt{v_m}} |\chi_m\rangle]$ and a sum of transmitted waves on the right of the form $\sum_{m=1}^{N'} \left(t_{mn} \frac{e^{ik'_mJ}}{\sqrt{v'_m}} \right) |\chi'_m\rangle$. Therefore a linear superposition of incident waves of the form $\sum_n A_n \frac{e^{ik_nJ}}{\sqrt{v_n}} |\chi_n\rangle$ produces a corresponding superposition of reflected and transmitted waves. As an example of such a wave function, one could choose $A_n = \frac{e^{-ik_nP}}{i\hbar\sqrt{v_n}}$. By analogy with figure 10.7 and equation (10.47), if P labels a cell in the left lead, the

Green's function $G_{JP}(E)$ in the left lead of the structure is obtained by introducing a modulus, as shown below

$$G_{JP}(E) = \sum_{n=1}^{N}\left[\frac{e^{ik_n|J-P|}}{i\hbar v_n}|\chi_n\rangle\langle\chi_n| + \sum_m r_{mn}\frac{e^{-ik_mJ}e^{-ik_nP}}{i\hbar\sqrt{v_nv_m}}|\chi_m\rangle\langle\chi_n|\right](P \leqslant 0,\ J \leqslant 0) \quad (10.86)$$

In this equation, both n and m sum over all channels in the left lead, because evanescent states originating from the source at P, which decay as $J \to -\infty$ are allowed.

Similarly the Green's function in the right lead is

$$G_{JP}(E) = \sum_{n=1}^{N}\sum_{m=1}^{N'} t_{mn}\frac{e^{ik_m'J}e^{-ik_nP}}{i\hbar\sqrt{v_nv_m'}}|\chi_m'\rangle\langle\chi_n|(P \leqslant 0,\ J \geqslant N+1) \quad (10.87)$$

where n sums over all channels in the left lead and m sums over all channels in the right lead.

Hence, the reflection and transmission coefficients associated with open channels n, m are obtained from

$$\boxed{\langle\chi_m|G_{00}(E)|\chi_n\rangle = \frac{\delta_{mn}}{i\hbar v_n} + \frac{r_{mn}}{i\hbar\sqrt{v_nv_m}}} \quad (10.88)$$

and

$$\boxed{\langle\chi_m|G_{0,N+1}(E)|\chi_n'\rangle = \frac{t_{mn}'e^{ik_n'(N+1)}}{i\hbar\sqrt{v_mv_n'}}} \quad (10.89)$$

Similarly, one finds

$$\boxed{\langle\chi_m'|G_{N+1,N+1}(E)|\chi_n'\rangle = \frac{\delta_{mn}}{i\hbar v_n} + \frac{r_{mn}'e^{i(k_m+k_n')(N+1)}}{i\hbar\sqrt{v_n'v_m'}}} \quad (10.90)$$

and

$$\boxed{\langle\chi_m|G_{0,N+1}(E)|\chi_n'\rangle = \frac{t_{mn}'e^{ik'_n(N+1)}}{i\hbar\sqrt{v_mv_n'}}} \quad (10.91)$$

10.12 Summary

In this chapter, analytic expressions were obtained for the Green's functions of linear chains with either free-end boundary conditions or periodic boundary conditions. Furthermore, by recognising that a Green's function of a linear chain is a wave function of a T-shaped structure, general equations relating Green's function to reflection and transmission amplitudes were obtained. The latter mean that if Dyson's equation of some other method is used to calculate Green's functions, then measurable properties such as electrical conductances and Seebeck coefficients can be predicted. Methods for computing Green's functions of complex structures are discussed in the following chapters.

Chapter 11

Connectivity theory revisited: heteroatom substitution, decimation and the Breit–Wigner formula

11.1 Introduction

In chapter 2, connectivity theory was introduced from an abstract viewpoint, simply by considering the relationship between sources and the waves they create. Without specifying the physical nature of the underlying structures, this led to the two-component noisy neighbour equations (NNEs) (2.34), equations for the Green's functions of coupled substructures, to NNEs for three-component and five-component structures.

In chapter 10, some simple examples of Green's functions were considered and analytic formulae were obtained for the Green's functions of isolated scatterers (i.e. closed systems) formed from a linear chain of N sites with free-end boundary conditions, or a chain of sites with periodic boundary conditions. At the simplest level of modelling, the former can be considered to be the Green's function of a cumulene and the latter the Green's function of an aromatic ring. In addition, an analytic formula for the Green's function of an open system formed from an infinite one-dimensional chain was obtained and the relationship between wave functions and Green's functions highlighted. This discussion led to formulae, which allow transmission and reflection amplitudes to be obtained directly from the Green's function G of an open system formed from a scatterer connected to crystalline leads.

In this chapter, connectivity theory is revisited, starting from Dyson's equations and working backwards towards the NNEs. This will provide a detailed step-by-step derivation of the three- and five-component NNEs and by completing the circle, will demonstrate that connectivity theory is a self-consistent theory. Connectivity theory aims to describe how the components of a quantum structure, such as a molecule, combine to yield the properties of the whole system. For example the structure shown in figure 11.1 is composed of two substructures, coupled together and

doi:10.1088/978-0-7503-3639-0ch11

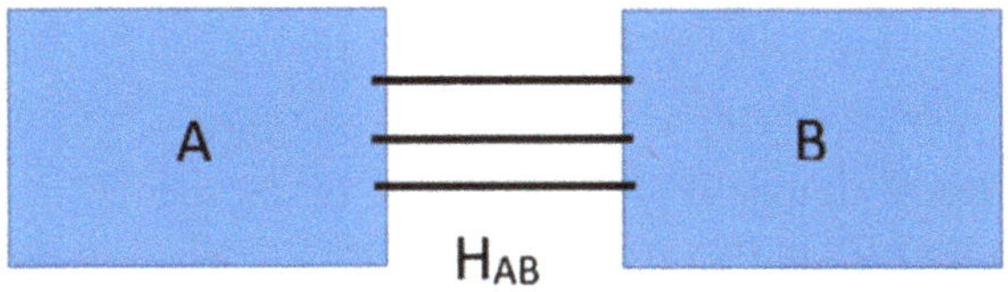

Figure 11.1. Two quantum substructures A and B, interacting via a coupling matrix H_{AB}.

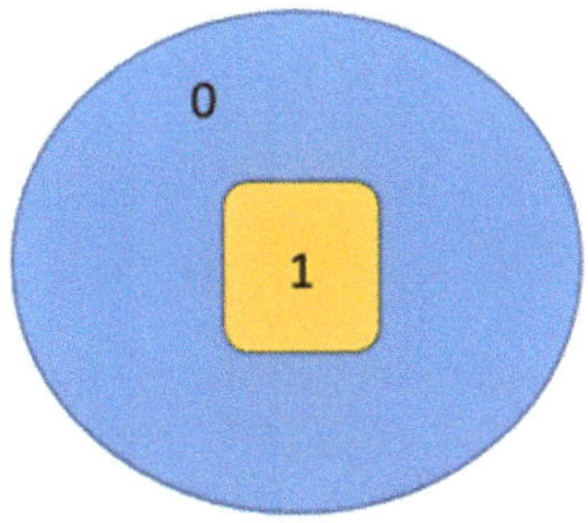

Figure 11.2. A quantum structure '0' containing a finite region '1' (shaded yellow), which is subject to a local perturbation.

connectivity theory describes how the presence of B affects the quantum properties of A and vice versa and how they both combine to determine the properties of the whole.

A closely related issue is shown in figure 11.2, in which one would like to describe the effect of perturbing a substructure (shaded yellow) of the whole, perhaps by performing a heteroatom substitution, imposing a local mechanical distortion, or applying a local electric field. To describe interference effects in these structures, this chapter will discuss three theoretical approaches, namely recursion, decimation, and perturbation.

11.2 Recursively describing the effect of changes to a molecular junction

The Green's function G of a structure described by a Hamiltonian H is defined to satisfy the equation

$$(EI - H)G = I \tag{11.1}$$

where I is the unit matrix. If H is a finite-size matrix, then the inverse of $(E - H)$ can be obtained using a computer and G is given by $G = (E - H)^{-1}$. On the other hand, for open systems such as a molecular junction containing electrodes which extend to infinity, H and G are infinite matrices and this matrix inversion cannot be carried out. Therefore in what follows, care will be taken to distinguish between infinite- and finite-size matrices.

In many cases, H may be the sum of two simpler matrices and can be written $= H_0 + H_1$, so the problem is to solve the equation

$$(EI - H_0 - H_1)G = I \tag{11.2}$$

This equation arises if one starts from a parent structure described by a Hamiltonian H_0 and then creates a 'daughter' by adding an extra term H_1 to H_0 to create the daughter Hamiltonian H. In this way, H_1 describes the effect of making an arbitrary change to the parental structure.

If the Green's function g_0 associated with the Hamiltonian H_0 is obtained by solving the equation

$$(EI - H_0)g_0 = I \tag{11.3}$$

then terms of g_0, the solution to equation (11.2) can be written

$$G = g_0 + g_0 H_1 G \tag{11.4}$$

To prove this, multiply equation (11.4) from the left by $(EI - H_0)$ to yield $(EI - H_0)G = (EI - H_0)g_0 + (EI - H_0)g_0 H_1 G$. Noting that $(E - H_0)g_0 = I$, yields $(E - H_0)G = I + H_1 G$, which is identical to equation (11.2).

The second term on the right-hand side of equation (11.4) reveals how the perturbation H_1 combines with the parental Green's function g_0 to create the Green's function of the daughter. Since the right-hand side of equation (11.4) contains the Green's function G, further steps are needed to evaluate it. Equation (11.4) can be written

$$(I - g_0 H_1)G = g_0 \tag{11.5}$$

Therefore for a closed system, where G is a finite matrix, one can write

$$G = (I - g_0 H_1)^{-1} g_0 \quad \text{(finite matrices only)} \tag{11.6}$$

However, if the matrices are infinite, as occurs in open systems, then this step is not permissible and further discussion is needed.

The expression (11.4) for G is only one of a number of ways to proceed. As an alternative, one could start from a parent described by a Hamiltonian H_1, whose Green's function g_1 satisfies $(E - H_1)g_1 = I$. Then repeating the above steps yields

$$G = g_1 + g_1 H_0 G \tag{11.7}$$

These two different approaches to computing G yield the same answer for the daughter, even though they take different viewpoints regarding the parent.

Consider now the more complicated problem of computing the Green's function $\tilde{G}$ of the Hamiltonian $\tilde{H} = H_0 + H_1 + H_2$, which satisfies $(E - \tilde{H})\tilde{G} = I$. In this case, there are even more routes to computing $\tilde{G}$. First would could write $H = H_0 + H_1$ and utilise the solutions (11.4), (11.5) or (11.7) to obtain G. Following the above steps, one then finds

$$\tilde{G} = G + G H_2 \tilde{G} \tag{11.8}$$

This shows that the Green's functions $\tilde{G}$ can be computed recursively, by first computing g_1 or g_0, then computing G via equations (11.4) or (11.7) and finally computing $\tilde{G}$ via equation (11.8).

Of course, as alternative recursive routes, one could write $H_4 = H_1 + H_2$, so that $\tilde{H} = H_0 + H_4$. Then from equation (11.4),

$$\tilde{G} = g_0 + g_0 H_4 \tilde{G} = g_0 + g_0(H_1 + H_2)\tilde{G} \tag{11.9}$$

and similarly

$$\tilde{G} = g_1 + g_1(H_0 + H_2)\tilde{G} = g_2 + g_2(H_0 + H_1)\tilde{G} \tag{11.10}$$

The route taken is a matter of choice. For example one might be interested in first understanding the effect of H_1 on the parental Green's function g_0 and then understanding the effect of H_2 on the daughter. Alternatively one might be interested in starting from a parental Green's function g_1 and then incorporating the effect of H_0 and H_1 etc. The final answer is independent of the route taken, but more detailed insight into how a molecule's functions derive from its components may be more obvious, if a particular route is chosen. As will be demonstrated in chapter 14, this flexibility can be used to iterate NNEs in different ways to produce a variety of NNEs, which clarify different aspects of quantum interference in coupled structures.

11.3 Decimation and the effect of pendant groups

The above discussion describes the effect of perturbing a parental structure, by adding an arbitrary perturbation H_1, which does not change the size of the Hamiltonian. Consider now the effect of adding a finite-size pendant substructure 'B' to the parental substructure 'A', which increases the size of the system, as shown in figure 11.1. In this case, the combined structure is described by a Hamiltonian of the form

$$H = \begin{pmatrix} H_{AA} & H_{AB} \\ H_{BA} & H_{BB} \end{pmatrix} \tag{11.11}$$

In principle, H could be obtained self-consistently from a mean-field theory such as density functional theory or Hartree–Fock theory, which ensures that the resulting Hamiltonian H is consistent with the ground-state geometry and charge distribution of the whole structure. In that case, the submatrix H_{AA} is the Hamiltonian of the substructure A in the presence of B and H_{BB} is the Hamiltonian of B in the presence of A. Typically, these differ from the corresponding Hamiltonians describing the isolated gas-phase substructures, since the latter would possess different geometries and charge distributions.

Starting from a self-consistent Hamiltonian H, it is of interest to determine how quantum properties are affected by the presence of the coupling matrix H_{AB}. To address this question, it is useful to first set the coupling H_{AB} to zero, so that the Hamiltonian and Green's function of the parent take the block diagonal forms

$\begin{pmatrix} H_{AA} & 0 \\ 0 & H_{BB} \end{pmatrix}$ and $\begin{pmatrix} g_{AA} & 0 \\ 0 & g_{BB} \end{pmatrix}$, respectively, where g_{AA} and g_{BB} are the Green's functions of the decoupled substructures, defined to satisfy

$$(EI - H_{AA})g_{AA} = I \tag{11.12}$$

and

$$(EI - H_{BB})g_{BB} = I \tag{11.13}$$

If the pendant substructure B is finite, H_{BB} is also finite and therefore equation (11.13) can be written

$$g_{BB} = (EI - H_{BB})^{-1} \tag{11.14}$$

If the substructure B is open, then g_{BB} should be obtained without an explicit matrix inversion, for example as discussed in sections 10.3 and 10.5.

When H_{AB} is non-zero, the Green's function of the whole coupled structure is

$$G = \begin{pmatrix} G_{AA} & G_{AB} \\ G_{BA} & G_{BB} \end{pmatrix} \tag{11.15}$$

and satisfies

$$\begin{pmatrix} EI - H_{AA} & -H_{AB} \\ -H_{BA} & EI - H_{BB} \end{pmatrix} \begin{pmatrix} G_{AA} & G_{AB} \\ G_{BA} & G_{BB} \end{pmatrix} = \begin{pmatrix} I & 0 \\ 0 & I \end{pmatrix} \tag{11.16}$$

Multiplying the two matrices on the left-hand side yields

$$\begin{pmatrix} (EI - H_{AA})G_{AA} - H_{AB}G_{BA} & (EI - H_{AA})G_{AB} - H_{AB}G_{BB} \\ -H_{BA}G_{AA} + (EI - H_{BB})G_{BA} & -H_{BA}G_{AB} + (EI - H_{BB})G_{BB} \end{pmatrix} = \begin{pmatrix} I & 0 \\ 0 & I \end{pmatrix} \tag{11.17}$$

The bottom left-hand corner equation yields $(EI - H_{BB})G_{BA} = H_{BA}G_{AA}$ and when H_{BB} is finite, this leads to the following NNE (cf. equations (2.34)):

$$G_{BA} = g_{BB}H_{BA}G_{AA} \tag{11.18}$$

where g_{BB} is given by equation (11.14). Substituting equation (11.18) into the top-right-hand corner equation (i.e. $(EI - H_{AA})G_{AA} - H_{AB}G_{BA} = I$) yields

$$(EI - H_{AA} - \Sigma_A)G_{AA} = I \tag{11.19}$$

where

$$\Sigma_A = H_{AB}g_{BB}H_{BA} \tag{11.20}$$

Alternatively equation (11.19) can be written

$$(EI - H'_{AA})G_{AA} = I \tag{11.21}$$

where

$$H'_{AA} = H_{AA} + \Sigma_A \tag{11.22}$$

In equations (11.20)–(11.22), Σ_A is called a 'self-energy' and H'_{AA} is called a 'renormalized Hamiltonian,' as first discussed in appendix A.17 of chapter 2. Comparing equation (11.12) with equation (11.21) reveals that the Green's function submatrix G_{AA} of the daughter satisfies the same equation as the Green's function g_{AA} of the parent, except that the Hamiltonian H_{AA} of the parent is replaced by the renormalized Hamiltonian H'_{AA} of the daughter. If only the submatrix G_{AA} is of interest, then this step simplifies the problem of computing G_{AA}. Apart from the presence of the self-energy Σ_A all other information about substructure B is lost from equation (11.21). This process of eliminating a substructure such as B, to obtain an equation for the Green's function of a coupled substructure such as A is called 'decimation'. Nowadays, the word 'decimate' means 'to kill a large portion of a population', whereas in ancient Roman culture it meant 'to kill one on every tenth soldier in a mutinous army'. Here the former usage is implied, because the substructure B has been completely eliminated to obtain the renormalized Hamiltonian H'_{AA} of substructure A.

11.4 Illustration of destructive quantum interference due to pendant atoms

Even though the Hamiltonian H_{AA} may be of infinite dimensionality, decimation can yield insight into the quantum transport properties of a substructure such as A. As a simple example of the utility of decimation, consider the case of a pendant atom attached to a linear chain, as shown in figure 11.3. In this case, substructure B is a single site of energy ϵ_B and therefore $g_B = 1/(E - \epsilon_B)$, whereas substructure A is a doubly infinite linear chain of identical sites, (with site energies ϵ_0), except that site p has a site energy ϵ_p. The coupling matrix H_{AB} has only a single non-zero element $(-\alpha)$ between site p of substructure A and the pendant site.

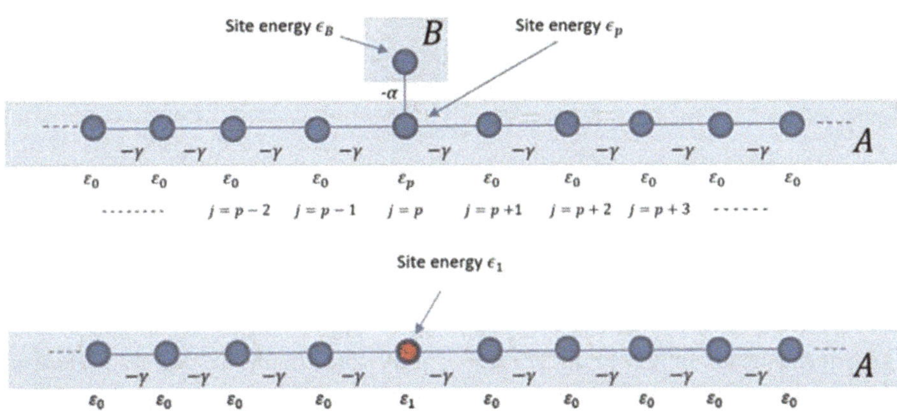

Figure 11.3. Single-site decimation. After decimating the pendant site, the impurity site energy becomes $\epsilon_1 = \epsilon_p + \alpha^2/(E - \epsilon_B)$.

Hence from equation (11.20), the only difference between H'_{AA} and H_{AA} is that the site energy of site p is changed from ϵ_p to a renormalized site energy $\epsilon_1 = \epsilon_p + \alpha^2/(E - \epsilon_B)$. Since $\epsilon_1 = \infty$ when $E = \epsilon_B$, the transmission coefficient of the linear chain must vanish when the energy E of an electron propagating along the linear chain coincides with the energy ϵ_B of the pendant site. In other words, perfect destruction quantum interference (DQI) is predicted to occur when $E = \epsilon_B$. This conclusion is reached without explicitly calculating the transmission coefficient of the chain and would clearly apply to any linear chain with such a pendant atom. For this particular example, the decimated structure is identical to that of figure 6.1 and therefore the transmission coefficient is given by (see equation (6.59))

$$T(E) = \frac{1}{\left(1 + \left|\frac{(\epsilon_1 - \epsilon_0)}{\gamma 2 \sin k(E)}\right|^2\right)} \tag{11.23}$$

As expected $T(E) = 0$ when $\epsilon_1 = \infty$.

11.5 The effect of a local perturbation

In the analysis of section 11.3, the subsystem A could be open, in which case it is not possible to invert the matrix $EI - H_{AA}$. On the other hand, as demonstrated by the example of figure 11.3, it is reasonable to assume that the coupling matrix H_{AB} contains only a finite number of non-zero elements. This arises, because the interactions contained in H_{AB} have a finite range and therefore if H_{AB} couples to a finite number of sites in B, then it will only couple to a finite number of sites in A, even if the substructure A is infinite. Before examining the consequences of the finite range of the coupling matrix H_{AB}, in this section, the effect of adding an arbitrary finite-range perturbation to a possibly infinite parental Hamiltonian is discussed.

Consider a parental system described by a Hamiltonian H_0 of dimensionality N_0, with a known Green's function g satisfying $(E - H_0)g = I$. In what follows, N_0 could be finite or infinite. Our task is to compute the new Green's function G of the daughter when a perturbation h_{11} changes the Hamiltonian matrix elements of a finite sub-set of N_1 sites, where $N_1 < N_0$. Since only matrices of the same dimensions can be added together, the daughter Hamiltonian is $H = H_0 + H_1$, where H_1 a $N_0 \times N_0$ perturbation of the form

$$H_1 = \begin{pmatrix} h_{11} & 0 \\ 0 & 0 \end{pmatrix} \tag{11.24}$$

In this expression, the finite-size submatrix h_{11} is a $N_1 \times N_1$ square matrix. Clearly the introduction of h_{11} creates a partitioning of the whole structure into $N_1 \times N_1$ and $(N_0 - N_1) \times (N_0 - N_1)$ substructures denoted 1 and 0 respectively and coloured yellow and blue in figure 11.2. Therefore it is convenient to express G and g in terms of submatrices as follows:

$$G = \begin{pmatrix} G_{11} & G_{10} \\ G_{01} & G_{00} \end{pmatrix} \quad \text{and} \quad g = \begin{pmatrix} g_{11} & g_{10} \\ g_{01} & g_{00} \end{pmatrix}, \tag{11.25}$$

where G_{11} and g_{11} are finite submatrices and G_{00} could be an infinite matrix. Our task is to solve the equation

$$(EI - H_0 - H_1)G = I \tag{11.26}$$

As discussed in section (11.2), the solution to this is

$$G = g + gH_1G \tag{11.27}$$

From equations (11.24) and (11.25), equation (11.27) is of the form

$$\begin{pmatrix} G_{11} & G_{10} \\ G_{01} & G_{00} \end{pmatrix} = \begin{pmatrix} g_{11} & g_{10} \\ g_{01} & g_{00} \end{pmatrix} + \begin{pmatrix} g_{11} & g_{10} \\ g_{01} & g_{00} \end{pmatrix}\begin{pmatrix} h_{11}G_{11} & h_{11}G_{10} \\ 0 & 0 \end{pmatrix}$$

That is

$$G = \begin{pmatrix} G_{11} & G_{10} \\ G_{01} & G_{00} \end{pmatrix} = \begin{pmatrix} g_{11} & g_{10} \\ g_{01} & g_{00} \end{pmatrix} + \begin{pmatrix} g_{11}h_{11}G_{11} & g_{11}h_{11}G_{10} \\ g_{01}h_{11}G_{11} & g_{01}h_{11}G_{10} \end{pmatrix}$$

Hence the submatrices G_{IJ} are given by

$$G_{IJ} = g_{IJ} + g_{I1}h_{11}G_{1J} \quad (I, J = 1 \text{ or } 0) \tag{11.28}$$

Choosing $I = 1$ yields $G_{1J} = g_{1J} + g_{11}h_{11}G_{1J}$. Hence $(1 - g_{11}h_{11})G_{1J} = g_{1J}$, i.e.

$$G_{1J} = (I - g_{11}h_{11})^{-1}g_{1J} \tag{11.29}$$

After substituting this into equation (11.28), one obtains

$$G_{IJ} = g_{IJ} + g_{I1}U_{11}g_{1J} \quad (I, J = 1 \text{ or } 0) \tag{11.30a}$$

where U_{11} is a renormalized perturbation, given by

$$U_{11} = h_{11}(I - g_{11}h_{11})^{-1} \tag{11.31a}$$

Equation (11.30a) is an expression for each of the four submatrices of G and therefore is also an expression for any of the individual matrix elements of G. When written in terms of individual matrix elements, it takes the form

$$G_{ij} = g_{ij} + \sum\nolimits_{p,q=1}^{N_1} g_{ip}(U_{11})_{pq}g_{qj} \tag{11.30b}$$

Note that from equation (11.29), choosing $J = 1$ or from equation (11.30a), choosing $I = J = 1$, yields for the finite-size submatrix G_{11},

$$G_{11} = \left(g_{11}^{-1} - h_{11}\right)^{-1} = g_{11} + g_{11}U_{11}g_{11} \tag{11.32}$$

Hence, comparing the first of equation (11.32) with equation (11.31) yields the following alternative expression for U_{11}

$$U_{11} = h_{11}G_{11}g_{11}^{-1} \tag{11.31b}$$

Furthermore, $h_{11}G_{11} = h_{11}g_{11}(1 - h_{11}g_{11})^{-1} = (1 - 1 + h_{11}g_{11})(1 - h_{11}g_{11})^{-1} = (1 - h_{11}g_{11})^{-1} - 1 = -1 + g_{11}^{-1}G_{11}$. Hence from equation (11.30a)

$$G_{I1} = g_{I1}g_{11}^{-1}G_{11} \quad \text{and} \quad G_{1J} = G_{11}g_{11}^{-1}g_{1J} \tag{11.33}$$

The fundamental result is that all submatrices G_{IJ} of the daughter Green's function G can be obtained from equation (11.30), once the finite matrices G_{11} and/or U_{11} are obtained from equations (11.31)–(11.33). Since the latter involve only $N_1 \times N_1$ matrices, the problem is considerably reduced. This reduction occurs, because the perturbation H_1 is localised and acts only on the substructure 1, rather than the whole (possibly infinite) structure.

11.6 The effect of introducing a single heteroatom

As an example of the use of equations (11.30) and (11.31), consider a simple description of heteroatom substitution into the conjugated π system of naphthalene (see figure 11.4), whose parental Green's function g is known. Our task is to compute the Green's function G of a daughter molecule, such as quinoline, obtained by substituting a nitrogen atom into a single site. This problem is closely related to the single-impurity problem considered in section 6.1, where the task was to compute the transmission coefficient of a doubly infinite chain containing a heteroatom at site $j = 0$.

The simplest model of heteroatom substitution is obtained by adding a term ϵ_l to the site energy of a single site (labelled l) of the parent, to yield a daughter molecule. As an example, consider a simple model of a naphthalene parent described by a Hamiltonian $h_0 = -C$, where C is given in table 2.3 (left) of chapter 2. In this model, all site energies are zero and all nearest neighbour couplings are -1, so that h_0 is simply a connectivity matrix. The mid-gap Green's function corresponding to $E = 0$ is $g = M/3$, where M is the magic number table given in table 2.3 (right). If a site energy ϵ_l is added to site l, as shown in figure 11.3, then the corresponding perturbation matrix h_{11} (in equation (11.24)) has matrix elements $(h_{11})_{ij} = \epsilon_l \delta_{il} \delta_{jl}$.

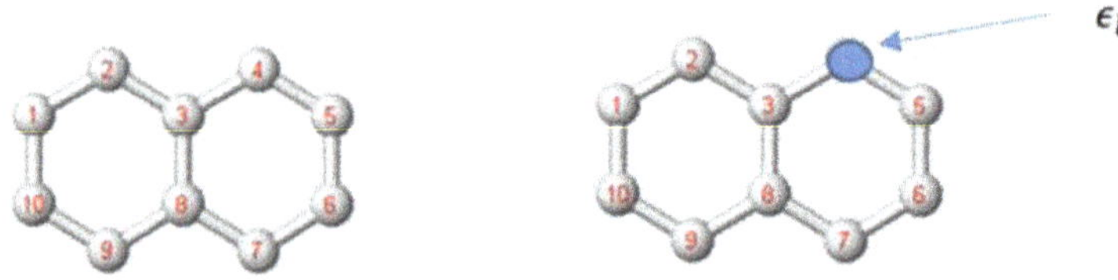

Figure 11.4. (Left) A lattice of sites representing the connectivity of naphthalene. (Right) A lattice of sites representing the connectivity of quinoline, in which the site energy of site $l = 4$ is perturbed by the presence of a nitrogen atom, coloured blue.

In this case, the matrices U_{11} and G_{11} are simply numbers, given by

$$U_{ll} = \frac{\epsilon_l}{1 - g_{ll}\epsilon_l}, \qquad G_{ll} = \frac{1}{g_{ll}^{-1} - \epsilon_l} = \frac{g_{ll}}{1 - g_{ll}\epsilon_l} \tag{11.34}$$

and equation (11.30b) yields for the Green's function matrix elements between any two sites i and j of the daughter,

$$G_{ij} = g_{ij} + \frac{g_{il}\epsilon_l g_{lj}}{1 - g_{ll}\epsilon_l} \tag{11.35}$$

Since the lattice is bipartite and H_0 is a connectivity matrix, the magic number table at $E = 0$ yields $g_{ll} = 0$ and therefore the mid-gap Green's function in the presence of the heteroatom is

$$G_{ij} = g_{ij} + g_{il}\epsilon_l g_{lj} \tag{11.36}$$

If i is odd and j is even or vice versa (corresponding to CQI for the parent), then either g_{il} or g_{lj} are zero, so the heteroatom does not affect the mid-gap Green's function and $G_{ij} = g_{ij}$. On the other hand, if i, j are both odd or both even (corresponding to DQI for the parent), then $g_{ij} = 0$ and $G_{ij} = g_{il}\epsilon_l g_{lj}$. This will be non-zero if l has the opposite parity to i, j, and therefore the heteroatom alleviates mid-gap DQI. However if l has the same parity as i, j there will be no effect and DQI of the parent persists.

11.7 The effect of introducing two heteroatoms

Consider the case where h_{11} describes the effect of adding a term ϵ_l to site l and ϵ_m to site m. In this case,

$$h_{11} = \begin{pmatrix} \epsilon_l & 0 \\ 0 & \epsilon_m \end{pmatrix} \tag{11.37}$$

and

$$U_{11} = \begin{pmatrix} \epsilon_l & 0 \\ 0 & \epsilon_m \end{pmatrix} \begin{pmatrix} 1 - g_{ll}\epsilon_l & -g_{lm}\epsilon_m \\ -g_{ml}\epsilon_l & 1 - g_{mm}\epsilon_m \end{pmatrix}^{-1} = \frac{1}{\Delta} \begin{pmatrix} \epsilon_l & 0 \\ 0 & \epsilon_m \end{pmatrix} \begin{pmatrix} 1 - g_{mm}\epsilon_m & g_{lm}\epsilon_m \\ g_{ml}\epsilon_l & 1 - g_{ll}\epsilon_l \end{pmatrix}$$

where $\Delta = (1 - g_{mm}\epsilon_m)(1 - g_{ll}\epsilon_l) - g_{ml}\epsilon_l g_{lm}\epsilon_m$.

That is

$$U_{11} = \frac{1}{\Delta} \begin{pmatrix} \epsilon_l(1 - g_{mm}\epsilon_m) & \epsilon_l g_{lm}\epsilon_m \\ \epsilon_m g_{ml}\epsilon_l & \epsilon_m(1 - g_{ll}\epsilon_l) \end{pmatrix} \tag{11.38}$$

In this case, equation (11.30) yields for the Green's function matrix elements between any two sites i and j of the daughter,

$$G_{ij} = g_{ij} + \sum\nolimits_{p,q=l,m} g_{ip} U_{pq} g_{qj} \tag{11.39}$$

As an example, consider the mid-gap Green's function of the bipartite lattice of figure 11.3 that occurs if l, m are both odd or both even. In this case, $g_{lm} = 0$ and $g_{mm} = g_{ll} = 0$.

In this case, $U_{11} = \begin{pmatrix} \epsilon_l & 0 \\ 0 & \epsilon_m \end{pmatrix}$ and equation (11.39) reduces to

$$G_{ij} = g_{ij} + g_{il}\epsilon_l g_{lj} + g_{im}\epsilon_m g_{mj} \tag{11.40}$$

As for a single heteroatom, if i and j have opposite parities (corresponding to CQI for the parent), then $G_{ij} = g_{ij}$ and the heteroatoms have no effect on the mid-gap Green's function. If they have the same parity (corresponding to DQI for the parent), then $g_{ij} = 0$ and

$$G_{ij} = g_{il}\epsilon_l g_{lj} + g_{im}\epsilon_m g_{mj} \tag{11.41}$$

As an example if i, j are both odd, then only heteroatoms on even-numbered sites contribute. If only a single heteroatom lies on an even site, then the single-heteroatom result of equation (11.35) is obtained. If both heteroatoms are on even sites, then equation (11.41) shows that G_{ij} is a sum of two terms. If these terms have the same sign, then they interfere constructively and DQI of the parent is alleviated. If they have opposite signs, then they interfere destructively are less effective in alleviating DQI.

11.8 The Green's function of a semi-infinite lead revisited

As a third example of using equation (11.30), consider the problem of computing the Green's function of a semi-infinite lead, starting from the known Green's function of a doubly infinite lead. The latter is (see equation (10.39)),

$$g_{jp}(E) = \frac{e^{ik(E)|j-p|}}{i\hbar v(E)} \tag{11.42}$$

where $\hbar v(E) = 2\gamma \sin k(E)$. Consider now introducing a single heteroatom on site $l+1$, described by a perturbation of the form $(h_1)_{mn} = \epsilon\delta_{ml+1}\,\delta_{nl+1}$. Taking the limit $\epsilon \to \infty$ has the effect of cutting the doubly infinite chain into two semi-infinite leads. The left lead terminates at site l and the right lead terminates at site $l+2$, as shown in figure 11.5.

From equation (11.35),

$$G_{ij} = g_{ij} + \frac{g_{il+1}\,\epsilon g_{l+1j}}{1 - g_{l+1,\,l+1}\epsilon} \tag{11.43}$$

As $\epsilon \to \infty$, equation (11.43) becomes

$$G_{ij} = g_{ij} - \frac{g_{il+1}\,g_{l+1j}}{g_{l+1,\,l+1}} \tag{11.44}$$

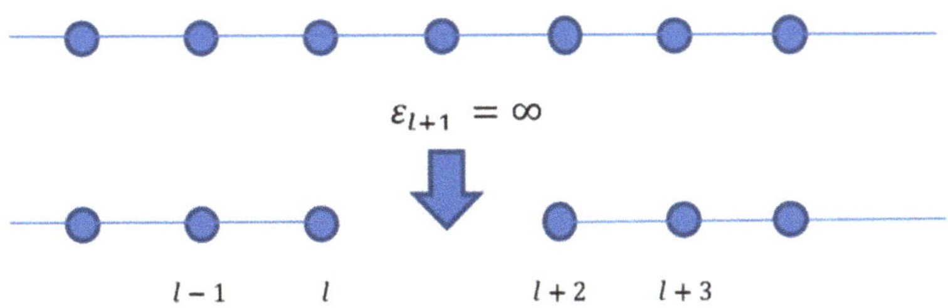

Figure 11.5. The top figure shows a double-infinite chain. After adding an infinite perturbation to site $l+1$, the chain is cut into two semi-infinite leads, as shown in the lower panel figure.

If i, j belong to the left semi-infinite lead, then they are both less than $l+1$, so after taking into account the modulus in equation (11.42), $g_{i,l+1}(E) = \frac{e^{ik(E)(l+1-i)}}{i\hbar v(E)}$ and $g_{j,l+1}(E) = \frac{e^{ik(E)(l+1-j)}}{i\hbar v(E)}$.

Hence the Green's function of the left semi-infinite lead is

$$G_{ij} = \frac{e^{ik(E)|i-j|} - e^{-ik(E)(i+j-2l-2)}}{i\hbar v(E)} \tag{11.45}$$

This vanishes at the missing site $i = l+1$ or $j = l+1$ and for $i = j = l$, it yields the surface Green's function

$$G_{ll}(E) = -\frac{e^{ik(E)}}{\gamma} \tag{11.46}$$

As expected, equations (11.45) and (11.46) are identical to equations (10.44) and (10.46) of chapter 10.

11.9 Fusion of substructures via Dyson's equation

Returning to the discussion of section 11.5, consider dividing the subspace 1 into two partitions labelled A and B, with dimensions N_A and N_B respectively, such that $N_1 = N_A + N_B$. Consider also the case where A and B are decoupled when $H_{11} = 0$, so that g_{11} has the structure

$$g_{11} = \begin{pmatrix} g_{AA} & 0 \\ 0 & g_{BB} \end{pmatrix} \tag{11.47}$$

Note that g_{11} and therefore g_{AA} and g_{BB} are submatrices of a larger (possibly infinite) matrix and could be non-Hermitian or Hermitian. In what follows, it is assumed that these submatrices are known. Starting from these decoupled subspaces, it is of interest to understand the effect of introducing a coupling between the subspaces A and B and therefore to choose h_{11} to have the form

$$h_{11} = \begin{pmatrix} 0 & H_{AB} \\ H_{BA} & 0 \end{pmatrix} \tag{11.48}$$

The task is to compute the Green's function G_{11} in the presence of h_{11}, where

$$G_{11} = \begin{pmatrix} G_{AA} & G_{AB} \\ G_{BA} & G_{BB} \end{pmatrix} \tag{11.49}$$

From equation (11.32) the Green's function G_{11}, satisfies $G_{11} = \left(g_{11}^{-1} - h_{11}\right)^{-1}$ or equivalently $(g_{11}^{-1} - h_{11})G_{11} = I$. With the above block structure of the matrices, this equation takes the form

$$\begin{pmatrix} g_{AA}^{-1} & -H_{BA} \\ -H_{AB} & g_{BB}^{-1} \end{pmatrix}\begin{pmatrix} G_{AA} & G_{AB} \\ G_{BA} & G_{BB} \end{pmatrix} = \begin{pmatrix} I & 0 \\ 0 & I \end{pmatrix} \tag{11.50}$$

Note that if two finite invertible matrices X and Y satisfy $XY = I$, where I is the unit matrix, then they also satisfy $YX = I$. So from equation (11.50), one can also write

$$\begin{pmatrix} G_{AA} & G_{AB} \\ G_{BA} & G_{BB} \end{pmatrix}\begin{pmatrix} g_{AA}^{-1} & -H_{AB} \\ -H_{BA} & g_{BB}^{-1} \end{pmatrix} = \begin{pmatrix} 1 & 0 \\ 0 & 1 \end{pmatrix} \tag{11.51}$$

Multiplying the above pairs of matrices and being careful to preserve the order of submatrices yields

$$\begin{pmatrix} g_{AA}^{-1}G_{AA} - H_{AB}G_{BA} & g_{AA}^{-1}G_{AB} - H_{AB}G_{BB} \\ -H_{BA}G_{AA} + g_{BB}^{-1}G_{BA} & -H_{BA}G_{AB} + g_{BB}^{-1}G_{BB} \end{pmatrix} = \begin{pmatrix} I & 0 \\ 0 & I \end{pmatrix} \tag{11.52}$$

and

$$\begin{pmatrix} G_{AA}g_{AA}^{-1} - G_{AB}H_{BA} & -G_{AA}H_{AB} + G_{AB}g_{BB}^{-1} \\ G_{BA}g_{AA}^{-1} - G_{BB}H_{BA} & -G_{BA}H_{AB} + G_{BB}g_{BB}^{-1} \end{pmatrix} = \begin{pmatrix} I & 0 \\ 0 & I \end{pmatrix} \tag{11.53}$$

From the top left of equation (11.52) and the top left of equation (11.53) we obtain two alternative expressions for G_{AA}

$$G_{AA} = g_{AA} + g_{AA}H_{AB}G_{BA} \tag{11.54a}$$

and

$$G_{AA} = g_{AA} + G_{AB}H_{BA}g_{AA} \tag{11.54b}$$

From the top-right of (11.52) and the top-right of (11.53) one obtains two alternative expressions for G_{AB}

$$G_{AB} = g_{AA}H_{AB}G_{BB} \tag{11.55a}$$

and

$$G_{AB} = G_{AA}H_{AB}g_{BB} \tag{11.55b}$$

From the bottom left of (11.52) and the bottom left of (11.53), or by interchanging A and B in equations (11.55), two alternative expressions for G_{BA} are obtained:

$$G_{BA} = g_{BB}H_{BA}G_{AA} \tag{11.56a}$$

and

$$G_{BA} = G_{BB}H_{BA}g_{AA} \tag{11.56b}$$

Substituting (11.55a) into (11.54b) yields

$$G_{AA} = g_{AA} + g_{AA}H_{AB}G_{BB}H_{BA}g_{AA} \tag{11.57}$$

Substituting equation (11.55b) into equation (11.54b) yields

$$G_{AA} = g_{AA} + G_{AA}H_{AB}g_{BB}H_{BA}g_{AA}$$

That is $G_{AA}(1 - H_{AB}g_{BB}H_{BA}g_{AA}) = g_{AA}$. Multiplying this from the right by g_{AA}^{-1} yields $G_{AA}(g_{AA}^{-1} - H_{AB}g_{BB}H_{BA}) = I$. Hence

$$G_{AA} = \left(g_{AA}^{-1} - \Sigma_A\right)^{-1} \tag{11.58}$$

where

$$\Sigma_A = H_{AB}g_{BB}H_{BA} \tag{11.59}$$

Corresponding equations for G_{BB} can be obtained by interchanging A and B in equations (11.57)–(11.60). For later reference, a summary of the resulting equations for these finite submatrices is provided below. Clearly these are the NNEs of chapter 2.

$$G_{BA} = g_{BB}H_{BA}G_{AA} = G_{BB}H_{BA}g_{AA} \tag{11.56}$$

$$G_{AA} = (g_{AA}^{-1} - \Sigma_A)^{-1} = g_{AA} + g_{AA}H_{AB}G_{BB}H_{BA}g_{AA} \tag{11.57/58}$$

where

$$\Sigma_A = H_{AB}g_{BB}H_{BA} \tag{11.59}$$

$$G_{BB} = (g_{BB}^{-1} - \Sigma_B)^{-1} = g_{BB} + g_{BB}H_{BA}G_{AA}H_{AB}g_{BB} \tag{11.60a}$$

where

$$\Sigma_B = h_{BA}g_{AA}h_{AB} \tag{11.60b}$$

In many structures of interest, the matrix H_{AB} does not couple every element of A to every element of B. Instead, it is often more 'local' and couples only a subspace a of

A to a subspace b of B, as shown in figure 11.4. In other words, H_{AB} and H_{BA} have the form

$$H_{AB} = \begin{pmatrix} h_{ab} & 0 \\ 0 & 0 \end{pmatrix} \quad \text{and} \quad H_{BA} = \begin{pmatrix} h_{ba} & 0 \\ 0 & 0 \end{pmatrix} \tag{11.61}$$

In view of this local structure of H_{AB}, it is convenient to write

$$g_{AA} = \begin{pmatrix} g_{aa} & g_{aa'} \\ g_{a'a} & g_{a'a'} \end{pmatrix}, \; g_{BB} = \begin{pmatrix} g_{bb} & g_{bb'} \\ g_{b'b} & g_{b'b'} \end{pmatrix},$$
$$G_{AA} = \begin{pmatrix} G_{aa} & G_{aa'} \\ G_{a'a} & G_{a'a'} \end{pmatrix}, \; G_{BB} = \begin{pmatrix} G_{bb} & G_{bb'} \\ G_{b'b} & G_{b'b'} \end{pmatrix} \text{etc}$$

With this notation, equation (11.57) becomes

$$G_{AA} = g_{AA} + g_{Aa}h_{ab}G_{bb}h_{ba}g_{aA} \tag{11.62}$$

where

$$g_{Aa} = \begin{pmatrix} g_{aa} \\ g_{a'a} \end{pmatrix} \quad \text{and} \quad g_{aA} = \begin{pmatrix} g_{aa} & g_{aa'} \end{pmatrix} \tag{11.63}$$

Similarly, equation (11.58) yields $G_{aa} = g_{aa} + G_{aa}h_{ab}g_{bb}h_{ba}g_{aa}$ and therefore following the steps leading to equation (11.59) one obtains

$$G_{aa} = \left(g_{aa}^{-1} - \Sigma_a\right)^{-1}, \quad \text{where } \Sigma_a = h_{ab}g_{bb}h_{ba} \tag{11.64}$$

The corresponding equations for G_{bb} and Σ_b are obtained by interchanging a and b in equations (11.62)–(11.64). After noting equations (11.55) and (11.56), a summary of the key equations is provided below.

$$G_{AA} = g_{AA} + g_{Aa}h_{ab}G_{bb}h_{ba}g_{aA} \tag{11.65}$$

$$G_{BA} = g_{Bb}h_{ba}G_{aA} = G_{Bb}h_{ba}g_{aA} \tag{11.66}$$

$$G_{bb} = (g_{bb}^{-1} - \Sigma_b)^{-1} \tag{11.67}$$

and

$$\Sigma_b = h_{ba}g_{aa}h_{ab} \tag{11.68}$$

Equations for other submatrices are obtained by interchanging a and b in the above equations.

Note that equation (11.33) yields

$$U_{11} = \begin{pmatrix} U_{AA} & U_{AB} \\ U_{BA} & U_{BB} \end{pmatrix} = \begin{pmatrix} 0 & H_{AB} \\ H_{BA} & 0 \end{pmatrix} \begin{pmatrix} G_{AA} & G_{AB} \\ G_{BA} & G_{BB} \end{pmatrix} \begin{pmatrix} g_{AA}^{-1} & 0 \\ 0 & g_{BB}^{-1} \end{pmatrix}$$

$$= \begin{pmatrix} H_{AB}G_{BA}g_{AA}^{-1} & H_{AB}G_{BB}g_{BB}^{-1} \\ H_{BA}G_{AA}g_{AA}^{-1} & H_{BA}G_{AB}g_{BB}^{-1} \end{pmatrix}$$

Using equation (11.56b) yields the following expression for U_{AA}

$$U_{AA} = H_{AB}G_{BB}H_{BA} \tag{11.69}$$

and therefore equation (11.57) can be written

$$G_{AA} = g_{AA} + g_{AA}U_{AA}g_{AA} \tag{11.70}$$

Alternatively using equation (11.56a) yields

$$U_{AA} = H_{AB}g_{BB}H_{BA}G_{AA}g_{AA}^{-1} = \Sigma_A(1 - g_{AA}\Sigma_A)^{-1} \tag{11.71}$$

In the above equations, the Green's functions g_{AA} and g_{BB} could be finite submatrices of the Green's function of a larger structure or the substructures A and B could combine to form the whole structure, as in figure 11.1. In the former case, once the equations are solved for the submatrices G_{AA}, G_{BB}, G_{AB} and G_{BA} of equation (11.49), the Green's function of the larger structure can be obtained from equations (11.30) and (11.33).

Equation (11.71) is similar to equation (11.31), in which Σ_A of the former plays the role of h_{11} in the latter. This is to be expected from the discussion in section 11.3, since if subsystem B is decimated, it renormalizes the Hamiltonian of subsystem A by adding a perturbation Σ_A to the Hamiltonian of A, as described in equation (11.22). Not surprisingly, equation (11.65) can also be written $G_{AA} = g_{AA} + g_{Aa}U_{aa}g_{aA}$, where $U_{aa} = h_{ab}G_{bb}h_{ba} = \Sigma_a(1 - g_{aa}\Sigma_a)^{-1}$. This connection to decimation can be made more explicit by noting that if the larger structure is closed and the whole structure is formed from A and B as in figure 11.1, then $g_{AA} = (E - H_{AA})^{-1}$ and $g_{BB} = (E - H_{BB})^{-1}$, where H_{AA} and H_{BB} are the diagonal submatrices in equation (11.11). In this case, equation (11.58) becomes $G_{AA} = (E - H_{AA} - \Sigma_A)^{-1} = (E - H'_{AA})^{-1}$, where H'_{AA} is given by equation (11.22).

On the other hand, if A and B are subsystems of a larger structure, then the equations $g_{AA} = (E - H_{AA})^{-1}$ and $g_{BB} = (E - H_{BB})^{-1}$ are not valid. For example, if subsystem A is formed from sites 1 and 2 of a doubly infinite linear chain, whose Green's function is given by equation (11.42), then g_{AA} would have the form $g_{AA} = \frac{1}{i\hbar v(E)}\begin{pmatrix} 1 & e^{ik(E)} \\ e^{ik(E)} & 1 \end{pmatrix}$ and therefore $g_{AA}^{-1} = \frac{i\hbar v(E)}{1 - e^{2ik(E)}}\begin{pmatrix} 1 & -e^{ik(E)} \\ -e^{ik(E)} & 1 \end{pmatrix} = \gamma\begin{pmatrix} -e^{-ik(E)} & 1 \\ 1 & -e^{-ik(E)} \end{pmatrix}$. Starting from g_{AA}^{-1}, one could define an 'effective Hamiltonian'

$\hat{H}_{AA}$ via the equation $E - \hat{H}_{AA} = g_{AA}^{-1}$, so that by definition, $g_{AA} = (E - \tilde{H}_{AA})^{-1}$. Clearly $\hat{H}_{AA}$ is a non-Hermitian, energy dependent matrix and should not be regarded as a true Hamiltonian. Nevertheless, if this notation is adopted, one obtains $G_{AA} = (E - \hat{H}_{AA} - \Sigma_A)^{-1} = (E - H'_{AA})^{-1}$, where $H'_{AA} = \hat{H}_{AA} + \Sigma_A$. With this notation, the relationship to decimation is clear, even in the case where A and B are subsystems of a larger structure, provided one notes that $\hat{H}_{AA}$ is obtained by inverting a Green's function and is not a Hamiltonian. Furthermore, the equations for submatrices of finite and infinite Green's functions can be made to resemble each other, which makes it easier to memorise them.

11.10 Constructing a lead-scatterer-lead system using Dyson's equation

In this section, the Green's function of a 3-component quantum system is obtained, which as a special case, reduces to the Green's function of a molecular junction composed of a left electrode (L), a molecule (B) and a right electrode (R). Alternatively it could describe two substructures L, R of a molecule, connected by a bridging moiety B. The system of interest is obtained by partitioning the subspace A into two smaller substructures L, R as shown in figure 11.6.

In this case, the matrices of section 11.9 take the form

$$\begin{pmatrix} g_{AA} & 0 \\ 0 & g_{BB} \end{pmatrix} = \begin{pmatrix} g_{LL} & 0 & 0 \\ 0 & g_{RR} & 0 \\ 0 & 0 & g_{BB} \end{pmatrix} \tag{11.72}$$

$$\begin{pmatrix} g_{aa} & 0 \\ 0 & g_{bb} \end{pmatrix} = \begin{pmatrix} g_{ll} & 0 & 0 \\ 0 & g_{rr} & 0 \\ 0 & 0 & g_{bb} \end{pmatrix} \tag{11.73}$$

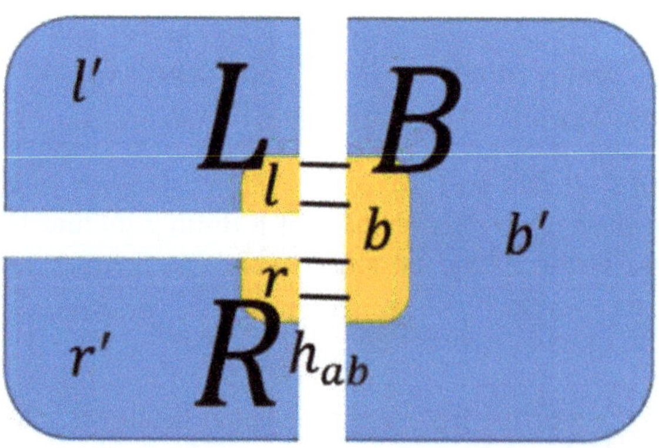

Figure 11.6. A division of the whole structure into three substructures L, B, R.

$$\begin{pmatrix} 0 & h_{AB} \\ h_{BA} & 0 \end{pmatrix} = \begin{pmatrix} 0 & 0 & h_{LB} \\ 0 & 0 & h_{RB} \\ h_{BL} & h_{BR} & 0 \end{pmatrix} \tag{11.74}$$

$$\begin{pmatrix} G_{AA} & G_{AB} \\ G_{BA} & G_{BB} \end{pmatrix} = \begin{pmatrix} G_{LL} & G_{LR} & G_{LB} \\ G_{RL} & G_{RR} & G_{RB} \\ G_{BL} & G_{BR} & G_{BB} \end{pmatrix} \tag{11.75}$$

The above submatrices are given by equations (11.65)–(11.68). For example from equation (11.65), the Green's function submatrix linking the left electrode to the right electrode is

$$G_{RL} = g_{Rr} h_{rb} G_{bb} h_{bl} g_{lL} \tag{11.76a}$$

and

$$G_{LR} = g_{Ll} h_{lb} G_{bb} h_{br} g_{rR} \tag{11.76b}$$

and from equation (11.66), the Green's function linking the left electrode to the region B is

$$G_{BL} = g_{Bb} h_{bl} G_{lL} = G_{Bb} h_{bl} g_{lL} \tag{11.77}$$

Furthermore from equation (11.68), the self-energies Σ_a and Σ_b are

$$\begin{aligned} \Sigma_a = \begin{pmatrix} \Sigma_{ll} & \Sigma_{lr} \\ \Sigma_{rl} & \Sigma_{rr} \end{pmatrix} &= h_{ab} g_{bb} h_{ba} = \begin{pmatrix} h_{lb} \\ h_{rb} \end{pmatrix} g_{bb} (h_{bl} \quad h_{br}) \\ &= \begin{pmatrix} h_{lb} g_{bb} h_{bl} & h_{lb} g_{bb} h_{br} \\ h_{rb} g_{bb} h_{bl} & h_{rb} g_{bb} h_{br} \end{pmatrix} \end{aligned} \tag{11.78}$$

and

$$\Sigma_b = (h_{bl} \quad h_{br}) \begin{pmatrix} g_{ll} & 0 \\ 0 & g_{rr} \end{pmatrix} \begin{pmatrix} h_{lb} \\ h_{rb} \end{pmatrix} = h_{bl} g_{ll} h_{lb} + h_{br} g_{rr} h_{rb} \tag{11.79}$$

$$G_{bb} = \left(g_{bb}^{-1} - \Sigma_b\right)^{-1} = \left(g_{bb}^{-1} - h_{bl} g_{ll} h_{lb} - h_{br} g_{rr} h_{rb}\right)^{-1} \tag{11.80}$$

Finally the renormalized perturbation takes the form

$$U_{aa} = \begin{pmatrix} U_{ll} & U_{lr} \\ U_{rl} & U_{rr} \end{pmatrix} = h_{ab} G_{bb} h_{ba} = \begin{pmatrix} h_{lb} G_{bb} h_{bl} & h_{lb} G_{bb} h_{br} \\ h_{rb} G_{bb} h_{bl} & h_{rb} G_{bb} h_{br} \end{pmatrix} \tag{11.81}$$

Clearly, equation (11.76a) is the Green's function in three-component NNE of equation (2.38) and is a consequence of the connectivity theory of chapter 2, which is based on the intimate relationship between a source and the wave it creates, from which Dyson's equation emerges.

11.11 From Green's functions to transmission functions

To obtain an expression for the transmission coefficient of a lead-molecule-lead junction, choose $L = l$ and $R = r$ in equation (11.81), where l and r are the surfaces of the left and right leads respectively. With such a choice, g_{ll} is a $N \times N$ surface Green's function of the form (10.82)

$$g_{ll} = \sum_{n=1}^{N} |\chi_n\rangle \, g_n \, \langle\chi_n| \tag{11.82}$$

where

$$g_n = \frac{-e^{ik_n(E)}}{\gamma_L} \tag{11.83}$$

In this expression, $k_n(E)$ is the wave vector of channel n of the left lead. Similarly, g_{rr} is a $M \times M$ surface Green's function of the right lead

That is

$$g_{rr} = \sum_{m=1}^{M} |\chi_m'\rangle \, g_m \, \langle\chi_m'| \tag{11.84}$$

In the above expression

$$g_m = \frac{-e^{ik_m'(E)}}{\gamma_R} \tag{11.85}$$

where $k_m'(E)$ is the wave vector of channel m of the right lead. Henceforth, n will be used to label channels in the left lead and m to label channels in the right lead.

From equation (11.81), the Green's function matrix element $(G_{rl})_{mn} = \langle\chi_m'|G_{lr}|\chi_n\rangle$ is given by

$$(G_{rl})_{mn} = g_m\langle\chi_m'|h_{rb}G_{bb}h_{bl}\,|\chi_n\rangle g_n = g_m\langle B_m'|G_{bb}|B_n\rangle g_n, \tag{11.86}$$

where

$$|B_m'\rangle = h_{br}|\chi_m'\rangle \text{ and } |B_n\rangle = h_{bl}|\chi_n\rangle \tag{11.87}$$

Hence

$$|(G_{RL})_{mn}|^2 = |g_m|^2\langle B_m'|G_{bb}|B_n\rangle|g_n|^2\langle B_n|\mathrm{G}_{bb}^{\dagger}|B_m'\rangle \tag{11.88}$$

If X is any square matrix with the same dimension as $|B_n\rangle$, cyclic invariance under the trace yields

$$\mathrm{Tr}|B_n\rangle\langle B_n|X = \langle B_n|X|B_n\rangle \tag{11.89}$$

Hence

$$|(G_{RL})_{mn}|^2 = \mathrm{Tr}\{|B_m'\rangle|g_m|^2\langle B_m'|G_{bb}|B_n\rangle|g_n|^2\langle B_n|\mathrm{G}_{bb}^{\dagger}\} \tag{11.90}$$

From the modulus squared of equation (10.89), this yields the transmission coefficient from an open channel n of the left lead to an open channel m of the right lead via the equation

$$T_{mn} = \hbar v_n \hbar v_m |(G_{rl})_{mn}|^2, \tag{11.91}$$

where

$$\hbar v_n = 2\gamma_L \sin k_n(E) \quad \text{and} \quad \hbar v_m = 2\gamma_R \sin k'_m(E) \tag{11.92}$$

To simplify this equation, note that to obtain the retarded Green's function, k_n (and k'_m) must be chosen such that:

(a) If k_n is real, the group velocity v_n is positive. In this case, channel n is said to be open and

That is

$$g_n = \frac{-e^{ik_n}}{\gamma_L} = \frac{-\cos k_n}{\gamma_L} - i\frac{\sin k_n}{\gamma_L} \tag{11.93}$$

Hence, for an open channel

$$\frac{\hbar v_n \, |g_n|^2}{2} = \frac{\sin k_n}{\gamma_L} \tag{11.94}$$

where k_n is chosen such that $\frac{\sin k_n}{\gamma_L}$ is positive.

(b) If k_n is imaginary, then $Im\ k_n$ is positive. In this case, channel n is closed and $g_n = \frac{-e^{ik_n}}{\gamma_L}$ is real.

Hence equations (11.90) and (11.91) combine to yield

$$T_{mn} = \mathrm{Tr}\{|B_n\rangle \hbar v_n |g_n|^2 \langle B_n|\, G_{bb} |B'_m\rangle \hbar v_m |g_m|^2 \langle B'_m|\, G^{\dagger}_{bb}\}$$

That is

$$T_{mn} = 4\mathrm{Tr}\{|B_n\rangle \frac{\sin k_n}{\gamma_L} \langle B_n| G_{bb} |B'_m\rangle \frac{\sin k'_m}{\gamma_R} \langle B'_m| G^{\dagger}_{bb}\} \tag{11.95}$$

Hence one obtains

$$T_{mn} = 4\mathrm{Tr}[\Gamma_L(n) G_{bb} \Gamma_R(m) G^{\dagger}_{bb}] \tag{11.96}$$

where

$$\Gamma_L(n) = |B_n\rangle \frac{\sin k_n}{\gamma_L} \langle B_n| = h_{bl}\, |\chi_n\rangle \frac{\sin k_n}{\gamma_L} \langle \chi_n| h_{lb} \tag{11.97}$$

and

$$\Gamma_R(m) = |B'_m\rangle \frac{\sin k'_m}{\gamma_R} \langle B'_m| = h_{br}\, |\chi'_m\rangle \frac{\sin k'_m}{\gamma_R} \langle \chi'_m| h_{rb} \tag{11.98}$$

Since, $|g_n|^2 = 1/[\gamma_L]^2$, and $|g_m|^2 = 1/[\gamma_R]^2$ so equations (11.56) and (11.59) yield,

$$T_{mn} = 4\frac{\sin k_n}{\gamma_L}\frac{\sin k'_m}{\gamma_R}|\langle\chi'_m|h_{rb}G_{bb}h_{bl}|\chi_n\rangle|^2 \tag{11.99}$$

Equations (11.91), (11.96) and (11.99) are three equivalent expressions for T_{mn}. To evaluate them, it is useful to simplify the expression for the Green's function G_{bb}.

11.12 Simplification of the Green's function of a scatterer in the presence of electrodes

To simplify equations (11.80) and (11.99), following the discussion in chapter 9, define

$$\Sigma_R = h_{br}g_{rr}h_{rb} = \sum\nolimits_{m=1}^{M}\Sigma_{\mathrm{R}}(\mathrm{m}) \quad \text{and} \quad \Sigma_L = h_{bl}g_{ll}h_{lb} = \sum\nolimits_{n=1}^{N}\Sigma_{\mathrm{L}}(\mathrm{n}) \tag{11.100}$$

where

$$\Sigma_{\mathrm{L}}(\mathrm{n}) = h_{bl}|\,\chi_n\rangle g_n\langle\chi_n|h_{lb} \quad \text{and} \quad \Sigma_{\mathrm{R}}(\mathrm{m}) = h_{br}|\chi'_m\rangle g_m\langle\chi'_m|h_{rb} \tag{11.101}$$

From equation (11.93), it is convenient to write

$$g_n = a_n - ib_n \tag{11.102}$$

If channel n is open, then $a_n = \frac{-\cos k_n}{\gamma_L}$ and $b_n = \frac{\sin k_n}{\gamma_L}$. On the other hand, if channel n is closed, then $a_n = \frac{-e^{ik_n}}{\gamma_L}$ and $b_n = 0$. Similarly, $g_m = a_m - ib_m$.

With this notation, it is convenient to write

$$\Sigma_{\mathrm{L}}(\mathrm{n}) = \sigma_{\mathrm{L}}(\mathrm{n}) - i\Gamma_{\mathrm{L}}(\mathrm{n}) \tag{11.103}$$

where

$$\sigma_{\mathrm{L}}(\mathrm{n}) = H_{BL}|n\rangle a_n\langle n|H_{LB} \quad \text{and} \quad \Gamma_{\mathrm{L}}(\mathrm{n}) = H_{BL}|n\rangle b_n\langle n|H_{LB} \tag{11.104}$$

Hence

$$\Sigma_L = \sigma_L - i\Gamma_L \tag{11.105}$$

where

$$\Gamma_{\mathrm{L}} = \sum\nolimits_{n=1}^{N}\Gamma_{\mathrm{L}}(\mathrm{n}), \tag{11.106}$$

$$\sigma_{\mathrm{L}} = \sum\nolimits_{n=1}^{N}\sigma_{\mathrm{L}}(\mathrm{n}), \tag{11.107}$$

then

$$G_{bb} = \left(g_{bb}^{-1} - \Sigma_b\right)^{-1} = \left(g_{bb}^{-1} - \sigma + \mathrm{i}\Gamma\right)^{-1} \tag{11.108}$$

An important point to note is that the $\Gamma_L(n)$ and $\Gamma_R(m)$ appearing in equations (11.96) and (11.104) are the same quantities. This is a consequence of the fact that the surface Green's function satisfies $g_n = -\frac{\hbar v_n |g_n|^2}{2}$.

11.13 Summary of Green's function equations for the transmission coefficient

For reference, the box below contains a summary of the most important equation derived above.

$$T_{mn} = 4\mathrm{Tr}[\Gamma_L(n)G_{bb}\Gamma_R(m)G_{bb}^{\dagger}] = 4\frac{\sin k_n}{\gamma_L}\frac{\sin k_m'}{\gamma_R}|\langle\chi_m'|h_{rb}G_{bb}h_{bl}|\chi_n\rangle|^2 \tag{11.109}$$

In these expressions,

$$G_{bb} = (g_{bb}^{-1} - \Sigma_b)^{-1} \text{ where } \Sigma_b = h_{bl}g_{ll}h_{lb} + h_{br}g_{rr}h_{rb} \tag{11.110}$$

The self-energy Σ_b can be written $\Sigma_b = \sigma - i\Gamma$, where σ and Γ are the Hermitian and anti-Hermitian contributions given by $\sigma = \sigma_L + \sigma_R$ and $\Gamma = \Gamma_L + \Gamma_R$. The contributions to the latter from the left and right leads are

$$\Gamma_L = \sum_{n=\text{open}}^{N}\Gamma_L(n),\ \sigma_L = \sum_{n=1}^{N}\sigma_L(n),\ \Gamma_R = \sum_{m=\text{open}}^{M}\Gamma_R(m),\ \sigma_R = \sum_{m=1}^{M}\sigma_R(m), \tag{11.111}$$

with

$$\Gamma_R(m) = h_{br}|\chi_m'\rangle\frac{\sin k_m'}{\gamma_R}\langle\chi_m'|h_{rb} \quad \Gamma_L(n) = h_{bl}|\chi_n\rangle\frac{\sin k_n}{\gamma_L}\langle\chi_n|h_{lb} \tag{11.112}$$

$$\sigma_R(m) = h_{br}|\chi_m'\rangle\frac{-\cos k_m'}{\gamma_R}\langle\chi_m'|h_{rb} \text{ and } \sigma_L(n) = h_{bl}|\chi_n\rangle\frac{-\cos k_n}{\gamma_L}\langle\chi_n|h_{lb} \tag{11.113}$$

11.14 The Breit–Wigner formula for an arbitrary scatterer connected to finite-width leads

The Breit–Wigner formula is immediately obtained from equation (11.110) by making the approximation (see appendix A for a detailed discussion)

$$g_{bb} \approx \frac{|\psi_p\rangle\langle\psi_p|}{E - E_p} \tag{11.114}$$

where $|\psi_p\rangle$ is an eigenstate of the isolated scatterer, corresponding to eigenvalue E_p.

In this case, equation (11.110) becomes
$G_{BB} \approx |\psi_p\rangle G_p\langle\psi_p|$, where

$$G_{BB} \approx \frac{|\psi_p\rangle\langle\psi_p|}{E - E_p - \sigma_p + i\Gamma_p} \tag{11.115}$$

Hence equation (11.109) yields

$$T_{mn} = \frac{4\Gamma_{\mathrm{L}}^{\mathrm{p}}(\mathrm{n})\Gamma_{\mathrm{R}}^{\mathrm{p}}(\mathrm{m})}{(E - E_p - \sigma_p)^2 + (\Gamma_p)^2} \tag{11.116}$$

where

$$\Gamma_L^p(n) = \langle\psi_p|\Gamma_L(n)|\psi_p\rangle \quad \text{and} \quad \Gamma_R^p(m) = \langle\psi_p|\Gamma_R(m)|\psi_p\rangle \tag{11.117}$$

The total transmission coefficient from all open channels on the left to all open channels on the right is therefore

$$T = \sum\nolimits_{m,n} T_{mn} = \frac{4\Gamma_L\Gamma_R}{(E - E_p - \sigma_p)^2 + (\Gamma_L + \Gamma_R)^2} \tag{11.118}$$

where

$$\Gamma_L = \sum\nolimits_{n\,(\mathrm{open})} \Gamma_L^p(n), \quad \Gamma_R = \sum\nolimits_{n\,(\mathrm{open})} \Gamma_R^p(m) \tag{11.119}$$

and it was noted that

$$\Gamma_p = \Gamma_L + \Gamma_R \tag{11.120}$$

This demonstrates that the Breit–Wigner formula (8.71) is not restricted to single-channel leads, but instead is valid in the presence of finite-dimensional leads, for an arbitrary scatterer, provided equation (11.114) is a good approximation when E is close to an eigenvalue of the isolated scatterer. In practice, this means that the level broadening Γ_p should be less than the energy level spacing of the isolated scatterer.

An extension of the Breit–Wigner formula in the case of a 2-level approximation to g_{bb} is also discussed in appendix A and is shown in equation (A.38).

11.15 Derivation of the five-component NNE

If the electrodes are compound electrodes, of the kind discussed in chapter 2, then in equation (11.76), g_{Rr} and g_{lL} are Green's functions of the right and left compound electrodes in the absence of the couplings h_{rb} and h_{bl}. They are the Green's function of a right contact group coupled to a right electrode surface and a left contact group coupled to a left electrode surface respectively. To obtain these quantities, note that equation (11.66) is a formula for the Green's function of an arbitrary subsystem B coupled to a subsystem A, in which g_{Bb} is the Green's function of B in the absence of A and G_{aA} is the Green's function of A in the presence of B, and similarly g_{aA} is the Green's function of A in the absence of B and G_{Bb} is the Green's function of B in the presence of A. They allow us to write the the Green's function of the left compound electrode as

$$g_{lL} = g_{ll}H_{lL}\tilde{g}_{LL} = \tilde{g}_{ll}H_{lL}g_{LL} \quad \text{and} \quad g_{Ll} = g_{LL}H_{Ll}\tilde{g}_{ll} = \tilde{g}_{LL}H_{Ll}g_{ll} \tag{11.121}$$

where g_{ll} is the Green's function of the left anchor in the absence of the left lead and $\tilde{g}_{LL}$ is the surface Green's function of the left lead in the presence of the left anchor and similarly g_{LL} is the surface Green's function of the left lead in the absence of the left anchor and $\tilde{g}_{ll}$ is the Green's function of the left anchor in the presence of left lead.

Similarly for the right compound electrode,

$$g_{rR} = g_{rr}H_{rR}\tilde{g}_{RR} = \tilde{g}_{rr}H_{rR}g_{RR} \quad \text{and} \quad g_{Rr} = g_{RR}H_{Rr}\tilde{g}_{rr} = \tilde{g}_{RR}H_{Rr}g_{rr} \tag{11.122}$$

Substituting these into equation (11.76) yields the following version of the 5-component NNE:

$$G_{RL} = g_{RR}H_{Rr}\tilde{g}_{rr}h_{rb'}G_{b'b'}h_{b'l}\tilde{g}_{ll}H_{lL}g_{LL} \tag{11.123a}$$

and

$$G_{LR} = g_{LL}H_{Ll}\tilde{g}_{ll}h_{lb'}G_{b'b'}h_{b'r}\tilde{g}_{rr}H_{rR}g_{RR} \tag{11.123b}$$

In these equations $G_{b'b'}$ is a submatrix of G_{BB} containing elements which couple to l and r.

For the case where $h_{rb'}$ and $h_{lb'}$ do not couple to all atoms of a central core and it is convenient to denote the submatrix of non-zero elements of $h_{rb'}$ by H_{rb}, where the index b labels all sites (i.e. basis vectors) in the molecular core which couple to right contact group. Similarly the submatrix of non-zero elements of $h_{lb'}$ will be denoted by H_{la}, where the index a labels all sites (i.e. basis vectors) in molecules core, which couple to left contact group. If the labels a and b are disjoint sets of indices, then only the submatrix G_{ab} of $G_{b'b'}$ is needed. In the NNE (2.38a) of chapter 2, this is denoted $(G_M)_{ab}$.

With this notation, equations (11.123) become

$$G_{RL} = g_{RR}H_{Rr}\tilde{g}_{rr}H_{rb}(G_M)_{ba}H_{al}\tilde{g}_{ll}H_{lL}g_{LL} \tag{11.124a}$$

and

$$G_{LR} = g_{LL}H_{Ll}\tilde{g}_{ll}H_{la}(G_M)_{ab}H_{br}\tilde{g}_{rr}H_{rR}g_{RR} \tag{11.124b}$$

The Green's functions of the contact groups in the presence of the electrode surfaces are obtained by introducing the self-energies

$$\Sigma_{rr} = h_{rR}g_{RR}h_{Rr} \quad \text{and} \quad \Sigma_{ll} = h_{lL}g_{LL}h_{Ll} \tag{11.125}$$

which yield

$$\tilde{g}_{ll} = \left(g_{ll}^{-1} - \Sigma_{ll}\right)^{-1} \quad \text{and} \quad \tilde{g}_{rr} = \left(g_{rr}^{-1} - \Sigma_{rr}\right)^{-1} \tag{11.126}$$

Similarly, the Green's function G_M of the molecular core in the presence of the contact groups is given by

$$G_M = \left(g_M^{-1} - \Sigma_M\right)^{-1} \tag{11.127}$$

where the self-energy Σ_M is obtained by writing

$$\tilde{g}_c = \begin{pmatrix} \tilde{g}_{ll} & 0 \\ 0 & \tilde{g}_{rr} \end{pmatrix},\ H_{CM} = \begin{pmatrix} H_{la} & 0 \\ 0 & H_{rb} \end{pmatrix},\ H_{MC} = \begin{pmatrix} H_{al} & 0 \\ 0 & H_{br} \end{pmatrix} \tag{11.128}$$

This yields

$$\Sigma_M = H_{MC}\tilde{g}_c H_{CM} = \begin{pmatrix} \Sigma_a & 0 \\ 0 & \Sigma_b \end{pmatrix} \tag{11.129}$$

where

$$\Sigma_a = H_{al}\tilde{g}_{ll}H_{la} \quad \text{and} \quad \Sigma_b = H_{br}\tilde{g}_{rr}H_{rb} \tag{11.130}$$

To evaluate equation (11.127), it is convenient to write the submatrix of G_M involving sites a, b as

$$G_M = \begin{pmatrix} (G_M)_{aa} & (G_M)_{ab} \\ (G_M)_{ba} & (G_M)_{bb} \end{pmatrix} \tag{11.131}$$

and the corresponding submatrix when $H_{CM} = 0$ as

$$g_M = \begin{pmatrix} g_{aa} & g_{ab} \\ g_{ba} & g_{bb} \end{pmatrix} \tag{11.132}$$

Note that to obtain the submatrices in equation (11.132), it is first necessary to compute the whole Green's function g_{BB} of the isolated core and then to read off the submatrices.

From equation (11.32), since only the sites a, b are affected by the perturbation Σ_M,

$$G_M = \left[\begin{pmatrix} g_{aa} & g_{ab} \\ g_{ba} & g_{bb} \end{pmatrix}^{-1} - \begin{pmatrix} \Sigma_a & 0 \\ 0 & \Sigma_b \end{pmatrix}\right]^{-1} \tag{11.133}$$

To evaluate the inverse, it is convenient to define the matrices A_{ab} etc by

$$\begin{pmatrix} g_{aa} & g_{ab} \\ g_{ba} & g_{bb} \end{pmatrix}^{-1} = \begin{pmatrix} A_{aa} & -A_{ab} \\ -A_{ba} & A_{bb} \end{pmatrix} \tag{11.134}$$

Then

$$G_M = \begin{pmatrix} A_{aa} - \Sigma_a & -A_{ab} \\ -A_{ba} & A_{bb} - \Sigma_b \end{pmatrix}^{-1} \tag{11.135}$$

The submatrices of G_M are given by equations (11.50), (11.55) and (11.59), which allow us to write

$$(G_M)_{ab} = (G_M)_{aa}A_{ab}\hat{g}_{bb} = \hat{g}_{aa}A_{ab}(G_M)_{bb} \tag{11.136}$$

where

$$\hat{g}_{aa} = (A_{aa} - \Sigma_a)^{-1} \quad \text{and} \quad \hat{g}_{bb} = (A_{bb} - \Sigma_b)^{-1} \tag{11.137}$$

$$(G_M)_{aa} = [A_{aa} - \Sigma_a - A_{ab}\hat{g}_{bb}A_{ba}]^{-1} \tag{11.138}$$

$$(G_M)_{bb} = [A_{bb} - \Sigma_b - A_{ba}\hat{g}_{aa}A_{ab}]^{-1} \tag{11.139}$$

Note that in the above equations,

$$A_{aa} = \left(g_{aa} - g_{ab}g_{bb}^{-1}g_{ba}\right)^{-1},\ A_{bb} = \left(g_{bb} - g_{ba}g_{aa}^{-1}g_{ab}\right)^{-1} \tag{11.140}$$

$$A_{ab} = A_{aa}g_{ab}g_{bb}^{-1} \quad \text{and} \quad A_{ba} = A_{bb}g_{ba}g_{aa}^{-1} \tag{11.141}$$

which can be verified by evaluating the product

$$\begin{pmatrix} \left(g_{aa} - g_{ab}g_{bb}^{-1}g_{ba}\right)^{-1} & -A_{aa}g_{ab}g_{bb}^{-1} \\ -A_{bb}g_{ba}g_{aa}^{-1} & \left(g_{bb} - g_{ba}g_{aa}^{-1}g_{ab}\right)^{-1} \end{pmatrix} \begin{pmatrix} g_{aa} & g_{ab} \\ g_{ba} & g_{bb} \end{pmatrix} = \begin{pmatrix} I & 0 \\ 0 & I \end{pmatrix}$$

Hence

$$(G_M)_{ab} = (G_M)_{aa}A_{aa}g_{ab}g_{bb}^{-1}\hat{g}_{bb} = \hat{g}_{aa}A_{aa}g_{ab}g_{bb}^{-1}(G_M)_{bb} \tag{11.142}$$

As an example of this general formula, for the simplest case, where Σ_a, Σ_b, g_{aa}, g_{ab}, g_{bb} are simply numbers, equation (11.142) reduces to

$$(G_M)_{ab} = \frac{g_{ab}}{\delta} \tag{11.143}$$

where

$$\delta = (1 - g_{aa}\Sigma_a)(1 - \Sigma_b g_{bb}) - g_{ab}\Sigma_b g_{ba}\Sigma_a \tag{11.144}$$

and equation (11.145) becomes

$$G_{LR} = (g_{LL}H_{Ll}\tilde{g}_{ll}H_{la}g_{ab}H_{br}\tilde{g}_{rr}H_{rR}g_{RR})/\delta \tag{11.145a}$$

and

$$G_{RL} = g_{RR}H_{Rr}\tilde{g}_{rr}H_{rb}g_{ba}H_{al}\tilde{g}_{ll}H_{lL}g_{LL}/\delta \tag{11.145b}$$

Note that only g_{ab} and δ are simple numbers in this formula; all other quantities could be matrices.

In the simplest case of a single orbital on each of the contact groups, with orbital energies ϵ_l and ϵ_r, $g_{ll}^{-1} = E - \epsilon_l$, $g_{rr}^{-1} = E - \epsilon_r$ and

$$\tilde{g}_{ll} = (E - \epsilon_l - \Sigma_{ll})^{-1} \quad \text{and} \quad \tilde{g}_{rr} = (E - \epsilon_r - \Sigma_{rr})^{-1} \tag{11.146}$$

$$\Sigma_a = H_{al}\tilde{g}_{ll}H_{la} = \frac{H_{al}H_{la}}{E - \epsilon_l - \Sigma_{ll}} \quad \text{and} \quad \Sigma_b = H_{br}\tilde{g}_{rr}H_{rb} = \frac{H_{br}H_{rb}}{E - \epsilon_r - \Sigma_{rr}} \tag{11.147}$$

If the leads are one-dimensional tight-binding chains,

$$\begin{aligned} \Sigma_{ll} &= H_{lL}g_{LL}H_{Ll} = H_{lL}\left(\frac{-e^{-k_L}}{\gamma_L}\right)H_{Ll} \quad \text{and} \\ \Sigma_{rr} &= H_{rR}g_{RR}H_{Rr} = H_{rR}\left(\frac{-e^{-k_R}}{\gamma_R}\right)H_{Rr} \end{aligned} \tag{11.148}$$

Hence

$$G_{LR} = \frac{g_{LL}H_{Ll}H_{la}g_{ab}H_{br}H_{rR}g_{RR}}{\Delta} \tag{11.149a}$$

and

$$G_{RL} = \frac{g_{RR}H_{Rr}H_{rb}g_{ba}H_{al}H_{lL}g_{LL}}{\Delta} \tag{11.149b}$$

where

$$\begin{aligned} \Delta = \frac{\delta}{\tilde{g}_{ll}\tilde{g}_{rr}} &= (E - \epsilon_l - \Sigma_{ll} - g_{aa}H_{al}^2)(E - \epsilon_r - \Sigma_{rr} - H_{br}^2 g_{bb}) \\ &- g_{ab}H_{br}^2 g_{ba}H_{al}^2 \end{aligned} \tag{11.150}$$

As an example, if the molecule is a linear chain and a, b are the end sites of the chain, then from equation (10.70)

$$g_{aa} = g_{bb} = \frac{-\sin kN}{\gamma \sin k(N+1)} \quad \text{and} \quad g_{ab} = g_{ba} = \frac{-1}{\gamma \sin k(N+1)} \tag{11.151}$$

11.16 Summary

In this chapter, Dyson's equation was used to obtained a variety of formulae for Green's functions of scatterers connected to finite-width, multi-channel leads. In doing so, the NNEs of chapter 2 were re-derived. Perturbation theory applied to infinite, open systems was also discussed, along with the decimation method, which can be used to eliminate unwanted information about substructures of more complex systems. The latter allows instances of DQI to be identified without explicitly solving the Schrödinger equation, while the former facilitates the description of changes which occur due to heteroatom substitution. In section 11.14, the Breit–Wigner formula was derived, which reveals how open channels contribute to the width of resonances in transmission coefficients. In chapters 12 and 13, these formulae will be used to analyse interference effects is a selection of multi-component quantum structures.

Appendix A Similarity transformations

If a matrix A is related to another matrix $\hat{A}$ by the equation

$$\hat{A} = C^{-1}AC \text{ or equivalently } A = C\hat{A}C^{-1} \tag{A.1}$$

Then $\hat{A}$ is said to be the similarity transform of A with respect to the transformation matrix C. Clearly, for this to make sense, the inverse of C must exist. If A is a product of two matrices, such as $A = XY$, then $\hat{A} = C^{-1}XYC = C^{-1}XCC^{-1}YC$. Hence $\hat{A} = \hat{X}\hat{Y}$. If, for some reason, the product XY is difficult to evaluate, whereas the product $\hat{X}\hat{Y}$ is easy to evaluate, then this allows A to be obtained by first evaluating the product $\hat{A} = \hat{X}\hat{Y}$ and then back transforming to A via the relation $A = C\hat{A}C^{-1}$. As another example, if

$$A = (I - XY)^{-1} \quad \text{then one finds } \hat{A} = (I - \hat{X}\hat{Y})^{-1} \tag{A.2}$$

This is proved by noting that the inverse of a product of two finite matrices P, Q satisfies $(PQ)^{-1} = Q^{-1}P^{-1}$. Hence $\hat{A} = C^{-1}(I - XY)^{-1}C = [C(I - XY)^{-1}C^{-1}]^{-1} = (I - CXYC^{-1})^{-1} = (I - CXC^{-1}CYC^{-1})^{-1}$.

As a further example, if

$$A = (X - Y)^{-1} \quad \text{then one finds } \hat{A} = (\hat{X} - \hat{Y})^{-1} \tag{A.3}$$

Similarly if $A = f(X)$, then since a function of a matrix is defined by a Taylor expansion of the form $f(X) = f(0) + f'(0)X + \frac{f''(0)X^2}{2!} + \dots$

On finds

$$\hat{A} = f(\hat{X}) \tag{A.4}$$

If the matrix C is unitary, then $C^{-1} = C^{\dagger}$, in which case, the transformation (A.1) is called a unitary transformation.

As an example of a unitary transformation, consider the Green's function g_{BB} of an isolated scatterer with N sites (i.e. N atomic orbitals or basis functions). g_{BB} is a $N \times N$ matrix, with matrix elements $(g_{BB})_{ij}$ of the form

$$g_{BB} = \begin{pmatrix} (g_{BB})_{11} & (g_{BB})_{12} & \cdots & (g_{BB})_{1N} \\ \vdots & \vdots & \ddots & \vdots \\ \vdots & \vdots & & \vdots \\ (g_{BB})_{N1} & (g_{BB})_{N2} & \cdots & (g_{BB})_{NN} \end{pmatrix} \tag{A.5}$$

An alternative notation for the matrix elements $(g_{BB})_{ij}$ is obtained by introducing N component vectors $|j\rangle$, as shown below,

$$|j> = \begin{pmatrix} 0 \\ \vdots \\ 1 \\ \vdots \\ 0 \end{pmatrix} \leftarrow \text{row } j \quad \updownarrow \text{ length } N \tag{A.6}$$

such that $|j\rangle$ is a column of zeros, except for a '1' at element j. These satisfy

$$\langle i|j\rangle = \delta_{ij} \tag{A.7}$$

and

$$\sum_{i=1}^{N} |i\rangle\langle i| = I_N, \tag{A.8}$$

where I_N is the $N \times N$ unit matrix. With this notation,

$$(g_{BB})_{ij} = \langle i|g_{BB}|j\rangle \tag{A.9}$$

and the full matrix can be written

$$g_{BB} = \sum_{i,j=1}^{N} |i\rangle(g_B)_{ij}\langle j| \tag{A.10}$$

Note that equation (A.10) follows from equation (A.8), because

$$g_{BB} = I_N g_{BB} I_N = \sum_{i,\, j=1}^{N} |i\rangle\langle i|g_B|j\rangle\langle j| = \sum_{i,\, j=1}^{N} |i\rangle(g_B)_{ij}\langle j|$$

On the other hand, if the Hamiltonian of the isolated scatterer is h_B, then the eigenvectors $|\psi_n\rangle$ of h_B satisfy

$$h_B|\psi_n\rangle = E_n|\psi_n\rangle \tag{A.11}$$

Choosing these to be orthonormal allows us to write

$$\langle\psi_m|\psi_n\rangle = \delta_{nm} \tag{A.12}$$

and

$$\sum_{n=1}^{N} |\psi_n\rangle\langle\psi_n| = I_N \tag{A.13}$$

Since $g_{BB} = (E - h_B)^{-1}$, g_{BB} can be written

$$g_{BB} = (E - h_B)^{-1} I_N = \sum_{n=1}^{N} (E - h_B)^{-1}\, |\psi_n\rangle\langle\psi_n| = \sum_{n=1}^{N} (E - E_n)^{-1}\, |\psi_n\rangle\langle\psi_n|$$

That is

$$g_{BB} = \sum_{n=1}^{N} \frac{|\psi_n\rangle\langle\psi_n|}{E - E_n} \tag{A.14}$$

Hence the components of g_{BB} satisfy

$$(g_{BB})_{ij} = \langle i|\, g_{BB}\, |j\rangle = \sum_{n=1}^{N} \frac{\psi_n(i)\psi_n^*(j)}{E - E_n} \tag{A.15}$$

where $\psi_n(i) = \langle i|\psi_n\rangle$ is the amplitude of the n th eigenvector of site i.

More generally, introducing any set of N orthonormal vectors $\{|\phi_n\rangle\}$ satisfying

$$\langle\phi_m|\phi_n\rangle = \delta_{nm} \tag{A.16}$$

and

$$\sum_{n=1}^{N} |\phi_n\rangle\langle\phi_n| = I_N \tag{A.17}$$

allows us to write

$$g_{BB} = I_N g_{BB} I_N = \sum_{n,m=1}^{N} |\phi_n\rangle\langle\phi_n|\, g_B\, |\phi_m\rangle\langle\phi_m| = \sum_{n,m=1}^{N} |\phi_n\rangle(\hat{g}_B)_{nm}\langle\phi_m| \tag{A.18}$$

where

$$(\hat{g}_{BB})_{nm} = \langle\phi_n|\, g_{BB}\, |\phi_m\rangle \tag{A.19}$$

Note that the matrix g_{BB} on the left-hand side of equation (A.14) is the same as that shown in equation (A.5). On the other hand, the matrix $\hat{g}_{BB}$ formed from the matrix elements $(\hat{g}_{BB})_{nm}$ is a different matrix, of the form

$$\hat{g}_{BB} = \begin{pmatrix} (\hat{g}_{BB})_{11} & (\hat{g}_{BB})_{12} & \ddots & (\hat{g}_{BB})_{1N} \\ \vdots & \vdots & & \vdots \\ (\hat{g}_{BB})_{N1} & (\hat{g}_{BB})_{N2} & \cdots & (\hat{g}_{BB})_{NN} \end{pmatrix} \tag{A.20}$$

$\hat{g}_{BB}$ is said to be the matrix representation of g_{BB} in the basis of $\{|\phi_n\rangle\}$. If $\hat{g}_{BB}$ and the set $\{|\phi_n\rangle\}$ are known, then g_{BB} can be obtained from equation (A.18), which yields

$$(g_{BB})_{ij} = \langle i|\, g_{BB}\, |j\rangle = \sum_{n,m=1}^{N} \phi_n(i)(\hat{g}_B)_{nm}\phi_m^*(j) \tag{A.21}$$

where $\phi_n(i) = \langle i|\phi_n\rangle$. If the basis $\{|\phi_n\rangle\}$ is chosen to be the set of eigenstates $\{|\psi_n\rangle\}$, then by comparing equations (A.15) and (A.21), one finds that the matrix $\hat{g}_{BB}$ is diagonal

$$\text{i. e.}\quad (\hat{g}_{BB})_{nm} = \frac{\delta_{nm}}{E - E_n} \tag{A.22}$$

or equivalently

$$\hat{g}_{BB} = \begin{pmatrix} \frac{1}{E - E_1} & 0 & 0 & 0 \\ 0 & \ddots & 0 & 0 \\ 0 & 0 & \ddots & 0 \\ 0 & 0 & 0 & \frac{1}{E - E_N} \end{pmatrix} \tag{A.23}$$

Note that equation (A.21) can be written

$$(g_{BB})_{ij} = \sum_{n,m=1}^{N} C^{\dagger}{}_{in}(\hat{g}_{BB})_{nm}(C)_{mj} \tag{A.24}$$

That is

$$g_{BB} = C^{\dagger}\hat{g}_{BB}C \tag{A.25}$$

where $C^{\dagger}{}_{in} = \phi_n(i) = \langle i|\phi_n\rangle$ and $(C)_{mj} = C^{\dagger}{}^{*}_{jm} = \phi^{*}_{m}(j) = \langle\phi_m|j\rangle$. Since completeness and orthonormality yield $\sum_n \langle i|\phi_n\rangle\langle\phi_n|j\rangle = \delta_{ij}$ and $\sum_j \rangle\langle\phi_n|j\rangle\langle j|\phi_m\rangle = \delta_{nm}$, one finds that

$$C^{\dagger}C = CC^{\dagger} = I_N \tag{A.26}$$

That is C is a unitary matrix and the transformation (A.25) between g_{BB} and $\hat{g}_{BB}$ is called a unitary transformation.

From equation (B.3), the corresponding transformation of the Green's function $G_{BB} = \left(g_{BB}^{-1} - \Sigma_B\right)^{-1}$ is

$$G_{BB} = \sum_{n,m=1}^{N} |\psi_n\rangle(\hat{G}_{BB})_{nm}\langle\psi_m| \tag{A.27}$$

where

$$\hat{G}_{BB} = \left(\hat{g}_{BB}^{-1} - \hat{\Sigma}_B\right)^{-1} \tag{A.28}$$

i.e.

$$\hat{G}_{BB} = \begin{pmatrix} E - E_1 - \left(\hat{\Sigma}_B\right)_{11} & -\left(\hat{\Sigma}_B\right)_{12} & \cdots & -\left(\hat{\Sigma}_B\right)_{1N} \\ -\left(\hat{\Sigma}_B\right)_{21} & E - E_2 - \left(\hat{\Sigma}_B\right)_{22} & \cdots & \vdots \\ \vdots & \cdots & \ddots & \vdots \\ -\left(\hat{\Sigma}_B\right)_{N1} & \cdots & 0 & E - E_N - \left(\hat{\Sigma}_B\right)_{NN} \end{pmatrix}^{-1} \tag{A.29}$$

This transformation is advantageous when E is close to an eigenvalue of h_B and the self-energy matrix elements are small compared to the energy level spacing. For example, if the off-diagonal elements in the matrix of equation (A.29) are set to zero, then

$$\hat{G}_{BB} \approx \begin{pmatrix} \left[E - E_1 - \left(\hat{\Sigma}_B\right)_{11}\right]^{-1} & 0 & \cdots & 0 \\ 0 & \left[E - E_2 - \left(\hat{\Sigma}_B\right)_{22}\right]^{-1} & \cdots & \vdots \\ \vdots & \cdots & \ddots & \vdots \\ 0 & \cdots & 0 & \left[E - E_N - \left(\hat{\Sigma}_B\right)_{NN}\right]^{-1} \end{pmatrix} \tag{A.30}$$

If $E \approx E_1$, then the matrix element $(\hat{G}_{BB})_{11} = [E - E_1 - (\hat{\Sigma}_B)_{11}]^{-1}$ in equation (A.29) will be much larger than all others and therefore

$$\hat{G}_{BB} \approx \frac{|\psi_1\rangle\langle\psi_1|}{E - E_1 - \left(\hat{\Sigma}_B\right)_{11}} \tag{A.31}$$

where

$$\left(\hat{\Sigma}_B\right)_{11} = \langle\psi_1|\Sigma_{\mathrm{B}}|\psi_1\rangle \tag{A.32}$$

Introducing the notation $(\hat{\Sigma}_B)_{11} = \sigma_1 - i\Gamma_1$ leads to the Breit–Wigner formula.

As a second example, consider the approximation of retaining two terms in equation (A.29). If E_2 is the closest energy level to E_1 and $E \approx E_1$, then

$$G_{BB} = \sum\nolimits_{n,m=1}^{2} |\psi_n\rangle\left(\hat{G}_{BB}\right)_{nm}\langle\psi_m| \tag{A.33}$$

In this expression, the 2×2 submatrix of $\hat{G}_{BB}$ is given by

$$\hat{G}_{BB} = \begin{pmatrix} E - E_1 - \left(\hat{\Sigma}_B\right)_{11} & -\left(\hat{\Sigma}_B\right)_{12} \\ -\left(\hat{\Sigma}_B\right)_{21} & E - E_2 - \left(\hat{\Sigma}_B\right)_{22} \end{pmatrix}^{-1} \tag{A.34}$$

where

$$\hat{\Sigma}_{nm} = \langle\psi_n|\Sigma_B\,|\psi_m\rangle \tag{A.35}$$

In this case, equation (11.98) becomes

$$T_{mn} = 4\mathrm{Tr}\left[\hat{\Gamma}_L(n)\hat{G}_{BB}\hat{\Gamma}_R(m)\hat{G}^{\dagger}_{BB}\right] \tag{A.36}$$

where $\hat{\Gamma}_L(n)$ is a 2×2 matrix with elements $[\hat{\Gamma}_L(n)]_{ij} = \left\langle\psi_i|\hat{\Gamma}_L(n)|\psi_j\right\rangle$ and similarly for $\hat{\Gamma}_R(m)$.

Inverting the 2×2 matrix on the right-hand side of equation (A.34) yields

$$\hat{G}_{BB} = \frac{1}{\Delta}\begin{pmatrix} E - E_2 - \left(\hat{\Sigma}_B\right)_{22} & \left(\hat{\Sigma}_B\right)_{12} \\ \left(\hat{\Sigma}_B\right)_{21} & E - E_1 - \left(\hat{\Sigma}_B\right)_{11} \end{pmatrix} \tag{A.37}$$

where

$$\Delta = \left[E - E_1 - \left(\hat{\Sigma}_B\right)_{11}\right]\left[E - E_2 - \left(\hat{\Sigma}_B\right)_{22}\right] - \left(\hat{\Sigma}_B\right)_{12}\left(\hat{\Sigma}_B\right)_{21} \quad \text{(A.38)}$$

As an exercise, it is interesting to verify that the same result is obtained from equations (11.56)–(11.60). To use the latter equations, compare equation (A.34) with equation (11.50) and define

$$g_{11}^{-1} = E - E_1 - \left(\hat{\Sigma}_B\right)_{11}, \quad g_{22}^{-1} = E - E_2 - \left(\hat{\Sigma}_B\right)_{22}, \quad \text{(A.39)}$$

$$\tilde{\Sigma}_1 = \left(\hat{\Sigma}_B\right)_{12} g_{22} \left(\hat{\Sigma}_B\right)_{21} \quad \text{and} \quad \tilde{\Sigma}_2 = \left(\hat{\Sigma}_B\right)_{21} g_{11} \left(\hat{\Sigma}_B\right)_{12} \quad \text{(A.40)}$$

Then (cf equations (11.58) and (11.60))

$$\left(\hat{G}_{BB}\right)_{11} = \left(g_{11}^{-1} - \tilde{\Sigma}_1\right)^{-1}, \quad \left(\hat{G}_{BB}\right)_{22} = \left(g_{22}^{-1} - \tilde{\Sigma}_2\right)^{-1} \quad \text{(A.41)}$$

and (cf equation (11.56))

$$\left(\hat{G}_{BB}\right)_{12} = g_{11}\left(\hat{\Sigma}_B\right)_{12}\left(\hat{G}_{BB}\right)_{22} = \left(\hat{G}_{BB}\right)_{11}\left(\hat{\Sigma}_B\right)_{12} g_{22} \quad \text{(A.42)}$$

The above expressions allow us to estimate the error in using the single-level Breit–Wigner formula. By writing $(\hat{\Sigma}_B)_{11} = \sigma_1 - i\Gamma_1$ and noting that when $E = E_1 + \sigma_1$, $g_{11} = 1/i\Gamma_1$ and $g_{22} = (\delta + \sigma_1 - (\hat{\Sigma}_B)_{22})^{-1}$, where $\delta = E_1 - E_2$ If the level spacing $|\delta|$ is much greater than $|\sigma_1 - (\hat{\Sigma}_B)_{22}|$, then $(\hat{G}_{BB})_{22} \approx g_{22} \approx 1/\delta$. Similarly, $(\hat{G}_{BB})_{12}$ is of order $1/\delta$, whereas $(\hat{G}_{BB})_{11} = (i\Gamma_1 - \tilde{\Sigma}_1)^{-1}$. Consequently in this limit, the error is of order the magnitude of the self-energy matrix element divided by the level spacing.

Appendix B An alternative derivation of the core Green's function

To evaluate equation (11.127), it is convenient to write it in the form

$$G_M = (1 - Y)^{-1} g_M \quad \text{(B.1)}$$

where

$$Y = g_M \Sigma_M = \begin{pmatrix} g_{aa} & g_{ab} \\ g_{ba} & g_{bb} \end{pmatrix}\begin{pmatrix} \Sigma_a & 0 \\ 0 & \Sigma_b \end{pmatrix} = \begin{pmatrix} g_{aa}\Sigma_a & g_{ab}\Sigma_b \\ g_{ba}\Sigma_a & g_{bb}\Sigma_b \end{pmatrix} = \begin{pmatrix} Y_{aa} & Y_{ab} \\ Y_{ba} & Y_{bb} \end{pmatrix} \quad \text{(B.2)}$$

Note that

$$\begin{pmatrix} 1 - Y_{aa} & -Y_{ab} \\ -Y_{ba} & 1 - Y_{bb} \end{pmatrix}^{-1} = \begin{pmatrix} X_{aa} & X_{ab} \\ X_{ba} & X_{bb} \end{pmatrix}, \quad \text{(B.3)}$$

where

$$X_{aa} = [1 - Y_{aa} - Y_{ab}(1 - Y_{bb})^{-1} Y_{ba}]^{-1} \quad \text{(B.4)}$$

and

$$X_{ba} = (1 - Y_{bb})^{-1} Y_{ba} X_{aa} = X_{bb} Y_{ba} (1 - Y_{aa})^{-1} \quad \text{(B.5)}$$

$$X_{bb} = [1 - Y_{bb} - Y_{ba}(1 - Y_{aa})^{-1} Y_{ab}]^{-1} \tag{B.6}$$

$$X_{ab} = (1 - Y_{aa})^{-1} Y_{ab} X_{bb} = X_{aa} Y_{ab}(1 - Y_{bb})^{-1} \tag{B.7}$$

Since

$$G_M = \begin{pmatrix} X_{aa} & X_{ab} \\ X_{ba} & X_{bb} \end{pmatrix} \begin{pmatrix} g_{aa} & g_{ab} \\ g_{ba} & g_{bb} \end{pmatrix} \tag{B.8}$$

One obtains

$$(G_M)_{ab} = X_{aa} g_{ab} + X_{ab} g_{bb} = X_{aa}(g_{ab} + Y_{ab}(1 - Y_{bb})^{-1} g_{bb})$$

That is

$$(G_M)_{ab} = X_{aa} g_{ab}(1 + \Sigma_b(1 - g_{bb}\Sigma_b)^{-1} g_{bb}) \tag{B.9}$$

where

$$X_{aa} = [1 - g_{aa}\Sigma_a - g_{ab}\Sigma_b(1 - g_{bb}\Sigma_b)^{-1} g_{ba}\Sigma_a]^{-1} \tag{B.10}$$

Finally, since $1 + \Sigma_b(1 - g_{bb}\Sigma_b)^{-1} g_{bb} = (1 - \Sigma_b g_{bb})^{-1}$, one obtains

$$(G_M)_{ab} = X_{aa} g_{ab}(1 - \Sigma_b g_{bb})^{-1} \tag{B.11}$$

The corresponding formula for $(G_M)_{ba}$ is obtained by interchanging a and b in the above equation.

Note that as expected, if $\Sigma_a = 0$, $(G_M)_{ab} = g_{ab}(1 - \Sigma_b g_{bb})^{-1}$ and if $\Sigma_b = 0$, $(G_M)_{ab} = (1 - g_{aa}\Sigma_a)^{-1} g_{ab}$.

IOP Publishing

Chapter 12

Linear molecules

12.1 Introduction

In this chapter, recursive use of the decimation method will be compared with iteration of the noisy neighbour equations (NNEs) and both will be utilised to obtain the transport properties of linear chains. In particular, when a molecule is composed of linear chains of sub-units, one finds that the electrical conductance of such a molecule decays with the number of sub-units N as $e^{-\beta N}$, provided the HOMO–LUMO gap remains finite for large N. For a diatomic chain or a chain with alternating bonds, it is found that β is proportional to the geometric mean of the inter-sub-unit couplings, divided by the HOMO–LUMO gap. When carrying decimation site-by-site, a negative eigenvalue theorem is noted, which allows rapid numerical evaluation of densities of states.

12.2 Single-site decimation and the variation of conductance with length

Before utilising decimation, it is enlightening to obtain the results of section 11.3 from a different viewpoint, by returning to the time-independent Schrödinger equation, $H|\psi\rangle = E|\psi\rangle$, which for a system of N sites takes the form

$$\sum\nolimits_{j=1}^{N} H_{ij}\psi_j = E\psi_i \qquad (12.1)$$

Imagine that the wave function of site l is of no interest. Then it is advantageous to eliminate it from the above equation, by writing the equation as

$$\sum\nolimits_{j=1\neq l}^{N} H_{ij}\psi_j + H_{il}\psi_l = E\psi_i \qquad (12.2)$$

Choosing $i = l$ yields $\sum\nolimits_{j=1\neq l}^{N} H_{lj}\psi_j + H_{ll}\psi_l = E\psi_l$, which can be solved to yield

doi:10.1088/978-0-7503-3639-0ch12

$$\psi_l = \sum\nolimits_{j=1\neq l}^{N} \frac{H_{lj}}{E - H_{ll}} \psi_j \tag{12.3}$$

Substituting this back into equation (12.2) yields

$$\sum\nolimits_{j=1\neq l}^{N} H_{ij}^{(1)} \psi_j = E\psi_i \tag{12.4}$$

where

$$H_{ij}^{(1)} = H_{ij} + \frac{H_{il}H_{lj}}{E - H_{ll}} \tag{12.5}$$

Clearly the term $\frac{H_{il}H_{lj}}{E - H_{ll}}$ is simply the self-energy to be added to the $(N - 1) \times (N - 1)$ sub-matrix of the parental Hamiltonian, to create the renormalized $(N - 1) \times (N - 1)$ Hamiltonian of the daughter. Equation (12.5) is an example of equation (11.22), obtained by eliminating a single site.

Since the aim is to use this Hamiltonian to compute Green's functions as a function of E, the fact that $H_{ij}^{(1)}$ is a function of E will present no difficulties. Having eliminated site l, one can now repeat the above procedure to eliminate a second site k, starting from equations (12.4) and (12.5). The result is

$$\sum\nolimits_{j=1\neq l,k}^{N} H_{ij}^{(2)} \psi_j = E\psi_i \tag{12.6}$$

where

$$H_{ij}^{(2)} = H_{ij}^{(l)} + \frac{H_{ik}^{(l)} H_{kj}^{(l)}}{E - H_{kk}^{(l)}} \tag{12.7}$$

In this equation, $H_{ij}^{(2)}$ is a matrix element of the renormalized Hamiltonian that arises after two decimation steps. Notice that the right-hand side of equation (12.7) involves matrix elements of the renormalized Hamiltonian $H_{ij}^{(1)}$, obtained after eliminating only site l. Since the summation in equation (12.6) does not involve sites l and k, the problem has been reduced in size to a $(N - 2) \times (N - 2)$ matrix equation. Furthermore, the wave function ψ_i and the energy E is not changed by the elimination step.

To see how this works in practice, consider the one-dimensional chain shown in figure 12.1, whose Hamiltonian is $H^{(0)}$, with matrix elements $H_{ij}^{(0)}$. After eliminating site 2, the renormalized matrix elements involving sites 1 and 3 are

$$H_{11}^{(1)} = H_{11}^{(0)} + \frac{H_{12}^{(0)} H_{21}^{(0)}}{E - H_{22}^{(0)}},\ H_{33}^{(1)} = H_{33}^{(0)} + \frac{H_{32}^{(0)} H_{23}^{(0)}}{E - H_{22}^{(0)}} \text{ and } H_{13}^{(1)} = \frac{H_{12}^{(0)} H_{23}^{(0)}}{E - H_{22}^{(0)}} \tag{12.8}$$

No other Hamiltonian matrix elements are changed, because in equation (12.5) the matrix element between sites i and j is only changed if both H_{i2} and H_{2j} are non-zero.

Similarly, after the second decimation step in which site 3 is eliminated, the renormalized matrix elements involving sites 1 and 4 are

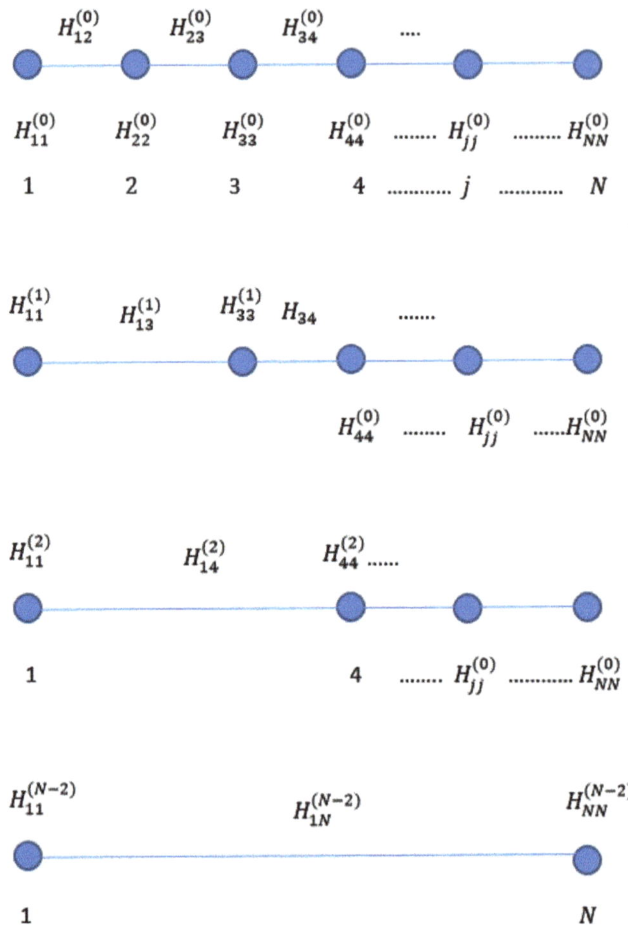

Figure 12.1. Decimation on a one-dimensional chain.

$$H_{11}^{(2)} = H_{11}^{(1)} + \frac{H_{13}^{(1)}H_{31}^{(1)}}{E - H_{33}^{(1)}},\ H_{44}^{(2)} = H_{44}^{(0)} + \frac{H_{43}^{(0)}H_{34}^{(0)}}{E - H_{33}^{(1)}} \text{ and } H_{14}^{(2)} = \frac{H_{13}^{(1)}H_{34}^{(0)}}{E - H_{33}^{(1)}} \tag{12.9}$$

where it has been noted that $H_{34}^{(1)} = H_{34}^{(0)}$, because this matrix element was unchanged during the previous step.

Repeating this procedure many times yields for the j th decimation step

$$\begin{gathered} H_{11}^{(j)} = H_{11}^{(j-1)} + \frac{H_{1,j+1}^{(j-1)}H_{j+1,1}^{(j-1)}}{D^{(j)}(E)},\ H_{j+2,j+2}^{(j)} = H_{j+2,j+2}^{(j-1)} + \frac{H_{j+2,j+1}^{(0)}H_{j+1,j+2}^{(0)}}{D^{(j)}(E)}, \\ H_{1,j+2}^{(j)} = \frac{H_{1,j+1}^{(j-1)}H_{j+1,j+2}^{(0)}}{D^{(j)}(E)} \end{gathered} \tag{12.10}$$

where the denominator at step j is

$$D^{(j)}(E) = E - H^{(j-1)}_{j+1,j+1} \tag{12.11}$$

After $N - 2$ steps, the renormalized Hamiltonian has become a 2×2 matrix

$$H^{(N-2)} = \begin{pmatrix} H^{(N-2)}_{11} & H^{(N-2)}_{1N} \\ H^{(N-2)}_{N1} & H^{(N-2)}_{NN} \end{pmatrix} \tag{12.12}$$

whose Green's function is obtained simply by inverting the 2×2 matrix $E - H^{(N-2)}$; i.e.

$$G = \begin{pmatrix} G_{11} & G_{1N} \\ G_{N1} & G_{NN} \end{pmatrix} = (E - H^{(N-2)})^{-1} = \frac{1}{\Delta}\begin{pmatrix} E - H^{(N-2)}_{NN} & H^{(N-2)}_{1N} \\ H^{(N-2)}_{N1} & E - H^{(N-2)}_{11} \end{pmatrix} \tag{12.13}$$

where $\Delta = \det(E - H^{(N-2)})$. This shows that

$$G_{1N} = \frac{H^{(N-2)}_{1N}}{E - H^{(N-2)}_{11}} G_{NN} \tag{12.14}$$

where G_{NN} is the surface Green's function of the chain of N sites, given by

$$G_{NN} = \frac{E - H^{(N-2)}_{11}}{\Delta} = \left[E - H^{(N-2)}_{NN} - \frac{H^{(N-2)}_{1N} H^{(N-2)}_{N1}}{E - H^{(N-2)}_{11}}\right]^{-1} \tag{12.15}$$

From equation (12.25), $H^{(N-2)}_{1,N} = \dfrac{H^{(N-3)}_{1,N-1} H^{(0)}_{N-1,N}}{D^{(N-2)}(E)}$. Hence

$$G_{1N} = \frac{H^{(N-3)}_{1,N-1} H^{(0)}_{N-1,N} G_{NN}}{D^{(N-2)}(E)\left(E - H^{(N-2)}_{11}\right)}$$

It is useful to note that

$$\begin{aligned} D^{(N-2)}(E)\left(E - H^{(N-2)}_{11}\right) &= \left(E - H^{(N-3)}_{N-1,N-1}\right)\left(E - H^{(N-3)}_{11} - \frac{H^{(N-3)}_{1,N-1} H^{(N-3)}_{N-1,1}}{E - H^{(N-3)}_{N-1,N-1}}\right) \\ &= \left(E - H^{(N-3)}_{11}\right)\left(E - H^{(N-3)}_{N-1,N-1} - \frac{H^{(N-3)}_{1,N-1} H^{(N-3)}_{N-1,1}}{E - H^{(N-3)}_{11}}\right) \\ &= \frac{\left(E - H^{(N-3)}_{11}\right)}{G_{N-1,N-1}} \end{aligned}$$

Hence

$$G_{1N} = \frac{H^{(N-3)}_{1,N-1} G_{N-1,N-1} H^{(0)}_{N-1,N} G_{NN}}{\left(E - H^{(N-3)}_{11}\right)}$$

where $G_{N-1,N-1}$ is the surface Green's function of the chain of $N - 1$ sites.

Repeating these steps yields finally

$$G_{1N} = G_{11}H^{(0)}_{1,2}G_{22}H^{(0)}_{2,3} \ldots\ldots H^{(0)}_{N-2,N-1}G_{N-1,N-1}H^{(0)}_{N-1,N}G_{NN} \tag{12.16}$$

where $G_{11} = \left(E - H^{(0)}_{11}\right)^{-1}$.

Equation (12.16) is rather general and can be applied to any linear chain of sites, including a random chain, in which the couplings $H^{(0)}_{j,j+1}$ or sites energies $H^{(0)}_{j,j}$ are random numbers. As a simpler example, note that for the periodic linear chain of N sites shown in figure 4.8, $H_{j,j+1} = -\gamma$ and $H^{(0)}_{jj} = \epsilon_0$, in which case equation (12.16) reduces to

$$G_{1N} = G_{11}(-\gamma)G_{22}(-\gamma) \ldots\ldots (-\gamma)G_{N-1,N-1}(-\gamma)G_{NN}$$

The Green's function for such a chain of N sites is given by equations (10.70) and (10.71). Choosing $j = p = 1$ in the former, or $j = p = N$ in the latter yields for the surface Green's function

$$G_{NN} = \frac{\sin kN}{(-\gamma)\sin k(N+1)}$$

As a quick check, note that $G_{11} = \frac{\sin k}{-\gamma \sin 2k} = \frac{\sin k}{-2\gamma \sin k \cos k} = \frac{1}{E - \epsilon_0}$, as expected.

After cancelling factors of $\sin 2k$, this yields for example $G_{11}(-\gamma)G_{22} = \frac{\sin k}{-\gamma \sin 3k}$ and after repeated cancellations,

$$G_{1N} = \frac{\sin k}{-\gamma \sin k(N+1)}, \tag{12.17}$$

which could have been obtained from equation (10.70) by choosing $j = 1$ and $p = N$.

In the above equation, $k = \text{acos}\left[\frac{\epsilon_0 - E}{2\gamma}\right]$. If $\epsilon_0 - 2|\gamma| \leqslant E \leqslant \epsilon_0 + 2|\gamma|$, then E lies within the energy band of the corresponding doubly infinite chain and k is real. In that case G_{NN} diverges when $k = n\pi/(N+1)$ (i.e. when $E = \epsilon_0 - 2\gamma \cos \frac{n\pi}{N+1}$), leading to Breit–Wigner resonances in the transmission coefficient, when such a chain is connected to electrodes. This is the simplest model of a metallic nanowire, or a conductive molecular wire such as cumulene.

On the other hand, if E lies outside the energy band, k is complex. Since E is related to k by $E = \epsilon_0 - 2\gamma \cos k$, it is clear that if $\gamma > 0$ and $E < \epsilon_0 - 2|\gamma|$, or if $\gamma < 0$ and $E > \epsilon_0 + 2|\gamma|$, then $k = \frac{i\beta}{2}$, where β is real. (Note that $\cos i\beta/2 = \cosh \beta/2 \geqslant 1$). Otherwise if $\gamma < 0$ and $E < \epsilon_0 - 2|\gamma|$, or if $\gamma > 0$ and $E > \epsilon_0 + 2|\gamma|$, then $k = \pi + \frac{i\beta}{2}$. (Note that $\cos(\pi + i\beta/2) = -\cosh \beta/2 \leqslant -1$.) In either case,

$$|G_{1N}| = \frac{\sinh\frac{\beta}{2}}{|\gamma|\sinh\beta(N+1)/2}, \text{ which for large } N \text{ tends to } |G_{1N}| \approx \frac{\sinh\frac{\beta}{2}}{2|\gamma|}e^{-\beta(N+1)/2} \tag{12.18}$$

From equation (8.41), this means that the electrical conductance will be proportional to $e^{-\beta(N+1)}$.

12.3 A negative eigenvalue theorem

The above procedure can be used to obtain information about the eigenvalues of a closed system. To demonstrate this, consider carrying out the above site-by-site decimation for an arbitrary finite system of N sites, using equations (12.15), (12.17), etc. The order in which the sites are decimated is not important, so let l_j be the site, which eliminated at the j th decimation step. If the denominator at step j is denoted $D^{(j)}(E) = E - H^{(j-1)}_{l_j,l_j}$ then after eliminating $N - 1$ sites, the renormalised Hamiltonian is reduced to a single number, $H^{(N-1)}_{p,p}(E)$, where p is the last remaining site. The Schrödinger equation for this last site is simply

$$H^{(N-1)}_{p,p}(E)\psi_p = E\psi_p$$

That is

$$D^{(N)}(E)\psi_p = 0, \quad \text{where } D^{(N)}(E) = E - H^{(N-1)}_{p,p}(E) \tag{12.19}$$

Hence the eigenvalues E_n of the N-site system are those values of E for which $D^{(N)}(E) = 0$. When carrying out the above decimation procedure, one first chooses an energy E and then decimates the $N \times N$ Hamiltonian to obtain $D^{(N)}(E)$. If this is repeated for many values of E, then a graph of $D^{(N)}(E)$ versus E can be plotted. The values of E at which the graph crosses the horizontal axis are the eigenvalues E_n.

One advantage of this approach is that it can be implemented on a computer without storing the $N \times N$ Hamiltonian and therefore it is a useful technique when H is a sparse matrix. On the other hand, if N is very large, the individual eigenvalues may not be of interest. Instead the density of states (see chapter 15) is often more interesting, which provides information about the number of eigenvalues between two energies. If $N(E_a)$ is the number of eigenvalues with energies less than chosen energy E_a then the number of eigenvalues between E_b and E_a (where $E_b > E_a$) is $N(E_b) - N(E_a)$. The quantity $N(E_a)$ is called the integrated density of states. It is equal to zero when E_a is less than the smallest eigenvalue of the system and equal to N when E_a is greater than the largest eigenvalue of the system.

During a complete series of decimation N steps, carried out at some chosen energy E_a, a series of denominators will be encountered, denoted $D^{(1)}(E_a)$, $D^{(2)}(E_a)$, ...$D^{(N)}(E_a)$., some of which will be positive and some will be negative. If the number of positive denominators is found to be $P(E_a)$, then the negative eigenvalue theorem tells us that the number of eigenvalues less than E_a is $P(E_a)$. In other words, the integrated density of states is given by

$$N(E_a) = P(E_a) \tag{12.20}$$

When implemented numerically, this is a fast method for computing $N(E_a)$, without computing individual eigenvalues. (Note: Vanishing denominators can be avoided by choosing irrational numbers for the energies E_a.)

12.4 Recurrence relations via the noisy neighbour equations

It should be noted that if the pink region of figure 12.2 is denoted A and the green region denoted B, then the noisy neighbour equation

$$G_{AB} = g_{AA}H_{AB}G_{BB}, \quad \text{where } G_{BB} = \left(g_{BB}^{-1} - \Sigma_B\right)^{-1} \text{ and } \Sigma_B = H_{BA}g_{AA}H_{AB} \tag{12.21}$$

yields

$$G_{1N} = G_{1,N-1}H^{(0)}_{N-1,N}G_{NN} \tag{12.22}$$

where,

$$G_{NN} = \left[E - H^{(0)}_{NN} - H^{(0)}_{N,N-1}G_{N-1,N-1}H^{(0)}_{N-1,N}\right]^{-1} \tag{12.23}$$

In these equations, G_{1N} denotes the sub-matrix of the Green's function of the coupled system, between sites in cell 1 and sites in cell N, $G_{1,N-1}$ denotes the sub-matrix of the Green's function of the isolated system B between cells 1 and $N - 1$, (denoted g_{AA} in equation (12.21)) $G_{N-1,N-1}$ denotes the surface Green's function of the isolated system A associated with cell $N - 1$ and G_{NN} is the Green's function of B in the presence of A (denoted G_{BB} in equation (12.21)). In other words, G_{NN} is the surface Green's function of the whole structure composed of N coupled cells and G_{1N} is the Green's function of the coupled structure between cells 1 and N.

Applying the same formula to a chain of $N - 1$ cells forming system B yields

$$G_{1,N-1} = G_{1,N-2}H^{(0)}_{N-2,N-1}G_{N-1,N-1} \tag{12.24}$$

where

$$G_{N-1,N-1} = \left[E - H^{(0)}_{N-1,N-1} - H^{(0)}_{N-1,N-2}G_{N-2,N-2}H^{(0)}_{N-2,N-1}\right]^{-1} \tag{12.25}$$

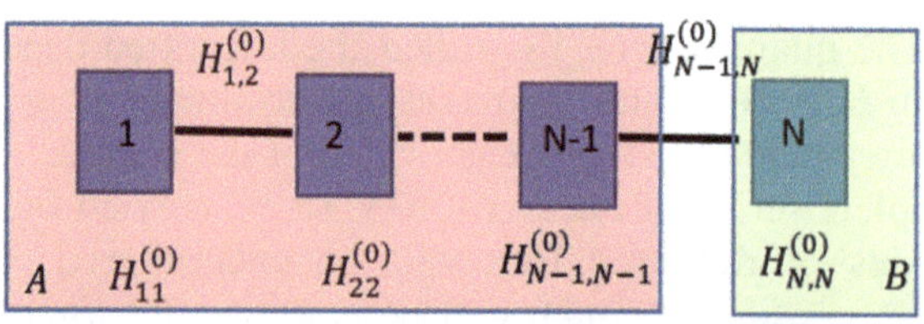

Figure 12.2. A chain of quantum cells connected in series.

where $G_{N-2,N-2}$ is the surface Green's function of a chain of the first $N - 2$ cells. Substituting this into equation (12.21) and repeating the iteration yields

$$G_{1N} = G_{11}H^{(0)}_{1,2}G_{22}H^{(0)}_{2,3} \ldots\ldots H^{(0)}_{N-2,N-1}G_{N-1,N-1}H^{(0)}_{N-1,N}G_{NN} \tag{12.26}$$

This is a generalisation of equation (12.16) to finite-width ribbons, in which all quantities such as $H^{(0)}_{n,n+1}$ and G_{nn} are now matrices.

This demonstrates that decimation and Dyson's equation yield the same results and therefore utilisation of one or the other is simply a matter of choice. Interestingly, equation (12.41) is a recurrence relation for surface Green's functions. As discussed in chapter 17, for a 1-dimensional tight-binding chain of figure 4.3, it reduces to

$$G_{NN} = [E - \epsilon_0 - \gamma^2 G_{N-1,N-1}]^{-1} \tag{12.27}$$

In the limit $N \to \infty$, if this converges, such that $G_{N-1,N-1} = G_{NN}$, then denoting the surface Green's function by g_s yields $g_s = [E - \epsilon_0 - \gamma^2 g_s]^{-1}$. Noting that $E = \epsilon_0 - 2\gamma \cos k$ yields a quadratic equation for g_s, with solution

$$g_s = -\frac{e^{\pm ik}}{\gamma} \tag{12.28}$$

Choosing the + sign yields the retarded surface Green's function of equation (10.31).

12.5 Decay of conductance with length of an oligoyne and a diatomic chain

As an example of the utility of decimation, consider the diatomic chain of $2N - 1$ sites, shown in the upper panel of figure 12.3, which for the choice $\epsilon_1 = \epsilon_2$ is a simple model of an oligoyne. After decimating the sites labelled 1', 2', … $N' - 1$, one obtained the chain of N site in the lower panel, where the renormalized couplings are denoted $-\gamma$. The renormalized parameters of the decimated lattice are

$$\begin{gathered} -\gamma = \gamma_1\gamma_2/(E - \epsilon_2),\ \epsilon_0 = \epsilon_1 + \frac{\gamma_1^2}{E - \epsilon_2} + \frac{\gamma_2^2}{E - \epsilon_2}, \\ \epsilon_a = -\frac{\gamma_2^2}{E - \epsilon_2} \text{ and } \epsilon_b = \frac{\gamma_1^2}{E - \epsilon_2} \end{gathered} \tag{12.29}$$

To obtain the Green's function G_{1N} of this chain, note that when $\epsilon_a = \epsilon_b = 0$, the Green's function g is given by equations (10.70) and (10.71), which yield

$$g_{11} = g_{NN} = \frac{\sin kN}{(-\gamma)\sin k(N+1)} \text{ and } g_{1N} = g_{N1} = \frac{\sin k}{-\gamma \sin k(N+1)} \tag{12.30}$$

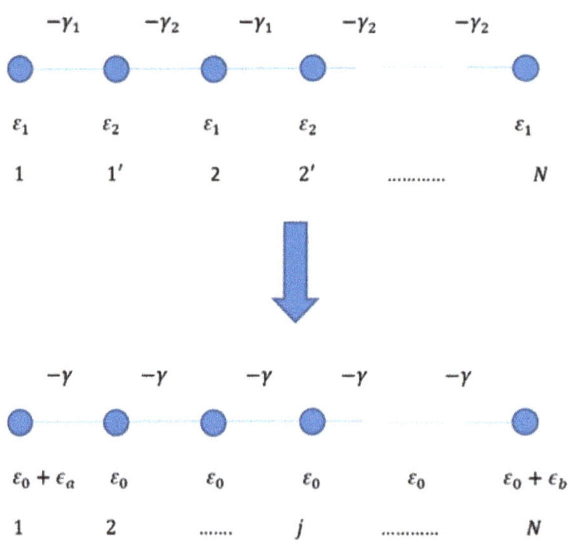

Figure 12.3. The upper panel shows a diatomic chain of $2N - 1$ sites. The lower panel shows the chain of N sites obtained by decimating the diatomic chain.

In other words, the Green's function between surface sites of the chain of N sites in figure 4.8 is

$$g = \begin{pmatrix} g_{11} & g_{1N} \\ g_{N1} & g_{NN} \end{pmatrix} \tag{12.31}$$

(Note that here a lower case g is used for this Green's function rather than upper case as in equation (12.16), since here, the linear chain in the lower panel of figure 4.8 represents the unperturbed Green's function appearing in Dyson's equation.) Starting from such a chain, Dyson's equation can be used to include the perturbations ϵ_a and ϵ_b and yields for the Green's function G of the chain in the lower panel of figure 12.3,

$$G = \begin{pmatrix} G_{11} & G_{1N} \\ G_{N1} & G_{NN} \end{pmatrix} = \left[\begin{pmatrix} g_{11} & g_{1N} \\ g_{N1} & g_{NN} \end{pmatrix}^{-1} - \begin{pmatrix} \epsilon_a & 0 \\ 0 & \epsilon_b \end{pmatrix}\right]^{-1} \tag{12.32}$$

Since $\begin{pmatrix} g_{11} & g_{1N} \\ g_{N1} & g_{NN} \end{pmatrix}^{-1} = \frac{1}{d}\begin{pmatrix} g_{NN} & -g_{1N} \\ -g_{N1} & g_{11} \end{pmatrix}$ where $d = g_{11}g_{NN} - g_{1N}g_{N1}$, this yields

$$G = d\begin{pmatrix} g_{NN} - d\epsilon_a & -g_{1N} \\ -g_{N1} & g_{11} - d\epsilon_b \end{pmatrix}^{-1}$$

Hence

$$G = \begin{pmatrix} G_{11} & G_{1N} \\ G_{N1} & G_{NN} \end{pmatrix} = \frac{1}{\mathrm{D}}\begin{pmatrix} g_{11} - d\epsilon_b & g_{1N} \\ g_{N1} & g_{NN} - d\epsilon_a \end{pmatrix} \tag{12.33}$$

where

$$D = 1 - \epsilon_b g_{NN} - \epsilon_a g_{11} + \epsilon_a \epsilon_b d \tag{12.34}$$

Finally, one obtains

$$G_{1N} = \frac{g_{1N}}{D} \tag{12.35}$$

As discussed in section 12.2, if k is complex, g_{1N} decays with increasing N as shown in equation (12.18), where β is the imaginary part of k. To obtain an expression for the decay constant β, note that since $E = \epsilon_0 - 2\gamma \cos k$, substituting the decimated parameters from equation (12.29) yields $E = \epsilon_1 + \frac{\gamma_1^2}{E-\epsilon_2} + \frac{\gamma_2^2}{E-\epsilon_2} + \frac{2\gamma_1\gamma_2}{E-\epsilon_2}\cos k$. Hence $(E - \epsilon_1)(E - \epsilon_2) = \gamma_1^2 + \gamma_2^2 + 2\gamma_1\gamma_2 \cos k$ and therefore

$$E = \bar{\epsilon} \pm |\Delta(\mathrm{k})| \tag{12.36}$$

where

$$|\Delta(\mathrm{k})| = \sqrt{\epsilon^2 + \gamma_1^2 + \gamma_2^2 + 2\gamma_1\gamma_2 \cos k} \tag{12.37}$$

$\bar{\epsilon} = \frac{\epsilon_1+\epsilon_2}{2}$ and $\epsilon = \frac{\epsilon_1-\epsilon_2}{2}$. The case $\epsilon = 0$ was discussed in section 5.1 (in section 5.1, $\bar{\epsilon}$ is denoted ϵ_0), where it was noted that there are two energy bands (see figure 5.2) with an energy gap $2\delta = 2|(|\gamma_1|-|\gamma_2|)|$. When $\epsilon \neq 0$, the energy gap is $2\delta_\epsilon$, where $\delta_\epsilon = \sqrt{\epsilon^2 + \delta^2}$. If E coincides with the gap centre (i.e. when $E = \bar{\epsilon}$), then the square root in equation (12.37) must vanish. That is $\cos k = -\frac{\epsilon^2+\gamma_1^2+\gamma_2^2}{2\gamma_1\gamma_2} = -s\frac{\epsilon^2+\gamma_1^2+\gamma_2^2}{2|\gamma_1\gamma_2|} = -s\left[1 + \frac{\epsilon^2+(|\gamma_1|-|\gamma_2|)^2}{2|\gamma_1\gamma_2|}\right]$ where s is the sign of the product $\gamma_1\gamma_2$. If $s = +1$, then $k = \pi + i\beta$, whereas if $s = -1$, $k = i\beta$, where

$$\cosh\beta = 1 + \frac{\epsilon^2 + (|\gamma_1| - |\gamma_2|)^2}{2|\gamma_1\gamma_2|}$$

For small β, $\cosh\beta \approx 1 + \frac{\beta^2}{2}$.

Hence

$$\beta \approx \delta_\epsilon / \sqrt{|\gamma_1\gamma_2|} \tag{12.38}$$

This shows that at the gap centre, β is approximately equal to the energy gap divided by the geometric mean of the nearest neighbour couplings.

More generally, for energies which differ from the gap centre, one finds

$$\cosh\beta = 1 + \frac{\epsilon^2 + (|\gamma_1| - |\gamma_2|)^2 - (E - \bar{\epsilon})^2}{2|\gamma_1\gamma_2|}$$

so that

$$\beta \approx \frac{\sqrt{\delta_\epsilon^2 - (E - \bar{\epsilon})^2}}{\sqrt{|\gamma_1\gamma_2|}} \tag{12.39}$$

This shows that the decay constant β is a maximum at the gap centre and decreases with energy, reaching zero when E coincides with the top of the lower band or the bottom of the upper band. For the reader interested in learning more about diatomic chains and bipartite lattices, reference [3] is recommended reading.

12.6 Quantum circuit rules for the conductance and Seebeck coefficient of linear molecules

When resistors are coupled in series, Ohms law tells us that the total resistance is the sum of the individual resistances. When molecular moieties are coupled in series, the Green's function of the resulting oligomer is given by equation (12.26), which is the quantum replacement of Ohm's law at a molecular scale. When the energy lies within the HOMO–LUMO gap, if the coupling matrices $H^{(0)}_{j,j+1}$ are sufficiently small, then equation (12.39) can be replaced by

$$G_{N-1,N-1} \approx G^{(0)}_{N-1,N-1}$$

where $G^{(0)}_{N-1,N-1}$ is the Green's function of moiety $N - 1$

$$G^{(0)}_{N-1,N-1} = \left[E - H^{(0)}_{N-1,N-1}\right]^{-1} \tag{12.40}$$

Hence in this weak coupling limit, equation (12.26) is approximated by

$$G_{1N} = G^{(0)}_{11}H^{(0)}_{1,2}G^{(0)}_{22}H^{(0)}_{2,3} \;\ldots\ldots\; H^{(0)}_{N-2,N-1}G^{(0)}_{N-1,N-1}H^{(0)}_{N-1,N}G^{(0)}_{N,N} \tag{12.41}$$

If the molecule is weakly coupled to electrodes, (so that G_{BB} can be replaced by G_{1N} in equations (11.54) and (11.59)) then the latter equations reveal that in a molecular junction, the transmission coefficient from an open channel n of the left lead to an open channel m of the right lead is proportional to

$$|(G_{1N})_{mn}|^2 = |\langle \chi'_m | H_{RB} G_{1N} H_{BL} | \chi_n \rangle|^2 \tag{12.42}$$

In general, these are complicated matrix equations, but is certain cases they simplify. For example, if the coupling matrices $H^{(0)}_{j,j+1}$ couple only a single site in moiety j to a single site I moiety $j + 1$, then equation (12.41) becomes simply a product of numbers.

For the case where all $H^{(0)}_{j,j+1}$ have the same magnitude,

$$G_{1N} \sim G^{(0)}_{11}G^{(0)}_{22} \ldots G^{(0)}_{N-1,N-1}G^{(0)}_{N,N} \tag{12.43}$$

As an example, consider the case of three moieties in series, for which

$$G_{13} \sim G^{(0)}_{11}G^{(0)}_{22}G^{(0)}_{33} \tag{12.44}$$

For convenience, denote the product on the right-hand side of equation (12.44) by G_{123}, so that

$$G_{123} = G^{(0)}_{11}G^{(0)}_{22}G^{(0)}_{33}$$

Now consider the effect of changing the moieties forming such oligomers. For an oligomer formed from moieties 1,2 and 1, the corresponding product is

$$G_{121} = G_{11}^{(0)} G_{22}^{(0)} G_{11}^{(0)}$$

for a third molecule formed from moieties 2,2,1

$$G_{212} = G_{22}^{(0)} G_{11}^{(0)} G_{22}^{(0)}$$

and for fourth and fifth molecules formed from moieties 2,2,2 and 1,1,1

$$G_{222} = G_{22}^{(0)} G_{22}^{(0)} G_{22}^{(0)} \text{ and } G_{111} = G_{11}^{(0)} G_{11}^{(0)} G_{11}^{(0)}$$

These satisfy the product rule [1, 2]

$$G_{222} G_{111} = G_{212} G_{121} \tag{12.45}$$

Hence for molecules formed from weakly coupled moieties in series, one expects their electrical conductances to satisfy a corresponding product rule.

In addition to the above circuit rule for electrical conductance, since in the low-temperature limit, the Seebeck coefficient is proportional to the slope of the logarithm of the transmission coefficient and since the logarithm of a product is the sum of theory logarithms one finds the following circuit rule for Seebeck coefficients of weakly coupled moieties in series

$$S_{222} + S_{111} = S_{212} + S_{121} \tag{12.46}$$

12.7 Summary

In this chapter, the decimation method was used to obtain Green's functions of linear chains. It was found that the electrical conductance of such chains decay exponentially, provided the Fermi energy lies within the HOMO–LUMO gap and that the gap persists in the long-chain limit. An efficient method for computing the integrated density of states of quantum structures was also discussed. Finally, it was noted that electrical conductances and Seebeck coefficients satisfy product rules and sum rules, provided substructures within linear molecules are weakly coupled.

References

[1] Manrique D Z *et al* 2015 A quantum circuit rule for interference effects in single-molecule electrical junctions *Nat. Commun.* **6** 6389

[2] Manrique D Z, Al-Galiby Q, Hong W and Lambert C J L 2016 A new approach to materials discovery for electronic and thermoelectric properties of single-molecule junctions *Nano Lett.* **16** 1308

[3] Asbóth J K, Oroszlány L and Pályi A 2016 *A Short Course on Topological Insulators* (*Lecture Notes in Physics* vol 919) (Berlin: Springer)

IOP Publishing

Quantum Transport in Nanostructures and Molecules
An introduction to molecular electronics
Colin John Lambert

Chapter 13

Quantum interference in molecules with parallel paths and pendant groups

13.1 Introduction

In chapter 12, Green's functions of linear molecules formed by coupling building blocks in series were presented. In this chapter, the Green's functions of molecules with bridging units and/or pendant groups will be discussed. In chapter 11, equations (11.109) to (11.113) show that for the structure of figure 11.5, the transmission coefficient from channel n of the right lead, to channel m of the left lead could be obtained, once the Green's function g_{bb} of the isolated scatterer is known. In what follows, g_{bb} will be computed for a number of model structures. Since the isolated scatterer is now the focus of attention, and its Green's function is the output of the calculations below, the notation will now change. The Green's function of the isolated scatterer formed by coupling together building blocks will be denoted either G, or $\hat{g}$, while the Green's functions of isolated building blocks will be denoted by a lower-case g. As an example, if external one-dimensional electrodes are coupled to single sites labelled a and b, (instead of 1 and N as shown in figure 8.1), then in this new notation, the equation (8.46) becomes

$$T = |t|^2 = 4\left[\frac{\alpha^2 \sin k_L}{\gamma_L}\right]\left[\frac{\beta^2 \sin k_R}{\gamma_R}\right]\left|\frac{G_{ab}}{\Delta}\right|^2 \tag{13.1}$$

where

$$\Delta = 1 - G_{aa}\Sigma_L - G_{bb}\Sigma_R + \Sigma_L\Sigma_R(G_{aa}G_{bb} - G_{ab}G_{ba}) \tag{13.2}$$

and the self-energies Σ_L, Σ_R are given by equation (8.33). Formulae for G_{ab} are obtained below to illustrate how quantum interference created by the building blocks of a molecule affect electron transport through the molecule. Quantum interference effects are more pronounced when a molecule or a molecular junction contains 'pinch points', which force the wave function of electrons passing through the molecule to superpose on particular atoms or orbitals. The examples below are chosen to illustrate this point. Section 13.2 describes how DQI arises when a pendant group interacts with a molecular orbital on the backbone of a molecule. Sections 13.3–13.5 describe quantum interference effects arising when a bridging group couples two substructures, which lie on the current path. Sections 13.6 and 13.7 describe QI between parallel paths created by the presence of two or more bridging groups.

13.2 The effect of pendant groups

Consider first the following structure, formed from building blocks (i.e. moieties) labelled 1 (coloured blue) and 2 (coloured green). The moiety 1 represents a subspace of the isolated scatterer, which for the moment could be arbitrary, but later will be chosen to be the subspace which connects to electrodes. Substructure 2 is a pendant group attached to subsystem 1 by a coupling matrix H_{12} (figure 13.1).

The Green's function of the decoupled structure, before coupling moieties 1 and 2 together is

$$g_0 = \begin{pmatrix} g_{11} & 0 \\ 0 & g_{22} \end{pmatrix} \tag{13.3}$$

When the coupling H_{12} between the two substructures is added, the Green's function of the combined structure becomes

$$\begin{pmatrix} G_{11} & G_{12} \\ G_{21} & G_{22} \end{pmatrix} = \begin{pmatrix} g_{11}^{-1} & -H_{12} \\ -H_{21} & g_{22}^{-1} \end{pmatrix}^{-1} \tag{13.4}$$

This is analogous to equation (11.49) (with A replaced by 1 and B replaced by 2) and therefore from equations (11.58) and (11.59), one obtains

$$G_{11} = \left(g_{11}^{-1} - \Sigma_2\right)^{-1} \qquad \text{where} \quad \Sigma_2 = H_{12} g_{22} H_{21} \tag{13.5}$$

To introduce a 'pinch point', consider the case where the energy E of an electron passing through the structure from a to b is close to an energy level E_p of the isolated moiety 1, so that g_{11} can be approximated by (cf equation (A.31) in chapter 11)

Figure 13.1. Two quantum coloured blue and green, connected by a coupling matrix H_{12}. The finite-width electrodes are shaded yellow.

$$g_{11} \approx \frac{|\psi_p\rangle\langle\psi_p|}{E - E_p} \tag{13.6}$$

and hence

$$G_{11} \approx \frac{|\psi_p\rangle\langle\psi_p|}{E - E_p - \Sigma_p} \tag{13.7}$$

where

$$\Sigma_p = \langle\psi_p|H_{12}g_{22}H_{21}|\psi_p\rangle \tag{13.8}$$

In this case, the pinch point is the molecular orbital $|\psi_p\rangle$ located on substructure 1, though which all electrons must pass.

In section 11.14, a scatterer was connected to finite-width electrodes and the Breit–Wigner formula was derived. Before connecting the scatterer to the electrodes, the Green's function was denoted g_{bb} and was of the form shown in equation (11.114). In the present case, the corresponding Green's function of the isolated scatterer is G_{11}, which has the same form as equation (11.114), except that E_p is replaced by $E_p + \Sigma_2$. Hence after making this replacement, in the analysis of section 11.14, equation (11.118) yields for the transmission coefficient,

$$T(E) = \sum_{m,n} T_{mn} = \frac{4\Gamma_L\Gamma_R}{(E - \epsilon_p - \sigma_p)^2 + (\Gamma_L + \Gamma_R)^2} \tag{13.9}$$

where

$$\epsilon_p = E_p + \Sigma_p \tag{13.10}$$

Equation (13.9) is identical to the Breit–Wigner formula, except that E_p is replaced by ϵ_p, due to the presence of the pendant group.

As noted in appendix A of chapter 11, (equation (A.14)) the Green's function of any finite-size isolated moiety such as the pendant group 2 can be written

$$g_{22} = \sum_{n=1}^{N} \frac{|\phi_n\rangle\langle\phi_n|}{E - \bar{E}_n} \tag{13.11}$$

where $|\phi_n\rangle$ is a molecular orbital of the isolated moiety 2, with orbital energy $\bar{E}_n$. Hence the self-energy due to decimation of the pendant moiety is

$$\Sigma_p = \sum_{n=1}^{N} \frac{A_n}{E - \bar{E}_n} \tag{13.12}$$

where the amplitudes A_n are positive numbers, given by

$$A_n = \langle\psi_p|H_{12}|\phi_n\rangle\langle\phi_n|H_{21}|\psi_p\rangle = |\langle\psi_p|H_{12}|\phi_n\rangle|^2 \tag{13.13}$$

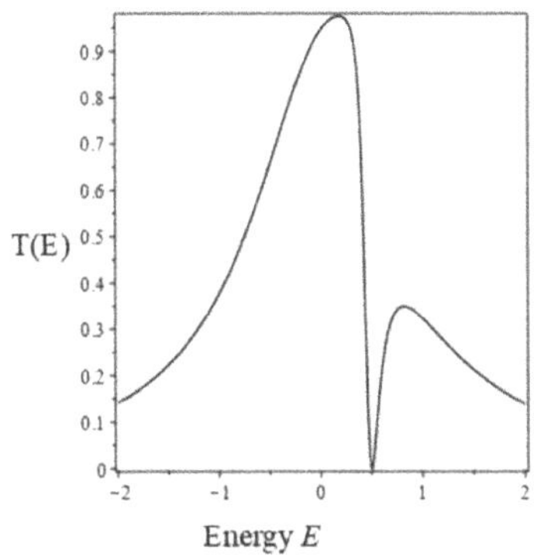

Figure 13.2. A plot of equation (13.17), with ϵ_p given by equation (13.23). In this plot, $\bar{E}_q = 0.5$ and the Breit–Wigner resonance occurs at $E = 0$.

This shows that $\Sigma_p \to \infty$ when E coincides with any of the energy levels $\bar{E}_n$ of the pendant group and therefore from equation (13.9) at these energies, DQI occurs and the transmission coefficient vanishes. In other words, $T(E)$ exhibits anti-resonances at $E = \bar{E}_n$.

As an example, if E is close to a molecular level $\bar{E}_q$ of the pendant group, then

$$\Sigma_p \approx \frac{A_q}{E - \bar{E}_q} \tag{13.14}$$

and

$$\epsilon_p \approx E_p + \frac{A_q}{E - \bar{E}_q} \tag{13.15}$$

From equation (13.9), the transmission coefficient shows a Breit–Wigner resonance at $E = \epsilon_p + \sigma_p$ and an anti-resonance at $E + \bar{E}_q$. A plot of $T(E)$ versus E is shown in figure 13.2. This characteristic line shape formed from a combination of a resonance and an anti-resonance is known as a Fano resonance.

It is interesting to note that the energy $\bar{E}_q$ at which the anti-resonance occurs does not depend on the connectivity of the moiety 1 to the electrodes, in contrast with DQI features associated with multi-path interference effects which do depend on connectivity to the electrodes.

13.3 The Green's function of an isolated molecule containing a bridging unit

As an example of a structure with a bridging unit, which creates a parallel transport path within a molecule, consider the upper structure in figure 13.3, in which single sites labelled 1 and 2 are pinch points, coupled by a matrix element δ and they are connected to a site labelled 3 of the green region by couplings v and w respectively.

Our task is to compute the Green's function matrix elements between sites a and b, assuming that when $\delta = v = w = 0$, the Green's function matrix elements g_{ij} of the decoupled blue and green regions are known.

After decimating the top left structure, the top-right structure is obtained, in which $\alpha = \delta + vg_{33}w$ and site energies $\epsilon_1 = v^2g_{33}$ and $\epsilon_2 = w^2g_{33}$ are added to sites 1 and 2 respectively. Since α is a local perturbation, the analysis of section 11.5 will now be used to compute G_{ab}.

Choosing $l = a$ and $m = b$ in equation (11.30b), and noting that for the decoupled structure in figure 13.1, $g_{ab} = 0$, yields

$$G_{ab} = \sum_{i,j=1}^{2} g_{ai}U_{ij}g_{jb} = g_{a1}U_{12}g_{2b} \tag{13.16}$$

where it was noted that $g_{a2} = g_{1b} = 0$. Furthermore, instead of the notation $(U_{11})_{ij}$ in equation (11.30b), the less cumbersome notation U_{ij} has now been adopted. From equation (11.31),

$$U = h_{11}(1 - gh_{11})^{-1}, \text{ where } g = \begin{pmatrix} g_{11} & 0 \\ 0 & g_{22} \end{pmatrix} \text{ and } h_{11} = \begin{pmatrix} \epsilon_1 & \alpha \\ \alpha & \epsilon_2 \end{pmatrix} \tag{13.17}$$

Hence

$$U = h_{11}\begin{pmatrix} 1 - g_{11}\epsilon_1 & -\alpha g_{11} \\ -\alpha g_{22} & 1 - g_{22}\epsilon_2 \end{pmatrix}^{-1} = \frac{1}{D}h_{11}\begin{pmatrix} 1 - g_{22}\epsilon_2 & \alpha g_{11} \\ \alpha g_{22} & 1 - g_{11}\epsilon_1 \end{pmatrix}$$

That is

$$U = \frac{1}{D}\begin{pmatrix} \epsilon_1(1 - g_{22}\epsilon_2) + \alpha^2 g_{22} & \alpha \\ \alpha & \epsilon_2(1 - g_{11}\epsilon_1) + \alpha^2 g_{11} \end{pmatrix} \tag{13.18}$$

where

$$D = (1 - g_{22}\epsilon_2)(1 - g_{11}\epsilon_1) - \alpha^2 g_{11}g_{22} \tag{13.19}$$

Hence

$$G_{ab} = \frac{g_{a1}\alpha g_{2b}}{D} = \frac{g_{a1}(\delta + vg_{33}w)g_{2b}}{D} \tag{13.20}$$

where

$$D = (1 - g_{22}w^2g_{33})(1 - g_{11}v^2g_{33}) - (\delta + vg_{33}w)^2 g_{11}g_{22} \tag{13.21}$$

Clearly $G_{ab} = 0$ if either $g_{a1} = 0$ or $g_{2b} = 0$. The former means that electrons cannot propagate between the pinch points a, 1 of the left-hand substructure, whereas the latter means that electrons cannot propagate between the pinch points b, 2. G_{ab} also vanishes if $\delta + vg_{33}w = 0$, which corresponds to DQI between waves passing from 1 to 2 either directly via the bond δ or indirectly via the pendant group. On the other hand if the pendant group is represented by a bipartite lattice, then in the middle of the HOMO–LUMO gap of the pendant group, $g_{33} = 0$ and such DQI cannot occur.

13.4 A molecular backbone coupled to two different atoms of a bridging unit

For the lower structure of figure 13.3, from equation (11.20), after decimating the green sub-unit, the following self-energy is added to the Hamiltonian of sites 1 and 2

$$\Sigma = \begin{pmatrix} \Sigma_{11} & \Sigma_{12} \\ \Sigma_{21} & \Sigma_{22} \end{pmatrix} = \begin{pmatrix} v & 0 \\ 0 & w \end{pmatrix} \begin{pmatrix} g_{33} & g_{34} \\ g_{43} & g_{44} \end{pmatrix} \begin{pmatrix} v & 0 \\ 0 & w \end{pmatrix} = \begin{pmatrix} v^2 g_{33} & v g_{34} w \\ w g_{43} v & w^2 g_{44} \end{pmatrix} \tag{13.22}$$

This shows that the perturbation is again of the form $h_{11} = \begin{pmatrix} \epsilon_1 & \alpha \\ \alpha & \epsilon_2 \end{pmatrix}$, but with

$$\epsilon_1 = v^2 g_{33}, \quad \epsilon_2 = w^2 g_{44} \quad \text{and} \quad \alpha = \delta + v g_{34} w \tag{13.23}$$

where is was noted that for real v and w, since g is a real Hermitian matrix, $v g_{34} w = w g_{43} v$.

Hence

$$G_{ab} = \frac{g_{a1} \alpha g_{2b}}{D} = \frac{g_{a1}(\delta + v g_{34} w) g_{2b}}{D} \tag{13.24}$$

where

$$D = (1 - g_{22} w^2 g_{44})(1 - g_{11} v^2 g_{33}) - (\delta + v g_{34} w)^2 g_{11} g_{22} \tag{13.25}$$

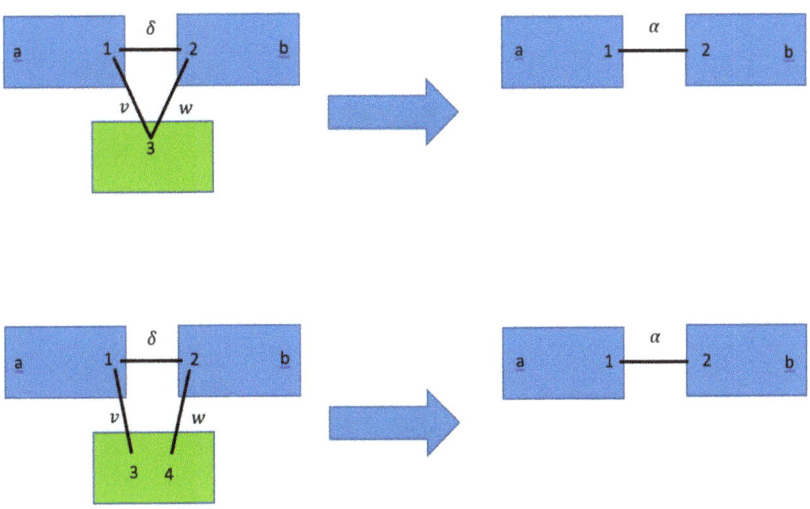

Figure 13.3. In the upper structure sites 1 and 2 connect to the same site 3 in the green region. In the lower structure, they connect to different sites labelled 3 and 4.

Again $G_{ab} = 0$ if either $g_{a1} = 0$ or $g_{2b} = 0$ and also if $\delta + vg_{34}w = 0$. Note that the subscripts 3,4 represent the labels of any two sites in the pendant substructure of figure 13.3. If the pendant group can be represented by a bipartite lattice, then the mid-gap g_{34} will be non-zero, if 3 and 4 label odd and even sites. In this case, as is clear from M-tables such as those of anthracene and pyrene (tables 2.3 and 2.4) g_{34} could be positive or negative and therefore the term $(\delta + vg_{34}w)$ could exhibit CQI or DQI, depending on the connectivity to the pendant group.

13.5 A molecule with three moieties in series

As a check on the above analysis, note that if $\delta = 0$, then the top structure of figure 13.3 is equivalent to the top structure shown in figure 13.4. In that case, the determinant D simplifies to

$$D = 1 - g_{22}w^2g_{33} - g_{11}v^2g_{33} \tag{13.26}$$

and therefore $G_{ab} = \frac{g_{a1}vg_{33}wg_{2b}}{g_{33}^{-1} - g_{22}w^2 - g_{11}v^2}$. If the blue regions are semi-infinite leads, then this is an example of the VUF. If the molecules are weakly coupled and the energy does not coincide with an eigenvalue of any of the isolated building blocks, so that $D \approx 1$, then this yields an example of the quantum circuit rule discussed in section 12.8.

Similarly, if $\delta = 0$, the lower structure of figure 13.1 reduces to that shown in lower panel of figure 13.4, for which

$$D = 1 - g_{22}w^2g_{44} - g_{11}v^2g_{33} + v^2w^2g_{11}g_{22}\left(g_{33}g_{44} - g_{34}^2\right) \tag{13.27}$$

and

$$G_{ab} = \frac{g_{a1}vg_{34}wg_{2b}}{D} \tag{13.28}$$

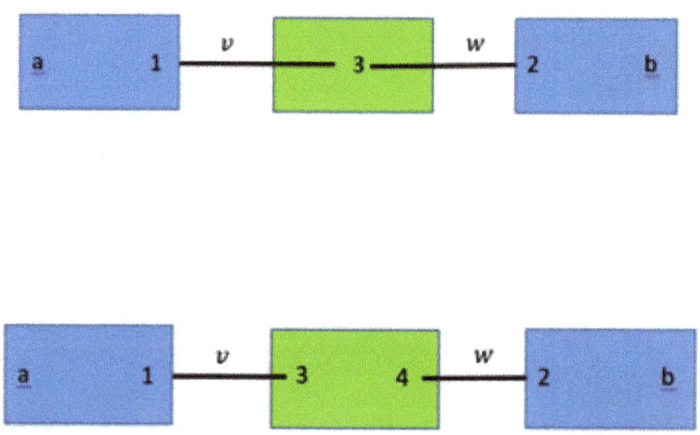

Figure 13.4. The top panel shows the top structure of figure 13.1, when $\delta = 0$. The lower panel shows the lower structure of figure 13.3, when $\delta = 0$.

If a and b are chosen to equal surface sites 1 and 2 of electrodes, then this again reproduces the VUF and when $D \approx 1$, the quantum circuit rule $G_{ab} \approx g_{a1}vg_{34}wg_{2b}$ is obtained.

13.6 A molecule with two bridging groups in parallel

Now consider the structure shown in figure 13.5, which contains two bridging groups in parallel.

In this case, after decimating the bridging groups, the structure on the left of figure 13.5 reduces to the structure on the right, where

$$\epsilon_1 = v^2 g_{33}, \quad \epsilon_2 = w^2 g_{44}, \quad \epsilon_1' = v'^2 g_{3'3'}, \quad \epsilon_2' = w'^2 g_{4'4'} \tag{13.29}$$

$$\alpha = \delta + v g_{34} w \quad \text{and} \quad \alpha' = \delta' + v' g_{3'4'} w' \tag{13.30}$$

An example of this structure is shown on the left in figure 13.6. If the blue regions represent 6-membered aryl rings, then after introducing the coupling α' and the pendant group, this is a model of a fluorene core, shown on the right of the figure. In this case, $v' = w' = \delta = 0$, $v = w$ and $g_{33} = g_{44}$.

Consequently, for a fluorene core,

$$\epsilon_1 = \epsilon_2 = \alpha = v^2 g_{33}, \ \epsilon_1' = \epsilon_2' = 0 \text{ and } \alpha' = \delta' \tag{13.31}$$

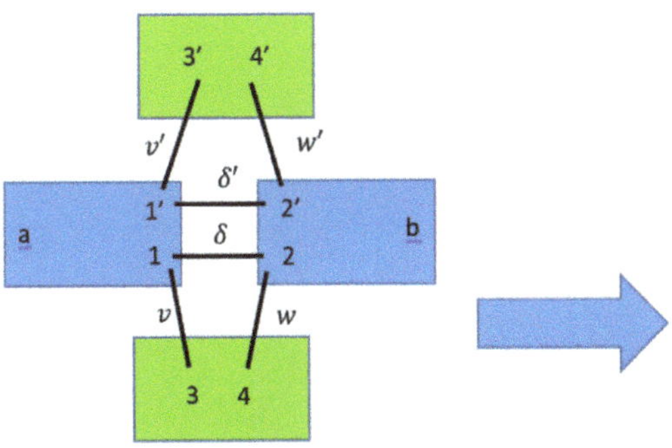

Figure 13.5. A structure with two bridging groups in parallel.

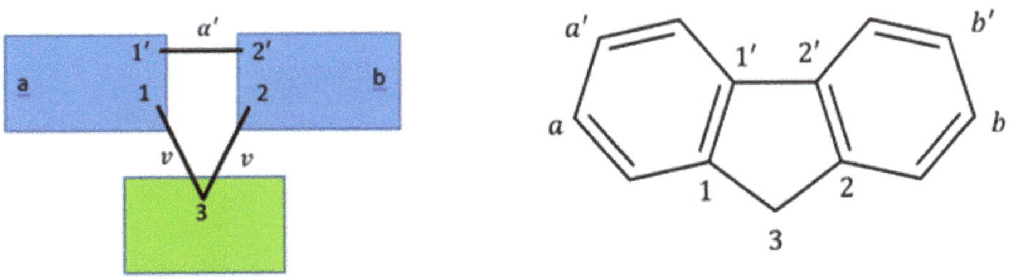

Figure 13.6. (Left) A structure with a single pendant group and a coupling α between sites $1'$ and $2'$. (Right) The structure of a fluorene core.

To solve for the Green's functions of the structures in figures 13.5 and 13.6, while avoiding 4×4 matrices, it is convenient to introduce the perturbation recursively, as discussed in section 11.2. Let g denote the Green's function of the decoupled structure on the right of figure 13.5, when $v' = w' = v = w = 0$; i.e. when $\alpha = \alpha' = \epsilon_1 = \epsilon_2 = \epsilon_1' = \epsilon_2' = 0$. Also let $\hat{g}$ denote the Green's function of the decimated structure on the right of figure 13.5, when $v = w = 0$, but $v', w' \neq 0$; i.e. when $\alpha = \epsilon_1 = \epsilon_2 = 0$, but $\alpha', \epsilon_1', \epsilon_2' \neq 0$. Finally, let G denote the Green's function of the structure on the right when all the perturbative parameters are non-zero. By solving the problem in two such steps, only 2×2 matrices are needed. The first step involves solving the 'two-impurity problem,' shown in figure 13.7.

To obtain G from $\hat{g}$, consider the structure in figure 13.7, in which sites energies ϵ_1 and ϵ_2 are added to sites 1 and 2 respectively and an extra coupling α between sites 1 and 2 is added.

The matrix element G_{ab} between any two sites a, b of the structure is given by

$$G_{ab} = \hat{g}_{ab} + \sum_{i,j=1}^{2} \hat{g}_{ai} U_{ij} \hat{g}_{jb} \tag{13.32}$$

where

$$U = h_{11}(1 - \hat{g}h_{11})^{-1} = (h_{11}^{-1} - \hat{g})^{-1} \tag{13.33}$$

In these equations, the perturbation is

$$h_{11} = \begin{pmatrix} \epsilon_1 & \alpha \\ \alpha & \epsilon_2 \end{pmatrix} \tag{13.34}$$

and g is the Green's function before adding the perturbation. Since

$$h_{11}^{-1} = \frac{1}{d}\begin{pmatrix} \epsilon_2 & -\alpha \\ -\alpha & \epsilon_1 \end{pmatrix} \tag{13.35}$$

where

$$d = \det h_{11} = \epsilon_1 \epsilon_2 - \alpha^2 \tag{13.36}$$

equation (13.32) yields $U = d\begin{pmatrix} \epsilon_2 - d\hat{g}_{11} & -\alpha - d\hat{g}_{12} \\ -\alpha - d\hat{g}_{21} & \epsilon_1 - d\hat{g}_{22} \end{pmatrix}^{-1}$. Hence

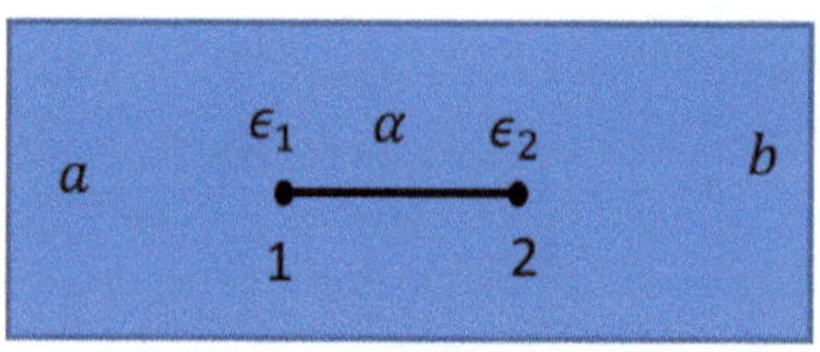

Figure 13.7. The two-impurity problem involves adding a two-site perturbation to a structure. In this case the sites are labelled 1 and 2. The Green's function before adding the perturbation (i.e. when $\alpha = \epsilon_1 = \epsilon_2 = 0$) is $\hat{g}$. The Green's function after adding the perturbation is G.

$$U = \frac{1}{\Delta}\begin{pmatrix} \epsilon_1 - d\hat{g}_{22} & \alpha + d\hat{g}_{12} \\ \alpha + d\hat{g}_{21} & \epsilon_2 - d\hat{g}_{11} \end{pmatrix} \quad (13.37)$$

where $D = [(\epsilon_1 - d\hat{g}_{22})(\epsilon_2 - d\hat{g}_{11}) - (\alpha + d\hat{g}_{12})(\alpha + d\hat{g}_{21})]/d$

That is

$$D = 1 - \epsilon_1\hat{g}_{11} - \epsilon_2\hat{g}_{22} - \alpha(\hat{g}_{12} + \hat{g}_{21}) + d(\hat{g}_{11}\hat{g}_{22} - \hat{g}_{12}\hat{g}_{21}) \quad (13.38)$$

Equations (13.32), (13.37) and (13.38) are the general solution to the two-site perturbation problem, in which site energies ϵ_1 and ϵ_2 and a coupling α are added to two sites labelled 1 and 2.

For the example of a fluorene core described by equation (13.31), for which $\epsilon_1 = \epsilon_2 = \alpha$, one obtains $d = 0$ and therefore equations (13.39) and (13.40) simplify to

$$D = 1 - \alpha(\hat{g}_{11} + \hat{g}_{22} + \hat{g}_{12} + \hat{g}_{21}) \qquad \text{(fluorene)} \quad (13.39)$$

$$U = \frac{\alpha}{D}\begin{pmatrix} 1 & 1 \\ 1 & 1 \end{pmatrix} \qquad \text{(fluorene)} \quad (13.40)$$

Hence from equation (13.32)

$$G_{ab} = \hat{g}_{ab} + \frac{\alpha}{D}(\hat{g}_{a1} + \hat{g}_{a2})(\hat{g}_{1b} + \hat{g}_{2b}) \qquad \text{(fluorene)} \quad (13.41)$$

This shows that the effect of the two-site perturbation in figure 13.6 depends on the outcome of quantum interference between the Green's functions $\hat{g}_{a1}$, $\hat{g}_{a2}$ and $\hat{g}_{1b}$, $\hat{g}_{2b}$. For example if $\hat{g}_{a1}$, $\hat{g}_{a2}$ have opposite signs, then $(\hat{g}_{a1} + \hat{g}_{a2})$ could vanish. In this case, the transport paths $a \to 1$ and $a \to 2$ interfere destructively and the perturbation has no effect.

Interestingly, for this simple model of fluorene, $\alpha = v^2 g_{33}$ and therefore if the energy E of an electron passing through the fluorene core coincides with an energy level of the pendant group, g_{33} and α diverge. Hence α/D becomes independent of α and therefore independent of the coupling v to the pendant group.

To complete the calculation of the Green's function G, the Green's function $\hat{g}$ of the structure on the left of figure 13.8 will now be computed. This is identical to the

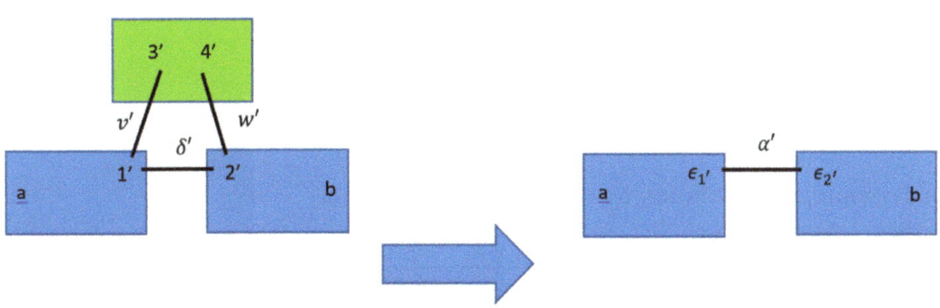

Figure 13.8. (Left) A structure with a perturbation involving sites $1'$ and $2'$. The Green's function before (after) adding the perturbation is $g(\hat{g})$. (Right) The equivalent structure after decimation. For fluorene, there is no pendant group and therefore $v' = w' = \epsilon_1' = \epsilon_2' = 0$, $\alpha' = \delta'$.

structure in figure 13.3 and therefore the result is contained in equation (13.18). Alternatively, if the Green's function in the absence of the perturbation is g, then repeating the above steps leading to equations (13.37) and (13.38) yields, after noting that for the unperturbed decoupled structure, $g_{1'2'} = g_{2'1'} = 0$

$$U' = \frac{1}{D'}\begin{pmatrix} \epsilon_1' - D'g_{2'2'} & \alpha' \\ \alpha' & \epsilon_2' - D'g_{1'1'} \end{pmatrix} \tag{13.42}$$

where

$$D' = \epsilon_1'\epsilon_2' - \alpha'^2 \tag{13.43}$$

and

$$\mathcal{D}' = 1 - \epsilon_1' g_{1'1'} - \epsilon_2' g_{2'2'} + D'(g_{1'1'}g_{2'2'})$$

i.e.

$$\mathcal{D}' = (g_{1'1'}g_{2'2'})\left[\left(g_{1'1'}^{-1} - \epsilon_1'\right)\left(g_{2'2'}^{-1} - \epsilon_2'\right) - \alpha'^2\right] \tag{13.44}$$

Furthermore, by comparison with equation (13.32)

$$\hat{g}_{ab} = g_{ab} + \sum_{i,j=1'}^{2'} g_{ai}U_{ij}'g_{jb} \tag{13.45}$$

Equations (13.42) to (13.45) are the general solution to the two-site perturbation problem, in which the two sites 1 and 2 belong to decoupled portions of the initial decoupled structure. $\hat{g}_{ab}$ is the Green's function matrix element between any two sites a, b. If a and b are located in different portions of the initially decoupled structure, as drawn on the right in figure 13.8, then $g_{ab} = g_{a2'} = g_{1'b} = 0$. If they are both on the left, then $g_{ab} \neq 0$, $g_{a2'} = g_{2'b} = 0$, whereas if they are both on the right, $g_{ab} \neq 0$, $g_{a1'} = g_{1'b} = 0$.

Choosing a, b to be on opposite sides of the structure, as drawn on the right in figure 13.8, yields

$$\hat{g}_{ab} = g_{a1'}U_{1'2'}'g_{2'b} = \frac{g_{a1'}\alpha' g_{2'b}}{\mathcal{D}'} \tag{13.46}$$

In contrast, if b is chosen to be a site a' on the left

$$\hat{g}_{aa'} = g_{aa'} + \frac{g_{a1}(\epsilon_1' - D'g_{2'2'})g_{1a'}}{\mathcal{D}'} \tag{13.47}$$

Furthermore, if a is chosen to be a site b' on the right

$$\hat{g}_{b'b} = g_{b'b} + \frac{g_{b'2}(\epsilon_2' - D'g_{1'1'})g_{2b}}{\mathcal{D}'} \tag{13.48}$$

As an example, for the fluorene core, equation (13.31) yields $D' = -\delta'^2$,

$$U' = \frac{\delta'}{D'}\begin{pmatrix} \delta' g_{2'2'} & 1 \\ 1 & \delta' g_{1'1'} \end{pmatrix} \tag{13.49}$$

$$D' = 1 - \delta'^2 g_{1'1'} g_{2'2'} \tag{13.50}$$

Choosing a, b to be on opposite sides of the structure, yields

$$\hat{g}_{ab} = g_{a1'} U'_{1'2'} g_{2'b} = \frac{g_{a1'} \delta' g_{2'b}}{1 - \delta'^2 g_{1'1'} g_{2'2'}} \tag{13.51}$$

On the other hand,

$$\hat{g}_{aa'} = g_{aa'} + \frac{g_{a1} \delta'^2 g_{1a'}}{g_{2'2'}^{-1} - \delta'^2 g_{1'1'}} \tag{13.52}$$

and

$$\hat{g}_{b'b} = g_{b'b} + \frac{g_{b'2} \delta'^2 g_{2b}}{g_{1'1'}^{-1} - \delta'^2 g_{2'2'}} \tag{13.53}$$

13.7 The electrical conductance of a molecule with weakly coupled bridges

From equation (13.37), in the weak coupling limit, keeping terms which are at most linear in the small quantities ϵ_1, ϵ_2, ϵ_1', ϵ_2', α' and α yields

$$U \approx \begin{pmatrix} \epsilon_1 & \alpha \\ \alpha & \epsilon_2 \end{pmatrix}$$

and therefore from equation (13.32),

$$G_{ab} \approx \hat{g}_{ab} + \hat{g}_{a1}\epsilon_1\hat{g}_{1b} + \hat{g}_{a2}\epsilon_2\hat{g}_{2b} + \alpha(\hat{g}_{a1}\hat{g}_{2b} + \hat{g}_{a2}\hat{g}_{1b}) \tag{13.54}$$

To lowest order in small quantities equation (13.46) yields $\hat{g}_{ab} \approx \alpha' g_{a1'} g_{2'b}$. Furthermore, in the last four terms of equation (13.54), all Green's function $\hat{g}_{ij}$ can be replaced by g_{ij}. Hence

$$G_{ab} \approx \alpha' g_{a1'} g_{2'b} + \alpha g_{a1} g_{2b} \tag{13.55}$$

If $\delta = \delta' = 0$, then

$$G_{ab} \approx g_{a1'} v' g_{3'4'} w' g_{2'b} + g_{a1} v g_{34} w g_{2b} \tag{13.56}$$

Reading these terms from right to left, shows that any current flowing through such a structure must flow through the bridging groups. In this case, from equation (13.30) one obtains

$$G_{ab} \approx A + A' \tag{13.57}$$

$$A' \approx g_{a1'}v'g_{3'4'}w'g_{2'b} \qquad \text{and} \qquad A \approx g_{a1}vg_{34}wg_{2b} \tag{13.58}$$

From the VUF, the electrical conductance is proportional to $(G_{ab})^2$. Therefore if the two branches are identical, such that $A = A'$, then

$$(G_{ab})^2 \approx 4A^2 \tag{13.59}$$

Since the conductance in the presence of only a single branch is proportional to A^2, this demonstrates that in the weak coupling limit, the conductance of two identical branches in parallel is four times higher than the conductance of a single branch. This is a purely quantum effect, since classically, the conductance of two identical conductors in parallel is only two times higher than that of a single conductor. This result is clearly valid in the weak coupling limit only, because for stronger coupling, all terms in equations (13.32), (13.37) and (13.38) must be taken into account. Furthermore, if the branches are different, then A and A' may have opposite signs and therefore may interfere destructively, leading to a conductance which is lower than the expected classical value. As discussed in chapter 1, this possibility arises, because the electrical conductance is proportional to $(A + A')^2$ rather than $A^2 + A'^2$.

13.8 The electrical conductance of a molecule with many parallel bridges

Consider now a molecule with many branches attached to sites 1′ and 2′, as shown on the left in figure 13.9, which after decimation, reduces to the figure on the right, where the following terms are added to sites 1′ and 2′:

$$\alpha' = \sum_{i=3}^{M} v_i g_{ii'} v_{i'}, \quad \epsilon_1' = \sum_{i=3}^{M} v_i g_{ii} v_i \quad \text{and} \quad \epsilon_2' = \sum_{i=3}^{M} v_{i'} g_{i'i'} v_{i'} \tag{13.60}$$

Since the two blue substructures in figure 13.9 are decoupled before adding the branches, the Green's function matrix element $\hat{g}_{ij}$ between any two sites i, j of the

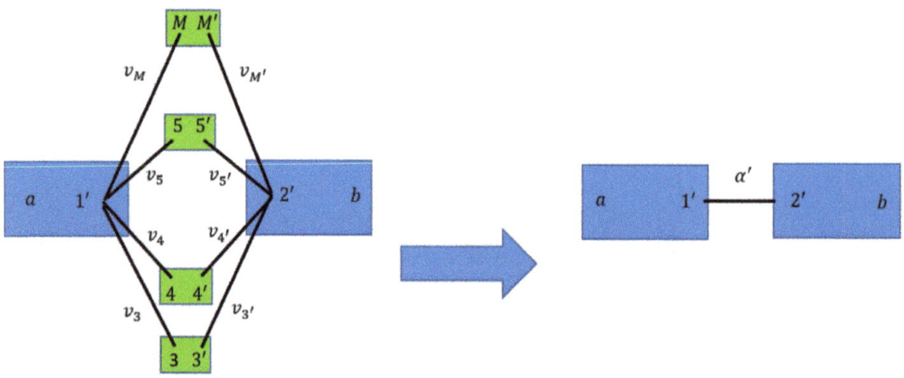

Figure 13.9. A molecule with many branches attached to sites 1′ and 2′.

perturbed structure is given by equations (13.46) to (13.48). In particular, for $\hat{g}_{1'2'}$, $\hat{g}_{1'1'}$ and $\hat{g}_{2'2'}$, these simplify to

$$\hat{g}_{1'2'} = \frac{\alpha'}{\bar{D}} \qquad \hat{g}_{1'1'} = \frac{g_{2'2'}^{-1} - \epsilon_2'}{\bar{D}} \qquad \hat{g}_{2'2'} = \frac{g_{1'1'}^{-1} - \epsilon_1'}{\bar{D}} \tag{13.61}$$

where

$$\bar{D} = \left(g_{1'1'}^{-1} - \epsilon_1'\right)\left(g_{2'2'}^{-1} - \epsilon_2'\right) - \alpha'^2 \tag{13.62}$$

Note that the above expressions could also be obtained from Dyson's equation:

$$\hat{g} = \begin{pmatrix} \hat{g}_{1'1'} & \hat{g}_{1'2'} \\ \hat{g}_{2'1'} & \hat{g}_{2'2'} \end{pmatrix} = (g^{-1} - \Sigma')^{-1} = \begin{pmatrix} g_{1'1'}^{-1} - \epsilon_1' & -\alpha' \\ -\alpha' & g_{2'2'}^{-1} - \epsilon_2' \end{pmatrix}^{-1} \tag{13.63}$$

where the self-energy arising from decimation of the bridging groups is $\Sigma' = \begin{pmatrix} \epsilon_1' & \alpha' \\ \alpha' & \epsilon_2' \end{pmatrix}$.

In the weak coupling limit, equation (13.46) reduces to

$$\hat{g}_{ab} \approx \sum_{i=3}^{M} A_i \tag{13.64}$$

where

$$A_i = g_{a1'} v_i g_{ii'} v_{i'} g_{2'b} \tag{13.65}$$

Again, since the conductance is proportional to $\left(\sum_{i=3}^{M} A_i\right)^2$ rather than $\sum_{i=3}^{M} A_i^2$, the final values of $(\hat{g}_{ab})^2$ and the transmission coefficient are determined by interference between the amplitudes A_i.

As a simpler example of the above structure, consider the case where the unperturbed system is composed of two sites $1' = a$ and $2' = b$, as shown in figure 13.10. In this case, if the site energies before adding the perturbations are ϵ_a and ϵ_b, then $g_{1'1'}^{-1} = E - \epsilon_a$ and $g_{2'2'}^{-1} = E - \epsilon_b$. Hence equations (13.61) and (13.62) simplify to

$$\hat{g}_{1'2'} = \frac{\alpha'}{\bar{D}} \qquad \hat{g}_{1'1'} = \frac{E - \epsilon_b - \epsilon_2'}{\bar{D}} \qquad \hat{g}_{2'2'} = \frac{E - \epsilon_a - \epsilon_1'}{\bar{D}} \tag{13.66}$$

where

$$\bar{D} = (E - \epsilon_a - \epsilon_1')(E - \epsilon_b - \epsilon_2') - \alpha'^2 \tag{13.67}$$

Furthermore, equation (13.63) simplifies to

$$\hat{g} = \begin{pmatrix} \hat{g}_{1'1'} & \hat{g}_{1'2'} \\ \hat{g}_{2'1'} & \hat{g}_{2'2'} \end{pmatrix} = (E - h)^{-1} \tag{13.68}$$

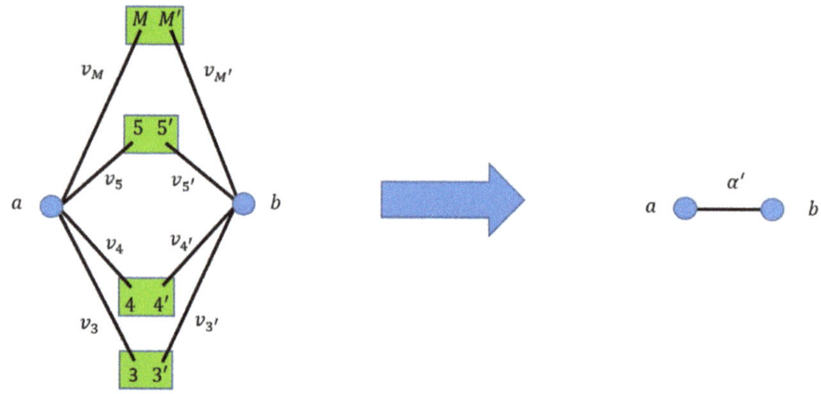

Figure 13.10. A molecule with many branches attached to sites 1 and b.

where

$$h = \begin{pmatrix} \epsilon_a + \epsilon_1' & -\alpha' \\ -\alpha' & \epsilon_b + \epsilon_2' \end{pmatrix} \tag{13.69}$$

These expressions highlight the fact that ϵ_1' and ϵ_2' renormalize the site energies and that $\hat{g}$ is the Green's function of the renormalized Hamiltonian h, obtained after decimating the branches.

When substituted into the VUF, equation (13.66) can be used to obtain the transmission function $T(E)$ of the structure of figure 13.10, when connected to external electrodes. From equation (8.45), $T(E) = B\,|\frac{\hat{g}_{1'2'}}{\Delta}|^2$, where the pre-factor B is a product of quantities associated with the leads and the couplings to the leads. i.e.

$$T(E) = B\left|\frac{\alpha'}{\bar{D}\Delta}\right|^2 \tag{13.70}$$

In this expression, from equations (8.43) and (13.67)

$$\Delta = 1 - \hat{g}_{1'1'}\Sigma_L - \hat{g}_{2'2'}\Sigma_R + \Sigma_L\Sigma_R(\hat{g}_{1'1'}\hat{g}_{2'2'} - \hat{g}_{1'2'}\hat{g}_{21'}) \tag{13.71}$$

where Σ_L and Σ_R are self-energies associated with the left and right leads.

Note that from equation (13.61), $\hat{g}_{1'1'}\hat{g}_{2'2'} - \hat{g}_{1'2'}\hat{g}_{21'} = \frac{1}{\bar{D}}$. Hence

$$\begin{aligned}\bar{D}\Delta = &(E - \epsilon_a - \epsilon_1')(E - \epsilon_b - \epsilon_2') - \alpha'^2 \\ &- (E - \epsilon_b - \epsilon_2')\Sigma_L - (E - \epsilon_a - \epsilon_1')\Sigma_R + \Sigma_L\Sigma_R\end{aligned}$$

and therefore

$$\bar{D}\Delta = (E - \epsilon_a - \Sigma_a)(E - \epsilon_b - \Sigma_b) - \alpha'^2 \tag{13.72}$$

where

$$\Sigma_a = \epsilon_1' + \Sigma_L \qquad \text{and} \qquad \Sigma_b = \epsilon_2' + \Sigma_R \tag{13.73}$$

This shows that the site energies of sites a and b are renormalized by adding the self-energies Σ_L and Σ_R associated with the left and right leads to the self-energies ϵ_1' and ϵ_2' associated with the bridging groups. From equation (13.60), the values of ϵ_1', ϵ_2' and α' are determined by quantum interference between waves travelling along the bridging groups.

13.9 Summary

In this chapter, formulae for Green's functions of isolated scatterers were obtained, which illustrate how quantum interference created by the building blocks of a molecule affect electron transport through the molecule. In particular, molecules with multiple parallel paths and pendant groups were analysed and conditions for the appearance of destructive and constructive quantum interference were discussed.

IOP Publishing

Quantum Transport in Nanostructures and Molecules
An introduction to molecular electronics
Colin John Lambert

Chapter 14

Connectivity theory and equations of motion

14.1 Introduction

The aim of this chapter is to extend the discussion of chapter 2, by utilising the time-dependent Schrödinger equation to clarify the concept of a source and to demonstrate how the active and passive viewpoints of connectivity theory lead to a variety of equivalent equations for the Green's function of connected substructures. The concept of a source and stationary states for a vibrating structure will also be discussed.

In this chapter, the time-dependent Schrödinger equation $i\hbar\frac{d}{dt}|\phi(t)\rangle = H|\phi(t)\rangle + |f(t)\rangle$ is used to analyze examine the wave function $|\phi(t)\rangle$ that arises from a source of the form $|f(t)\rangle = e^{-i(\omega_0+i\eta/\hbar)t}|S\rangle$, where $|S\rangle$ is independent of time t, ω_0 is a frequency and η is a positive real number. For such a source, the time dependence is contained in the pre-factor $e^{-i(\omega_0+i\eta/\hbar)t}$, whose modulus is $e^{\eta t}$. Since this modulus tends to zero as t tends to $-\infty$, the source vanishes at $t = -\infty$ and is equal to $|S\rangle$ at $t = 0$. Since we are seeking the wave function $|\phi(t)\rangle$ which arises from the source, we insist that $|\phi(t)\rangle$ also vanishes at $t = -\infty$.

It will be found that $|\phi(t)\rangle = e^{-i\left(\omega_0+\frac{i\eta}{\hbar}\right)t}g(E_0 + i\eta)|S\rangle$, where $E_0 = \hbar\omega_0$. Note that η does not need to be small. Also note that in general, any solution to the equation $i\hbar\frac{d}{dt}|\phi(t)\rangle = H|\phi(t)\rangle$ could be added to $|\phi(t)\rangle$, but the boundary condition $|\phi(-\infty)\rangle = 0$ prevents this, because if a solution to the latter equation is non-zero at some time t, then it remains non-zero for all t, including $t = -\infty$.

The solution to the above problem is particularly useful, because the principle of superposition (equations (2.3) and (2.5)) implies that for a source of the form $|f(t)\rangle = \sum_{\omega_0} e^{-i(\omega_0+i\eta/\hbar)t}|S(\omega_0)\rangle$, which is simply a superposition of the above sources, the resulting wave function will be a superposition of the above solutions.

doi:10.1088/978-0-7503-3639-0ch14

14.2 Stationary states and Green's functions

To understand the relationship between Green's functions and stationary wave amplitudes, it is useful to recall what is meant by a 'stationary wave'. To this end, consider the time-dependent wave function $|\phi(t)\rangle = a(t)|\psi\rangle$, where $a(t)$ is an arbitrary function of time and the vector $|\psi\rangle$ is independent of time. The time dependence of such a state is simple, because the amplitudes on all sites are moving in unison, with the same time dependence $a(t)$. Clearly the amplitude $\phi_j(t) = a(t)\psi_j$ on site j changes with time. However the ratio of two amplitudes is time independent, because for any two sites j and l,

$$\frac{\phi_j(t)}{\phi_l(t)} = \frac{\psi_j}{\psi_l} \tag{14.1}$$

This means that if $\phi_l(t)$ is non-zero, then if all amplitudes are divided by $\phi_l(t)$, the resulting ratios are time independent. In other words, if the amplitude ratios $\frac{\phi_j(t)}{\phi_l(t)}$ are plotted as functions of j, then the resulting wave pattern will be stationary.

To demonstrate that Green's functions are stationary wave amplitudes due to a source, the term 'source' will now be clarified. To this end, consider the system shown below, in which a substructure A described by a Hamiltonian H_{AA} is coupled to a substructure B described by a Hamiltonian H_{BB}.

If the $N \times M$ coupling matrix is $H_{AB} = (H_{BA})^{\dagger}$, then the full Hamiltonian is

$$H = \begin{pmatrix} H_A & H_{AB} \\ H_{BA} & H_B \end{pmatrix} \tag{14.2}$$

and the time-dependent Schrödinger equation is

$$i\hbar\frac{d}{dt}\begin{pmatrix} |\varphi_A(t)\rangle \\ |\varphi_B(t)\rangle \end{pmatrix} = \begin{pmatrix} H_A & H_{AB} \\ H_{BA} & H_B \end{pmatrix}\begin{pmatrix} |\varphi_A(t)\rangle \\ |\varphi_B(t)\rangle \end{pmatrix} \tag{14.3}$$

In particular,

$$i\hbar\frac{d}{dt}|\varphi_A(t)\rangle = H_A|\varphi_A(t)\rangle + |f(t)\rangle \tag{14.4}$$

where

$$|f(t)\rangle = H_{AB}|\phi_B(t)\rangle \tag{14.5}$$

If $|f(t)\rangle$ is known, then $|f(t)\rangle$ can be regarded as an external source, which drives the wave function of region A.

To obtain a relationship between $|\varphi_A(t)\rangle$ and $|f(t)\rangle$, one could attempt to solve equation (14.4) for the wave function $|\varphi_A(t)\rangle$ arising from a known source $|f(t)\rangle$. In fact it is much easier to consider the reverse question and ask 'what source would be needed to create a known wave function $|\varphi_A(t)\rangle$'. As a starting point, it is of interest to determine what source $|f(t)\rangle$ would be needed to create a stationary wave pattern in region A of the form $|\varphi_A(t)\rangle = a(t)|\psi\rangle$, where $a(t) = e^{-i(\omega_0+i\eta)t}$. In this expression, η is chosen to be a positive real number, so that $|\varphi_A(t)\rangle$ vanishes as $t \to -\infty$. In other words one would like to create a wave function of the form

$$|\varphi_A(t)\rangle = e^{-i(\omega_0+i\eta/\hbar)t}|\psi\rangle \tag{14.6}$$

Since $|e^{-i\left(\omega_0+\frac{i\eta}{\hbar}\right)t}| = e^{(\eta/\hbar)t}$, which changes by e when t changes by $\hbar/\eta$, the quantity $\tau = \hbar/\eta$ is the time constant controlling the switching on of the wave function with time. Since $|\varphi_A(t)\rangle$ vanishes as $t \to -\infty$, if the term $|f(t)\rangle$ were absent from equation (14.4), then $|\varphi_A(t)\rangle$ would remain zero for all time. Therefore one can regard the driving term $|f(t)\rangle$ as a source, which creates $|\varphi(t)\rangle$ at finite times.

To obtain the source that produces the desired wave pattern (14.4), note that

$$\begin{aligned} i\hbar\frac{d}{dt}|\varphi_A(t)\rangle &= i\hbar\frac{d}{dt}e^{-i(\omega_0+i\eta/\hbar)t}|\psi\rangle = (\hbar\omega_0 + i\eta)e^{-i(\omega_0+i\eta/\hbar)t}|\psi\rangle \\ &= (E_0 + i\eta)|\varphi_A(t)\rangle \end{aligned}$$

where $E_0 = \hbar\omega_0$. Adding and subtracting $H_A|\varphi_A(t)\rangle$ on the right-hand side, yields

$$i\hbar\frac{d}{dt}|\varphi_A(t)\rangle = H_A|\varphi_A(t)\rangle + (E_0 - H_A + i\eta)|\varphi_A(t)\rangle \tag{14.7}$$

Comparing equation (14.5) with (14.4) shows that the required source is

$$|f(t)\rangle = (E_0 - H_A + i\eta)|\varphi_A(t)\rangle \tag{14.8}$$

In other words, the required source is

$$|f(t)\rangle = e^{-i(\omega_0+i\eta/\hbar)t}|S\rangle \tag{14.9}$$

where $|S\rangle$ is a time-independent source vector given by

$$|S\rangle = (E_0 - H_A + i\eta)|\psi\rangle \tag{14.10}$$

This equation reveals that the source $|f(t)\rangle$ with a spatial dependence $|S\rangle$ yields the desired stationary state $|\varphi_A(t)\rangle$ with a spatial dependence $|\psi\rangle$.

Notice that both the source $|f(t)\rangle = e^{-i(\omega_0+i\eta/\hbar)t}|S\rangle$ and the state $|\varphi_A(t)\rangle = e^{-i(\omega_0+i\eta/\hbar)t}|\psi\rangle$ vanish as $t \to -\infty$, because of the presence of the term $+i\eta/\hbar$ and as a consequence the operator $E_0 - H_A + i\eta$ which maps the vector $|\psi\rangle$ onto the vector $|f\rangle$ contains a factor $+i\eta$.

To answer the question 'what stationary wave function $|\varphi_A(t)\rangle$ arises from a known source $|f(t)\rangle$', it is necessary to invert equation (14.10). If the Hamiltonian H_A is a finite-size matrix, which would be the case when region is closed, then the answer to this question is

$$|\psi\rangle = g(E_0 + i\eta)|S\rangle \tag{14.11}$$

where $g(E_0 + i\eta)$ is the Green's functions given by

$$g(E_0 + i\eta) = (E_0 - H_A + i\eta)^{-1} \tag{14.12}$$

In the literature, $g(E_0 + i\eta)$ is called a retarded Green's function.

Clearly $|\varphi(t)\rangle$ is a stationary state with a spatial wave pattern $|\psi\rangle$, created by the source ($|f(t)\rangle$) which vanishes at $t = -\infty$. The amplitude of this wave pattern on site i is $\psi_i = \langle i|\psi\rangle$ and the amplitude of the source on site i is $f_i = \langle i|f\rangle$. For the special case where $f_i = \delta_{il}$ (i.e. when the source acts with unit amplitude on site l only), one obtains $\psi_i = g_{il}(E_0 + i\eta)$, where $g_{il}(E_0 + i\eta)$ is the i, l th element of $g(E_0 + i\eta)$. In other words, the l th column of the matrix $g(E_0 + i\eta)$ contains the stationary wave amplitudes due to a source of unit amplitude acting on site l.

More generally, as mentioned in the introduction to this chapter, for an arbitrary sum of sources of the form

$$|f(t)\rangle = \sum_{\omega_0} e^{-i(\omega_0 + i\eta/\hbar)t}|S(\omega_0)\rangle \tag{14.13a}$$

the resulting wave is

$$|\phi(t)\rangle = \sum_{\omega_0} e^{-i(\omega_0 + i\eta/\hbar)t} g(E_0 + i\eta)|S(\omega_0)\rangle \tag{14.13b}$$

14.3 Green's function of a 2-component system revisited

In this section, the two-component noisy neighbour equations (NNEs) (2.34) of chapter 2 are re-derived, starting from equation (14.3). In what follows, all matrices are of finite dimensions, either because the system is closed, or because they are sub-matrices of infinite matrices associated with an open system, as discussed in appendix B.

Equations (14.6) and (14.5) show that the wave functions in sub-systems A and B of figure 14.1 are related by

$$(E - H_A)|\varphi_A(t)\rangle = H_{AB}|\phi_B(t)\rangle$$

where $E = E_0 + i\eta$. Therefore the wave function in A, due to the source $|f_B\rangle = H_{AB}|\phi_B(t)\rangle$ is

$$|\varphi_A(t)\rangle = g_{AA}|f_B\rangle$$

where $g_{AA} = (E - H_A)^{-1}$. Consider now the three coupled substructures labelled 1, 2 and 3, shown in figure 14.2. In this case, equation (14.3) generalises to

$$i\hbar\frac{d}{dt}\begin{pmatrix} |\varphi_1(t)\rangle \\ |\varphi_2(t)\rangle \\ |\varphi_3(t) \end{pmatrix} = \begin{pmatrix} H_{11} & H_{12} & 0 \\ H_{21} & H_{22} & H_{23} \\ 0 & H_{32} & H_{33} \end{pmatrix}\begin{pmatrix} |\varphi_1(t)\rangle \\ |\varphi_2(t)\rangle \\ |\varphi_3(t) \end{pmatrix} \tag{14.14}$$

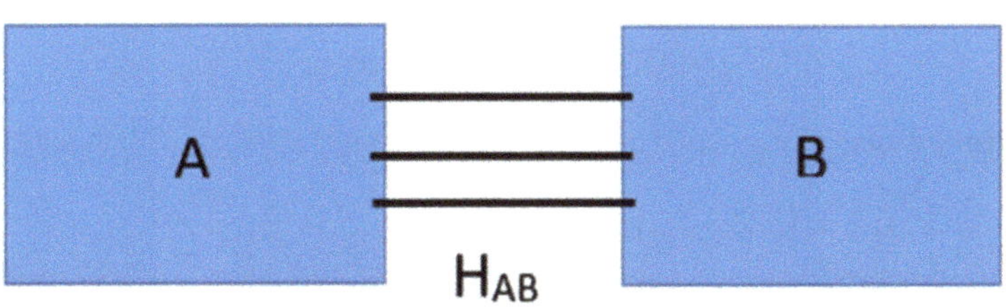

Figure 14.1. Two coupled substructures labelled A and B, interacting via a coupling matrix H_{AB}.

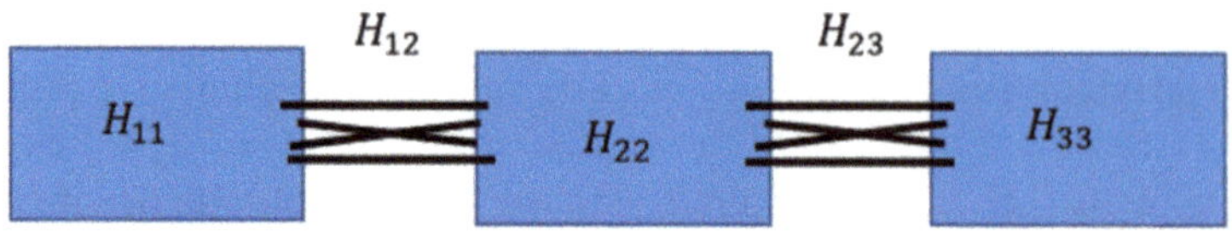

Figure 14.2. Three coupled substructures labelled 1, 2 and 3.

After introducing the time-dependent factor $e^{-i(\omega_0+i\eta/\hbar)t}$, by writing $|\varphi_A(t)\rangle = e^{-i(\omega_0+i\eta/\hbar)t}|\psi_A\rangle$ etc, this yields

$$\begin{pmatrix} E - H_{11} & -H_{12} & 0 \\ -H_{21} & E - H_{22} & -H_{23} \\ 0 & -H_{32} & E - H_{33} \end{pmatrix} \begin{pmatrix} |\psi_1\rangle \\ |\psi_2\rangle \\ |\psi_3\rangle \end{pmatrix} = 0 \tag{14.15}$$

Written out in full, equation (14.15) yields

$$|\psi_1\rangle = g_{11}H_{12}|\psi_2\rangle \tag{14.16}$$

and

$$|\psi_2\rangle = g_{22}[H_{21}|\psi_1\rangle + |S_2'\rangle] \tag{14.17}$$

In this equation, the source $|S_2'\rangle$ is given by

$$|S_2'\rangle = H_{23}|\psi_3\rangle \tag{14.18}$$

and the Green's functions of the isolated substructures are

$$g_{11} = (E - H_{11})^{-1} \text{ and } g_{22} = (E - H_{22})^{-1}. \tag{14.19}$$

The term $|S_2'\rangle$ will be regarded as a controlled external source, which is independent of the waves in substructures 1 and 2. Since $|\varphi_j(t)\rangle = e^{-i(\omega_0+i\eta/\hbar)t}|\psi_j\rangle$, and the time-dependent factors cancel everywhere, it is clear that equations (14.16) and (14.17) are identical to equations (2.10) and (2.16) of chapter 2 and therefore all equations derived there are reproduced. In particular the NNEs are obtained:

$$|\psi_1\rangle = G_{12}|S_2'\rangle \tag{14.20}$$

and

$$|\psi_2\rangle = G_{22}|S_2'\rangle \tag{14.21}$$

where

$$G_{12} = G_{11}H_{12}g_{22} = g_{11}H_{12}G_{22} \tag{14.22}$$

In these expressions,

$$G_{11} = \left(g_{11}^{-1} - \Sigma_1\right)^{-1} \text{ and } G_{22} = \left(g_{22}^{-1} - \Sigma_2\right)^{-1} \tag{14.23}$$

$$\Sigma_1 = H_{12}g_{22}H_{21} \text{ and } \Sigma_2 = H_{21}g_{11}H_{12} \tag{14.24}$$

The first of the NNEs (14.22) is the 'active viewpoint' (2.26), which incorporates the back-action back-action of $|\psi_1\rangle$ onto itself via G_{11}. The second is the 'passive viewpoint' (2.14), which incorporates the back-action of $|\psi_2\rangle$ via G_{22}. It should be noted that the derivation of these equations in chapter 2 did not rely on equation (14.19) and therefore if substructures 1 and 2 are themselves connected to other substructures, g_{11} and g_{22} can be regarded as Green's function obtained by decimating those substructures. If substructures 1 and 2 are connected to open substructures, then g_{11} and g_{22} will not be Hermitian.

When utilising the above equations, it should be noted that they are solutions to the equation

$$\begin{pmatrix} G_{11} & G_{12} \\ G_{21} & G_{22} \end{pmatrix} = \begin{pmatrix} g_{11}^{-1} & -H_{12} \\ -H_{21} & g_{22}^{-1} \end{pmatrix}^{-1}$$

Therefore with a change of notation, the solution to the equation

$$\begin{pmatrix} X_{\alpha\alpha} & X_{\alpha\beta} \\ X_{\beta\alpha} & X_{\beta\beta} \end{pmatrix} = \begin{pmatrix} g_{\alpha\alpha}^{-1} & -H_{\alpha\beta} \\ -H_{\beta\alpha} & g_{\beta\beta}^{-1} \end{pmatrix}^{-1} \tag{14.25}$$

is

$$X_{\alpha\beta} = X_{\alpha\alpha}H_{\alpha\beta}g_{\beta\beta} = g_{\alpha\alpha}H_{\alpha\beta}X_{\beta\beta} \tag{14.26}$$

where

$$X_{\alpha\alpha} = \left(g_{\alpha\alpha}^{-1} - \Sigma_{\alpha\alpha}(\beta)\right)^{-1} \text{ and } X_{\beta\beta} = \left(g_{\beta\beta}^{-1} - \Sigma_{\beta\beta}(\alpha)\right)^{-1} \tag{14.27}$$

$$\Sigma_{\alpha\alpha}(\beta) = H_{\alpha\beta}g_{\beta\beta}H_{\beta\alpha} \text{ and } \Sigma_{\beta\beta}(\alpha) = H_{\beta\alpha}g_{\alpha\alpha}H_{\alpha\beta} \tag{14.28}$$

Equations (14.25)–(14.28) will be used below to obtain expressions for Green's functions of more complex structures.

14.4 Connectivity equations from combinations of active and passive viewpoints

In this section, the aim is to show that by taking both passive and active viewpoints when connecting substructures to form larger structures, the NNEs (14.26) can be used to obtain collections of equivalent formulae, which allow wave functions in complex structures to be understood from different viewpoints. These formulae are equations for off-diagonal sub-matrices of the Green's function of the whole structure, expressed in terms of diagonal sub-matrices belonging to individual substructures. The 2-component NNEs (14.22) are the simplest examples of such formulae, because they express the off-diagonal Green's function G_{12} in terms of the diagonal Green's functions G_{11}, G_{22}, g_{11} and g_{22}.

As a starting point, to demonstrate how the 2-component NNEs (14.22) can be iterated to yield multi-component NNEs, the 3-component structure of figure 14.3 will

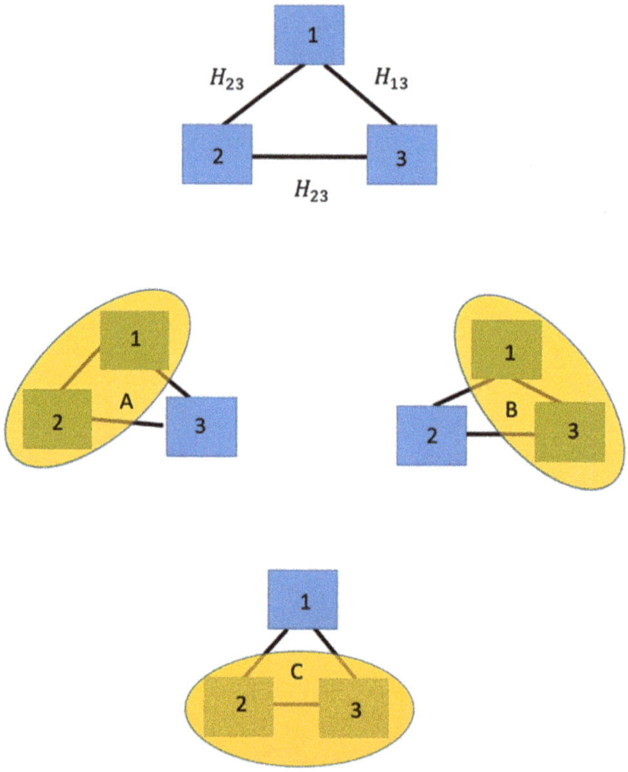

Figure 14.3. The top structure comprising three substructures coloured blue, can be regarded as either of the three structures labelled A, B or C.

first be analysed, which is composed of three substructures coloured blue and labelled 1, 2 and 3. This structure can be partitioned in three ways, as shown by the lower three figures. The structure on the left is composed of two substructures labelled A and 3, coloured orange and brown respectively, in which substructure A is composed of substructure 1 and 2. Similarly, the structure on the right is composed of substructures B and 2, while the lower structure is composed of substructures C and 1.

When analysing such structures, it may be useful to describe their quantum properties in terms of the quantum properties of substructures 1, 2 and 3, along with the couplings between them. On the other hand, one may be interested in describing their properties in terms of the quantum properties of substructures A and 3, or B and 2 or C and 3. The choice is a matter of taste, in much the same way as the language used to describe a work of art is a matter of preference. In what follows, all of these options will be explored.

The task is to obtain expressions for the off-diagonal Green's functions in terms of the various diagonal Green's functions. In what follows, expressions for G_{13} will be obtained. Corresponding expressions for G_{12} and G_{23} can be obtain by simply renumbering the substructures.

Before seeking such expressions, it is useful to identify the various Green's function which arise within such descriptions, starting with the Green's function G of the whole structure, which can be written

$$G = \begin{pmatrix} G_{11} & G_{12} & G_{13} \\ G_{21} & G_{22} & G_{23} \\ G_{31} & G_{32} & G_{33} \end{pmatrix} = \begin{pmatrix} g_{11}^{-1} & -H_{12} & -H_{13} \\ -H_{21} & g_{22}^{-1} & -H_{23} \\ -H_{31} & -H_{32} & g_{33}^{-1} \end{pmatrix}^{-1}, \tag{14.29}$$

or

$$G = \begin{pmatrix} G_{AA} & G_{A3} \\ G_{3A} & G_{33} \end{pmatrix} = \begin{pmatrix} A^{-1} & -H_{A3} \\ -H_{3A} & g_{33}^{-1} \end{pmatrix}^{-1}, \tag{14.30}$$

or

$$G = \begin{pmatrix} G_{BB} & G_{B2} \\ G_{2B} & G_{22} \end{pmatrix} = \begin{pmatrix} B^{-1} & -H_{B2} \\ -H_{2B} & g_{22}^{-1} \end{pmatrix}^{-1}, \tag{14.31}$$

or

$$G = \begin{pmatrix} G_{CC} & G_{C1} \\ G_{1C} & G_{11} \end{pmatrix} = \begin{pmatrix} C^{-1} & -H_{C1} \\ -H_{1C} & g_{11}^{-1} \end{pmatrix}^{-1} \tag{14.32}$$

In these equations,

$$G_{AA} = \begin{pmatrix} G_{11} & G_{12} \\ G_{21} & G_{22} \end{pmatrix}, \; G_{BB} = \begin{pmatrix} G_{11} & G_{13} \\ G_{31} & G_{33} \end{pmatrix} \text{ and } G_{CC} = \begin{pmatrix} G_{22} & G_{23} \\ G_{32} & G_{33} \end{pmatrix} \tag{14.33}$$

$$G_{A3} = G_{3A}^{\dagger} = \begin{pmatrix} G_{13} \\ G_{23} \end{pmatrix}, \; G_{B2} = G_{2B}^{\dagger} = \begin{pmatrix} G_{12} \\ G_{32} \end{pmatrix} \text{ and } G_{C1} = G_{1C}^{\dagger} = \begin{pmatrix} G_{21} \\ G_{31} \end{pmatrix} \tag{14.34}$$

and the coupling matrices are

$$H_{A3} = H_{3A}^{\dagger} = \begin{pmatrix} H_{13} \\ H_{23} \end{pmatrix}, \; H_{B2} = H_{2B}^{\dagger} = \begin{pmatrix} H_{12} \\ H_{32} \end{pmatrix}, \; H_{C1} = H_{1C}^{\dagger} = \begin{pmatrix} H_{21} \\ H_{31} \end{pmatrix} \tag{14.35}$$

Furthermore, A, B and C are the Green's functions of the three isolated orange structures shown in figure 14.4, given by

$$A = \begin{pmatrix} A_{11} & A_{12} \\ A_{21} & A_{22} \end{pmatrix} = \begin{pmatrix} g_{11}^{-1} & -H_{12} \\ -H_{21} & g_{22}^{-1} \end{pmatrix}^{-1}, \tag{14.36}$$

$$B = \begin{pmatrix} B_{11} & B_{13} \\ B_{31} & B_{33} \end{pmatrix} = \begin{pmatrix} g_{11}^{-1} & -H_{13} \\ -H_{31} & g_{33}^{-1} \end{pmatrix}^{-1} \tag{14.37}$$

and

$$C = \begin{pmatrix} C_{22} & C_{23} \\ C_{32} & C_{33} \end{pmatrix} = \begin{pmatrix} g_{22}^{-1} & -H_{23} \\ -H_{32} & g_{33}^{-1} \end{pmatrix}^{-1} \tag{14.38}$$

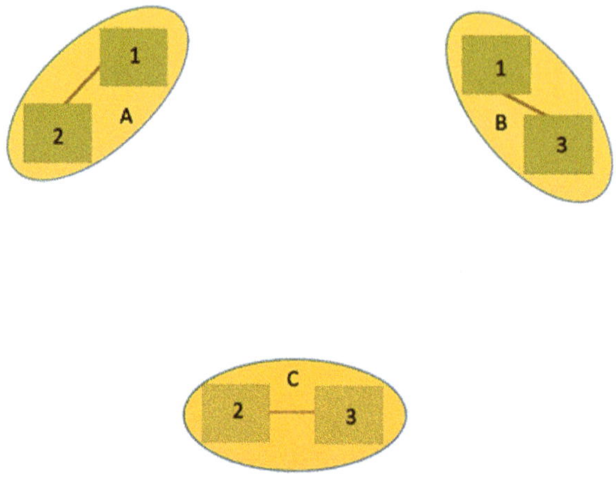

Figure 14.4. Three structures labelled A, B and C, each comprising two substructures.

Equations (14.29)–(14.38) merely define the language that will be used to describe quantum transport properties of the structure shown in figure 14.3. Expressions for G_{13} will be obtained by utilising the above equations. For example, with an appropriate change of notation, (i.e. $X \to A$, $\alpha \to 1$, $\beta \to 2$) comparing equation (14.25) with equation (14.36), yields

$$A_{12} = A_{11}H_{12}g_{22} = g_{11}H_{12}A_{22} \tag{14.39}$$

where

$$A_{11} = \left(g_{11}^{-1} - H_{12}g_{22}H_{21}\right)^{-1}, \tag{14.40}$$

and

$$A_{22} = \left(g_{22}^{-1} - H_{21}g_{11}H_{12}\right)^{-1} \tag{14.41}$$

Appendix D shows how a variety of equivalent expressions for G_{13} can be obtained by utilising the above equations. For the simplest case, when $H_{13} = 0$, these reduce to

$$G_{13} = A_{11}H_{12}g_{22}H_{23}G_{33} \tag{D.17a}$$

$$G_{13} = g_{11}H_{12}A_{22}H_{23}G_{33} \tag{D.31}$$

$$G_{13} = G_{11}H_{12}C_{22}H_{23}g_{33} \tag{D.24}$$

$$G_{13} = G_{11}H_{12}g_{22}H_{23}C_{33} \tag{D.17b}$$

$$G_{13} = g_{11}H_{12}G_{22}H_{23}g_{33} \tag{D.25}$$

In these equations, there are two in which G_{33} appears on the right, with active and passive NNEs for A_{12} on the left, two in which G_{11} appears on the left, with active and passive NNEs for C_{23} on the right and one in which G_{22} appears in the middle, with Green's functions of isolated substructures 1 and 2 as book ends. Like the two-component NNEs, they provide alternative viewpoints for describing G_{13} and they can all be shown to be equivalent.

The choice of which equation to use is of course a matter of convenience. For example, if E is real and coincides with an eigenvalue of H_{11}, then $(E - H_{11})^{-1}$ diverges and it is convenient to avoid using equations (14.55a) and (14.55b). If $H_{33} = H_{11}$, then it would also be convenient to avoid equation (14.56b). Of course, with careful handling, one would find that these divergences cancel in the overall expressions, but if one chooses to use equation (14.56a) this issue is avoided.

As discussed in chapter 11, and appendix E, the 5-component NNE is obtained by a further iteration of equation (D.25),

14.5 Stationary states of vibrating structures

To understand the relationship between Green's functions and stationary wave amplitudes of vibrating systems, it is useful to repeat the discussion of section 14.2 for a driven vibrating structure described by the equation of motion

$$-\frac{d^2}{dt^2}|\phi(t)\rangle = H|\phi(t)\rangle + |f(t)\rangle \tag{14.42}$$

In this equation $|\phi(t)\rangle$ is a vector containing the displacements from equilibrium of a vibrating set of masses and therefore is necessarily real. Again, to understand the relationship between $|\phi(t)\rangle$ and a real source $|f(t)\rangle$, it is useful to determine the source needed to create a known set of wave amplitudes of the form

$$|\phi(t)\rangle = e^{\eta t}\cos\omega_o t|\psi\rangle \tag{14.43}$$

To find the source $|f(t)\rangle$ that will produce the desired wave pattern, note that

$$\frac{d}{dt}|\phi(t)\rangle = [-\omega_o\sin\omega_o t + \eta\cos\omega_o t]e^{\eta t}|\psi\rangle$$

and therefore

$$\begin{aligned}-\frac{d^2}{dt^2}|\phi(t)\rangle &= \left[(\omega_o^2 - \eta^2)\cos\omega_o t + 2\eta\omega_o\sin\omega_o t\right]e^{\eta t}|\psi\rangle \\ &= H|\phi(t)\rangle + |f(t)\rangle,\end{aligned} \tag{14.44}$$

where

$$|f(t)\rangle = \left[(\omega_o^2 - \eta^2 - H)\cos\omega_o t + 2\eta\omega_o\sin\omega_o t\right]e^{\eta t}|\psi\rangle \tag{14.45}$$

This shows that the source $|f(t)\rangle$ that produces the wave amplitude (14.43) involves not just a time variation $\cos\omega_o t$, but also has a contribution proportional to $\sin\omega_o t$. Since $\sin(\omega_o t + 90°) = \cos\omega_o t$, the contribution $\sin\omega_o t$ is 90° out of phase with the contribution $\cos\omega_o t$. The presence of both the in-phase contribution $\cos\omega_o t$ and the out-of-phase contribution $\sin\omega_o t$ arises, because the left-hand side of equation (14.42) involves a second order time derivative.

The above argument shows that to produce a wave with a time dependence proportional to $\cos\omega_o t$ only, the source must have both an in-phase and an out-of-phase component. It will now be demonstrated that if a source has only a $\cos\omega_o t$ time dependence, then the resulting wave will have both an in-phase and an out-of-phase component.

To demonstrate this result, note that since $\cos\omega_o t = \text{Re}\, e^{-i\omega_o t}$ and $\sin\omega_o t = \text{Re} ie^{-i\omega_o t}$, equation (14.45) can be written

$$|f(t)\rangle = \text{Re}\Big[(\omega_o^2 - \eta^2 - H) + i2\eta\omega_o\Big]e^{-i\omega_o t}e^{\eta t}|\psi\rangle$$

That is

$$|f(t)\rangle = \text{Re}[e^{-i(\omega_o+i\eta)t}(\omega_o + i\eta)^2 - H]|\psi\rangle \tag{14.46}$$

From equation (14.43), the real vector $|\phi(t)\rangle$ is given by

$$|\phi(t)\rangle = \text{Re}[e^{-i(\omega_o+i\eta)t}]|\psi\rangle \tag{14.47}$$

From a mathematical viewpoint, it is also interesting to note that if the complex vector

$$|\phi_c(t)\rangle = [e^{-i(\omega_o+i\eta)t}]|\psi\rangle \tag{14.48}$$

is substituted into the equation $-\frac{d^2}{dt^2}|\phi_c(t)\rangle = H|\phi_c(t)\rangle + |f_c(t)\rangle$, one obtains

$$|f_c(t)\rangle = e^{-i(\omega_o+i\eta)t}[(\omega_o + i\eta)^2 - H]|\psi\rangle \tag{14.49}$$

This means that, the complex source $|f_c(t)\rangle$ produces the complex wave $|\phi_c(t)\rangle$ and the real part of $|f_c(t)\rangle$ produces the physically relevant real part of $|\phi_c(t)\rangle$. From equations (14.48) and (14.49), writing

$$|S_c\rangle = [(\omega_o + i\eta)^2 - H]|\psi\rangle \tag{14.50}$$

demonstrates that the source

$$|f_c(t)\rangle = e^{-i(\omega_0+i\eta/\hbar)t}|S_c\rangle \tag{14.51}$$

produces the wave $|\phi_c(t)\rangle = [e^{-i(\omega_o+i\eta)t}]|\psi\rangle$, where

$$|\psi\rangle = [(\omega_o + i\eta)^2 - H]^{-1}|S_c\rangle \tag{14.52}$$

Similarly, if $|S\rangle$ is any source vector, then the source

$$|f_c(t)\rangle = e^{-i(\omega_0+i\eta/\hbar)t}|S\rangle \tag{14.53}$$

will produce the wave

$$|\phi_c(t)\rangle = [e^{-i(\omega_o+i\eta)t}]|\psi_c\rangle, \tag{14.54}$$

where

$$|\psi_c\rangle = [(\omega_o + i\eta)^2 - H]^{-1}|S\rangle \tag{14.55}$$

Choosing $|S\rangle$ to be real, shows that the real source

$$|f(t)\rangle = \mathrm{Re}|f_c(t)\rangle = e^{\eta t}\cos\omega_o t|S\rangle \tag{14.56}$$

produces a real wave

$$|\phi(t)\rangle = \mathrm{Re}|\phi_c(t)\rangle = \mathrm{Re}[e^{-i(\omega_o+i\eta)t}[(\omega_o + i\eta)^2 - H]^{-1}]|S\rangle \tag{14.57}$$

Writing

$$[[(\omega_o + i\eta)^2 - H]^{-1}]|S\rangle = |A\rangle + i|B\rangle \tag{14.58}$$

where $|A\rangle$ and $|B\rangle$ are real vectors, yields

$$|\phi(t)\rangle = e^{\eta t}[\cos\omega_o t|A\rangle + \sin\omega_o t|B\rangle] \tag{14.59}$$

This demonstrates that if a source has only a $\cos\omega_o t$ time dependence, then the resulting wave will have both an in-phase and an out-of-phase component. Clearly $|A\rangle$ and $|B\rangle$ are the amplitudes of the in-phase and out-of-phase components of $|\phi(t)\rangle$ respectively. These can be measured experimentally and since the ratios of in-phase components at different sites are independent of time (and similarly for the out-of-phase components), the wave is said to be stationary.

14.6 Connectivity theory, equality of currents and unitarity of the scattering matrix

In this section, the aim is to show that unitarity of the S-matrix describing currents passing from a substructure (labelled 2) to another (substructure labelled 1), derives from the Hermiticity of the Green's function of substructure 1, which in turn depends on whether or not substructure 1 is open or closed.

Consider the structure of figure 14.5, formed from two coupled substructures. From appendix B in chapter 6, the current passing from substructure 2 to substructure 1 is

$$J_{1\to 2} = -\frac{2}{\hbar}\mathrm{Im}\left[\langle\phi_1|H_{12}|\phi_2\rangle\right] \tag{14.60}$$

Also from equations (2.5) and (2.8) of the noisy neighbour discussion in chapter 2, the stationary wave functions $|\phi_1\rangle$ and $|\phi_2\rangle$ in the two substructures are related by

$$|\phi_1\rangle = g_{11}|S_1\rangle \tag{14.61}$$

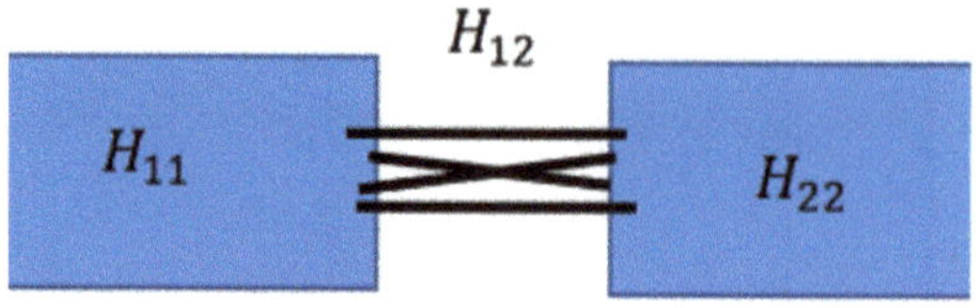

Figure 14.5. Two coupled substructures.

where

$$|S_1\rangle = H_{12}|\phi_2\rangle \tag{14.62}$$

Hence

$$H_{12}|\phi_2\rangle = g_{11}^{-1}|\phi_1\rangle \tag{14.63}$$

and therefore equation (14.60) becomes

$$J_{1\to2} = -\frac{2}{\hbar}\mathrm{Im}\left[\langle\phi_1|g_{11}^{-1}|\phi_1\rangle\right] \tag{14.64}$$

Consider the number $z = \langle\phi_1 \mid A \mid \phi_2\rangle$. The complex conjugate of z is $z^* = \langle\phi_1 \mid A \mid \phi_2\rangle^* = \langle\phi_1 \mid A \mid \phi_2\rangle^\dagger = |\phi_2\rangle^\dagger A^\dagger \langle\phi_1|^\dagger = \langle\phi_2 \mid A^\dagger \mid \phi_1\rangle$, where it was noted that the Hermitian conjugate of a product is the product of the Hermitian conjugates in reverse order. Hence if $z = \langle\phi_1|g_{11}^{-1}|\phi_1\rangle$, then $z^* = \langle\phi_1|g^{\dagger}{}_{11}^{-1}|\phi_1\rangle$. This means that if $g_{11} = g_{11}^\dagger$, then z is real and $\mathrm{Im}\, z = 0$. From equation (14.64), one obtains the fundamental result that if the Green's function g_{11} of substructure 1 is Hermitian, then no current can flow between substructure 1 and substructure 2.

Since

$$(EI - H_{11})g_{11} = I, \tag{14.65}$$

then provided H_{11} is a finite-size matrix, one can multiply from the left by $(EI - H_{11})^{-1}$ to obtain $g_{11} = (EI - H_{11})^{-1}$.

As discussed in appendix E, the above discussion can be extended to shows that a state initially localised on a substructure decays with time, with a rate determined by the non-Hermitian part of the appropriate self-energy.

14.7 The limit $\eta \to 0$

In the limit $\eta \to 0$, since H_{11} is Hermitian, g_{11} is Hermitian and no current can flow between the two substructure. If H_{11} is of finite size, then the substructure 1 is said to be closed and therefore stationary states cannot carry a current from substructure 1 to substructure 2. This means that any current entering substructure 1 must be reflected back into substructure 2. If substructure 2 represents a collection of electrodes attached to substructure 1, as discussed for example in chapter 9, then the associated scattering matrix must be unitary.

On the other hand, as discussed above and in appendix B, if H_{11} is infinite, then the matrix $(EI - H_{11})$ cannot be inverted and subsystem 1 is said to be open. Taking the Hermitian conjugate of equation (14.65) yields

$$g_{11}^{\dagger}(EI - H_{11}) = I \tag{14.66}$$

Since $g_{11}^{\dagger}$ satisfies a different equation from g_{11}, one cannot conclude that g_{11} is Hermitian. Indeed, as an example, equation (10.39) shows that the retarded Green's function of a doubly infinite tight-binding chain is not Hermitian, because $G_{jp}(E) \neq G_{pj}(E)*$. As another example, the Green's function (11.115) of a scatterer connected to one or more external leads is not Hermitian, due to the presence of a complex self-energy. For these reasons, one concludes that if substructure 1 is open, then a steady-state current can flow between substructure 1 to substructure 2. In other words, a current flowing into substructure 1 from substructure 2 may not be fully reflected back into substructure 2 and a scattering matrix describing wave propagation between these substructures will not be unitary.

The above argument can also be applied to structure of figure 14.2, to obtain the conditions for which the current from substructure 1 to substructure 2 is equal to the current from substructure 2 to substructure 3. Taking the Hermitian conjugate of equation (14.63) and noting that $H_{21}^{\dagger} = H_{21}$ yields $\langle\phi_2|H_{21}g_{11}^{\dagger} = \langle\phi_1|$. Substituting this into equation (14.60) yields

$$J_{1\to 2} = -\frac{2}{\hbar}\mathrm{Im}\left[\langle\phi_2|H_{21}g_{11}^{\dagger}H_{12}|\phi_2\rangle\right] \tag{14.67}$$

On the other hand, the equivalent of equation (14.60) applied to substructures 2 and 3 is

$$J_{2\to 3} = -\frac{2}{\hbar}\mathrm{Im}\left[\langle\phi_2|H_{23}|\phi_3\rangle\right] \tag{14.68}$$

From the NNE (2.12) $|\phi_2\rangle = G_{22}|S_2'\rangle$, where $|S_2'\rangle = H_{23}|\phi_3\rangle$ and therefore $H_{23}|\phi_3\rangle = G_{22}^{-1}|\phi_2\rangle$. Hence

$$J_{2\to 3} = -\frac{2}{\hbar}\mathrm{Im}\left[\langle\phi_2|G_{22}^{-1}|\phi_2\rangle\right] \tag{14.69}$$

From equations (2.22) and (2.19), this yields

$$J_{2\to 3} = -\frac{2}{\hbar}\mathrm{Im}\left[\langle\phi_2|g_{22}^{-1} - H_{21}g_{11}H_{12}|\phi_2\rangle\right] \tag{14.70}$$

That is

$$J_{2\to 3} = -\frac{2}{\hbar}\mathrm{Im}[\langle\phi_2|g_{22}^{-1}|\phi_2\rangle] + \frac{2}{\hbar}\mathrm{Im}[\langle\phi_2|H_{21}g_{11}H_{12}|\phi_2\rangle] \tag{14.71}$$

Since $\langle\phi_2|H_{21}g_{11}H_{12}|\phi_2\rangle = \langle\phi_2|H_{21}g_{11}^{\dagger}H_{12}|\phi_2\rangle*$, $\mathrm{Im}\langle\phi_2|H_{21}g_{11}H_{12}|\phi_2\rangle = -\mathrm{Im}\langle\phi_2|H_{21}g_{11}^{\dagger}H_{12}|\phi_2\rangle$

Hence comparing equation (14.71) with equation (14.67) yields

$$J_{2\to 3} = J_{1\to 2} - \frac{2}{\hbar}\mathrm{Im}[\langle\phi_2|g_{22}^{-1}|\phi_2\rangle] \tag{14.72}$$

If g_{22} is Hermitian, then $[\langle\phi_2|g_{22}^{-1}|\phi_2\rangle]$ is real and $\mathrm{Im}[\langle\phi_2|g_{22}^{-1}|\phi_2\rangle] = 0$. Therefore if substructure 2 is closed, so that g_{22} is Hermitian $J_{2\to3} = J_{1\to2}$.

On the other hand if substructure 2 is open and g_{22} is not Hermitian, then $J_{2\to3} \neq J_{1\to2}$, because current can flow to infinity via substructure 2.

14.8 The case $\eta \neq 0$

When $\eta \neq 0$, it is convenient to write $|\phi_j\rangle = e^{-i\omega t}e^{\eta t/\hbar}|\psi_j\rangle$, so that equation (14.64) becomes

$$J_{1\to2} = -\frac{2}{\hbar}e^{2\eta t/\hbar}\mathrm{Im}\left[\langle\psi_1|\,g_{11}^{-1}\,|\psi_1\rangle\right] \tag{14.73}$$

For a closed system, $g_{11}^{-1} = ((\hbar\omega + i\eta)I - H_{11})$ and therefore

$$J_{1\to2} = -\frac{2\eta}{\hbar}e^{\frac{2\eta t}{\hbar}}\langle\psi_1|\psi_1\rangle = -\frac{2\eta}{\hbar}\langle\phi_1|\phi_1\rangle \tag{14.74}$$

As expected, this is equal to $-\frac{d\langle\phi_1\,|\,\phi_1\rangle}{dt}$

14.9 Comparison between advanced and retarded Green's functions

The Green's functions discussed so far are called retarded Green's functions and are waves resulting from a source, which start from zero at $t = -\infty$. One can also ask, if it is possible to create a wave of the form

$$|\varphi_A(t)\rangle = e^{-i(\omega_0 - i\eta/\hbar)t}|\psi\rangle \tag{14.75}$$

which vanishes as $t \to \infty$. In this case, $\eta > 0$ and one finds that

$$\begin{aligned} i\hbar\frac{d}{dt}|\varphi_A(t)\rangle &= i\hbar\frac{d}{dt}e^{-i(\omega_0 - i\eta/\hbar)t}|\psi\rangle = (\hbar\omega_0 - i\eta)e^{-i(\omega_0 - i\eta/\hbar)t}|\psi\rangle \\ &= (E_0 - i\eta)|\varphi_A(t)\rangle \end{aligned}$$

where $E_0 = \hbar\omega_0$. Adding and subtracting $H_A|\varphi_A(t)\rangle$ on the right-hand side, yields

$$i\hbar\frac{d}{dt}|\varphi_A(t)\rangle = H_A|\varphi_A(t)\rangle + |f(t)\rangle \tag{14.76}$$

where

$$|f(t)\rangle = (E_0 - H_A - i\eta)|\varphi_A(t)\rangle \tag{14.77}$$

In this case, $|f(t)\rangle$ is referred to as a 'sink', since it causes $|\varphi_A(t)\rangle$ to vanish at large times.

In other words, the required sink is

$$|f(t)\rangle = e^{-i(\omega_0 - i\eta/\hbar)t}|S\rangle \tag{14.78}$$

where $|S\rangle$ is a time-independent sink vector given by

$$|S\rangle = (E_0 - H_A - i\eta)|\psi\rangle \tag{14.79}$$

This equation tells us that the sink $|f(t)\rangle$ with a spatial dependence $|S\rangle$ yields the desired retarded Green's function $|\varphi_A(t)\rangle$ with a spatial dependence $|\psi\rangle$, which decays to zero at large times.

To answer the question 'what stationary wave function $|\varphi_A(t)\rangle$ tends to zero at $t \to \infty$ under the influence of a known sink $|f(t)\rangle$', it is necessary to invert equation (14.79). If the Hamiltonian H_A is a finite-size matrix, which would be the case when the region is closed, then the answer to this question is

$$|\psi\rangle = g(E_0 - i\eta)|S\rangle \tag{14.80}$$

where $g(E_0 - i\eta)$ is the advanced Green's functions given by

$$g(E_0 - i\eta) = (E_0 - H_A - i\eta)^{-1} \tag{14.12}$$

Clearly $|\varphi(t)\rangle$ is a stationary state with a spatial wave pattern $|\psi\rangle$, which under the influence of the sink ($|f(t)\rangle$), vanishes at $t = +\infty$. The amplitude of this wave pattern on site i is $\psi_i = \langle i|\psi\rangle$ and the amplitude of the sink on site i is $S_i = \langle i|S\rangle$. For the special case where $S = \delta_{il}$ (i.e. when the sink acts with unit amplitude on site l only), one obtains $\psi_i = g_{il}(E_0 - i\eta)$, where $g_{il}(E_0 - i\eta)$ is the i, l th element of $g(E_0 + i\eta)$. In other words, $g_{il}(E_0 - i\eta)$ is the stationary wave amplitude at i due to a sink of unit amplitude acting on site l. As an example, the advanced Green's function of a one-dimensional tight-binding chain is presented in appendix C. Symmetry relations between Green's functions are discussed in appendix B of chapter 17 and are presented in equation (B.46). In appendix B of chapter 17, the role of time reversal symmetry is discussed in section B.9.

14.10 Summary

In this chapter, the concept of stationary states for quantum structures and vibrating systems was placed on a firm footing, by analysing their equations of motion in the presence of an external source. Furthermore, the NNEs of chapter 2 were re-derived starting from these equations of motion. By iterating the passive and active NNEs in different sequences, a variety of equations for Green's functions of coupled quantum structures were obtained, which allow one to examine interference effects within such structures from different viewpoints.

Appendix A Explicit integration of equation (14.4)

Note that for a closed system, the solution to equation (14.4) is

$$|\varphi(t)\rangle = e^{-iH(t-t_0)/\hbar}\left[|\varphi(t_0)\rangle + \frac{1}{i\hbar}\int_{t_0}^{t} dt' e^{\frac{iH(t'-t_0)}{\hbar}}|f(t')\rangle\right] \tag{A.1}$$

as can be verified by differentiating both sides with respect to t. In the above expression, t_0 is an arbitrary constant. As an example, consider the case where the source has the form

$$|f(t)\rangle = e^{-i\omega_0 t}e^{\eta t/\hbar}|\psi\rangle \tag{A.2}$$

where $|\psi\rangle$ is independent of t and η is a positive real number. In this case,

$$|\varphi(t)\rangle = e^{-iH(t-t_0)/\hbar}|\varphi(t_0)\rangle + I(\eta, \omega_0, t)|\psi\rangle - e^{-iH(t-t_0)/\hbar}I(\eta, \omega_0, t_0)|\psi\rangle \quad \text{(A.3)}$$

where

$$I(\eta, \omega_0, t) = \frac{e^{\frac{-iHt}{\hbar}}}{i\hbar}\int^t dt' e^{\frac{iHt'}{\hbar}} e^{-i\omega_0 t'} e^{\eta t'/\hbar} = g(E_0 + i\eta)e^{-i\omega_0 t} e^{\eta t/\hbar} \quad \text{(A.4)}$$

In this equation, $g(E_0 + i\eta)$ is the following Green's function

$$g(E_0 + i\eta) = (E_0 - H + i\eta)^{-1} \quad \text{(A.5)}$$

where $E_0 = \hbar\omega_0$. As a further simplification, consider the case where $t_0 \to -\infty$ and $|\varphi(t_0)\rangle = 0$. In this case $I(\eta, \omega_0, t_0) = 0$ and equation (A.3) reduces to

$$|\varphi(t)\rangle = e^{-i\omega_0 t} e^{\eta t/\hbar}|\psi\rangle \quad \text{(A.6)}$$

Appendix B Green's functions of open systems

The Green's functions discussed is section 14.2 are defined by

$$g(E) = (E - H_A)^{-1} \quad \text{(B.1)}$$

where $E = E_0 + i\eta$. Note that the presence of the term $i\eta$ means that $g(E)$ is not Hermitian. The right-hand side of equation (B.1) is well defined for a closed region A with a finite-size Hamiltonian H_A, but for an open system with an infinitely large Hamiltonian, the inverse is not defined. To address this issue, first note the intuitively obvious fact that for an open system, the wave amplitude ψ_i of equation (14.13), which arises from a source at site l decays as the distance between l and i increases, because on a finite time scale $\tau = \hbar/\eta$, the wave due to a source can only travel a finite distance from the source. Appendix C discusses an example of such a decay. Therefore the amplitudes ψ_i are non-negligible on a finite number N_1 of sites only, when f_l is non-zero on a finite number $N_2 < N_1$ of sites only. In this case, equation (14.10) becomes

$$f_l = \sum_{i=1}^{N_1}(E_0 - H_A + i\eta)_{li}\psi_i \quad \text{(B.2)}$$

In this equation, although the matrix H_A is infinite, the range of i in the sum is restricted by the finite number of non-negligible elements. Therefore if one constructs a N_1-component vector $|\tilde{f}\rangle$ containing the non-zero elements, (and some zeros), then equation (B.2) becomes

$$\tilde{f}_l = \sum_{i=1}^{N_1}(E_0 - H_A + i\eta)_{li}\psi_i \quad \text{(B.3)}$$

where l is restricted to the range $l = 1, 2, \ldots N_1$.

Hence equation (B.2) can be replaced by

$$|\tilde{f}\rangle = (E_0 - \tilde{H} + i\eta)|\tilde{\psi}\rangle \quad \text{(B.4)}$$

where $\tilde{H}$ is a submatrix of H_A with matrix elements $\tilde{H}_{ij} = H_{ij}$, where $i, j = 1, 2, \ldots N_1$ and $|\tilde{f}\rangle$, $|\tilde{\psi}\rangle$ are N_1-component vectors. Since the matrix $(E_0 - \tilde{H} + i\eta)$ is now a $N_1 \times N_1$ finite matrix, this equation can be inverted to yield

$$|\tilde{\psi}\rangle = \tilde{g}|\tilde{f}\rangle \tag{B.5}$$

where

$$\tilde{g} = (E_0 - \tilde{H} + i\eta)^{-1} \tag{B.6}$$

The above steps are permissible, provided η is non-zero, because then the non-negligible amplitudes ψ_i are spread over only a finite number of sites.

Provided $\eta \neq 0$, the fact that sub-matrices of Green's functions of infinite systems can be obtained from equation (B.6) means that equations involving Green's functions do not need to treat open systems differently from closed systems. One should bear in mind however that in the limit $\eta \to 0$, the Green's functions of closed systems are Hermitian, whereas the retarded Green's functions of open systems are not. Furthermore for open systems, when $\eta = 0$, equation (B.6) is not valid. If $\eta \to 0$, one should return to equations (14.11) and (14.12), which have the form

$$f_j = \underset{\eta \to 0}{Lt} \sum_{i=1}^{\infty} (E_0 - H_A + i\eta)_{ji}\psi_i \tag{B.7}$$

where j and i run over the infinite number of sites in region A. In this case, if the driving term has the form $f_j = \delta_{jl}$ (i.e. if the driving term acts with unit amplitude on site l only), then the amplitude ψ_i of the stationary wave pattern on site i of region A is given by

$$\delta_{jl} = \underset{\eta \to 0}{Lt} \sum_{i=1}^{\infty} (E_0 - H_A + i\eta)_{ji}\psi_i \tag{B.8}$$

Obviously this wave pattern will depend on the choice of l and therefore one should attach a label l to symbol used to denote it. Therefore it will be denoted $g_{il}(E_0 + i\eta)$. In other words, for an infinite region A, $g_{il}(E_0 + i\eta)$ is defined by the equation

$$\chi_i = \underset{\eta \to 0}{Lt}\, g_{il}(E_0 + i\eta) \tag{B.9}$$

That is the Green's function $g_{il}(E_0 + i\eta)$ is defined to be the stationary wave pattern on site i created by a driving term of unit amplitude acting on site l.

With this notation, equation (B.8) becomes

$$\delta_{jl} = \underset{\eta \to 0}{Lt} \sum_{i=1}^{\infty} (E_0 - H_A + i\eta)_{ji} g_{il}(E_0 + i\eta) \tag{B.10}$$

or more concisely,

$$I = \underset{\eta \to 0}{Lt} (E - H_A) g(E) \tag{B.11}$$

where I is the unit matrix and $E = E_0 + i\eta$. Equations (B.10) and (B.11) do not involve the inverse of the matrix $(E_0 - H_A + i\eta)$ and are valid even when the matrix is of infinite size. Starting from equation (B.10) it may be possible to solve directly for the Green's function g_{il}, without inverting a matrix, as discussed in chapter 10, where the Green's function of a doubly infinite chain (equation (10.39)) was obtained for the case $\eta = 0$. The Green's function of a doubly infinite chain when $\eta \neq 0$ is discussed in appendix C.

Appendix C Green's functions of a double infinite tight-binding chain revisited

As an example, which demonstrates the decay of Green's functions in open systems, consider the retarded Green's function of a doubly infinite chain, when $\eta > 0$. From equation (10.39), this is

$$g_{jl}(E) = \frac{e^{ik(E)|j-l|}}{2i\gamma \sin k(E)} \tag{C.1}$$

where $k(E)$ is obtained by solving $E = E_0 + i\eta = \epsilon_0 - 2\gamma \cos k(E)$ or equivalently, $\epsilon_0 - E_0 - i\eta = \gamma[e^{ik(E)} + e^{-ik(E)}]$. Writing $k = k_0 + ik_1$, where k_0 and k_1 are real yields

$$\begin{aligned} e^{ik(E)} + e^{-ik(E)} &= e^{ik_0}e^{-k_1} + e^{-ik_0}e^{k_1} = (e^{-k_1} + e^{k_1}) \cos k_0 \\ &+ i(e^{-k_1} - e^{k_1}) \sin k_0 \epsilon_0 - E_0 - i\eta = (e^{-k_1} + e^{k_1})\gamma \cos k_0 \\ &+ i(e^{-k_1} - e^{k_1})\gamma \sin k_0 \end{aligned} \tag{C.2}$$

The requirement that the wave is outgoing from the source at l means that $\gamma \sin k_0 > 0$. Therefore, since the imaginary part of the left-hand side of equation (C.2) is negative, the term $(e^{-k_1} - e^{k_1})$ on the right-hand side must be negative, which means that k_1 is positive. Hence, since equation (C.1) is of the form

$$g_{jl}(E) = \left[\frac{e^{ik_0(E)|j-l|}}{2i\gamma \sin k(E)}\right] e^{-k_1(E)|j-l|} = \left[\frac{e^{ik_0(E)|j-l|}}{2i\gamma \sin k(E)}\right] e^{-|j-l|/L_\eta}, \tag{C.3}$$

This shows that the retarded Green's function $g_{jl}(E)$ decays exponentially with distance $|j - l|$, with a decay length $L_\eta = 1/k_1$.

From equation (C.2)

$$(e^{k_1} - e^{-k_1}) = \eta/\gamma \sin k_0$$

Writing $x = e^{k_1}$ allows this to be written as a quadratic equation $x^2 - bx - 1 = 0$, where $b = \eta/\gamma \sin k_0$. Hence $x = e^{k_1} = \frac{b \pm \sqrt{b^2+4}}{2}$. Since $k_1 > 0$, only the + sign is acceptable. Hence

$$k_1 = \ln\left[\frac{b + \sqrt{b^2 + 4}}{2}\right] \tag{C.4}$$

If b is small compared with unity, then

$$k_1 \approx \ln\left[1 + \frac{b}{2}\right] \approx \frac{b}{2} \tag{C.5}$$

Hence $k_1 \approx \frac{\eta}{2\gamma \sin k_0} = \eta/\hbar v(E)$, where $v(E)$ is the group velocity. If a time T_η is defined by $T_\eta = \hbar/\eta$, then the length $L_\eta = \frac{1}{k_1} \approx v(E)T_\eta$. Clearly T_η is the time scale over which the source is switched on and L_η is the distance that an electron can travel ballistically over that time. As noted below equation (C.3) L_η is also the decay length of the wave $g_{jl}(E)$. The condition of small b corresponds to small η and large T_η and therefore corresponds to a source that is switched on slowly. In the opposite limit of large η, equation (C.5) yields

$$k_1 \approx \ln[b] \tag{C.6}$$

and therefore $L_\eta < v(E)T_\eta$, meaning that the wave spreads less quickly than in the small η limit.

Similarly, the advanced Green's function $g^A_{jl}(E)$ is of the form

$$g^A_{jl}(E) = \frac{e^{-ik(E)|j-l|}}{-2i\gamma \sin k(E)} \tag{C.7}$$

where $k(E)$ is obtained by solving $E = E_0 - i\eta = \epsilon_0 - 2\gamma \cos k(E)$. Writing $k = k_0 - ik_1$, where k_0 and k_1 are real yields

$$\epsilon_0 - E_0 + i\eta = (e^{k_1} + e^{-k_1})\gamma \cos k_0 + i(e^{k_1} - e^{-k_1})\gamma \sin k_0 \tag{C.8}$$

The requirement that the wave is incoming towards the sink at l means that $\gamma \sin k_0 > 0$. Therefore, since the imaginary part of the left-hand side of equation (C.8) is positive, the term $(e^{k_1} - e^{-k_1})$ on the right-hand side must be positive and k_1 must be positive. Hence, since equation (C.7) is of the form

$$g^A_{jl}(E) = \left[\frac{e^{-ik_0(E)|j-l|}}{-2i\gamma \sin k(E)}\right] e^{-k_1(E)|j-l|} = \left[\frac{e^{-ik_0(E)|j-l|}}{-2i\gamma \sin k(E)}\right] e^{-|j-l|/L_\eta} \tag{C.9}$$

$g^A_{jl}(E)$ is again found to decay exponentially with distance $|j - l|$, with a decay length $L_\eta = 1/k_1$.

Appendix D Iterative analysis of the 3-component structure of figure 14.3

In this appendix, 2-component NNEs will be iterated to obtain a variety of expressions for G_{13}. For example, in section 14.4, it was noted that for the Green's function (14.36),

$$A_{12} = A_{11}H_{12}g_{22} = g_{11}H_{12}A_{22} \tag{14.39}$$

$$A_{11} = \left(g_{11}^{-1} - \Sigma_{11}(2)\right)^{-1}, \; A_{22} = \left(g_{22}^{-1} - \Sigma_{22}(1)\right)^{-1} \tag{14.40}$$

where

$$\Sigma_{11}(2) = H_{12}g_{22}H_{21} \text{ and } \Sigma_{22}(1) = H_{21}g_{11}H_{12} \tag{D.1}$$

Similarly with the notation change ($X \to B$, $\alpha \to 1$, $\beta \to 3$) comparing equation (14.25) with equation (14.37) yields,

$$B_{13} = B_{11}H_{13}g_{33} = g_{11}H_{13}B_{33} \tag{D.2}$$

where

$$B_{11} = \left(g_{11}^{-1} - \Sigma_{11}(3)\right)^{-1},\ B_{33} = \left(g_{33}^{-1} - \Sigma_{33}(1)\right)^{-1} \tag{D.3}$$

$$\Sigma_{11}(3) = H_{13}g_{33}H_{31} \text{ and } \Sigma_{33}(1) = H_{31}g_{11}H_{13} \tag{D.4}$$

Furthermore from equation (14.38),

$$C_{23} = C_{22}H_{23}g_{33} = g_{22}H_{23}C_{33} \tag{D.5}$$

where

$$C_{22} = \left(g_{22}^{-1} - \Sigma_{22}(3)\right)^{-1},\ C_{33} = \left(g_{33}^{-1} - \Sigma_{33}(2)\right)^{-1} \tag{D.6}$$

$$\Sigma_{22}(3) = H_{23}g_{33}H_{32} \text{ and } \Sigma_{33}(2) = H_{32}g_{22}H_{23} \tag{D.7}$$

From equations (14.33) and (14.34), it is clear that the off-diagonal Green's function G_{13} can be obtained by obtaining expressions for either G_{BB}, G_{A3} or G_{1C}, all of which contain G_{13} as a submatrix. Therefore these offer three distinct routes to obtaining equations for G_{13}.

For the first route, an expression for G_{BB} is obtained by noting that with the change of notation, (i.e. $X \to G$, $\alpha \to B$, $\beta \to 2$) comparing equation (14.25) with equation (14.31), yields

$$G_{BB} = [B^{-1} - H_{B2}g_{22}H_{2B}]^{-1} \tag{D.8}$$

In the above expression,

$$B^{-1} - H_{B2}g_{22}H_{2B} = \begin{pmatrix} g_{11}^{-1} & -H_{13} \\ -H_{31} & g_{33}^{-1} \end{pmatrix} - \begin{pmatrix} \Sigma_{11}(2) & \Sigma_{13}(2) \\ \Sigma_{31}(2) & \Sigma_{33}(2) \end{pmatrix}, \tag{D.9}$$

where

$$\Sigma_{ij}(2) = H_{i2}g_{22}H_{2j} \tag{D.10}$$

After taking note of equations (D.0) and (D.6), one finds

$$G_{BB} = \begin{pmatrix} G_{11} & G_{13} \\ G_{31} & G_{33} \end{pmatrix} = \begin{pmatrix} A_{11}^{-1} & -\tilde{H}_{13} \\ -\tilde{H}_{31} & C_{33}^{-1} \end{pmatrix}^{-1} \tag{D.11}$$

where

$$\tilde{H}_{13} = H_{13} + \Sigma_{13}(2) \text{ and } \tilde{H}_{31} = H_{31} + \Sigma_{31}(2) \tag{D.12}$$

As an aside, for later use, note that a similar argument yields

$$G_{AA} = \begin{pmatrix} G_{11} & G_{12} \\ G_{21} & G_{22} \end{pmatrix} = \begin{pmatrix} B_{11}^{-1} & -\tilde{H}_{12} \\ -\tilde{H}_{21} & C_{22}^{-1} \end{pmatrix}^{-1}, \text{ where } \tilde{H}_{12} = H_{12} + \Sigma_{12}(3) \tag{D.13}$$

and

$$G_{CC} = \begin{pmatrix} G_{22} & G_{23} \\ G_{32} & G_{33} \end{pmatrix} = \begin{pmatrix} A_{22}^{-1} & -\tilde{H}_{23} \\ -\tilde{H}_{32} & B_{33}^{-1} \end{pmatrix}^{-1}, \text{ where } \tilde{H}_{23} = H_{23} + \Sigma_{23}(1) \tag{D.14}$$

As discussed in appendix A.13 in chapter 2, it should also be noted that, equation (D.8) also implies[1]

$$G_{BB} = B + BH_{B2}G_{22}H_{2B}B \tag{D.15}$$

(To see this, note that $G_{BB} = [B^{-1} - \Sigma_B(2)]^{-1} = [B^{-1} - \Sigma_B(2)]^{-1}[B^{-1}]B = [B^{-1} - \Sigma_B(2)]^{-1}$ $[B^{-1} - \Sigma_B(2) + \Sigma_B(2)]B$. That is $G_{BB} = B + [B^{-1} - \Sigma_B(2)]^{-1}\Sigma_B(2)B = B + G_{BB}\Sigma_B(2)B = B + BH_{B2}G_{22}H_{2B}B$. In the last step, the NNE $G_{BB}H_{B2}g_{22} = BH_{B2}G_{22}$ was used. This is obtained by comparing equation (14.25) with equation (14.31), using the notation change ($X \to G$, $\alpha \to B$, $\beta \to 2$).)

Hence

$$G_{13} = B_{13} + \sum_{\substack{i=1,3 \\ j=1,3}} B_{1i}H_{i2}G_{22}H_{2j}B_{j3} \tag{D.16}$$

Continuing with the first route, applying the two-component NNEs (14.26) to equation (D.11) (with the change of notation, (i.e. $X \to G$, $\alpha \to 1$, $\beta \to 3$)) yields the following pair of expressions for G_{13}

$$G_{13} = A_{11}\tilde{H}_{13}G_{33} = G_{11}\tilde{H}_{13}C_{33} \tag{D.17}$$

Equations (D.12) are the first pair of equations for G_{13}. Clearly equation (D.9) could be written $B^{-1} = EI - \tilde{H}_{BB}$, where $\tilde{H}_{BB}$ is the renormalized Hamiltonian obtained after decimating substructure 2, given by

$$\tilde{H}_{BB} = \begin{pmatrix} \tilde{H}_{11} & \tilde{H}_{13} \\ \tilde{H}_{31} & \tilde{H}_{33} \end{pmatrix} = \begin{pmatrix} \mathrm{H}_{11} + \Sigma_{11}(2) & \mathrm{H}_{13} + \Sigma_{13}(2) \\ \mathrm{H}_{31} + \Sigma_{31}(2) & \mathrm{H}_{33} + \Sigma_{33}(2) \end{pmatrix} \tag{D.18}$$

Therefore, equations (D.13) are expressions for G_{13}, obtained by adopting a 'decimation approach' to solving the problem.

For the second route to obtaining a pair of expressions for G_{13}, it is useful to note that applying the two-component NNEs (14.26) to equation (14.30), (with the change of notation, (i.e. $X \to G$, $g_{\alpha\alpha} \to A$, $\alpha \to A$, $\beta \to 3$)) yields

[1] To see this, note that

$$\begin{aligned} G_{BB} &= [B^{-1} - \Sigma_B(2)]^{-1} = [B^{-1} - \Sigma_B(2)]^{-1}[B^{-1}]B \\ &= [B^{-1} - \Sigma_B(2)]^{-1}[B^{-1} - \Sigma_B(2) + \Sigma_B(2)]B \end{aligned}$$

$$G_{A3} = G_{AA}H_{A3}g_{33} = AH_{A3}G_{33} \tag{D.19}$$

That is from equations (14.33) and (14.34), when written out in full, these become

$$G_{A3} = \begin{pmatrix} G_{13} \\ G_{23} \end{pmatrix} = \begin{pmatrix} G_{11} & G_{12} \\ G_{21} & G_{22} \end{pmatrix}\begin{pmatrix} H_{13} \\ H_{23} \end{pmatrix} g_{33} = \begin{pmatrix} A_{11} & A_{12} \\ A_{21} & A_{22} \end{pmatrix}\begin{pmatrix} H_{13} \\ H_{23} \end{pmatrix} G_{33} \tag{D.20}$$

Hence a second pair of equations is

$$G_{13} = (G_{11}H_{13} + G_{12}H_{23})g_{33} = (A_{11}H_{13} + A_{12}H_{23})G_{33} \tag{D.21}$$

The story is not complete, because the right-hand sides of these equations contain off-diagonal Green's functions G_{12} and A_{12}. Applying the two-component NNEs (14.26) to equation (14.36) (with the change of notation, (i.e. $X \to A$, $\alpha \to 1$, $\beta \to 2$)) yields the following expression for A_{12}

$$A_{12} = A_{11}H_{12}g_{22} = g_{11}H_{12}A_{22} \tag{D.22}$$

Similarly, an expression for G_{12} is obtained by applying the two-component NNEs (14.26) to equation (D.13), to yield

$$G_{12} = G_{11}H_{12}C_{22} = B_{11}H_{12}G_{22} \tag{D.23}$$

Substituting the first of equations (D.22) into the second of equations (D.21) reproduces the first of equations (D.17), while substituting the first of equations (D.23) into the first of equations (D.21) yields

$$G_{13} = G_{11}(H_{13} + H_{12}C_{22}H_{23})g_{33} \tag{D.24}$$

The remaining two substitutions yield

$$G_{13} = (G_{11}H_{13} + B_{11}H_{12}G_{22}H_{23})g_{33} = (A_{11}H_{13} + g_{11}H_{12}A_{22}H_{23})G_{33} \tag{D.25}$$

For the third route, based on obtaining an expression for G_{1C}, applying the two-component NNEs (14.26) to equation (14.32), (with the change of notation, (i.e. $X \to G$, $g_{\alpha\alpha} \to C$, $\alpha \to C$, $\beta \to 1$)) yields

$$G_{1C} = G_{11}H_{1C}C = g_{11}H_{1C}G_{CC} \tag{D.26}$$

From equations (14.34), (14.35) and (14.38), these become

$$\begin{aligned} G_{1C} = (G_{12} \;\; G_{13}) &= G_{11}(H_{12} \;\; H_{13})\begin{pmatrix} C_{22} & C_{23} \\ C_{32} & C_{33} \end{pmatrix} \\ &= Gg_{11}(H_{12} \;\; H_{13})\begin{pmatrix} G_{22} & G_{23} \\ G_{32} & G_{33} \end{pmatrix} \end{aligned} \tag{D.27}$$

Hence the third pair of equations for G_{13} is

$$G_{13} = G_{11}(H_{12}C_{23} + H_{13}C_{33}) = g_{11}(H_{12}G_{23} + H_{13}G_{33}) \tag{D.28}$$

Again the story is not complete, because the right-hand sides of these equations contain off-diagonal Green's functions G_{23} and C_{23}. Applying the two-component

NNEs (14.26) to equation (14.38) (with the change of notation, (i.e. $X \to C$, $\alpha \to 2$, $\beta \to 3$)) yields

$$C_{23} = C_{22}H_{23}g_{33} = g_{22}H_{23}C_{33} \tag{D.29}$$

Similarly, an expression for G_{23} is obtained by applying the two-component NNEs (14.26) to equation (D.14), to yield

$$G_{23} = G_{22}H_{23}B_{33} = A_{22}H_{23}G_{33} \tag{D.30}$$

Substituting the second of equations (D.29) into the first of equations (D.28) reproduces the second of equations (D.17), while substituting the second of equations (D.30) into the second of equations (D.28) yields

$$G_{13} = g_{11}(H_{12}A_{22}H_{23} + H_{13})G_{33} \tag{D.31}$$

The remaining two substitution yields

$$G_{13} = G_{11}(H_{12}C_{22}H_{23}g_{33} + H_{13}C_{33}) = g_{11}(H_{12}G_{22}H_{23}B_{33} + H_{13}G_{33}) \tag{D.32}$$

In summary, the following equivalent equations for G_{13} have been obtained:

$$G_{13} = A_{11}\tilde{H}_{13}G_{33} = G_{11}\tilde{H}_{13}C_{33} \tag{D.17}$$

where

$$\tilde{H}_{13} = H_{13} + H_{12}g_{22}H_{23} \tag{D.12}$$

$$G_{13} = G_{11}\hat{\mathrm{H}}_{13}(C)g_{33} \tag{D.24}$$

where

$$\hat{\mathrm{H}}_{13}(C) = H_{13} + H_{12}C_{22}H_{23} \tag{D.33}$$

$$G_{13} = g_{11}\hat{\mathrm{H}}_{13}(A)G_{33} \tag{D.31}$$

where

$$\hat{\mathrm{H}}_{13}(A) = H_{13} + H_{12}A_{22}H_{23} \tag{D.34}$$

$$G_{13} = (G_{11}H_{13} + B_{11}H_{12}G_{22}H_{23})g_{33} = (A_{11}H_{13} + g_{11}H_{12}A_{22}H_{23})G_{33} \tag{D.25}$$

$$G_{13} = G_{11}(H_{12}C_{22}H_{23}g_{33} + H_{13}C_{33}) = g_{11}(H_{12}G_{22}H_{23}B_{33} + H_{13}G_{33}) \tag{D.32}$$

Also

$$G_{BB} = B + BH_{B2}G_{22}H_{2B}B \tag{D.15}$$

i.e.

$$G_{13} = B_{13} + \sum_{\substack{i=1,3 \\ j=1,3}} B_{1i}H_{i2}G_{22}H_{2j}B_{j3} \tag{D.16}$$

As an example, consider the case where $H_{13} = 0$, in which case, $B_{11} = g_{11}$, $B_{33} = g_{33}$, $B_{13} = 0$ and the above equations simplify to the following three-component NNEs, where equations (D.32) have been omitted, because they became identical to the first of equations (D.25) and to equation (D.31), while the second of equations (D.25) has been omitted, because it became identical to equation (D.31).

$$G_{13} = A_{11}H_{12}g_{22}H_{23}G_{33} \tag{D.17a}$$

$$G_{13} = g_{11}H_{12}A_{22}H_{23}G_{33} \tag{D.31}$$

$$G_{13} = G_{11}H_{12}C_{22}H_{23}g_{33} \tag{D.24}$$

$$G_{13} = G_{11}H_{12}g_{22}H_{23}C_{33} \tag{D.17b}$$

$$G_{13} = g_{11}H_{12}G_{22}H_{23}g_{33} \tag{D.25b}$$

In the above equations $G_{22} = \left(g_{22}^{-1} - \Sigma_{22}(1) - \Sigma_{22}(3)\right)^{-1}$

Appendix E The Green's function of a 5-component system

For completeness, the steps leading to the 5-component NNE are noted in this appendix. As discussed in chapter 11, to obtain the 5-component NNE, consider the case where

$$g_{11} = \begin{pmatrix} \tilde{g}_{LL} & \tilde{g}_{Ll} \\ \tilde{g}_{lL} & \tilde{g}_{ll} \end{pmatrix} = \begin{pmatrix} g_{LL}^{-1} & -H_{Ll} \\ -H_{lL} & g_{ll}^{-1} \end{pmatrix}^{-1} \tag{E.1}$$

and

$$g_{33} = \begin{pmatrix} \tilde{g}_{RR} & \tilde{g}_{Rr} \\ \tilde{g}_{rR} & \tilde{g}_{rr} \end{pmatrix} = \begin{pmatrix} g_{RR}^{-1} & -H_{Rr} \\ -H_{rR} & g_{rr}^{-1} \end{pmatrix}^{-1} \tag{E.2}$$

Consequently,

$$\tilde{g}_{Ll} = g_{LL}H_{Ll}\tilde{g}_{ll}, \text{ where } \tilde{g}_{ll} = (E - H_{ll} - H_{lL}g_{LL}H_{Ll})^{-1} \tag{E.3}$$

and

$$\tilde{g}_{rR} = \tilde{g}_{rr}H_{rR}g_{RR}, \text{ where } \tilde{g}_{rr} = (E - H_{rr} - H_{rR}g_{RR}H_{Rr})^{-1} \tag{E.4}$$

Then for the case where $H_{12} = \begin{pmatrix} 0 \\ H_{l2} \end{pmatrix}$ and $H_{23} = (0\ H_{2r})$, equation (D.25b) yields

$$G_{LR} = g_{LL} H_{Ll} \tilde{g}_{ll} H_{l2} G_{22} H_{2r} \tilde{g}_{rr} H_{rR} g_{RR} \tag{E.5}$$

Note that

$$\Sigma_{22}(3) = H_{23} g_{33} H_{32} = \begin{pmatrix} 0 & H_{2r} \end{pmatrix} \begin{pmatrix} \tilde{g}_{RR} & \tilde{g}_{Rr} \\ \tilde{g}_{rR} & \tilde{g}_{rr} \end{pmatrix} \begin{pmatrix} 0 \\ H_{r2} \end{pmatrix} = H_{2r} \tilde{g}_{rr} H_{r2} \tag{E.6}$$

and similarly

$$\Sigma_{22}(1) = H_{2l} \tilde{g}_{ll} H_{l2} \tag{E.7}$$

IOP Publishing

Quantum Transport in Nanostructures and Molecules

An introduction to molecular electronics

Colin John Lambert

Chapter 15

Relationship between Green's functions, molecular orbitals and densities of states

15.1 Introduction

In what follows, it will be shown that qualitative indicators of QI in isolated molecules can be obtained by examining their molecular orbitals. A second aim of this chapter is to highlight a relationship between Green's functions and the density of states of a quantum object, building on the discussion about Green's functions of closed systems in chapter 10.

15.2 The Coulson–Rushbrooke theorem

In section 10.2, it was noted that the Green's function of a finite-size Hamiltonian H is $g = (EI - H)^{-1}$, and f a real Hamiltonian describing a bipartite lattice is block off-diagonal and has form

$$H = \begin{pmatrix} 0 & h \\ h^t & 0 \end{pmatrix} \tag{15.1}$$

If $|\psi^{(E_n)}\rangle$ is a vector of amplitudes on even-numbered sites and $|\phi^{(E_n)}\rangle$ is a vector of amplitudes on odd-numbered sites, then the eigenstate satisfies the equation

$$\begin{pmatrix} 0 & h \\ h^t & 0 \end{pmatrix}\begin{pmatrix} |\psi^{(E_n)}\rangle \\ |\phi^{(E_n)}\rangle \end{pmatrix} = E_n\begin{pmatrix} |\psi^{(E_n)}\rangle \\ |\phi^{(E_n)}\rangle \end{pmatrix}$$

The top and bottom rows of this equation are

$$h|\phi^{(E_n)}\rangle = E_n|\psi^{(E_n)}\rangle \text{ and } h^t|\psi^{(E_n)}\rangle = E_n|\phi^{(E_n)}\rangle \tag{15.2}$$

doi:10.1088/978-0-7503-3639-0ch15

From equation (15.2)

$$\begin{pmatrix} 0 & h \\ h^t & 0 \end{pmatrix} \begin{pmatrix} |\psi^{(E_n)}\rangle \\ -|\phi^{(E_n)}\rangle \end{pmatrix} = E_n \begin{pmatrix} -|\psi^{(E_n)}\rangle \\ |\phi^{(E_n)}\rangle \end{pmatrix} = -E_n \begin{pmatrix} |\psi^{(E_n)}\rangle \\ -|\phi^{(E_n)}\rangle \end{pmatrix} \tag{15.3}$$

This demonstrates that eigenvalues of bipartite structures come in +/− pairs and the eigenstate belonging to $-E_n$ is related to the eigenstate belonging to E_n by

$$\begin{pmatrix} |\psi^{(-E_n)}\rangle \\ |\phi^{(-E_n)}\rangle \end{pmatrix} = \begin{pmatrix} |\psi^{(E_n)}\rangle \\ -|\phi^{(E_n)}\rangle \end{pmatrix} \tag{15.4}$$

Consequently eigenvectors come in +/− pairs, with eigenvalues $\pm E_n$ such that

$$|\psi^{(E_n)}\rangle = |\psi^{(-E_n)}\rangle \text{ and } |\phi^{(E_n)}\rangle = -|\phi^{(-E_n)}\rangle \tag{15.5}$$

Hence the amplitudes on sites a and b satisfy the relations

$$\psi_a^{(E_n)}\psi_b^{(E_n)} = \psi_a^{(-E_n)}\psi_b^{(-E_n)} \tag{15.6}$$

$$\phi_a^{(E_n)}\phi_b^{(E_n)} = \phi_a^{(-E_n)}\phi_b^{(-E_n)} \tag{15.7}$$

and

$$\psi_a^{(E_n)}\phi_b^{(E_n)} = -\psi_a^{(-E_n)}\phi_b^{(-E_n)} \tag{15.8}$$

Equations (15.6)–(15.8) are known as the Coulson–Rushbrooke (CR) theorem [1].

When a and b are both even the corresponding Green's function element is

$$g_{ab}(E) = \sum_{E_n>0} \left(\frac{\psi_a^{(E_n)}\psi_b^{(E_n)}}{E - E_n} + \frac{\psi_a^{(-E_n)}\psi_b^{(-E_n)}}{E + E_n} \right) \tag{15.9}$$

Hence at the middle of the HOMO–LUMO gap, $E = 0$,

$$g_{ab}(0) = \sum_{E_n>0} \left(\frac{\psi_a^{(-E_n)}\psi_b^{(-E_n)} - \psi_a^{(E_n)}\psi_b^{(E_n)}}{E_n} \right) = 0$$

where the last step follows from equation (15.6). Similarly, when a and b are both odd, equation (15.7) yields, $g_{ab}(0) = 0$. Hence the CR theorem reproduces the trivial zeros found in the diagonal blocks of the magic number tables of bipartite lattices. From the above viewpoint, these trivial zeros arise, because orbitals corresponding to energies $\pm E_n$ are guaranteed to interfere destructively in pairs.

On the other hand, when a is even and b is odd,

$$g_{ab}(0) = \sum_{E_n>0} \left(\frac{\psi_a^{(-E_n)}\phi_b^{(-E_n)} - \psi_a^{(E_n)}\phi_b^{(E_n)}}{E_n} \right)$$

That is

$$g_{ab}(0) = -2\sum_{E_n>0}\left(\frac{\psi_a^{(E_n)}\phi_b^{(E_n)}}{E_n}\right) \tag{15.10}$$

In general there is no particular relationship between terms such as $\frac{\psi_a^{(E_n)}\phi_b^{(E_n)}}{E_n}$. Some of these could be positive and some negative. If they add up to zero, then this would lead to a non-trivial zero in the corresponding magic number table and corresponds to a connectivity exhibiting destructive quantum interference (DQI). If they add up to a non-zero value, then the corresponding magic number is non-zero and this corresponds to a connectivity exhibiting constructive quantum interference (CQI).

The CR theorem is illustrated in figure 15.1, which shows the frontier orbitals of pyrene. (Blue indicates a positive amplitude and red indicates a negative amplitude). As an example, consider the 1,2 connectivity between even and odd sites. For the LUMO, the amplitudes on sites 1 and 2 have opposite signs (i.e. colours) and therefore their product is negative, whereas for the HOMO, the amplitudes on sites 1 and 2 have the same opposite signs (the same red colour) and therefore their product is positive. Hence the HOMO and LUMO products have opposite signs, in agreement with equation (15.8). Similarly, for the 1,6 connectivity, the HOMO and LUMO products have opposite signs, in agreement with equation (15.8), whereas for the 2,6 connectivity, the HOMO and LUMO products have the same sign, in agreement with equation (15.6).

15.3 An orbital symmetry rule

As mentioned in chapter 2, magic ratio theory is based on utilising the exact core Green's function, defined by $g(E_F) = (IE_F - H)^{-1}$. In the literature, various approximations to $g(E)$ are discussed, one of which involves the approximation of including only the contributions to $g(E)$ from the HOMO and LUMO.

If the amplitudes of the HOMO on sites a and b are denoted $\psi_a^{(E_H)}$ and $\psi_b^{(E_H)}$ and the amplitudes of the LUMO are $\psi_a^{(E_L)}$ and $\psi_b^{(E_L)}$, then if the contributions from all other orbitals are ignored, then as discussed in chapter 10, a crude approximation to the Green's function $g_{ba}(E)$ is

$$g_{ab}(E) \approx \frac{\psi_a^{(E_H)}\psi_b^{(E_H)}}{E - E_H} + \frac{\psi_a^{(E_L)}\psi_b^{(E_L)}}{E - E_L} \tag{15.11}$$

where E_H and E_L are the energies of the HOMO and LUMO respectively.

If the product HOMO product $\psi_b^H\psi_a^H$ has the same sign as the LUMO product $\psi_b^L\psi_a^L$, then the right-hand side of this equation will vanish at some energy E in the range $E_H \leqslant E \leqslant E_L$. That is for some energy E within the HOMO–LUMO gap. In this case, one can say that the HOMO and the LUMO interfere destructively. On the other hand, if the HOMO and LUMO products have opposite signs then the right-hand side of this equation will not vanish within the HOMO–LUMO gap and one

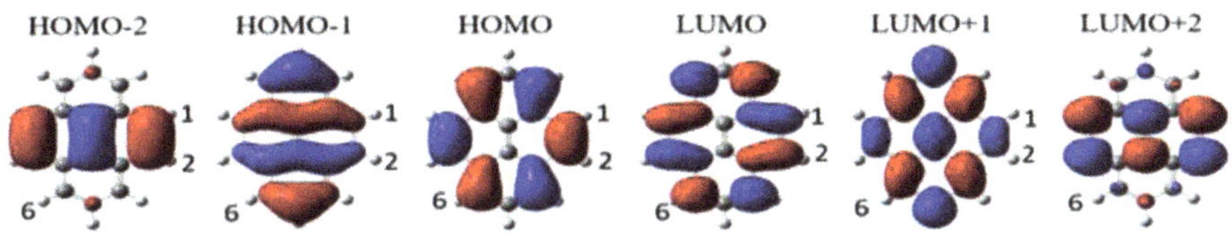

Figure 15.1. The wave functions of HOMO−2, HOMO−1, HOMO, LUMO, LUMO+1 and LUMO+2 levels in pyrene.

can say that the HOMO and LUMO interfere constructively within the gap. (They could of course interfere destructively at some other energy E outside the gap.)

When the right-hand side of (15.11) vanishes, the main contribution to $g_{ba}(E)$ comes from all other orbitals, so in general equation (15.11) could be a poor approximation. One exception to this occurs when the lattice is bipartite, because the CR theorem tells us that if a and b are both even or both odd, then the orbital products on opposite sides of equations (15.6) and (15.7) have the same sign. Consequently when the HOMO and LUMO interfere destructively, all other pairs of orbitals interfere destructively, leading to the trivial zeros in the magic number table, for which $g_{ab}(0) = 0$.

Obviously, this exact cancellation is a property of bipartite lattices only, but based on its success for bipartite lattices, one might suppose that equation (15.11) is a reasonable approximation, for other lattices. As discussed in section 10.3, this is not generally the case. Nevertheless, as pointed out by Yoshizawa and co-workers [2–5], since orbitals such as those in figure 15.1 are often available from DFT calculations, it can be helpful to examine the question of whether or not the HOMO and LUMO (or indeed any other pair of orbitals) interfere destructively or constructively, by examining the colours of orbitals. This is simplified by writing equation (15.11) in the form

$$g_{ab}(E) \approx \frac{a_{\rm H}}{E - E_{\rm H}} + \frac{a_{\rm L}}{E - E_{\rm L}} \tag{15.12}$$

where $a_{\rm H} = \psi_a^{(E_{\rm H})}\psi_b^{(E_{\rm H})}$ and $a_{\rm L} = \psi_a^{(E_{\rm L})}\psi_b^{(E_{\rm L})}$. If the HOMO product $a_{\rm H}$ has the same sign as the LUMO product $a_{\rm L}$ then the right-hand side of equation (15.12) will vanish for some energy E in the range $E_{\rm H} \leqslant E \leqslant E_{\rm L}$. In other words, the HOMO and LUMO will interfere destructively at some energy within the HOMO–LUMO gap. However this does not mean that the exact $g_{ab}(E)$ will vanish. Indeed, if the right-hand side of (15.12) vanishes, then the contributions from all other orbitals become the dominant terms [6]. Nevertheless, this is an appealing method of identifying QI effects in molecules and describing their qualitative features.

15.4 Densities of states

In the following sections, the notion of density of states will be discussed, followed by a derivation of their relationships to Green's functions. When the number of eigenvalues belonging to a quantum object is large, it is of interest to discuss their collective properties. For example consider the spectrum of eigenvalues λ_1, λ_2, ...λ_N, where for convenience, they are ordered, such that $\lambda_1 \leqslant \lambda_2 \leqslant \ldots \leqslant \lambda_N$. To describe their collective properties, one could for example compute their mean value $\langle \lambda \rangle$ defined by

$$\langle \lambda \rangle = \left(\frac{1}{N}\right)\sum_{n=1}^{N} \lambda_n \tag{15.13}$$

Their level spacings s_j may also be of interest. These are defined by $s_1 = \lambda_2 - \lambda_1$, $s_2 = \lambda_3 - \lambda_2$, ... etc, and the average level spacing

$$\langle s \rangle = \left(\frac{1}{N-1}\right)\sum_{n=1}^{N-1} s_n \tag{15.14}$$

More generally, one would like to describe the distribution of eigenvalues via a 'density of states' function $\rho(E)$, which is defined such that the number of eigenvalues between energies E_1 and E_2 (where $E_1 < E_2$) is given by the integral $\int_{E_1}^{E_2} dE\rho(E)$. The following qualitative discussion shows how the idea of a density of states emerges, when one considers building histograms of eigenvalues.

First note that eigenvalues could be displayed by simply plotting them along a horizontal line, as shown in the top panel of figure 15.2. Clearly the eigenvalues in this figure are more closely spaced at the lower and upper ends of the spectrum and have a larger spacing in the middle of the spectrum. An alternative way of showing how the eigenvalues are distributed is to create a histogram of eigenvalues, as shown in the lower panel of figure 15.2. To construct the histogram, a minimum energy $E_{\min}$ and a maximum energy $E_{\max}$ should be identified, such that $E_{\min} \leqslant \lambda_1$ and $E_{\max} \geqslant \lambda_N$. Next the number of histogram bins M should be chosen and the width of each bin defined to be $\Delta E = (E_{\max} - E_{\min})/M$. For example, in figure 15.2, $M = 5$ and centre of the jth bin is located at an energy $E_j = E_{\min} + \left(j - \frac{1}{2}\right)\Delta E$, with $j = 1, 2, \ldots M$. In figure 15.2, $E_{\min} = -2$, $E_{\max} = 2$, $\Delta E = 0.4$, $E_1 = -1.6$, $E_2 = -0.8$, ... etc.

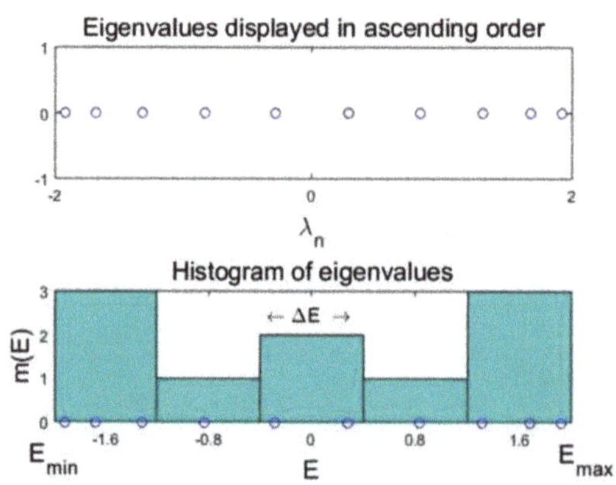

Figure 15.2. The top panel shows 10 eigenvalues, plotted along a horizontal line. The bottom panel shows a histogram of these eigenvalues.

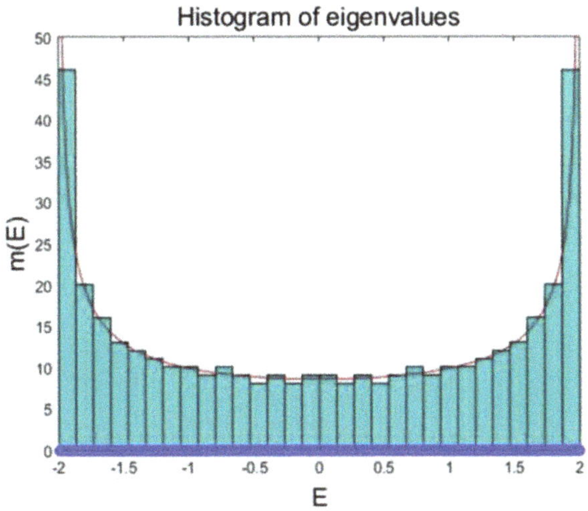

Figure 15.3. A histogram of the eigenvalues of equation (15.15).

If $m(E_j)$ is defined to be the number of eigenvalues contained within the bin centred on energy E_j, then in figure 15.3, $m(E_1) = 3$, $m(E_2) = 1$, ... etc. A histogram of this type is rather coarse grained, because the number of eigenvalues in each bin depends on the width of the bin and it contains no information about the fine-grained location of eigenvalues within a bin. The shape of the histogram is also somewhat arbitrary, because it depends on the choice of the number of bins. Ideally, one would like to choose the bin width to be as small as possible, to increase the resolution of the histogram. However, if the number of bins is chosen to be too large and their width too small, then each bin would either contain 1 or 0 eigenvalues and the histogram would become a series of isolated rectangles of height 1. To avoid this, one usually adopts a 'Goldilocks principle,' in which the bin widths are large enough to capture a reasonable number of eigenvalues in each bin, but small enough to achieve a reasonable resolution. Clearly these criteria are rather subjective.

Now consider what happens to $m(E)$ as the number of eigenvalues becomes very large. As an example, consider a linear chain of N tight-binding sites, with site energies $\epsilon_0 = 0$ and nearest neighbour couplings $-\gamma = -1$. The eigenvalues of the chain are given by $E = \epsilon_0 - 2\gamma \cos k$ and therefore with this choice of ϵ_0 and γ, lie in the range -2 to $+2$. To avoid clashes of notation, these eigenvalues will be denoted λ_n, so that

$$\lambda_n = -2 \cos k_n \tag{15.15}$$

where

$$k_n = n\pi/(N + 1), \tag{15.16}$$

with $n = 1, 2, \ldots .N$. With this notation, the group velocity associated with electrons in such a chain is $\hbar v = \frac{d\lambda}{dk} = 2 \sin k$. For large N, the histogram of figure 15.3 evolves into the histogram shown in figure 15.3. Following the 'Goldilocks principle', the histogram has been constructed by increasing the number of bins and decreasing their width, to improve the resolution of the histogram, whilst keeping the bin widths large enough to capture several eigenvalues in each bin.

In the limit of large N, the histogram tends to the smooth red curve shown in figure 15.3. To obtain an analytic expression for this curve, note that the number of k values between k and $k + \Delta k$ is $\Delta n = \frac{N+1}{\pi}\Delta k$. Furthermore in the large N limit, the spacing between eigenvalues tends to zero, so the number of eigenvalues between λ and $\lambda + \Delta\lambda$ is

$$\Delta n = \frac{N+1}{\pi}\left(\frac{dk}{d\lambda}\right)\Delta\lambda = \frac{N+1}{\pi}\left(\frac{1}{d\lambda/dk}\right)\Delta\lambda \tag{15.17}$$

Hence

$$\Delta n = \frac{N+1}{\pi}\left(\frac{1}{2\sin k}\right)\Delta\lambda = \frac{N+1}{\pi}\left(\frac{1}{2\sqrt{1-\cos^2 k}}\right)\Delta\lambda$$

That is $\Delta n = \frac{N+1}{\pi}\left(\frac{1}{\sqrt{4-\lambda^2}}\right)\Delta\lambda$. This means that the number of eigenvalues in a bin between E_j and $E_j + \Delta E$ (which in the limit of small ΔE is approximately the same as the number between $E_j - \Delta E/2$ and $E_j + \Delta E/2$) is

$$m(E_j) = \frac{N+1}{\pi}\left(\frac{1}{\sqrt{4-E_j^2}}\right)\Delta E \tag{15.18}$$

In the limit of large N, ΔE can be made very small, whilst keeping a non-zero percentage of eigenvalues in each bin finite, resulting in the red curve in figure 15.3, which is a plot of equation (15.18). Note that in the large N limit, this shows that the number of eigenvalues in a histogram bin is proportional to the width ΔE of the bin. To remove this dependence, it is useful to define the quantity $\rho(E_j)$ defined by

$$\rho(E_j) = \frac{m(E_j)}{\Delta E} \tag{15.19}$$

$\rho(E_j)$ is the number of eigenvalues per unit energy and in the limit $\Delta E \to 0$, is referred to as 'the density of states'.

15.5 van Hove singularities

Note that the dispersion relation in equation (15.15) is $\lambda = -2\cos k$ and has an upper band edge of $\lambda = 2$ at $k = \pi$ and a lower band edge of $\lambda = -2$ at $k = 0$. Since the group velocity $\frac{d\lambda}{dk} = 2\sin k = 0$ at these two band edges, i.e. when $\lambda^2 = 4$, this shows that $m(E_j)$ and $\rho(E_j)$ diverge at the band edges. These divergences are called 'van Hove singularities' and are a necessary feature of the densities of states of one-dimensional chains, because the dispersion relation $\lambda(k)$ of an infinite chain is a periodic function of k, with period 2π. Hence a graph of $\lambda(k)$ versus k has maxima and minima, at which $\frac{d\lambda}{dk} = 0$. Since the right-hand side of equation (15.17) contains $\frac{d\lambda}{dk}$ in the denominator, this means that the density of states must diverge at the band edges of one-dimensional chains.

15.6 A more precise form of the density of states

Having considered how the Goldilocks principle leads to a smooth function $\rho(E)$ in the large N limit, the precise form of $\rho(E)$ will now be discussed, when the number of eigenvalues is finite. So far, the discussion has focussed on the number of eigenvalues in bin j, namely $m(E_j)$, where E_j denotes the centre of bin j. To obtain a more precise definition of the density of states, let us go back to figure 15.2 and consider an alternative method of creating a histogram, using rectangular functions such as $f_n(E)$, shown in figure 15.4, where $f_n(E)$ is defined to be a rectangular function of height 1 and width ΔE centred on eigenvalue λ_n.

Now define a function $m(E)$ of the continuous variable E to be a sum of such rectangular functions:

$$m(E) = \sum_{n=1}^{N} f_n(E) \qquad (15.20)$$

Figure 15.5 shows plots of $m(E)$ for various rectangle widths ΔE. As ΔE increases, the lower figure begins to resemble figure 15.3. Indeed, for a large number of eigenvalues and an appropriate choice of bin width, figure 15.3 would be obtained.

Similarly, the coarse-grained density of states denoted $\hat{\rho}(E)$ can be written

$$\hat{\rho}(E) = \sum_{n=1}^{N} \delta_n(E), \qquad (15.21)$$

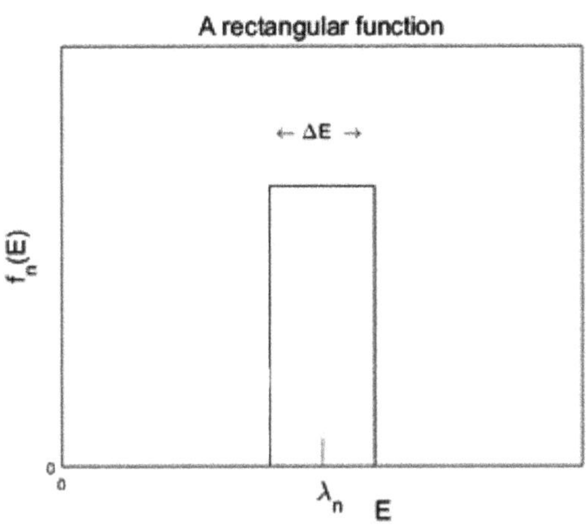

Figure 15.4. A rectangular function of width ΔE centred on eigenvalue λ_n.

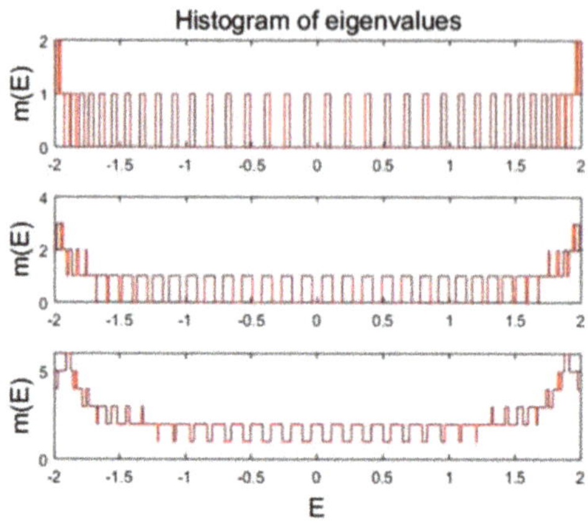

Figure 15.5. Plots of the function $m(E)$ for various rectangle widths ΔE.

where

$$\delta_{\mathrm{n}}(E) = \frac{f_n(E)}{\Delta E} \tag{15.22}$$

$\delta_{\mathrm{n}}(E)$ is a rectangular function of height $1/\Delta E$ and width ΔE centred on λ_n. The area under such a rectangle is base x height = 1. Therefore

$$\int_{\lambda_n - \epsilon}^{\lambda_n + \epsilon} dE\ \delta_{\mathrm{n}}(E) = 1 \tag{15.23}$$

where ϵ is any number greater than $\Delta E/2$.

The exact density of states $\rho(E)$ is obtained from the coarse-grained density of states $\hat{\rho}(E)$ by taking the limit of infinitesimally small bin widths. The top panel of figure 15.5 shows that for small enough ΔE, the function $m(E)$ is a series of rectangles of height 1 and area ΔE, centred on the eigenvalues λ_n. Similarly, for small enough ΔE, $\rho(E)$ is a series of rectangles of height $1/\Delta E$, area 1, centred on the eigenvalues.

In the limit $\Delta E \to 0$, $\delta_{\mathrm{n}}(E)$ is an infinitely thin rectangle of unit area, centred on λ_n. Such a function is called a delta function and denoted $\delta(E - \lambda_n)$. That is

$$\lim \Delta E \to 0\ \delta_{\mathrm{n}}(E) = \delta(E - \lambda_n) \tag{15.24}$$

From equation (15.23)

$$\int_{\lambda_n - \epsilon}^{\lambda_n + \epsilon} dE\ \delta(E - \lambda_n) = 1 \tag{15.25}$$

where ϵ is any finite positive number, which could be chosen to be infinitesimally small.

In this limit, combining equations (15.21) and (15.23) and taking the limit $\Delta E \to 0$, yields

$$\int_{E_1}^{E_2} dE\rho(E) = \sum_{n=1}^{N} \int_{E_1}^{E_2} dE\ \delta(E - \lambda_n), \tag{15.26}$$

If the eigenvalue λ_n lies within the interval E_1 to E_2, $\int_{E_1}^{E_2} dE\ \delta(E - \lambda_n) = 1$. Hence the right-hand side of this equation is simply equal to the number of eigenvalues in the interval E_1 to E_2. This shows that, in contrast with the coarse-grained density of states of equation (15.17), the exact density of states, is given by

$$\boxed{\rho(E) = \sum_{n=1}^{N} \delta(E - \lambda_n)} \tag{15.27}$$

Equation (15.27) is the precise definition of the density of states and does not rely on an imprecise Goldilocks principle for its construction. Clearly the quantity

$$\boxed{N(E) = \int_{-\infty}^{E_2} dE\rho(E)} \tag{15.28}$$

is equal to the number of eigenvalues between $-\infty$ and E_2. The quantity $N(E_2)$ is the number of eigenvalues less than E_2 and is referred to as the 'integrated density of states'.

15.7 Properties of delta functions

Since the definition (15.27) involves delta functions, these quantities will now be considered in more detail. A delta function is any quantity that satisfies equation (15.28).i.e.

$$\int_{\lambda-\epsilon}^{\lambda+\epsilon} dE\ \delta(E - \lambda) = 1 \tag{15.29}$$

where ϵ is any finite positive number, which could be chosen to be infinitesimally small.

In addition to (15.29), one can also show that for any smooth function $f(E)$,

$$\int_{\lambda-\epsilon}^{\lambda+\epsilon} dE\ f(E)\delta(E - \lambda) = f(\lambda) \tag{15.30}$$

This is proved by approximating $\delta(E - \lambda)$ by a rectangle of unit area and width ΔE, centred on λ so that (taking $\epsilon > \Delta E$) the left-hand side of (15.29) becomes

$$\frac{1}{\Delta E} \int_{\lambda-\Delta E/2}^{\lambda+\Delta E/2} dE\ f(E)$$

Writing $f(E) = f(\lambda) + \frac{df(\lambda)}{d\lambda}(E - \lambda) + \frac{1}{2}\frac{d^2 f(\lambda)}{d\lambda^2}(E - \lambda)^2 \ldots$ and noting that $\frac{1}{\Delta E}\int_{\lambda-\Delta E/2}^{\lambda+\Delta E/2} dE(E - \lambda)^n$ is equal to 1 for $n = 1$ and is of order $(\Delta E)^n$ for $n > 1$. Since the latter tends to zero as $\Delta E \to 0$, equation (15.30) is obtained.

From the discussion above, one might be tempted to believe that a delta function such as $\delta(E - \lambda)$ is a function which equals ∞ at $E = \lambda$ and equals zero for $E \neq \lambda$. However, because $\delta(E - \lambda)$ is defined to be any function which satisfies equation (15.29), the latter suggestion is not correct and in fact $\delta(E - \lambda)$ could take any value when $E \neq \lambda$. Consequently $\delta(E - \lambda)$ is not a function, because its value for $E \neq \lambda$ is not defined.

To demonstrate this point, consider the function shown in figure 15.7(a), which is formed from a rectangle of unit area, width η, height $1/\eta$ centred on $E = \lambda$ and also two other rectangles of width η and heights 1 and 2 respectively. Figure 15.7(b) shows the same function, but plotted with a smaller value of η. The integral with respect to E (i.e. the area) of the first rectangle is 1, independent of η, whereas the areas of the latter two rectangles are η and 2η respectively. Therefore their integrals over E (i.e. their areas) vanish as η decreases towards zero. Consequently, as $\eta \to 0$, this function satisfies equation (15.29), because the integral of the first rectangle remains at unity, whereas the other rectangles become infinitely narrow and do not contribute to the integral. Figure 9.5(c) shows another example, in which two further rectangles of heights 4 and 6 and width η have been added. Again these do not contribute to integral in equation (15.29) as $\eta \to 0$, so this function is a delta function.

These examples demonstrates that the value of $\delta(E - \lambda)$ tends to ∞ at $E = \lambda$. On the other hand, when E coincides with the centres of the four other rectangles in figure 15.7(c), $\delta(E - \lambda)$ is equal to 1, 4, 6 or 2, even in the limit $\eta = 0$. Clearly an arbitrarily large number of finite-height rectangles of width η could be included and they would not change the validity of equation (15.29). On the other hand, when E coincides with the centre of one of these arbitrary rectangles, the value of $\delta(E - \lambda)$ would be equal to an arbitrary non-zero number. Consequently the numerical value of $\delta(E - \lambda)$ is not defined by equation (15.29) and therefore $\delta(E - \lambda)$ is not a function. Instead $\delta(E - \lambda)$ is called a 'distribution'. Indeed any entity which is defined only by its action under an integral, as in equations (15.29) and (15.30), is called a distribution.

Since $\delta(E - \lambda)$ is a distribution, rather than a function, equation (15.27) should not be taken literally, because the right-hand side is not a function and does not have a value at any chosen E. Equation (15.27) simply means that the integral with respect to E of the left-hand side is equal to the integral with respect to E of the right-hand side. For this reason, the equality sign should be replaced by some other symbol, such as $\doteq$, to remind us that this is not an equation. With this notation, equation (15.27) should be replaced by

$$\rho(E) \doteq \sum_{n=1}^{N} \delta(E - \lambda_n) \tag{15.31}$$

In the literature, this replacement is not usually made; one simply remembers that any equation containing a delta function is not an equation and merely implies that the integral of the left-hand side is equal to the integral of the right-hand side.

15.8 Examples of delta functions

A delta function $\delta(E - \lambda)$ is a distribution, which satisfies equations (15.29) and (15.30) and can be considered to be the limit of a broadened delta function such as that in figure 15.6.

Other examples of broadened delta functions are

$$\delta_\eta(E - \lambda) = \frac{1}{\pi}\frac{\eta}{((E - \lambda)^2 + \eta^2)} \tag{15.32}$$

$$p_\eta(E - \lambda) = \frac{1}{\sqrt{2\pi\eta^2}}\,\exp-\frac{(E - \lambda)^2}{2\eta^2} \tag{15.33}$$

$$q_\eta = \frac{1}{\pi}\frac{\sin[(E - \lambda)/\eta]}{(E - \lambda)} \tag{15.34}$$

These are shown in figure 15.7. They all have unit area under the curve and all tend to $\delta(E - \lambda)$ as $\eta \to 0$.

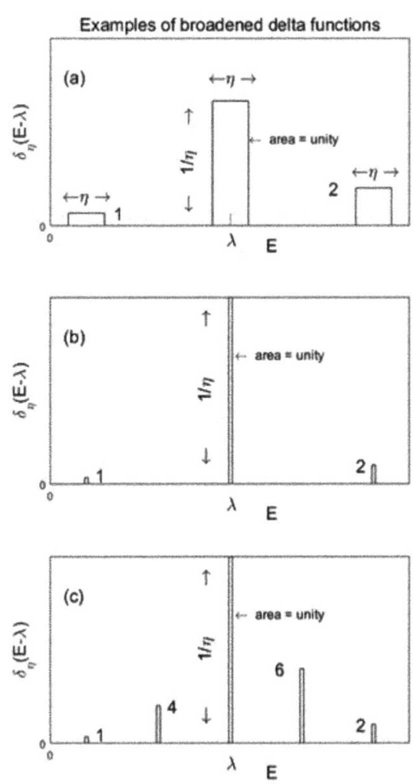

Figure 15.6. Examples of broadened delta functions.

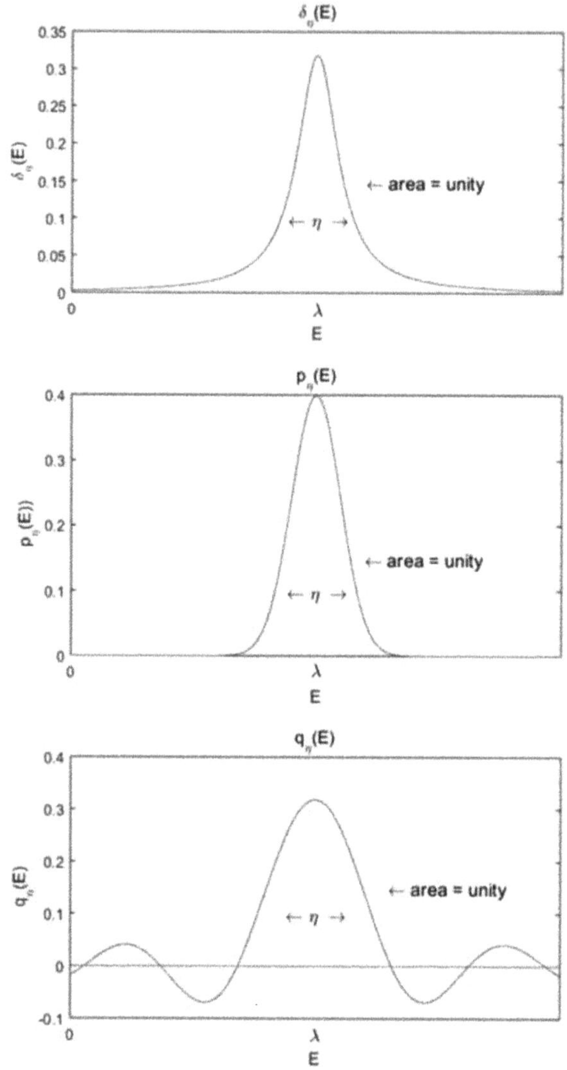

Figure 15.7. Examples of broadened delta functions.

Another example of a broadened delta function, obtained by differentiating the Fermi function, as follows:

$$\mathrm{F}_{k_\mathrm{B}T}(E - E_\mathrm{F}) = -\frac{df(E)}{dE} \tag{15.35}$$

where $f(E)$ is the Fermi function given by

$$f(E) = \frac{1}{\exp(E - E_\mathrm{F})/k_\mathrm{B}T + 1} \tag{15.36}$$

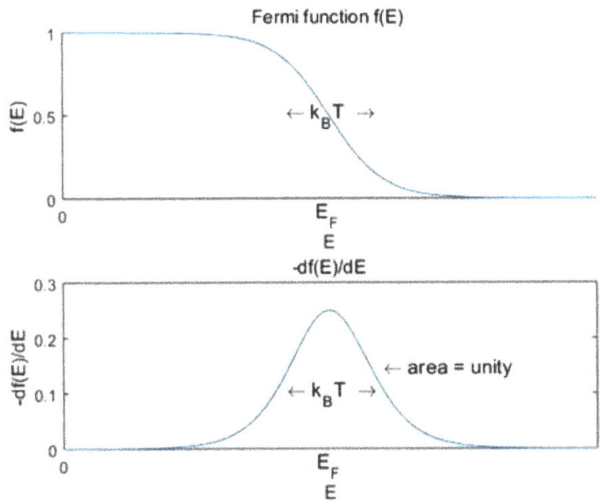

Figure 15.8. The Fermi function and its derivative.

In this case, the broadened delta function is

$$F_{k_BT}(E - E_F) = \frac{4}{k_BT\left(\cosh\frac{x}{2}\right)^2} \tag{15.37}$$

where $x = (E - E_F)/k_BT$. The functions (15.36) and (15.37) are shown in figure 15.8. Again the latter has unit area under the curve and tends to $\delta(E - \lambda)$ as $k_BT \to 0$.

Any of these broadened delta functions can be used to approximate the right-hand side of equation (15.27). For example, one could write

$$\rho(E) \approx \rho_\eta(E) \tag{15.38}$$

where

$$\rho_\eta(E) = \sum_{n=1}^{N} \delta_\eta(E - \lambda_n) \tag{15.39}$$

In this equation $\rho_\eta(E)$ is a sum of broadened density of states, which tends to $\rho(E)$ as $\eta \to 0$.

15.9 Relationship between Green's functions and densities of states

To obtain a relationship between Green's functions and densities of states, it is useful to note that

$$\frac{1}{E - \lambda + i\eta} = \frac{E - \lambda - i\eta}{(E - \lambda)^2 + \eta^2}$$

Comparing this with equation (15.32) yields

$$-\frac{1}{\pi}\text{Im}\left[\frac{1}{E-\lambda+i\eta}\right]=\delta_\eta(E-\lambda) \tag{15.40}$$

The Green's function $G(z)$ of a closed system is

$$G_{jl}(z)=\sum_{n=1}^{N}\frac{\psi_n(j)\psi_n^*(l)}{z-\lambda_n} \tag{15.41}$$

Therefore if z is chosen to be $z = E + i\eta$ and $j = l$, one finds

$$-\frac{1}{\pi}\text{Im }G_{jj}(E+i\eta)=\sum_{n=1}^{N}|\psi_n(j)|^2\delta_\eta(E-\lambda_n) \tag{15.42}$$

Since the eigenstates are normalised,

$$\sum_{j=1}^{N}|\psi_n(j)|^2=1$$

and one obtains

$$-\frac{1}{\pi}\text{Im }\sum_{j=1}^{N}G_{jj}(E+i\eta)=\sum_{n=1}^{N}\delta_\eta(E-\lambda_n)=\rho_\eta(E) \tag{15.43}$$

This relates the broadened density of states of equation (15.26) to the imaginary part of the Green's function. Hence one finds that the density of states is given by

$$\rho(E)=\text{Lim }\eta\to 0\left\{-\frac{1}{\pi}\text{Im }\sum_{j=1}^{N}G_{jj}(E+i\eta)\right\} \tag{15.44}$$

An alternative way of writing this is

$$\rho(E)=\sum_{j=1}^{N}\rho_{jj}(E) \tag{15.45}$$

where

$$\rho_{jj}(E)=\text{Lim }\eta\to 0\left\{-\frac{1}{\pi}\text{Im }G_{jj}(E+i\eta)\right\} \tag{15.46}$$

$\rho_{jj}(E)$ is called the partial density of states and is defined by

$$\rho_{jj}(E)=\sum_{n=1}^{N}|\psi_n(j)|^2\delta(E-\lambda_n) \tag{15.47}$$

Clearly $\int_{E_1}^{E_2} dE\rho_{jj}(E) = \sum_n |\psi_n(j)|^2$, where the sum includes only those eigenstates whose eigenvalues lie in the range E_1 to E_2. If all of these states are occupied by electrons, then it is equal to the probability that an electron from these states could be found of site j.

15.10 Summary

In this chapter, a detailed discussion about densities of states was presented and the relationship between Green's functions and densities of states was examined. The Coulson–Rushbrooke pairing theorem was also discussed, and an orbital symmetry rule highlighted.

References

[1] Coulson C A and Rushbrooke G S 1940 Note on the method of molecular orbitals *Math. Proc. Cambridge Philos. Soc.* **36** 193

[2] Yoshizawa K 2012 An orbital rule for electron transport in molecules *Acc. Chem. Res.* **45** 1612–21

[3] Tada T and Yoshizawa K 2015 Molecular design of electron transport with orbital rule: toward conductance-decay free molecular junctions *Phys. Chem. Chem. Phys.* **17** 32099

[4] Tsuji Y and Yoshizawa K 2017 Frontier orbital perspective for quantum interference in alternant and nonalternant hydrocarbons *J. Phys. Chem.* C **121** 9621–6

[5] Okazawa K, Tsuji Y, Yuta and Yoshizawa K 2020 Understanding single-molecule parallel circuits on the basis of Frontier orbital theory *J. Phys. Chem.* C **124** 3322

[6] Zhao X, Geskin V and Stadler R 2017 Destructive quantum interference in electron transport: A reconciliation of the molecular orbital and the atomic orbital perspective *J. Chem. Phys.* **146** 092308

IOP Publishing

Quantum Transport in Nanostructures and Molecules
An introduction to molecular electronics
Colin John Lambert

Chapter 16

Solving the time-dependent Schrödinger equation and the theory of representations

16.1 Introduction

In chapter 3, equation (3.2) the following matrix representation of the time-dependent Schrödinger equation was introduced

$$i\hbar\frac{d}{dt}|\varphi(t)\rangle = H|\varphi(t)\rangle \tag{16.1}$$

In section 3.9, this was solved by writing

$$|\varphi(t)\rangle = \sum\nolimits_{n=1}^{N} d_n(t)|\psi_n\rangle \tag{16.2}$$

where $|\psi_n\rangle$ is a N-dimensional vector of the form

$$|\psi_n\rangle = \begin{pmatrix} \psi_1^{(n)} \\ \psi_2^{(n)} \\ \vdots \\ \psi_N^{(n)} \end{pmatrix} \tag{16.3}$$

chosen to satisfy

$$H|\psi_n\rangle = E_n|\psi_n\rangle \tag{16.4}$$

This led to the solution

$$d_n(t) = d_n(0)\exp(-iE_nt/\hbar) \tag{16.5}$$

where the constant of integration $d_n(0)$ is determined by the wave function at time $t = 0$, via the equation

doi:10.1088/978-0-7503-3639-0ch16

$$d_n(0) = \langle\psi_n|\varphi(0)\rangle \tag{16.6}$$

As discussed in chapter 3, the eigenvectors $|\psi_n\rangle$ can be chosen to form a complete orthonormal set, satisfying

$$\langle\psi_m|\psi_n\rangle = \delta_{nm} \text{ and } \sum_{n=1}^{N}|\psi_n\rangle\langle\psi_n| = I \tag{16.7}$$

where I is the $N \times N$ unit matrix.

The aim of this chapter is to discuss the origin of equation (16.1), starting from the time-dependent Schrödinger equation written as a partial differential equation, namely

$$i\hbar\frac{\partial\Psi(x, t)}{\partial t} = H(x)\Psi(x, t) \tag{16.8}$$

where $H(x)$ is a differential operator.

To solve equation (16.8), it is useful start with a brief introduction to representation theory.

16.2 The theory of representations

Consider solving a mathematical problem for some function $\psi(x)$ and obtaining an answer of the form

$$\psi(x) = a_1 \cos x + a_2 x^2 \tag{16.9}$$

In this equation, a_1 is the amplitude of the term $\cos x$ and a_2 is the amplitude of the term x^2. If a_1 and a_2 are presented in the form of a vector, as follows

$$|\psi\rangle = \begin{pmatrix} a_1 \\ a_2 \end{pmatrix}, \tag{16.10}$$

then one says that the vector $|\psi\rangle$ is a representation of the function $\psi(x)$. Of course to obtain the function $\psi(x)$ from the vector $|\psi\rangle$, equation (16.9) is needed. In other words, one needs to know that a_1 is multiplied into $\cos x$ and a_2 is multiplied into x^2. The functions $\cos x$ and x^2 are known as the basis functions for this representation of $\psi(x)$. More generally, an arbitrary function $\psi(x)$ could be written as a linear superposition of N linearly-independent functions

$$\psi(x) = \sum_{j=1}^{N} a_j\rho_j(x) \tag{16.11}$$

In this case, the vector representing $\psi(x)$ is an N-dimensional vector of the form

$$|\psi\rangle = \begin{pmatrix} a_1 \\ a_2 \\ \vdots \\ a_N \end{pmatrix} \tag{16.12}$$

and the linearly-independent functions $\rho_j(x)$ are the basis functions.

16.3 Representations involving orthonormal basis functions

Consider introducing an arbitrary orthonormal set of basis functions $\{f_j(x)\}$ satisfying

$$\int_{-\infty}^{\infty} dx\, f_i^*(x) f_j(x) = \delta_{ij} \tag{16.13}$$

and writing

$$\Psi(x,\, t) = \sum_{j=1}^{N} \varphi_j(t) f_j(x) \tag{16.14}$$

Substituting this into equation (16.8) yields

$$\sum_{j=1}^{N} i\hbar \frac{d\varphi_j(t)}{dt} f_j(x) = \sum_{j=1}^{N} \varphi_j(t) H(x) f_j(x)$$

Multiplying from the left by $f_i^*(x)$, integrating with respect to x and taking advantage of equation (16.13) yields

$$i\hbar \frac{d\varphi_i(t)}{dt} = \sum_{j=1}^{N} H_{ij} \varphi_j(t) \tag{16.15}$$

where

$$H_{ij} = \int_{-\infty}^{\infty} dx\, f_i^*(x) H(x) f_j(x) \tag{16.16}$$

That is

$$i\hbar \frac{d}{dt} |\varphi(t)\rangle = H |\varphi(t)\rangle \tag{16.17}$$

where

$$|\varphi(t)\rangle = \begin{pmatrix} \varphi_1(t) \\ \varphi_2(t) \\ \vdots \\ \varphi_N(t) \end{pmatrix} \tag{16.18}$$

Equation (16.17) is identical to equation (16.1), with H given by equation (16.16). In the above equations, $|\varphi(t)\rangle$ is a representation of $\Psi(x,\, t)$ and $\{f_j(x)\}$ is the set of basis functions.

One particularly useful set of basis function is the set of linearly-independent, orthonormal eigenstates $\{\psi_n(x)\}$ of $H(x)$, obtained by solving the differential equation

$$H(x)\psi_n(x) = E_n \psi_n(x) \tag{16.19}$$

and

$$\int_{-\infty}^{\infty} dx\ \psi_n^*(x)\psi_m(x) = \delta_{nm} \tag{16.20}$$

When $\Psi(x, t)$ is written as a linear superposition of these eigenstates, it takes the form

$$\Psi(x, t) = \sum_{n=1}^{N} d_n(t)\psi_n(x) \tag{16.21}$$

where the vector $|d(t)\rangle = \begin{pmatrix} d_1(t) \\ d_2(t) \\ \vdots \\ d_N(t) \end{pmatrix}$ is a representation of $\Psi(x, t)$ and the functions $\psi_n(x)$ are the basis functions. In this representation, the Hamiltonian matrix will be denoted D, with matrix elements D_{nm} given by

$$D_{nm} = \int_{-\infty}^{\infty} dx\ \psi_n^*(x)H(x)\psi_m(x) = E_n\delta_{nm} \tag{16.22}$$

That is D is a diagonal matrix, with the eigenvalues of H along the diagonal and takes the form

$$D = \begin{pmatrix} E_1 & \dots & 0 \\ \vdots & E_n & \vdots \\ 0 & \dots & E_N \end{pmatrix} \tag{16.23}$$

In this representation, equation (16.17) is replaced by

$$i\hbar\frac{d}{dt}\begin{pmatrix} d_1(t) \\ d_2(t) \\ \vdots \\ d_N(t) \end{pmatrix} = \begin{pmatrix} E_1 & \dots & 0 \\ \vdots & E_n & \vdots \\ 0 & \dots & E_N \end{pmatrix}\begin{pmatrix} d_1(t) \\ d_2(t) \\ \vdots \\ d_N(t) \end{pmatrix} = \begin{pmatrix} E_1d_1(t) \\ E_2d_2(t) \\ \vdots \\ E_Nd_N(t) \end{pmatrix}$$

The nth row of the above equation is

$$i\hbar\frac{d}{dt}d_n(t) = E_nd_n(t) \tag{16.24}$$

The solution to equation (16.24) is given by equation (16.5), where from equations (16.20) and (16.21), $d(0)$ is a constant of integration, given by

$$d_n(0) = \int_{-\infty}^{\infty} dx\ \psi_n^*(x)\Psi(x, 0) \tag{16.25}$$

This equation shows that the values of the amplitudes $d_n(t)$ at $t = 0$ are determined by the wave function $\Psi(x, 0)$ at $t = 0$.

To complete the story, the time-independent differential equation (16.19) must be solved and for this purpose, it is useful to now introduce the above set of arbitrary basis functions $\{f_j(x)\}$ and express the solution to equation (16.19) as a linear superposition of these functions by writing

$$\psi_n(x) = \sum_{j=1}^{N} \psi_j^{(n)} f_j(x) \tag{16.26}$$

Substitute equation (16.26) into (16.19), then multiply both sides from the left by $f_i^*(x)$ and integrate over x to yield

$$\sum_{j=1}^{N} H_{ij} \psi_j^{(n)} = E_n \psi_i^{(n)} \tag{16.27}$$

where

$$H_{ij} = \int_{-\infty}^{\infty} dx\, f_i^*(x) H(x) f_j(x) \tag{16.28}$$

Equation (16.27) is equivalent to

$$H|\psi_n\rangle = E_n\, |\psi_n\rangle \tag{16.29}$$

where H is the $N \times N$ square matrix, whose matrix elements are given by equation (16.28) and $|\psi_n\rangle$ is a vector of amplitudes:

$$|\psi_n\rangle = \begin{pmatrix} \psi_1^{(n)} \\ \psi_2^{(n)} \\ \vdots \\ \psi_N^{(n)} \end{pmatrix} \tag{16.30}$$

The matrix H, with matrix elements H_{ij} is the Hamiltonian matrix in the representation $\{f_j(x)\}$.

Similarly, the matrix D with matrix elements

$$D_{nm} = \int_{-\infty}^{\infty} dx\, \psi_n^*(x) H(x) \psi_m(x) \tag{16.31}$$

is the Hamiltonian matrix in the representation $\{\psi_n(x)\}$.

From equations (16.26), (16.28) and (16.30)

$$D_{nm} = \sum_{i,j=1}^{N} \psi^{*}{}_i^{(n)} \psi_j^{(m)} \int_{-\infty}^{\infty} dx\, f_j^*(x) H(x) f_j(x) = \sum_{i,j=1}^{N} \psi^{*}{}_i^{(n)} H_{ij} \psi_j^{(m)} \tag{16.32}$$

Hence defining the matrix U with matrix elements $U_{in} = \psi_i^{(n)}$ and the Hermitian conjugate $U^\dagger$ with matrix elements $(U^\dagger)_{ni} = U_{in}^* = \psi^{*}{}_i^{(n)}$, yields

$$D_{nm} = \sum_{i,j=1}^{N} (U^\dagger)_{ni} H_{ij} U_{jm} \tag{16.33}$$

That is

$$D = U^\dagger H U \tag{16.34}$$

where

$$U = \begin{pmatrix} \psi_1^{(1)} & \dots & \psi_1^{(N)} \\ \vdots & \vdots & \vdots \\ \psi_N^{(1)} & \dots & \psi_N^{(N)} \end{pmatrix} \quad \text{and} \quad U^\dagger = \begin{pmatrix} \psi^{*}{}_1^{(1)} & \dots & \psi^{*}{}_N^{(1)} \\ \vdots & \vdots & \vdots \\ \psi^{*}{}_1^{(N)} & \dots & \psi^{*}{}_N^{(N)} \end{pmatrix} \tag{16.35}$$

Using the matrix U, the transformation between the bases given in equation (16.23) can be written

$$\langle\psi(x)| = \langle f(x)|U \tag{16.36}$$

where the following notation has been introduced for row vectors of functions

$$\langle f(x)| = (f_1(x) \dots f_N(x)) \text{ and } \langle\psi(x)| = (\psi_1(x) \dots \psi_N(x)) \tag{16.37}$$

Note that

$$|\psi(x)\rangle\langle\psi(x)| = \begin{pmatrix} \psi_1^*(x)\psi_1(x) & \dots & \psi_1^*(x)\psi_N(x) \\ \vdots & \vdots & \vdots \\ \psi_N^*(x)\psi_1(x) & \dots & \psi_N^*(x)\psi_N(x) \end{pmatrix} \tag{16.38}$$

and in view of the orthonormality condition (16.20),

$$\int_{-\infty}^{\infty} dx|\psi(x)\rangle\langle\psi(x)| = \begin{pmatrix} 1 & \dots & 0 \\ \vdots & \vdots & \vdots \\ 0 & \dots & 1 \end{pmatrix} = I \tag{16.39}$$

Similarly

$$\int_{-\infty}^{\infty} dx|f(x)\rangle\langle f(x)| = I \tag{16.40}$$

where I is the unit matrix. Since the Hermitian conjugate of equation (16.31) is $|\psi(x)\rangle = U^\dagger|f(x)\rangle$, multiplying this into equation (16.31) yields

$$|\psi(x)\rangle\langle\psi(x)| = U^\dagger|f(x)\rangle\langle f(x)|U$$

Integrating both sides with respect to x and utilising equations (16.39) and (16.40) yields

$$U^\dagger U = I \tag{16.41}$$

i.e. U is a unitary matrix. This shows that the transformation (16.36) between orthogonal bases is a unitary transformation. Since D is diagonal, equation (16.34) shows that H is diagonalised by a unitary transformation.

16.4 Representations involving non-orthogonal basis functions

In many cases of interest, it may not be convenient to utilise an orthonormal basis set and therefore this section examines how the above equations change when the differential equation (16.19) is solved using a non-orthogonal basis.

Consider introducing an arbitrary non-orthogonal basis $\{\chi_i(x)\}$ satisfying

$$\int_{-\infty}^{\infty} dx\, \chi_i^*(x)\chi_j(x) = S_{ij} \tag{16.42}$$

where the matrix S is called the overlap matrix. From equation (16.38), $S_{ij} = S_{ij}^*$ and therefore S is Hermitian.

As discussed in section 16.3, equation (16.19) is solved by writing

$$\psi_n(x) = \sum_{j=1}^{N} a_j^{(n)}\chi_j(x) \tag{16.43}$$

Substituting equation (16.43) into (16.19), then multiplying both sides from the left by $\phi_i^*(x)$ and integrating over x yields

$$\sum_{j=1}^{N} H'_{ij}a_j^{(n)} = E_n\sum_{j=1}^{N} S_{ij}a_j^{(n)} \tag{16.44}$$

where

$$H'_{ij} = \int_{-\infty}^{\infty} dx\, \chi_i^*(x)H(x)\chi_j(x) \tag{16.45}$$

Equation (16.44) is equivalent to

$$H'|a_n\rangle = E_nS|a_n\rangle \tag{16.46}$$

where H' is the $N \times N$ square matrix whose matrix elements are given by equation (16.46) and $|a_n\rangle$ is a vector of amplitudes:

$$|a_n\rangle = \begin{pmatrix} a_1^{(n)} \\ a_2^{(n)} \\ \vdots \\ a_N^{(n)} \end{pmatrix} \tag{16.47}$$

The matrix H' is the Hamiltonian matrix in the representation $\{\chi_i(x)\}$. Note that if the number of basis functions N is chosen to be finite, then equation (16.43) is an approximation. Clearly the $N \times N$ matrix equation (16.46) has only N linearly-independent solutions and therefore can provide at most N eigenstates of $H(x)$. During the past decades, when developing computational codes for quantum chemistry, a great deal of effort has been devoted to the question of choosing sets of basis functions, which deliver the most accurate solution to equation (16.19) for the smallest value of N. Clearly equation (16.29) looks simpler than equation (16.46), because the overlap matrix does not appear in the former. This is one of the reasons why orthonormal basis functions are often used to convert differential equations into matrix equations.

16.5 Transformations between representations

Sometimes it is useful to solve equation (16.19) using one basis and then to transform the solution to another basis. For example, having solved (16.19) for the amplitudes $a_j^{(n)}$ in equations (16.43) and (16.47), one may wish to obtain the amplitudes $\psi_j^{(n)}$ of equations (16.26) and (16.29), or vice versa.

To obtain a relationship between $|\psi_n\rangle$ and $|a_n\rangle$ write the basis functions $\chi_j(x)$ as a linear superposition of the basis functions $f_j(x)$:

$$\chi_i(x) = \sum_{j=1}^{N} f_j(x) c_{ji} \tag{16.48}$$

Multiplying both sides from the left by $f_l^*(x)$ and integrating over x yields

$$c_{li} = \int_{-\infty}^{\infty} dx\, f_l^*(x)\chi_i(x) \tag{16.49}$$

Similarly, substituting equation (16.48) into equation (16.42) yields

$$S_{ij} = \int_{-\infty}^{\infty} dx\, \chi_i^*(x)\chi_j(x) = \int_{-\infty}^{\infty} dx \sum_{l,m=1}^{N} f_l^*(x) c_{li}^* f_m(x) c_{mj}$$

In view of the orthonormality equation (16.13), this yields

$$S_{ij} = \sum_{l=1}^{N} c_{li}^* c_{lj} = (C^\dagger C)_{ij} \tag{16.50}$$

where C is a square matrix, with elements c_{ij}:

That is

$$C = \begin{pmatrix} c_{11} & \dots & c_{1N} \\ \vdots & \vdots & \vdots \\ c_{N1} & \dots & c_{NN} \end{pmatrix} \tag{16.51}$$

In terms of the matrices S and C, equation (16.50) can be written

$$S = C^\dagger C \tag{16.52}$$

If the basis $\chi_i(x)$ is chosen to be orthonormal, then $S = I$, and equation (16.48) would be a unitary transformation. In the present case, $S \neq I$ and the transformation between bases is referred to as a similarity transformation.

To complete the transformation, note that equations (16.26) and (16.43) yield

$$\sum_{j=1}^{N} \psi_j^{(n)} f_j(x) = \sum_{j=1}^{N} a_j^{(n)} \chi_j(x)$$

Multiplying both sides from the left by $f_l^*(x)$ and integrating over x yields

$$\psi_l^{(n)} = \sum_{j=1}^{N} c_{lj} a_j^{(n)} \tag{16.53}$$

That is

$$|\psi_n\rangle = C|a_n\rangle \tag{16.54}$$

16.6 A more compact notation for the transformation equation

Following the notation of equation (16.37), a neater way to present the above transformation is to write $\langle\chi(x)| = (\chi_1(x) \dots \chi_N(x))$, $|\chi(x)\rangle = \begin{pmatrix} \chi_1^*(x) \\ \vdots \\ \chi_N^*(x) \end{pmatrix}$ and therefore

$$|\chi(x)\rangle\langle\chi(x)| = \begin{pmatrix} \chi_1^*(x)\chi_1(x) & \dots & \chi_1^*(x)\chi_N(x) \\ \vdots & \vdots & \vdots \\ \chi_N^*(x)\chi_1(x) & \dots & \chi_N^*(x)\chi_N(x) \end{pmatrix} \tag{16.55}$$

Hence

$$\int_{-\infty}^{\infty} dx|\chi(x)\rangle\langle\chi(x)| = \begin{pmatrix} S_{11} & \dots & S_{1N} \\ \vdots & \vdots & \vdots \\ S_{N1} & \dots & S_{NN} \end{pmatrix} = S \tag{16.56}$$

Similarly equation (16.44) becomes

$$\langle\chi(x)| = \langle f(x)|C \tag{16.57}$$

where

$$C = \int_{-\infty}^{\infty} dx|f(x)\rangle\langle\chi(x)| \tag{16.58}$$

Taking the Hermitian conjugate of equation (16.56) yields $|\chi(x)\rangle = C^\dagger|f(x)\rangle$ and multiplying this into equation (16.56) yields

$$|\chi(x)\rangle\langle\chi(x)| = C^\dagger|f(x)\rangle\langle f(x)|C \tag{16.59}$$

As expected, integrating over x and noting equations (16.55) and (16.40) yields equation (16.52). With this notation, equations (16.26) and (16.43) become

$$\psi_n(x) = \langle\chi(x)|a_n\rangle = \langle f(x)|\psi_n\rangle \tag{16.60}$$

From equation (16.57), this yields

$$\langle f(x)|C|a_n\rangle = \langle f(x)|\psi_n\rangle \tag{16.61}$$

Again as expected, since the functions forming $\langle f(x)|$ are linearly independent, this yields equation (16.50).

This notation also makes sections 16.3 and 16.4 look neater. For example the Schrödinger equation (16.19) can be written

$$H(x)\langle\psi(x)| = \langle\psi(x)|D \tag{16.62}$$

With this notation, equation (16.60) becomes

$$\langle\psi(x)| = \langle\chi(x)|A = \langle f(x)|U \tag{16.63}$$

and equation (16.54) becomes

$$U = CA \tag{16.64}$$

Substituting $\langle\psi(x)| = \langle\chi(x)|A$ into the Schrödinger equation (16.62) yields, after multiplying by $|\chi(x)\rangle$ and integrating with respect to x,

$$H'A = SAD \tag{16.65}$$

whereas substituting $\langle\psi(x)| = \langle f(x)|U$ the Schrödinger equation (16.62) yields

$$HU = UD \tag{16.66}$$

Clearly if $U = CA$, then the last equation yields

$$HCA = CAD \tag{16.67}$$

Hence, $C^\dagger HCA = C^\dagger CAD$, so writing $H' = C^\dagger HC$ yields equation (16.65).

A summary of the above discussion and a comparison between notations is provided in the table below.

Transformation of functions	Transformation of functions
If $\chi_i(x) = \sum_{j=1}^N f_j(x)c_{ji}$	If $\langle\chi(x)| = \langle f(x)|C$
where $c_{ji} = \int_{-\infty}^{\infty} dx\, f_j^*(x)\chi_i(x)$	where $C = \int_{-\infty}^{\infty} dx|f(x)\rangle\langle\chi(x)|$
and $\psi_n(x) = \sum_{j=1}^N \psi_j^{(n)} f_j(x) = \sum_{j=1}^N a_j^{(n)}\chi_j(x)$	and $\langle\psi(x)| = \langle\chi(x)|A = \langle f(x)|U$
then $|\psi_n\rangle = C|a_n\rangle$	then $U = CA$
Orthogonality	Orthogonality
If $\int_{-\infty}^{\infty} dx\, f_i^*(x)f_j(x) = \delta_{ij}$	If $\int_{-\infty}^{\infty} dx|f(x)\rangle\langle f(x)| = I$
and $\int_{-\infty}^{\infty} dx\, \chi_i^*(x)\chi_j(x) = S_{ij}$	and $\int_{-\infty}^{\infty} dx|\chi(x)\rangle\langle\chi(x)| = S$
then $S_{ij} = \sum_{l=1}^N c_{li}^* c_{lj}$	then $S = C^\dagger C$
Transformation of the Schrödinger equation	Transformation of the Schrödinger equation
If $H(x)\psi_n(x) = E_n\psi_n(x)$	If $H(x)\langle\psi(x)| = \langle\psi(x)|D$
then $H|\psi_n\rangle = E_n|\psi_n\rangle$	then $HU = UD$
and $H'|a_n\rangle = E_nS|a_n\rangle$	and $H'A = SAD$
where $H' = C^\dagger HC$	where $H' = C^\dagger HC$

16.7 Atomic orbitals versus pseudo-atomic orbitals

So far the basis functions have not been specified. When describing molecules, the atomic orbitals belonging to each atom form a natural basis, since they are embedded within the language of chemistry. Unfortunately, although atomic orbitals belonging to the same atom can be chosen to be orthogonal, those belong to different atoms may not be orthogonal. Consider for example the π_z orbitals of benzene, shown in figure 2.2, or a linear chain of π_z orbitals, such as that shown in figure 4.3. If the normalised π_z orbital on atom j is denoted $\chi_j(x)$, then starting from these atomic orbitals, one can define orthogonal pseudo-atomic orbitals $f_i(x)$ via equation (16.48), or equivalently (16.58). The simplest choice of the transformation matrix C is to choose C to be real and symmetric, so that equation (16.52) yields $S = C^2$ or equivalently, $C = S^{1/2}$. From equation (16.57), this yields

$$\langle f(x)| = \langle \chi(x)| S^{-1/2} \tag{16.68}$$

or equivalently

$$f_i(x) = \sum\nolimits_{j=1}^{N} \chi_j(x)(S^{-1/2})_{ji} \tag{16.69}$$

Furthermore, if the Hamiltonian matrix H' is obtained from the integrals (16.45), then the Hamiltonian matrix H is

$$H = S^{-1/2} H' S^{-1/2} \tag{16.70}$$

Since the π_z orbitals are normalised, the diagonal matrix elements are $S_{ii} = 1$. If the nearest-neighbour matrix elements are denoted $S_{i,i\pm1} = \alpha$, then $|\alpha| < 1$. Since overlap integrals decay with distance, the next-nearest-neighbour matrix elements $S_{i,i\pm2}$ will be significantly smaller than α and therefore in the spirit of a tight-binding model, where only nearest-neighbour Hamiltonian matrix elements are retained, it is reasonable to consider the case where $S_{ii} = 1$, $S_{i,i+1} = S_{i,i-1} = \alpha$ and all other matrix elements are zero. Writing $S = I + X$, where $X_{i,i+1} = X_{i,i-1} = \alpha$ and all other matrix elements are zero, keeping only linear terms in α yields $S^{-1/2} = (1 + X)^{-1/2} \approx 1 - \frac{X}{2}$. Hence

$$f_i(x) \approx \chi_i(x) - \frac{\alpha}{2}(\chi_{i+1}(x) + \chi_{i-1}(x)) \tag{16.71}$$

Note that keeping only linear terms in α and only nearest-neighbour overlaps yields

$$\int_{-\infty}^{\infty} dx\, f_i(x) f_{i+1}(x) \approx \int_{-\infty}^{\infty} dx \left[\chi_i(x)\chi_{i+1}(x) - \frac{\alpha}{2}\left(\chi_i^2(x) + \chi_{i+1}^2(x)\right)\right] = 0$$

Hence to lowest order, the pseudo-atomic orbitals form an orthogonal basis.

Clearly, from equation (16.71), pseudo-atomic orbital $f_i(x)$ has a large amplitude on $\chi_i(x)$ and also incorporates a small contribution from the neighbouring π_z orbitals, via the small parameter α.

16.8 Contribution of overlap integrals to nearest-neighbour couplings

Note that keeping only linear terms in α and only nearest-neighbour overlaps yields

$$\int_{-\infty}^{\infty} dx\, f_i(x) f_{i+1}(x) \approx \int_{-\infty}^{\infty} dx \left[\chi_i(x)\chi_{i+1}(x) - \frac{\alpha}{2}\left(\chi_i^2(x) + \chi_{i+1}^2(x)\right)\right] = 0$$

Hence, as expected, to lowest order, the pseudo-atomic orbitals form an orthogonal basis.

From equation (16.70),

$$H \approx \left(1 - \frac{X}{2}\right)H'\left(1 - \frac{X}{2}\right) \approx H' - \frac{XH' + H'X}{2}$$

Since $(XH')_{ij} = X_{ii+1}H'_{i+1,j} + X_{ii-1}H'_{i-1,j} = \alpha\left(H'_{i+1,j} + H'_{i-1,j}\right)$ and $(H'X)_{ij} = H'_{i,j+1}X_{j+1,j} + H'_{i,j-1}X_{j-1,j} = \alpha(H'_{i,j+1} + H'_{i,j-1})$, one obtains

$$H_{ij} \approx H'_{i,j} - \frac{\alpha}{2}(H'_{i+1,j} + H'_{i-1,j} + H'_{i,j+1} + H'_{i,j-1})$$

For a one-dimensional tight-binding chain, if the on-site energies are $H'_{ii} = \epsilon_0$ and the nearest-neighbour couplings are $H'_{i,i\pm1} = -\beta$, and all other matrix elements are zero, then after keeping terms which are linear in α and β only, the on-site energies of H are $H_{ii} \approx \epsilon_0$ and the nearest-neighbour coupling of H are $H_{i,i\pm1} = -\gamma$, where

$$\gamma = \beta + \alpha\epsilon_0 \tag{16.72}$$

As noted by Löwdin [1], this shows that after transforming to a pseudo-atomic-orbital basis, the main effect is that the nearest neighbour overlap integral of the π_z orbitals contributes to the nearest-neighbour couplings of H.

Note that the solutions to the differential equation (16.19) are molecular orbitals $\psi_n(x)$, given by

$$\psi_n(x) = \sum_{j=1}^{N} \psi_j^{(n)} f_j(x) = \sum_{j=1}^{N} a_j^{(n)} \chi_j(x)$$

If the amplitudes $\psi_j^{(n)}$ are obtained by solving the matrix equation $H|\psi_n\rangle = E_n \mid \psi_n\rangle$, then the amplitude $a_j^{(n)}$ on atomic orbital $\chi_j(x)$ is obtained from the relation $|\psi_n\rangle = C|a_n\rangle$, or equivalently $|a_n\rangle = C^{-1}|\psi_n\rangle \approx S^{-\frac{1}{2}}|\psi_n\rangle \approx \left(1 - \frac{X}{2}\right)|\psi_n\rangle$. i.e. $a_j^{(n)} \approx \psi_j^{(n)} - \frac{\alpha}{2}\left[\psi_{j+1}^{(n)} + \psi_{j-1}^{(n)}\right]$.

16.9 Dyson's equation in a non-orthogonal basis

In an orthonormal basis $\{f_j(x)\}$, the Green's function equation is

$$(EI - H)G = I. \tag{16.73}$$

To express this in the transformed basis, multiply from the left by $C^\dagger$ to yield $(EC^\dagger - C^\dagger H)G = C^\dagger$. That is $(EC^\dagger - C^\dagger H)G(C^\dagger)^{-1} = I$ or equivalently i.e. $(EC^\dagger C - C^\dagger HC)C^{-1}G(C^\dagger)^{-1} = I$. Hence, after defining

$$G' = C^{-1}G(C^\dagger)^{-1} \text{ or equivalently } G = CG'C^\dagger, \tag{16.74}$$

the transformed equation is

$$(ES - H')G' = I, \tag{16.75}$$

where

$$H' = C^{\dagger}HC \text{ and } S = C^{\dagger}C \tag{16.76}$$

The above result could have been obtained from the real-space representation of Dyson's equation, which takes the form

$$(E - H(x))G(x, x') = \delta(x - x') \tag{16.77}$$

where $\delta(x - x')$ is a delta function. To convert this to a matrix equation, write

$$G(x, x') = \sum_{i,j=1}^{N} \chi_i(x)\, G'_{ij}\, \chi_j^*(x') = \sum_{i,j=1}^{N} f_i(x) G'_{ij} f_j^*(x') \tag{16.78}$$

and then multiply from the left by $f_l^*(x)$ and from the right by $f_m(x')$. After integrating with respect to x and x' one obtains

$$\int_{-\infty}^{\infty} dx dx' f_l^*(x) G(x, x') f_m(x') = G_{lm} = \sum_{i,j=1}^{N} c_{lj} G'_{ij} c^*_{mj} \tag{16.79}$$

where the last expression was obtained using equation (16.49). In matrix form, equation (16.79) is $G = CG'C^{\dagger}$, which is identical to equation (16.74).

Starting from the real-space equation (16.77), the matrix equation (16.74) is also obtained by substituting equation (16.78) into equation (16.77), multiplying from the left by $f_l^*(x)$ and from the right by $f_m(x')$ and integrating with respect to x and x'.

16.10 Summary

In this chapter, the solutions $\psi_n(x)$ to the differential equation (16.19) were discussed using a non-orthogonal basis and an orthogonal basis. If equation (16.19) describes a molecule, then the solutions $\psi_n(x)$ are the molecular orbitals. If the non-orthogonal basis is composed of atomic orbitals, then this is rather natural from the point of view of chemistry, but the resulting matrix equation $H'|a_n\rangle = E_n S|a_n\rangle$ contains the overlap matrix S. On the other hand, if an orthogonal pseudo atomic orbital basis is used, such as (16.71), which includes a slight 'leakage' onto nearest-neighbour atomic orbitals, then the resulting matrix equation $H|\psi_n\rangle = E_n|\psi_n\rangle$ is simpler. If the latter is solved for the eigenvectors $|\psi_n\rangle$, with amplitudes $\psi_j^{(n)}$, then the amplitudes of a molecular orbital $\psi_n(x)$ on the atomic orbitals are obtained from the relation $|a_n\rangle = C^{-1}|\psi_n\rangle$.

Reference

[1] Löwdin P-O 1950 On the non-orthogonality problem connected with the use of atomic wave functions in the theory of molecules and crystals *J. Chem. Phys.* **18** 365

Chapter 17

Scattering theory in the presence of material-specific leads

17.1 Introduction

The finite-width leads discussed in previous chapters were rather simple and therefore there is a need to generalise the arguments presented so far to leads described by arbitrary, material-specific mean-field Hamiltonians, which could be generated, for example, by density functional theory. It transpires that many of the concepts developed in previous can be applied in the general case, but certain crucial modifications are needed. The aim of this chapter is to develop a thorough understanding of scattering theory in the presence of arbitrary finite-width leads, and to derive general formulae for transmission and reflection coefficients.

The main theoretical steps needed to make the transition from simple leads, such as those shown in chapter 9, to arbitrary leads are as follows:

1. Understanding that the number of open channels is not necessarily equal to the number of sites in a unit cell.
2. Implementing the strategy of appendix A to deal with this 'H_1 problem'.
3. Devising a method for computing channel eigenstates and wave vectors.
4. Proving that there are no off-diagonal contributions to the current carried by a lead, even when the channel eigenstates are not orthogonal.
5. Obtaining expressions for the Green's functions of an arbitrary semi-infinite lead.
6. Obtaining expressions for inter-channel scattering amplitudes.

As discussed in chapter 5, a doubly infinite, finite-width lead, such as that shown in figure 17.1(a), is described by the equation

$$H_0|\psi_J\rangle + H_1|\psi_{J+1}\rangle + H_{-1}|\psi_{J-1}\rangle = E|\psi_J\rangle \tag{17.1}$$

doi:10.1088/978-0-7503-3639-0ch17

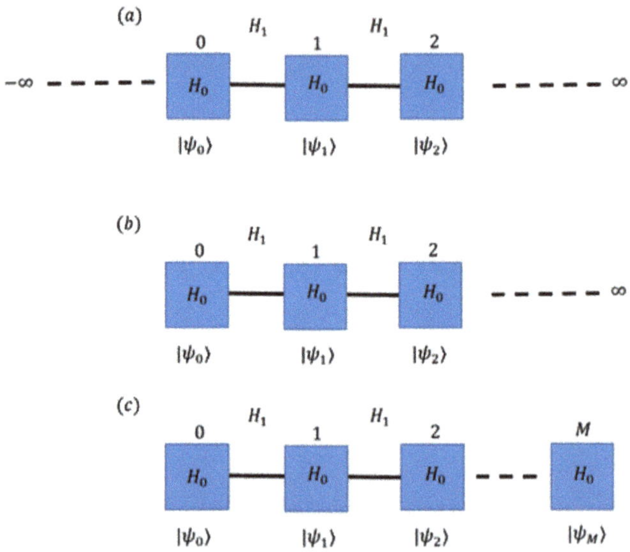

Figure 17.1. (a) A doubly infinite lead comprising a series of identical unit cells. (b) A semi-infinite lead. (c) A finite-length, crystalline lead formed from M identical unit cells.

where H_0, H_1, and H_{-1} are $N \times N$ square matrices satisfying $H_0 = (H_0)^\dagger$, $H_{-1} = (H_1)^\dagger$ and J labels the unit cells of the lead. In this chapter, the general solutions to this equation are discussed, without imposing further restrictions on these matrices.

Starting from equation (17.1), the band structure of this crystalline lead is obtained by writing

$$|\psi_J\rangle = |\chi\rangle e^{ikJ} \tag{17.2}$$

where to ensure linear independence, the wave vector k must be restricted to a range of width 2π. Substituting the solution (17.2) into equation (17.1) leads to

$$H(k)|\chi\rangle = E(k)|\chi\rangle \tag{17.3}$$

where

$$H(k) = H_0 + H_1 e^{ik} + H_{-1} e^{-ik} \tag{17.4}$$

In equation (17.4), k is a parameter, which labels the matrix $H(k)$. Clearly $H(k) \neq H(k')$ unless k' differs from k by an integral multiple of 2π.

For a given choice of k, since $H(k)$ is a $N \times N$ Hermitian matrix, there are N linearly independent orthonormal solutions to equation (17.3). If these are denoted $|\chi_n(k)\rangle$ and the corresponding energies are denoted $E_n(k)$, then equation (17.3) takes the form

$$H(k)|\chi_n(k)\rangle = E_n(k)|\chi_n(k)\rangle \tag{17.5}$$

When the N eigenvalues $E_n(k)$ are plotted against k, one obtains N mini-bands, such as those shown in figure 5.8.

For the simple choice considered in chapter 5, where $H_1 = -\gamma I$, one finds $H(k) = H_0 - 2\gamma I \cos k$ and therefore the Hamiltonian $H(k)$ differs from the Hamiltonian $H(k')$ by a multiple of the unit matrix. This means that if the solutions to the equation $H_0|\chi_n\rangle = \epsilon_n|\chi_n\rangle$ are known, then the solutions to equation (17.5) are

$$|\chi_n(k)\rangle = |\chi_n\rangle, \tag{17.6}$$

with energies

$$E_n(k) = \epsilon_n - 2\gamma \cos k \tag{17.7}$$

Equation (17.6) shows that for this simple choice of H_1, the states $|\chi_n(k)\rangle$ do not depend on k. Examples of the solutions $|\chi_n\rangle$ are discussed in sections 5.7 and 5.8. Since $\langle\chi_m|\chi_n\rangle = \delta_{nm}$, this means that for the simple choice $H_1 = -\gamma I$, eigenstates $|\chi_n(k)\rangle$ and $|\chi_n(k')\rangle$ belonging to different wave vectors are mutually orthonormal. i.e.

$$\langle\chi_m(k)|\chi_n(k')\rangle = \delta_{nm} \tag{17.8}$$

Clearly the above results are restricted to a very special case, because in general the eigenstates of two different Hermitian matrices have no particular relationship to each other. That is if A and B are Hermitian matrices satisfying $A|\phi_n\rangle = \lambda_n|\phi_n\rangle$ and $B|\psi_n\rangle = E_n|\psi_n\rangle$, then although one can choose $\langle\phi_m|\phi_n\rangle = \delta_{nm}$ and $\langle\psi_m|\psi_n\rangle = \delta_{nm}$, the product $\langle\phi_m|\psi_n\rangle$ has no particular value. The special case (17.8) occurs, because if $H_1 = -\gamma I$, then $H(k)$ and $H(k')$ commute (i.e. $H(k)H(k') = H(k')H(k)$). and matrices which commute share the same eigenvectors. The aim of this chapter is to discuss issues, which arise when $H(k)$ and $H(k')$ do not commute.

17.2 The most general solution to the Schrödinger equation for an arbitrary crystalline lead

In the band structure problem defined by equation (17.3), a value of k is chosen and the band energies $E_n(k)$ are computed by solving equation (17.5). Choosing real values of k in a range of width 2π, allows graphs of $E_n(k)$ versus k to be plotted. As discussed in chapter 5, to find the most general solution to the Schrödinger equation for a given energy E, the inverse problem must be solved, in which E is fixed and for each chosen value of E, the wave vectors $k_n(E)$ and channel eigenvectors $|\chi_n(E)\rangle$ are computed. In contrast with the band structure problem, in which k is real, the wave vectors $k(E)$ obtained by solving the inverse problem may be real or complex.

For the simple case, where $H_1 = -\gamma I$, the solutions can be obtained analytically, as shown in equation (5.52). As discussed in section 5.10, for this simple choice of H_1, there are $2N$ solutions, which can divided into two sets. When $H(k)$ and $H(k')$ do not commute, the problem of finding the allowed values of k is less straightforward. For example, as shown in section 17.9, when H_1 is arbitrary there are only $2r$

solutions, where r is the rank of H_1. (i.e. r is the number of linearly independent rows of H_1.) When $r < N$, H_1 does not possess an inverse. In this case, a decimation procedure described in appendix A, this chapter, should be adopted. This procedure converts H_1 into a renormalized rxr matrix, which possesses an inverse and the channel eigenstates $|\chi_n\rangle$ become r-dimensional vectors. As shown in section 17.9 and discussed in detail in section B.2 of appendix B, for a given E, when solving for the allowed wave vectors k_n, one finds that they divide into four sets. The first set is composed of M real wave vectors k_n, labelled $n = 1, 2, \ldots M$, with positive group velocities in the direction of increasing J. For $M < r$, the second set of $r - M$ wave vectors labelled $n = M + 1, \ldots r$ are complex wave vectors with positive imaginary parts, for which e^{ik_nJ} decays exponentially in the direction of increasing J. When discussing the semi-infinite lead shown in figure 17.1b, these wave vectors will be called 'outgoing'. The third set of M wave vectors labelled $n = r + 1, \ldots r + M$, correspond to real wave vectors with negative group velocities, while the fourth set, labelled $n = r + M + 1, \ldots .\ 2r$, are complex wave vectors with negative imaginary parts, for which e^{ik_nJ} increases exponentially in the direction of increasing J. When discussing the semi-infinite lead shown in figure 17.1b, these wave vectors will be called 'incoming'. For a given choice of E, this means that the most general solution to the Schrödinger equation can be written

$$|\psi_J\rangle = |I_J\rangle + |O_J\rangle \tag{17.9}$$

where

$$|O_J\rangle = \sum\nolimits_{n=1}^{r} A_n e^{ik_nJ} |\chi_n\rangle \text{ and } |I_J\rangle = \sum\nolimits_{n=r+1}^{2r} B_n e^{ik_nJ} |\chi_n\rangle \tag{17.10}$$

In the above equation, A_n and B_n are arbitrary plane wave amplitudes.

This general solution is appropriate for the finite-length chain of figure 17.1(c). However for the semi-infinite lead of figure 17.1(b), the 'incoming' sum in $|I_J\rangle$ must be restricted to real wave vectors only, because the states corresponding to complex wave vectors diverge a $J \to \infty$. Furthermore for the doubly infinite lead of figure 17.1(a) the sums in both $|I_J\rangle$ and $|O_J\rangle$ must be restricted to real wave vectors only, to avoid divergences as $J \to \pm\infty$.

In the simple case, where $H_1 = -\gamma I$, it was found in chapter 5 that the real wave vectors came in $\pm$ pairs, so that for $n \leqslant M$, $k_n = -k_{n+r}$. In contrast, when $H_1 \neq H_{-1}$, this is no longer the case. Instead, as discussed in section 17.9, one finds that to each real wave vector k_n with a positive group velocity, there corresponds a real wave vector k_{n+r} with a negative group velocity, but in general, $k_n \neq -k_{n+r}$.

An important step in understanding the properties of arbitrary crystalline leads is to obtain an expression for the current carried by a general solution. For example, in appendix B in chapter 6, it was shown that for the simple choice $H_1 = -\gamma I$, the current carried by the above sum of plane waves (equation (6.60)) is simply a sum over open channels of the form

$$I_{J\to J+1} = \sum_{n=1}^{M}(|A_n|^2 - |B_n|^2)$$

where v_n is the group velocity of channel n. This shows that only diagonal terms contribute to the current and there is no interference between scattering channels with different wave vectors. However the proof in appendix B in chapter 6 made use of the orthonormality condition (17.8), which is not generally satisfied. Therefore there is a need for a more general proof of this result, which is provided in the section 17.3.

17.3 Matrix elements between channel eigenvectors

Before demonstrating that there are no off-diagonal contributions to the current, it is useful to examine properties of matrix elements between channel vectors $|\chi_n\rangle$.

For some chosen value of E, consider two solutions to the Schrödinger equation (17.3), denoted $|\chi_n\rangle$ and $|\chi_m\rangle$, which satisfy

$$(E - H_0)|\chi_n\rangle = H_1 e^{ik_n}|\chi_n\rangle + H_{-1}e^{-ik_n}|\chi_n\rangle \tag{17.11}$$

$$\langle\chi_m|(E - H_0) = \langle\chi_m|H_{-1}e^{-ik_m^*} + \langle\chi_m|H_1 e^{ik_m^*} \tag{17.12}$$

In these equations E is fixed and k_n, k_m are functions of E, which could be real or complex.

Multiplying equation (17.11) from the left by $\langle\chi_m|$ and equation (17.12) from the right by $|\chi_n\rangle$ and subtracting the results yields

$$\langle\chi_m|H_1|\chi_n\rangle(e^{ik_n} - e^{ik_m^*}) + \langle\chi_m|H_{-1}|\chi_n\rangle\left(e^{-ik_n} - e^{-ik_m^*}\right) = 0$$

This is equivalent to

$$\langle\chi_m|H_1|\chi_n\rangle e^{ik_n}(1 - e^{i(k_m^*-k_n)}) = \langle\chi_m|H_{-1}|\chi_n\rangle e^{-ik_m^*}\left(1 - e^{i(k_m^*-k_n)}\right)$$

i.e.

$$z_{mn}(1 - e^{i(k_m^*-k_n)}) = z_{nm}^*\left(1 - e^{i(k_m^*-k_n)}\right) \tag{17.13}$$

where

$$z_{mn} = \langle\chi_m|H_1|\chi_n\rangle e^{ik_n} \tag{17.14}$$

For $k_m^* \neq k_n$, equation (17.13) implies that the two states $|\chi_n\rangle$ and $|\chi_m\rangle$ satisfy

$$z_{mn} = z_{nm}^* \quad (k_m^* \neq k_n) \tag{17.15}$$

Hence if k_n is complex, so that $k_n^* \neq k_n$, equation (17.15) yields

$$z_{nn} = z_{nn}^* \quad (\text{complex } k_n) \tag{17.16}$$

To accommodate the condition ($k_m^* \neq k_n$) in equation (17.15), it is useful to define the symmetric matrix L, with matrix elements

$$L_{mn} = 1 \text{ if } k_m^* = k_n \text{ and } L_{mn} = 0 \text{ if } k_m^* \neq k_n \tag{17.17}$$

The matrix L has the following properties:

Properties of L.

If there are no degeneracies, such that all wave vectors are different, then

1. If $n = 1 \ldots M$ then $L_{mn} = \delta_{m,n}$
2. If $n = r + 1, \ldots r + M$ then $L_{mn} = \delta_{m,n}$
3. If $n = M + 1, \ldots r$ then $L_{mn} = \delta_{m,n+r}$
4. $n = r + M + 1, \ldots 2r$ then $m = M + 1, \ldots r$ $L_{mn} = \delta_{m+r,n}$

More generally, in the presence of degeneracies, such that some wave vectors are repeated (for example if two mini bands cross at the chosen E), then L_{nm} is non-zero, only if one of the following conditions is satisfied[1]:

1. If $n \in \{1..M\}$ and $m \in \{1, \ldots M\} \cup \{r + 1, \ldots r + M\}$;
2. If $n \in \{r + 1, \ldots r + M\}$ and $m \in \{1, \ldots M\} \cup \{r + 1, \ldots r + M\}$;
3. If $n \in \{M + 1, \ldots r\}$ and $m \in \{r + M + 1, \ldots 2r\}$;
4. If $n \in \{r + M + 1, \ldots 2r\}$ and $m \in \{M + 1, \ldots r\}$.

Properties 5 and 6 recognise that if two mini bands cross, then a positive-current real wave vector could coincide with a negative-current real wave vector, or positive-current real wave vector.

Utilising the matrix L_{nm} allows equation (17.15) to be written

$$z_{mn} = z_{nm}^* - i\hbar w_{mn} L_{mn} \tag{17.18}$$

where

$$w_{mn} = \frac{z_{nm}^* - z_{mn}}{i\hbar} = \frac{\langle \chi_m | H_{-1} | \chi_n \rangle e^{-ik_m^*} - \langle \chi_m | H_1 | \chi_n \rangle e^{ik_n}}{i\hbar} \tag{17.19}$$

As discussed in section (17.9) (cf equation (17.86)), for real k_n, w_{nn} is the group velocity associated with channel n and therefore w_{mn} can be regarded as a 'generalised group velocity'.

[1] Here standard set notation is used, in which the intersection of two sets $\{A\}$ and $\{B\}$ is denoted $\{A\} \cap \{B\}$ and defined to be the largest set, which contains all the elements that are common to both the sets. For example if $\{A\} = \{1,2,7\}$ and $\{B\} = \{8,3,1,9,7\}$, then $\{A\} \cap \{B\} = \{1,7\}$. Similarly the union $\{A\} \cup \{B\}$ of two given sets is the smallest set, which contains all the elements of both the sets. In this example, $\{A\} \cup \{B\} = \{2,8,3,1,9,7\}$.

17.4 Cancellation of off-diagonal contributions to currents

From equation (A.5) of chapter 6 appendix A, the bond current from cell J to cell $J+1$ is $I_{J\to J+1}$ where

$$I_{J\to J+1} = -\frac{2}{\hbar}\mathrm{Im}\langle\psi_J|H_1|\psi_{J+1}\rangle \tag{17.20}$$

To analyse the currents carried by a semi-infinite lead, consider a general eigenstate of the form

$$|\psi_J\rangle = \sum_{n=1}^{2r} a_n e^{ik_n J}|\chi_n\rangle \tag{17.21}$$

this yields

$$I_{J\to J+1} = -\frac{2}{\hbar}Im\sum_{n,m=1}^{2r} a_m^* a_n\langle\chi_m|H_1|\chi_n\rangle e^{ik_n}e^{i(k_n-k_m^*)J}$$

i.e.

$$I_{J\to J+1} = -\frac{2}{\hbar}Im\sum_{n,m=1}^{2r} a_m^* a_n z_{mn} e^{i(k_n-k_m^*)J} \tag{17.22}$$

where z_{mn} is given by equation (17.14).

To show that only diagonal terms (for which $m = n$) contribute, note that by interchanging the dummy indices n, m, equation (17.22) becomes

$$I_{J\to J+1} = -\frac{2}{\hbar}Im\sum_{n,m=1}^{2r} a_n^* a_m z_{nm} e^{i(k_m-k_n^*)J} \tag{17.23}$$

Adding this to the right-hand side of equation (17.22) and dividing by 2 yields

$$I_{J\to J+1} = -\frac{1}{\hbar}Im\sum_{n,m=1}^{2r}\left[a_m^* a_n z_{mn} e^{i(k_n-k_m^*)J} + a_n^* a_m z_{nm} e^{i(k_m-k_n^*)J}\right] \tag{17.24}$$

From equation (17.12), the term in the square brackets is

$$a_m^* a_n z_{mn} e^{i(k_n-k_m^*)J} + a_n^* a_m z_{mn}^* e^{i(k_m-k_n^*)J} - i\hbar w_{nm} L_{nm} a_n^* a_m e^{i(k_m-k_n^*)J}$$

Since the second term is the complex conjugate of the first, the sum is real and therefore the imaginary part is zero, i.e.

$$\mathrm{Im}\left[a_m^* a_n z_{mn} e^{i(k_n-k_m^*)J} + a_n^* a_m z_{nm} e^{i(k_m-k_n^*)J}\right] = -\hbar L_{nm}\mathrm{Re}\left[w_{nm} a_n^* a_m e^{i(k_m-k_n^*)J}\right] \tag{17.25}$$

From the definition (17.17), this is

$$\mathrm{Im}\left[a_m^* a_n z_{mn} e^{i(k_n - k_m^*)J} + a_n^* a_m z_{nm} e^{i(k_m - k_n^*)J}\right] = -\hbar L_{nm} \mathrm{Re}\left[w_{nm} a_n^* a_m\right]$$

Hence

$$I_{J\to J+1} = \mathrm{Im} \sum_{n,m=1}^{2r} L_{nm} \mathrm{Re}\left[w_{nm} a_n^* a_m\right] \tag{17.26}$$

In what follows the aim is to compute $I_{J\to J+1}$ for a semi-infinite lead, which means that if complex wave vectors, with states which decay in the direction of increasing J are allowed, then those which increase in the direction of increasing J are not allowed, or vice versa. Consequently, from the above properties of L, conditions 3 and 4 or 7 and 8 are never satisfied. Therefore the terms corresponding to complex wave vectors are zero and taking into account properties 5 and 6, equation (17.26) simplifies to

$$I_{J\to J+1} = \mathrm{Im} \sum\nolimits_{(k_n,\, k_m \mathrm{real})} L_{nm} \mathrm{Re}\left[w_{nm} a_n^* a_m\right] \tag{17.27}$$

If there are no degeneracies, then from properties 1 and 2 of L, equation (17.27) reduces to

$$I_{J\to J+1} = \sum\nolimits_{(k_n \mathrm{real})} |a_n|^2 w_{nn} \tag{17.28}$$

This shows that only open channels contribute to the current and furthermore, there are no off-diagonal contributions. In other words, there is no interference between currents belonging to different channels.

17.5 Note about degeneracies

For $n \neq m$, there may be a degeneracy, such that the wave vectors of two or more linearly independent states $|\chi_n\rangle$ and $|\chi_m\rangle$ satisfy $k_m = k_n$. Two real wave vectors satisfy $k_m = k_n$, when their mini-bands cross. (If E is equal to the energy of a band edge, a degeneracy can also occur. However this will always be considered from the viewpoint of E approaching the band edge from above or below, in which case the wave vectors are considered to differ infinitesimally. Such states carry no current and therefore such a limiting procedure does not affect physical transport properties.) Degeneracies could also arise when a lead is composed of two or more identical non-interacting leads.

When $k_m = k_n$, solving the conventional band structure problem of equation (17.3) yields two linearly independent states $|\chi_n\rangle$ and $|\chi_m\rangle$. Alternatively, one notes that $|\chi_n\rangle$ and $|\chi_m\rangle$ satisfy equation (17.11) with the same k_n , so that in addition to equation (17.11), $|\chi_m\rangle$ satisfies

$$(E - H_0)|\chi_m\rangle = H_1 e^{ik_n}|\chi_m\rangle + H_{-1} e^{-ik_n}|\chi_m\rangle \tag{17.29}$$

Since $|\chi_n\rangle$ and $|\chi_m\rangle$ are linearly independent, one can define two new vectors $|y_n\rangle = a|\chi_m\rangle + b|\chi_n\rangle$ and $|y_m\rangle = c|\chi_m\rangle + d|\chi_n\rangle$, and choose the coefficients such that

$$\langle y_n|H_1|y_m\rangle e^{ik_n} = \langle y_n|H_{-1})|y_m\rangle e^{-ik_n} \tag{17.30}$$

This means that in the case of a degeneracy, if one adopts the convention of replacing $|\chi_m\rangle$ and $|\chi_n\rangle$ by $|y_m\rangle$ and $|y_n\rangle$, then $w_{nm} = 0$ for $n \neq m$ Consequently, only diagonal terms appear in equation (17.27) and equation (17.28) remains valid. The presence of degeneracies is considered in more detail in appendix C.

17.6 Analysis of a general scatterer connected to a periodic structure

In this section, a scatterer connected to a collection of scattering channels is discussed. The scattering channels may belong to different physical leads or to a single physical lead. In either case, they can be regarded as belonging to a semi-infinite chain of identical 'super cells', as shown in figure 17.2.

Figure 17.2(a) shows a collection of leads interacting with a scattering region. The Hamiltonian of the isolated scattering region is h. Figure 17.2(b) shows the same structure, but with the leads depicted as being in parallel. This conceptual construct will be called a 'super lead' and helps to simplify the notation below. Provided the Hamiltonian is the same for both structures (figure 17.2(a) and (b)), their scattering properties are identical. This means that from a conceptual viewpoint, the structure in figure 17.2(a) can be represented by the structure in figure 17.2(c), in which each cell is a 'super cell' containing multiple cells from the original leads; i.e. they are cells

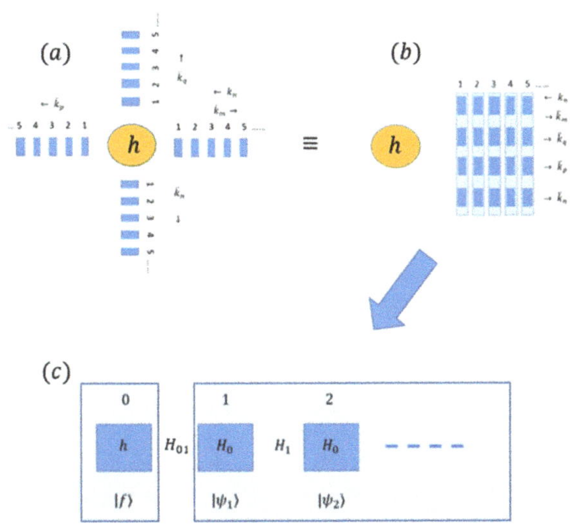

Figure 17.2. (a) Many leads connected to a scattering region. The Hamiltonian of the isolated scattering region is h. (b) A super lead composed of many leads depicted in parallel. (c) A scattering region labelled 0, coupled to a semi-infinite, periodic structure formed from cells labelled 1, 2, ... This represents the super lead of figure (b).

of the super lead. Here cells of the super lead are numbered by labels J, which increase in a direction moving away from the scattering region.

For such a structure, if the wave function is cell J is denoted $|\psi_J\rangle$ and the wave function within the scattering region is $|f\rangle$ the Schrödinger equation takes the form

$$g_0^{-1}|\psi_1\rangle = H_1|\psi_2\rangle + H_{10}|f\rangle \tag{17.31}$$

$$g^{-1}|f\rangle = H_{01}|\psi_1\rangle \tag{17.32}$$

where $H_{01}^{\dagger} = H_{10}$. In these equations, g_0 is the Green's function of an isolated unit cell, when $H_{10} = H_1 = 0$ and g is the Green's function of the scatterer when $H_{10} = 0$. As discussed in section 9.10, if the unit cells or the scatterer are in contact with an open system, then these Green's functions would be non-Hermitian. If the unit cells or the scatterer are closed systems, as in figure 17.2, then they are Hermitian and can be written

$$g_0 = (E - H_0)^{-1} \tag{17.33}$$

and

$$g = (E - h)^{-1} \tag{17.34}$$

where H_0 is the Hamiltonian of an isolated unit cell and h is the Hamiltonian of the isolated scatterer.

Eliminating $|f\rangle$ from equations (17.32) and (17.31) yields

$$\left(g_0^{-1} - \Sigma\right)|\psi_1\rangle = H_1|\psi_2\rangle \tag{17.35}$$

or equivalently

$$|\psi_1\rangle = GH_1|\psi_2\rangle \tag{17.36}$$

where

$$\Sigma = H_{10}gH_{01} \tag{17.37}$$

and

$$G = \left(g_0^{-1} - \Sigma\right)^{-1} \tag{17.38}$$

Equation (17.35) is of course equivalent to the result of decimating the scattering region.

So far, the property of translational invariance has not been used, which is captured by the equation

$$g_0^{-1}|\psi_J\rangle = H_{-1}|\psi_{J-1}\rangle + H_1|\psi_{J+1}\rangle \tag{17.39}$$

where $H_{-1} = H_1^{\dagger}$. In what follows, $|\psi_J\rangle$ will be chosen to satisfy equation (17.39) for all J. Choosing $J = 1$ and comparing the result with equation (17.31) yields

$$H_{10}|f\rangle = H_{-1}|\psi_0\rangle \tag{17.40}$$

Combining this with equation (17.32) yields

$$\Sigma|\psi_1\rangle = H_{-1}|\psi_0\rangle \tag{17.41}$$

Equation (17.41) is a general solution to the scattering problem. In section 17.8, it will be used to obtain a relationship between incoming and outgoing states, but first it is of interest to examine the currents carried by these states and to demonstrate that conservation of probability is embedded in the above equations.

17.7 Analysis of currents

Equations (17.35) and (17.41) will now be used to examine the probability currents flowing between cells in the superlead and between the scattering region and the superlead.

The most general form of $|\psi_J\rangle$ which satisfies equation (17.39) is

$$|\psi_J\rangle = \sum_k a_k e^{ik(J-1)}|\chi_k\rangle \tag{17.42}$$

where the sum is over all allowed wave vectors (both incoming and outgoing).The probability current carried by such a state is obtained from a sum over all real wave vectors and is given by

$$I_{1\to 2} = -\frac{2}{\hbar}\mathrm{Im}\ \langle\psi_1|H_1|\psi_2\rangle = \sum_{k\ (\mathrm{real})} |a_k|^2 v(k) \tag{17.43}$$

where

$$v(k) = -\frac{2}{\hbar}\mathrm{Im}\ \langle\chi_k|e^{ik}H_1|\chi_k\rangle \tag{17.44}$$

In this equation, the $v(k)$ has units of $(\mathrm{time})^{-1}$ and is the group velocity measured in unit cells per unit time.

Since equation (17.36) yields $G^{-1}|\psi_1\rangle = H_1|\psi_2\rangle$, equation (17.43) can be written

$$I_{1\to 2} = -\frac{2}{\hbar}\mathrm{Im}\ \langle\psi_1|G^{-1}|\psi_1\rangle \tag{17.45}$$

and therefore

$$I_{1\to 2} = -\frac{2}{\hbar}\mathrm{Im}\ \langle\psi_1|g_0^{-1}|\psi_1\rangle + \frac{2}{\hbar}\mathrm{Im}\ \langle\psi_1|\Sigma|\psi_1\rangle \tag{17.46}$$

Similarly, the current from the scattering region to cell 1 is

$$I_{0\to 1} = -\frac{2}{\hbar}\mathrm{Im}\ \langle f|H_{01}|\psi_1\rangle \tag{17.47}$$

After multiplying equation (17.32) by g and taking the Hermitian conjugate, one obtains, $\langle f| = \langle\psi_1|H_{10}g^\dagger$ and therefore $I_{0\to 1} = -\frac{2}{\hbar}\mathrm{Im}\ \langle\psi_1|H_{10}g^\dagger H_{01}|\psi_1\rangle$. Since

$\text{Im}\,\langle\psi_1|H_{10}g^\dagger H_{01}|\psi_1\rangle = -\text{Im}\,\langle\psi_1|H_{10}g^\dagger H_{01}|\psi_1\rangle^* = -\text{Im}\,\langle\psi_1|H_{10}gH_{01}|\psi_1\rangle = -\text{Im}\,\langle\psi_1|\Sigma|\psi_1\rangle$, (where the last step follows from equation (17.37)), one obtains

$$I_{0\to1} = \frac{2}{\hbar}\text{Im}\,\langle\psi_1|\Sigma|\psi_1\rangle \tag{17.48}$$

Comparing equations (17.46) and (17.48) yields

$$I_{1\to2} = I_{0\to1} - \frac{2}{\hbar}\text{Im}\,\langle\psi_1|g_0^{-1}|\psi_1\rangle \tag{17.49}$$

Since $\langle\psi_1|g_0^{-1}|\psi_1\rangle^* = \langle\psi_1|g_0^{\dagger-1}|\psi_1\rangle$, this shows that if cell 1 is closed, so that g_0 is Hermitian, then $\langle\psi_1|g_0^{-1}|\psi_1\rangle$ is real. Therefore, as expected, equation (17.49) yields

$$I_{1\to2} = I_{0\to1} \tag{17.50}$$

This is a mathematical proof of the intuitive property that if cell 1 is closed, the current entering cell 1 from the scattering region is equal to the current entering cell 2 from cell 1. On the other hand, if cell 1 is not closed and g_0 is not Hermitian, then these currents are not equal, because the non-hermiticity of g_0 means that cell 1 is connected to an open system, which can receive a non-zero current. Furthermore, if the scattering region is closed, so that g is Hermitian, the self-energy (17.37) is Hermitian and therefore $\langle\psi_1|\Sigma|\psi_1\rangle$ is real. From equation (17.48), this means that $I_{0\to1} = 0$ and therefore as expected, any current entering a closed scattering region is cancelled by an outgoing current, such that the net current is zero.

Writing

$$|\psi_J\rangle = |I_J\rangle + |O_J\rangle \tag{17.51}$$

and recalling that only diagonal terms contribute to the current, yields

$$I_{1\to2} = -\frac{2}{\hbar}\text{Im}\,\langle\psi_1|H_1|\psi_2\rangle = -\frac{2}{\hbar}\text{Im}\,\langle I_1|H_1|I_2\rangle - \frac{2}{\hbar}\text{Im}\,\langle O_1|H_1|O_2\rangle \tag{17.52}$$

If $|I_j\rangle$ carries an incoming current and $|O_j\rangle$ carries an outgoing current, then equations (17.50) and (17.52) show that the incoming and outgoing currents sum to zero. This demonstration that current conservation is built into the above equations guarantees unitarity of the resulting S-matrix.

17.8 Expressions for transmission amplitudes

In this section expressions for transmission amplitudes are obtained by solving equation (17.41), which was obtained by wave function matching at the interface between the scattering region and the surface of the super lead. To solve equation (17.41), a relationship between $|O_J\rangle$ and $|I_J\rangle$ is needed. In what follows, it will be convenient to distinguish different types of wave vectors by adopting the notation defined in section B.2 of appendix B of this chapter.

Consider the case where the incoming state $|I_J\rangle$ corresponds to an incoming open channel n:

$$|I_J\rangle = \frac{e^{ik_n^-(J-1)}}{\sqrt{|v(k_n^-)|}}|\chi_{k_n^-}\rangle \tag{17.53}$$

where $v(k_n^-)$is the group velocity given by equation (17.44). The most general outgoing state resulting from such a source is then of the form

$$|O_J\rangle = \sum_{m=1}^{r} t_{mn} \frac{e^{ik_m^{++}(J-1)}}{\sqrt{|w(k_m^{++}, k_m^{++})|}}|\chi_{k_m^{++}}\rangle \tag{17.54}$$

In this expression, following the notation of section 17.2, k_m^{++} is the wave vector of an outgoing plane wave or a state which decays in an outgoing direction from the scattering region. For real wave vectors, $w(k_m^{++}, k_m^{++})$ is chosen to be the group velocity $v(k_m^{++})$. For complex wave vectors, $w(k_m^{++}, k_m^{++})$ is arbitrary, but optionally (to allow a common notation for contributions from open and closed channels) could be chosen to be a 'generalised group velocity' defined in equation (B.32) of appendix B of this chapter. Note that t_{mn} is used to denote the amplitudes of all outgoing waves. Of course, outgoing waves belonging to the same lead as the incoming wave could be described as reflected waves, but there is no need to distinguish them from other outgoing waves and therefore there is no need to use a different notation (such as r_{mn}) for their amplitudes. Using the above notation, all outgoing waves are treated on the same footing.

Since only diagonal terms with real wave vectors contribute to the current, substituting these expressions into equation (17.52) yields the unitarity condition

$$\sum_{m=1}^{M} |t_{mn}|^2 = 1 \tag{17.55}$$

where the sum over wave vectors is restricted to terms for which k_m^{++} is real, because only these terms contribute to the current.

To obtain transmission amplitudes from equation (17.41), a relationship between $|\psi_0\rangle$ and $|\psi_1\rangle$ is needed. Note that from equations (17.53) and (17.54)

$$|I_0\rangle = \frac{e^{-ik_n^-}}{\sqrt{|v(k_n^-)|}}|\chi_{k_n^-}\rangle, \qquad |O_0\rangle = \sum_{m=1}^{r} t_{mn} \frac{e^{-ik_m^{++})}}{\sqrt{|w(k_m^{++}, k_m^{++})|}}|\chi_{k_m^{++}}\rangle, \tag{17.56}$$

$$|I_1\rangle = \frac{1}{\sqrt{|v(k_n^-)|}}|\chi_{k_n^-}\rangle \quad \text{and} \quad |O_1\rangle = \sum_{m=1}^{r} t_{mn} \frac{1}{\sqrt{|w(k_m^{++}, k_m^{++})|}}|\chi_{k_m^{++}}\rangle \tag{17.57}$$

To obtain relationships between these states, it is useful to introduce translation operators. These are constructed by noting that since states $|\chi_{k_n^{++}}\rangle$ belonging to different wave vectors are not necessarily orthogonal, it is convenient to introduce their dual vectors $|\chi_{k_n^{++}})$, which satisfy the orthonormality conditions

$$(\chi_{k_m^{++}}|\chi_{k_m^{++}}\rangle = \delta_{nm} \quad \text{and} \quad (\chi_{k_m^{--}}|\chi_{k_m^{--}}\rangle = \delta_{nm} \tag{17.58}$$

and the completeness conditions

$$\sum_{n=1}^{r} |\chi_{k_n^{++}}\rangle\langle\chi_{k_n^{++}}| = \sum_{n=1}^{r} |\chi_{k_n^{--}}\rangle\langle\chi_{k_n^{--}}| = I \tag{17.59}$$

These dual vectors are discussed in sections B.1 and B.3 of appendix B at the end of this chapter, where it is noted that they can be used to define translation operators denoted $\eta_{\sigma\sigma'}(J)$. (To clarify the notation, it is recommended that the reader follows the discussion in sections B.1–B.3 of appendix B.) Here, the translation operators $\eta_{++}(J)$ and $\eta_{-\sigma}(J)$ will be utilised, defined by

$$\eta_{++}(J) = \sum_{n=1}^{r} e^{ik_n^{++}J} |\chi_{k_n^{++}}\rangle\langle\chi_{k_n^{++}}| \quad \text{and} \quad \eta_{-\sigma'}(J) = \sum_{n=1}^{r} e^{ik_n^{-\sigma'}J} |\chi_{k_n^{-\sigma'}}\rangle\langle\chi_{k_n^{-\sigma'}}| \tag{17.60}$$

where the set of wave vectors $\{k_n^{\sigma\sigma'}\}$ is formed from the real wave vectors $\{k_n^{\sigma}\}$ and the complex wave vectors $\{c_n^{\sigma'}\}$.

From equations (17.56) and (17.57), these relate $|O_0\rangle$ to $|O_1\rangle$ and $|I_0\rangle$ to $|I_1\rangle$ via the equations

$$|O_0\rangle = \eta_{++}(-1)|O_1\rangle \quad \text{and} \quad |I_0\rangle = \eta_{-\sigma'}(-1)|I_1\rangle \tag{17.61}$$

In the second of equations (17.60), the wave vectors $k_n^{-\sigma'}$ run over the real incoming wave vectors k_n^{-} with channel vectors $|\chi_{k_n^-}\rangle$ and over the complex wave vectors $c_n^{\sigma'}$ with channel vectors $|\chi_{c_n^{\sigma'}}\rangle$. For the case where $k_n^{-\sigma'}$ is complex, $\langle\chi_{c_n^{\sigma'}}|I_1\rangle = \langle\chi_{c_n^{\sigma'}}|\chi_{k_n^-}\rangle = 0$, which means that in equation (17.61), σ' could be chosen arbitrarily to be either + or −.

Since equation (17.41) is of the form

$$\Sigma\left[|I_1\rangle + |O_1\rangle\right] = H_{-1}\left[|I_0\rangle + |O_0\rangle\right], \tag{17.62}$$

equations (17.61) yield

$$\Sigma\left[|I_1\rangle + |O_1\rangle\right] = H_{-1}\,\eta_{-\sigma'}(-1)|I_1\rangle + H_{-1}\,\eta_{++}(-1)|O_1\rangle \tag{17.63}$$

This can be written

$$\left[g_{++}^{-1} - \Sigma\right]|O_1\rangle = -\left[g_{-\sigma'}^{-1} - \Sigma\right]|I_1\rangle \tag{17.64}$$

where g_{++}^{-1} and $g_{-\sigma'}^{-1}$ are defined by

$$g_{++}^{-1} = H_{-1}\eta_{++}(-1) \quad \text{and} \quad g_{-\sigma'}^{-1} = H_{-1}\eta_{-\sigma'}(-1) \tag{17.65}$$

As shown in appendix B at the end of this chapter, g_{++} and g_{-+} are retarded and advanced surface Green's functions of the semi-infinite lead of figure 17.2c, arising when $H_{01} = 0$.

Equation (17.64) is a relationship between the outgoing state $|O_1\rangle$ and the incoming state $|I_1\rangle$ and is therefore the solution to the scattering problem.

Starting from the fundamental equation (17.64) one can derive an alternative form of this solution (equation (17.71) below), which leads to a convenient expression for transmission amplitudes. This can be derived directly, as shown in section B.12 of appendix B. Alternatively, note that with the above notation, equation (17.32) is

$$g^{-1}|f\rangle = H_{01}\big[\,|I_1\rangle + |O_1\rangle\big] \tag{17.66}$$

and equation (17.40) is

$$H_{10}|f\rangle = g_{-\sigma'}^{-1}|I_1\rangle + g_{++}^{-1}|O_1\rangle \tag{17.67}$$

$$|O_1\rangle = g_{++}H_{10}|f\rangle - g_{++}g_{-\sigma'}^{-1}|I_1\rangle \tag{17.68}$$

Equations (17.68) and (17.66) combine to yield

$$|f\rangle = G_{00}H_{01}\Big[1 - g_{++}g_{-\sigma'}^{-1}\Big]|I_1\rangle \tag{17.69}$$

where

$$\Sigma_0 = H_{01}g_r H_{10} \quad \text{and} \quad G_{00} = [g^{-1} - \Sigma_0]^{-1} \tag{17.70}$$

Substituting equation (17.69) into equation (17.68) yields

$$|O_1\rangle = g_{++}H_{10}G_{00}H_{01}g_{++}\Big[g_{++}^{-1} - g_{-\sigma'}^{-1}\Big]|I_1\rangle - g_{++}g_{-\sigma'}^{-1}|I_1\rangle \tag{17.71}$$

Equation (17.71) is simply an alternative form of equation (17.64).

Provided the procedure described in section A.3 of appendix A (below) is followed, the matrix H_{-1} is invertible and therefore

$$g_{++} = \sum_{k_n^{++}} e^{ik_n^{++}}|\chi_{k_n^{++}}\rangle\langle\chi_{k_n^{++}}|(H_{-1})^{-1} \tag{17.72}$$

(The reader should check that when the right-hand side of equation (17.72) is multiplied into the right-hand side of equation (17.65), after using equation (17.59), the result is the unit matrix.)·

Hence

$$\langle\chi_{k_p^+}|g_{++}g_{-\sigma'}^{-1}|I_1\rangle = \frac{1}{\sqrt{|v(k_n^-)|}}e^{i(k_p^+ - k_n^-)}\langle\chi_{k_p^+}|\chi_{k_n^-}\rangle$$

From equation (17.57), $\langle\chi_{k_p^+}|O_1\rangle = t_{pn}\frac{1}{\sqrt{|v((k_p^+))|}}$ and therefore from equation (17.70),

$$t_{pn} = \sqrt{\frac{|v(k_p^+)|}{|v(k_n^-)|}}\left[\langle\chi_{k_p^+}|g_{++}H_{10}G_{00}H_{01}g_{++}[g_{++}^{-1} - g_{-\sigma'}^{-1}]|\chi_{k_n^-}\rangle - e^{i(k_p^+ - k_n^-)}\langle\chi_{k_p^+}|\chi_{k_n^-}\rangle\right] \tag{17.73a}$$

In appendix B, equation (B.93), it is proved that $\left(g_{++}^{-1} - g_{-\sigma'}^{-1}\right)|\chi_{k_n^\sigma}\rangle = i\hbar|v(k_n^\sigma)||\chi_{k_n^\sigma}\rangle$. Hence the above expression simplifies to

$$\boxed{t_{pn} = i\hbar\sqrt{|v(k_p^+)||v(k_n^-)|}\left[(\chi_{k_p^+}|g_{++}H_{10}G_{00}H_{01}g_{++}|\chi_{k_n^-}\rangle - e^{i(k_p^+-k_n^-)}(\chi_{k_p^+}|\chi_{k_n^-}\rangle\right]} \quad (17.73b)$$

Equation (17.73) will be re-derived in section 17.16 using Green's functions alone. In the meantime, it is useful to clarify the physical meaning of the various terms in this general equation by applying it to the multi-lead structure of chapter 9. In the notation of chapter 9, $H_{10} \to V$, $H_{01} \to W$ and $H_{-1} \to -\gamma$, where γ, V and W are defined below equation (9.8). Also, for the simple one-dimensional leads of chapter 9, $k_p^- = -k_p^+$, and in terms of the unit vector defined above equation (9.9), $(\chi_{k_p^+}| = \langle p|$, $|\chi_{k_n^-}\rangle = |n\rangle$ and therefore $(\chi_{k_p^+}|\chi_{k_n^-}\rangle = \delta_{pn}$. Hence from equations (17.60) and (17.65), $(\chi_{k_p^+}|g_{++} = -\gamma_p^{-1}e^{ik_p^+}\langle p|$, $g_{++}|\chi_{k_n^-}\rangle = -\gamma_n^{-1}e^{-ik_n^+}|n\rangle$ and $g_{-\sigma'}|\chi_{k_n^-}\rangle = -\gamma_n^{-1}e^{ik_n^+}|n\rangle$. Therefore $g_{++}\left[g_{++}^{-1} - g_{-\sigma'}^{-1}\right]|\chi_{k_n^-}\rangle = \left(1 - e^{2ik_n^+}\right)|n\rangle = -2i\sin k_n^+e^{ik_n^+}|n\rangle$. With this notation, equation (17.73a) becomes $t_{pn} = i\sqrt{\frac{|v(k_p^+)|}{|v(k_n^+)|}}\left[\gamma_p^{-1}e^{i\bar{k}_p}\langle p|VG_{00}W|n\rangle 2\sin k_n^+e^{ik_n^+} - e^{i2k_p^+}\delta_{pn}\right]$ $= i\frac{\sqrt{\hbar|v(k_p^+)|}}{\gamma_p}e^{ik_p^+}\langle p|VG_{00}W|n\rangle e^{ik_n^+}\frac{\sqrt{\hbar|v(k_n^+)|}}{\gamma_n} - e^{i2k_p^+}\delta_{pn}$. Since G_{00} is defined by equation (17.70), which is the same as equation (9.22), the above result for t_{pn} is identical to equation (9.24).

17.9 Determining the allowed wave vectors for a given energy *E*

To make use of the above expressions and when constructing Green's functions in later sections, a method for computing the allowed wave vectors for a given energy will now be discussed. For a fixed choice of E, these are obtained by solving the equation

$$H(k)|\chi\rangle = E|\chi\rangle \quad (17.74)$$

where

$$H(k) = H_0 + H_1e^{ik} + H_{-1}e^{-ik}$$

If the inverse of H_1 exists, then the Schrödinger equation (17.3) can be written

$$H_1^{-1}(E - H_0)|\chi_n\rangle = e^{ik_n}|\chi_n\rangle + H_1^{-1}H_{-1}e^{-ik_n}|\chi_n\rangle \quad (17.75)$$

If the inverse of H_1 does not exist, then one should first adopt the decimation procedure discussed in appendix A and replace H_0 and H_1 by renormalised matrices, such that the inverse of the renormalized H_1 does exist. As a consequence of this procedure, the number of incoming (and outgoing) wave vectors is equal to the rank of H_1. In what follows, it is assumed that where necessary, such a replacement has been made and that H_0 and H_1 denote the renormalized matrices.

Writing

$$\lambda_n = e^{ik_n} \qquad \text{and} \qquad |\theta_n\rangle = \lambda_n^{-1}|\chi_n\rangle \tag{17.76}$$

yields

$$\hat{H}|y_n\rangle = \lambda_n|y_n\rangle \tag{17.77}$$

where

$$\hat{H} = \begin{pmatrix} H_1^{-1}(E - H_0) & -H_1^{-1}H_{-1} \\ I & 0 \end{pmatrix} \tag{17.78}$$

and

$$|y_n\rangle = \begin{pmatrix} |\chi_n\rangle \\ |\theta_n\rangle \end{pmatrix} \tag{17.79}$$

For a fixed value of E, the eigenvalue problem (17.77) can be solved numerically to yield the allowed values of k_n and the corresponding vectors $|\chi_n\rangle$.

Note that from equation (17.4), for a given E,

$$\det(\mathrm{E} - H(k)) = \det\left(E - H_0 - H_1\lambda_n - \frac{H_{-1}}{\lambda_n}\right) = p(\lambda_n) \tag{17.80}$$

and therefore the secular equation is $p(\lambda_n) = 0$, where $p(\lambda)$ is a polynomial in λ of degree $2N$ or $2r$ if $r<N$). Taking the Hermitian conjugate of equation (17.80) yields

$$\det(\mathrm{E} - H^{\dagger}(k)) = \det\left(E - H_0 - H_{-1}\lambda_n^* - \frac{H_1}{\lambda_n^*}\right) = p(1/\lambda_n^*) \tag{17.81}$$

and therefore the secular equation can also be written $p(1/\lambda_n^*) = 0$. This shows that if λ_n is a solution, then $1/\lambda_n^*$ is also a solution. If k_n is real, then $1/\lambda_n^* = \lambda_n$ and no extra information is obtained. On the other hand if $k_n = a_n - ib_n$, where a_n and b_n are real, then this shows that if $e^{ik_n} = e^{ia_n}e^{b_n}$ is a solution, then $e^{ia_n}e^{-b_n}$ is also a solution. That is, if k_n is a solution, then k_n^* is also a solution. Therefore to each eigenstate $|\chi_n\rangle e^{ik_nJ}$, which increases with increasing J, there is a partner eigenstate, which decreases with increasing J.

Similarly, if k_n is real, then the following argument shows that to each eigenstate $|\chi_n\rangle e^{ik_nJ}$, which carries a positive current in the direction of decreasing J, there is a partner eigenstate, which carries a positive current in the direction of increasing J. To demonstrate this, note that from equation (17.21), if k_n is real, the current carried by the corresponding eigenstate state is

$$I_{J\to J+1} = -\frac{2}{\hbar}\mathrm{Im}\,\langle\chi_n|H_1|\chi_n\rangle e^{ik_n} = \frac{i}{\hbar}[\langle\chi_n|H_1|\chi_n\rangle e^{ik_n} - \langle\chi_n|H_{-1}|\chi_n\rangle e^{-ik_n}] \tag{17.82}$$

On the other hand, from equation (17.3), (where the subscript n is omitted for convenience)

$$\langle\chi_k|H(k)|\chi_k\rangle = \langle\chi_k|H_0 + H_1e^{ik} + H_{-1}e^{-ik}|\chi_k\rangle = E(k) \tag{17.83}$$

Differentiating with respect to k yields

$$\frac{\partial E(k)}{\partial k} = \left[\frac{\partial\langle\chi_k|}{\partial k}\right]H(k)|\chi_k\rangle + \langle\chi_k|\left[\frac{\partial H(k)}{\partial k}\right]|\chi_k\rangle + \langle\chi_k|H(k)\left[\frac{\partial|\chi_k\rangle}{\partial k}\right]$$

Since $H(k)|\chi_k\rangle = E(k)|\chi_k\rangle$ and $\langle\chi_k|H(k) = E(k)\langle\chi_k|$, one finds $\left[\frac{\partial\langle\chi_k|}{\partial k}\right]H(k)|\chi_k\rangle + \langle\chi_k|H(k)\left[\frac{\partial|\chi_k\rangle}{\partial k}\right] = E(k)\left(\left[\frac{\partial\langle\chi_k|}{\partial k}\right]|\chi_k\rangle + \langle\chi_k|\left[\frac{\partial|\chi_k\rangle}{\partial k}\right]\right) = E(k)\frac{\partial\langle\chi_k|\chi_k\rangle}{\partial k} = 0$, where the last step follows from the normalisation condition $\langle\chi_k|\chi_k\rangle = 1$ for all k. Hence

$$\frac{\partial E(k)}{\partial k} = \langle\chi_k|\left[\frac{\partial H(k)}{\partial k}\right]|\chi_k\rangle = i\langle\chi_k|[H_1e^{ik} - H_{-1}e^{-ik}]|\chi_k\rangle \tag{17.84}$$

Comparing this with equation (17.82) shows that the current carried by such a state is,

$$I_{J\to J+1} = \frac{1}{\hbar}\frac{\partial E(k)}{\partial k} = v(k) \tag{17.85}$$

where $v(k)$ is the group velocity given by

$$\hbar v(k) = \frac{\partial E(k)}{\partial k} \tag{17.86}$$

This shows that equation (17.86), which was derived for a rather simple Hamiltonian in chapter 6, holds for a finite-width crystalline lead described by an arbitrary mean-field Hamiltonian. Furthermore, since $E(k)$ is a periodic function satisfying $E(k + 2\pi) = E(k)$, then a line of constant energy (such as the dashed red line in figure 5.8) must intersect the curve $E(k)$ at an even number of values of k, such that $\frac{\partial E(k)}{\partial k}$ is positive at half of these values of k (denoted k_n^+) and $\frac{\partial E(k)}{\partial k}$ is negative at the other half (denoted k_n^-). (An exception to this pairing rule occurs when E coincides with the upper or lower mini-band edge, at which $\frac{\partial E(k)}{\partial k} = 0$. However such states do not contribute to the current and therefore as noted earlier, they can be described using a limiting procedure in which the band edge is approached from above or below.) For the Hamiltonians considered in earlier chapters, where $E(k)$ is an even function of k satisfying $E(k) = E(-k)$, it was found that $k_n^- = -k_n^+$, but more generally, this will not be the case.

17.10 The Green's function of an arbitrary material-specific, finite-width, doubly infinite lead

So far, the transmission amplitudes have been obtained by wave function matching at the interface between the scattering region and the super lead. In this section expressions for transmission amplitudes are obtained using Green's functions. The first step is to obtain an expression for the retarded Green's function $g(L, L')$ of a doubly infinite lead, which satisfies

$$(EI - H_0)g(J, J') - H_1 g(J+1, J') - H_{-1}g(J-1, J') = I\delta_{IJ} \tag{17.87}$$

where H_0 and H_1 are NxN matrices and (if necessary following the procedure in appendix A, this chapter) H_1 is invertible. Since $g(J, J')$ is an outgoing wave from the source at J', the most general solution must have the form $g(J, J') = g^<(J, J')$ for $(J \leqslant J')$ and $g(J, J') = g^>(J, J')$ for $(J \geqslant J')$, where, after utilising the wave vector notation of appendix B,

$$g^<(J, J') = \sum_{n=1}^{r} e^{ik_n^{--}(J-J')}|\chi_{k_n^{--}}\rangle\langle A_n| \quad (J \leqslant J') \tag{17.88}$$

$$g^>(J, J') = \sum_{n=1}^{r} e^{ik_n^{++}(J-J')}|\chi_{k_n^{++}}\rangle\langle \overline{A}_n| \quad (J \geqslant J') \tag{17.89}$$

In these expressions, the vectors $\langle A_n|$ and $\langle \overline{A}_n|$ are yet to be determined and to ensure the outgoing nature of g, (ie to ensure that g is a retarded Green's function) k_n^{++} and k_n^{--} are chosen such that the contributions to the current $I_{J'\to J}$ from each term in the sums are outgoing from the source at J'.

To obtain expressions for the vectors $\langle A_n|$ and $\langle \overline{A}_n|$, note that for $J = J'$, the following continuity condition must be satisfied

$$g(J, J) = \sum_{n=1}^{r} |\chi_{k_n^{--}}\rangle\langle A_n| = \sum_{\overline{k}_n} |\chi_{k_n^{++}}\rangle\langle \overline{A}_n| \tag{17.90}$$

As is section 17.7, to solve this for $\langle A_n|$ and $\langle \overline{A}_n|$, introduce the dual vectors $|\chi_{k_m^{--}})$ and $|\chi_{k_m^{++}})$ which satisfy the orthonormality conditions

$$(\chi_{k_m^{++}}|\chi_{k_n^{++}}\rangle = \delta_{nm} \quad \text{and} \quad (\chi_{k_m^{--}}|\chi_{k_n^{--}}\rangle = \delta_{nm} \tag{17.91}$$

and the completeness conditions

$$\sum_{n=1}^{r} |\chi_{k_n^{++}}\rangle(\chi_{k_n^{++}}| = \sum_{n=1}^{r} |\chi_{k_n^{--}}\rangle(\chi_{k_n^{--}}| = I \tag{17.92}$$

Equation (17.90) is then satisfied by writing $\langle A_n| = (\chi_{k_n^{--}}|(i\hbar V)^{-1}$ and $\langle \overline{A}_n| = (\chi_{k_n^{++}}|(i\hbar V)^{-1}$, where V is an invertible square NxN matrix, which is yet to be determined[2]. This yields

$$g^<(J, J') = \sum_{n=1}^{r} e^{ik_n^{--}(J-J')}|\chi_{k_n^{--}}\rangle(\chi_{k_n^{--}}|(i\hbar V)^{-1} \tag{17.93}$$

$$g^>(J, J') = \sum_{n=1}^{r} e^{ik_n^{++}(J-J')}|\chi_{k_n^{++}}\rangle(\chi_{k_n^{++}}|(i\hbar V)^{-1} \tag{17.94}$$

[2] The notation $i\hbar V$ is used, because as shown later, V is a velocity operator, dimensions $(\text{time})^{-1}$.

After noting equations (17.92), this clearly satisfies equation (17.90). To determine the matrix V, note that for $J = J'$, equation (17.87) yields

$$(EI - H_0)g(J, J) - H_1 g(J + 1, J) - H_{-1} g(J - 1, J) = I \qquad (17.95a)$$

i.e.

$$(EI - H_0)g^>(J, J) - H_1 g^>(J + 1, J) - H_{-1} g^<(J - 1, J) = I \qquad (17.95b)$$

On the other hand $(EI - H_0)g^>(J, J) - H_1 g^>(J + 1, J) - H_{-1} g^>(J - 1, J) = 0$ and therefore equation (17.95b) simplifies to

$$H_{-1} g^>(J - 1, J) - H_{-1} g^<(J - 1, J) = I \qquad (17.95c)$$

Substituting equations (17.93) and (17.94) into equation (17.95c) yields

$$\sum_{n=1}^{r} e^{-ik_n^{++}} H_{-1}|\chi_{k_n^{++}}\rangle\langle\chi_{k_n^{++}}|(i\hbar V)^{-1} - \sum_{n=1}^{r} e^{-ik_n^{--}} H_{-1}|\chi_{k_n^{--}}\rangle\langle\chi_{k_n^{--}}|(i\hbar V)^{-1} = I$$

Hence,

$$i\hbar V = H_{-1}\big(\eta_{++}(- 1) - \eta_{--}(- 1)\big) \qquad (17.96)$$

where $\eta_{++}(- 1)$ and $\eta_{--}(- 1)$ are defined in appendix B, of this chapter, section B.3. As discussed in appendix B, section B.9, V can be regarded as a velocity operator, since expectation values of V yield the modulus of group velocities $|v(k_n^{\pm})|$, defined in equation (17.86).

17.11 The Green's function of an arbitrary finite-width, semi-infinite lead

Consider a crystalline lead with cells labelled $J = J_0 + 1, J_0 + 2, \ldots \infty$. Since $g(J_0, J') = 0$, one must subtract an outgoing wave from equation (17.93) and (17.94) to satisfy this boundary condition.

Such a wave is

$$W_J = \sum_{m=1}^{r} |\chi_{k_n^{++}}\rangle e^{ik_m^{++}(J-J_0)}\Delta_m \qquad (17.97)$$

where the matrix Δ_m is to be determined. This yields for the retarded Green's function $\tilde{g}(J, J')$ of a semi-infinite lead

$$\tilde{g}(J, J') = \sum_{n=1}^{r} e^{ik_n^{--}(J-J')}|\chi_{k_n^{--}}\rangle\langle\chi_{k_n^{--}}|(i\hbar V)^{-1} - W_J(J \leqslant J') \qquad (17.98)$$

and

$$\tilde{g}(J, J') = \sum_{n=1}^{r} e^{ik_n^{++}(J-J')}|\chi_{k_n^{++}}\rangle\langle\chi_{k_n^{++}}|(i\hbar V)^{-1} - W_J \quad (J \geqslant J') \qquad (17.99)$$

For $J = J_0 \leqslant J'$, setting $\tilde{g}(J_0, J') = 0$ yields from equations (17.97) and (17.98),

$$\sum_{n=1}^{r} e^{ik_n^{--}(J_0-J')}|\chi_{k_n^{--}}\rangle(\chi_{k_n^{--}}|(i\hbar V)^{-1} - \sum_{m=1}^{r}|\chi_{k_m^{++}}\rangle\Delta_m \tag{17.100}$$

Hence from equation (17.92), Δ_m is given by

$$\Delta_m = \sum_{n=1}^{r} e^{ik_n^{--}(J_0-J')}(\chi_{k_m^{++}}|\chi_{k_n^{--}}\rangle(\chi_{k_n^{--}}|(i\hbar V)^{-1} \tag{17.101}$$

and therefore

$$W_J = W'_J(i\hbar V)^{-1} \tag{17.102}$$

where

$$W'_J = \sum_{m,n=1}^{r} |\chi_{k_m^{++}}\rangle e^{ik_m^{++}(J-J_0)}(\chi_{k_m^{++}}|\chi_{k_n^{--}}\rangle e^{ik_n^{--}(J_0-J')}(\chi_{k_n}| \tag{17.103}$$

17.12 The surface Green's function of an arbitrary finite-width, semi-infinite lead

Setting $J = J' = J_0 + 1$ yields for the (retarded) surface Green's function

$$g_r = \tilde{g}(J_0 + 1, J_0 + 1) = \left[\sum_{n=1}^{r}|\chi_{k_n^{--}}\rangle(\chi_{k_n^{--}}| - W'_{J_0+1}\right](i\hbar V)^{-1} \tag{17.104}$$

Utilising equation (17.92), reduces this to

$$g_r = \left[1 - W'_{J_0+1}\right](i\hbar V)^{-1} \tag{17.105}$$

where, after using the notation in section B.3 of appendix B,

$$W'_{J_0+1} = \sum_{m,n}|\chi_{k_n^{++}}\rangle e^{ik_m^{++}}(\chi_{k_m^{++}}|\chi_{k_n^{--}}\rangle e^{-ik_n^{--}}(\chi_{k_n^{--}}| = \eta_{++}(1)\eta_{--}(-1) \tag{17.106}$$

Note that from equation (17.96),

$$i\hbar V = H_{-1}\eta_{++}(-1)(1 - \eta_{++}(1)\eta_{--}(-1)) \tag{17.107}$$

Hence

$$g_r = \left[1 - \eta_{++}(1)\eta_{--}(-1\right](i\hbar V)^{-1} = \eta_{++}(1)(H_{-1})^{-1} \tag{17.108}$$

and therefore

$$g_r = \sum_{n=1}^{r} e^{ik_n^{++}}|\chi_{k_n^{++}}\rangle(\chi_{k_n^{++}}|(H_{-1})^{-1} \tag{17.109}$$

Clearly, g_r is identical to the Green's function g_{++} of equation (17.72).

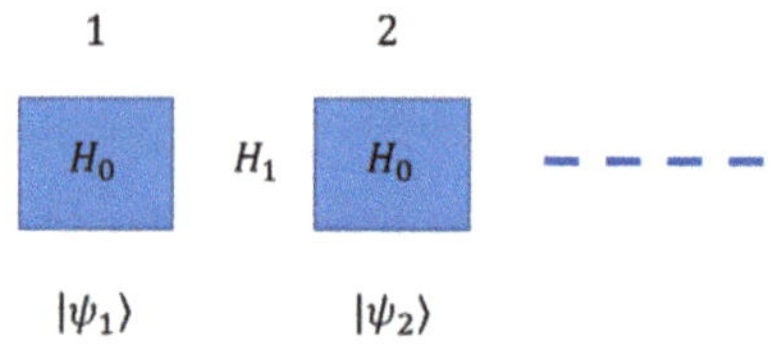

Figure 17.3. A semi-infinite lead terminating at cell $J = 1$.

As an example, for a simple one-dimensional tight binding shown in figure 10.3, where $H_{-1} = -\gamma$ and $k_n^{++} = k$, this reproduces equation (10.46).

17.13 A simple derivation of the surface Green's function of an arbitrary finite-width, semi-infinite lead

In section 17.11, equations (17.98) and (17.99) are equations for the Green's function of a semi-infinite lead, for all values of J and J', which as a special case yields the surface Green's function. It is interesting to note that the above result for the surface Green's function could be obtained more directly using continuity, as follows. Consider the semi-infinite lead shown in figure 17.3, for which $J_0 = 0$.

For this structure, the Green's function $g(J, 1)$ satisfies equation (17.95) for $J \neq 1$)

$$(EI - H_0)g(J, 1) - H_1 g(J + 1,1) - H_{-1} g(J - 1,1) = 0 \tag{17.110}$$

whereas for $J = 1$, since cell $J = 0$ is absent,

$$(EI - H_0)g(1,1) - H_1 g(2,1) = I \tag{17.111}$$

Since $g(J, 1)$ is a wave function due to a source in cell 1, it must be a sum of outgoing wave functions of the form

$$g(J, 1) = \sum_n e^{ik_n^{++}J} |\chi_{k_n^{++}}\rangle a_n, \tag{17.112}$$

where a_n is to be determined.

Since the vector $|\psi_J\rangle = e^{ik_n^{++}J} |\chi_{k_n^{++}}\rangle$ satisfies $(EI - H_0)|\psi_J\rangle - H_1|\psi_{J+1}\rangle - H_{-1}|\psi_{J-1}\rangle = 0$ for all J, it satisfies $(EI - H_0)|\psi_1\rangle - H_1|\psi_2\rangle = H_{-1}|\psi_0\rangle$. Therefore when the right-hand side of equation (17.112) is substituted into equation (17.111), one obtains

$$H_{-1} g(0,1) = I \tag{17.113}$$

i.e. $\sum_n |\chi_{k_n^{++}}\rangle a_n = (H_{-1})^{-1}$. Comparing this with equation (17.92) yields $a_n = \langle \chi_{k_n^{++}}|(H_{-1})^{-1}$ and therefore

$$g(1,1) = \sum_n e^{ik_n^{++}} |\chi_{k_n^{++}}\rangle\langle\chi_{k_n^{++}}|(H_{-1})^{-1}, \tag{17.114}$$

which is identical to equation (17.109), with the notation $g_r = g(1,1)$.

17.14 Dyson's equation for the surface Green's function of an arbitrary finite-width, semi-infinite lead

The aim of this section is to derive an expression for the surface Green's function using Dyson's equation, which avoids the need to solve explicitly for channel wave vectors. To obtain an equation for the surface Green's function using Dyson's equation, consider the structure shown in figure 17.4, composed of a substructure A (coloured green) interacting with a substructure B (coloured pink), with nearest neighbour inter-cell interactions H_1.

Dyson's equation for such a structure is

$$G_{AA} = \left(g_A^{-1} - \Sigma_A\right)^{-1} \tag{17.115}$$

where the self-energy is

$$\Sigma_A = H_1 g_B H_{-1}. \tag{17.116}$$

In this expression, since H_1 is coupled only to the surface cell 2, g_B is the Green's function of the pink region on cell 2. That is, g_B is the surface Green's function of a chain of $N - 1$ cells, which is conveniently denoted G_{N-1}. Similarly, in equation (17.119) G_{AA} is the surface Green's function of a chain of N cells, which is conveniently denoted G_N. Furthermore, g_A is the Green's function of an isolated cell, which is conveniently denoted g_0 and is given by

$$g_0 = (EI - H_0)^{-1} \tag{17.117}$$

Using this notation, equation (17.115) becomes

$$G_N = \left[g_0^{-1} - H_1 G_{N-1} H_{-1}\right]^{-1} \tag{17.118}$$

In the limit $N \to \infty$, if the surface Green's function of a N-cell chain tends to the surface Green's function of a $(N - 1)$-cell chain, then equation (17.119) becomes

$$G_\infty = \left[g_0^{-1} - H_1 G_\infty H_{-1}\right]^{-1} \tag{17.119}$$

When solving this equation, one finds that the solution is not unique and therefore further criteria are needed to determine which solution to choose. This non-uniqueness is evident if one replaces G_∞, g_0, H_1 and H_{-1} by numbers, as would be

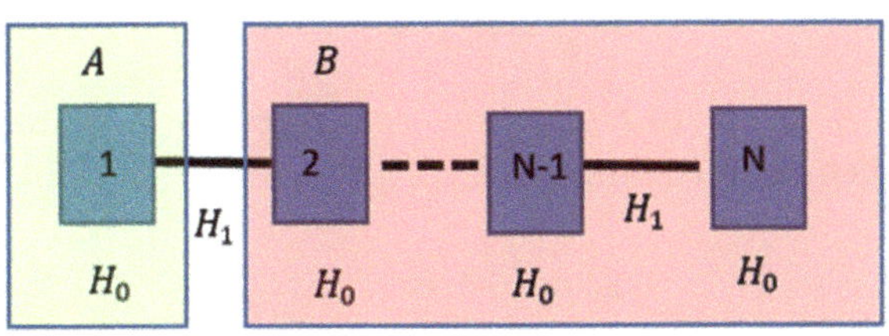

Figure 17.4. A chain of N identical cells.

the case for a one-dimensional, tight-binding chain. In this case, writing $H_1 = -\gamma$, $z = G_\infty\gamma$ and $g_0^{-1} = E - \epsilon_0 = -2\gamma\cos k$ yields

$$z = [-2\cos k - z]^{-1} \tag{17.120}$$

i.e. $z^2 + 2z\cos k + 1 = 0$ and therefore

$$z = -\cos k + is\sqrt{1 - \cos^2 k} = -\cos k + is\sin k \tag{17.121}$$

where $s = \pm 1$. Hence $G_\infty = -\frac{e^{-isk}}{\gamma}$. The presence of the undetermined sign s means that the solution is not unique and there is more than one solution to equation (17.120), even for a simple one-dimensional chain. Comparing this with equation (10.46) reveals that to obtain the retarded Green's function, one must choose $s = -1$. However this choice is not determined by equation (17.119). By choosing $s = -1$, one is simply selecting the retarded Green's function from a choice of two possible solutions.

In general, equation (17.119) cannot be solved analytically. To solve it numerically, it is useful to return to equation (17.118), and make a guess for $G_{N-1} = G_{\text{guess}}$. Then equation (17.118) yields

$$G_N = \left[g_0^{-1} - H_1 G_{\text{guess}} H_{-1}\right]^{-1} \tag{17.122}$$

Utilising equation (17.118) yields

$$G_{N+1} = \left[g_0^{-1} - H_1 G_N H_{-1}\right]^{-1}$$

Then again

$$G_{N+2} = \left[g_0^{-1} - H_1 G_{N+1} H_{-1}\right]^{-1}$$
$$\vdots$$
$$\vdots$$

This iterative procedure can be repeated until convergence is achieved.

Of course there is no guarantee that the series of iterations does converge. In fact, since the solution to equation (17.119) is not unique, the iteration cannot possibly converge unless an extra condition is imposed, such as the requirement that the wave is outgoing. This can be achieved by adding an imaginary part $i\eta$ to E, where η is a real number. To demonstrate this for a one-dimensional chain, consider the equation (17.118), which takes the form

$$z_N = [b - z_{N-1}]^{-1}$$

where $b = E - \epsilon_0 + i\eta$. If this iteration converges to a solution z_∞ satisfying $z_\infty = [b - z_\infty]^{-1}$, then as in equation (17.121),

$$z_\infty = \frac{b}{2} + isr \tag{17.123}$$

where $r = \sqrt{1 - \left(\frac{b}{2}\right)^2}$. In the theory of iterative non-linear maps, z_∞ is called a 'fixed point' of the recurrence relation. If successive iterations converge to the fixed point, then the fixed point is said to be stable. This stability of the fixed point means that if z_{N-1} is close to z_∞, then after the next iteration, z_N should be even closer. That is $|z_N - z_\infty|$ should be smaller than $|z_{N-1} - z_\infty|$. Therefore the fixed point is stable if

$$\left| \frac{z_N - z_\infty}{z_{N-1} - z_\infty} \right| < 1 \tag{17.124}$$

Since $[b - z_{N-1}]^{-1} \approx [b - z_\infty]^{-1} + [b - z_\infty]^{-2}(z_{N-1} - z_\infty) = z_\infty + [b - z_\infty]^{-2}(z_{N-1} - z_\infty)$, the stability condition (17.124) becomes

$$|b - z_\infty|^2 > 1 \tag{17.125}$$

From equation (17.123) this becomes

$$\left| \frac{b}{2} - isr \right|^2 > 1$$

Writing $b = b_0 + i\eta$ and Taylor expanding in β yields to first order in β

$$|b - z_\infty|^2 \approx 1 - \frac{s\eta}{r_0}$$

where $r_0 = \sqrt{1 - \left(\frac{b_0}{2}\right)^2}$. Hence the stability condition (17.125) becomes

$$s\eta < 0 \tag{17.126}$$

Therefore since $\eta > 0$, one finds that $s < 0$. Since $b_0 = E - \epsilon_0 = -2\cos k + i\eta$ and $r_0 = \sin k$, the iteration converges to the retarded Green's function. On the other hand, if η is chosen to be negative, then the iteration converges to the solution with $s > 0$, which corresponds to the advanced Green's function.

Clearly, if $\eta = 0$, then $|\frac{z_N - z_\infty}{z_{N-1} - z_\infty}| = 1$ and the fixed point is neither stable nor unstable, which means that the iteration does not converge and the surface Green's function will simply oscillate with increasing N. This is evident in the analytic solution for the Green's function of a one-dimensional chain presented in equation (10.70), which yields for the surface Green's function (i.e. choosing $j = p = 1$ in equation (10.70)),

$$G_N = \frac{-\sin kN}{\gamma \sin k(N+1)} \tag{17.127}$$

On the other hand, choosing $\eta > 0$ causes the Green's function $g(J, 1)$ to decay with increasing values of J, as discussed in appendix C in chapter 14, which means that for finite η, the Green's function of a large enough finite system approaches that of an infinite system.

It is interesting to note that channel wave vectors are not explicitly present in equation (17.119) and therefore if a numerical solution can be obtained, for example by iteration, their introduction is not needed. However, if channel eigenvectors are not introduced, then it would not be possible to obtain transmission amplitudes between individual channels. Furthermore, if analytic progress towards a solution is sought, then channel wave vectors are needed. To illustrate this feature, it is first interesting to recognise that a recurrence relation for $g(1,1)$ can be obtained from equation (17.111), by noting that from equation (17.114)

$$g(1,1)H_{-1}g(1,1) = \sum_n e^{i2k_n^{++}}|\chi_{k_n^{++}}\rangle\langle\chi_{k_n^{++}}|(H_{-1})^{-1} = g(2,1), \tag{17.128}$$

Therefore equation (17.111) becomes

$$(EI - H_0)g(1,1) - H_1 g(1,1)H_{-1}g(1,1) = I \tag{17.129}$$

or equivalently, after writing $g(1,1) = g_r$

$$g_r = \left[g_0^{-1} - H_1 g_r H_{-1}\right]^{-1} \tag{17.130}$$

After writing $G_\infty = g_r$, this becomes identical to equation (17.119). This means that the solution to equation (17.119) is given by equation (17.114). As discussed in sections 17.9 and 17.10, to obtain the retarded Green's function, the wave vectors in equation (17.114) must be chosen to be outgoing. Other choices could be made, which again highlights the non-uniqueness of the solution to equation (17.119). In summary, if the retarded Green's function is sought, then one can either choose $\eta = 0$ and selected outgoing wave vectors only in equation (17.114), or choose $\eta > 0$ and obtain the surface Green's function by iteration.

17.15 Expressions for transmission and reflection amplitudes of a scatterer connected to arbitrary, material-specific leads

In section 17.15, the retarded Green's functions of a doubly infinite and a semi-infinite lead were discussed. More generally, consider a scatterer connected to left and right leads, such that the super lead of figure 17.2(b) is composed of two leads only, labelled left and right. Since the cells of the left lead are labelled $J = 0,1,2,\ldots\infty$ and those of the right lead are labelled $J = 0,1,\ldots.\infty$, a scatterered wave function takes the form

$$|\psi_J\rangle = \sqrt{\frac{1}{|v(k_n^-)|}}e^{ik_n^-J}|\chi_{k_n^-}\rangle + \sum_{m=1}^{r}\sqrt{\frac{1}{|v(k_m^{++})|}}r_{mn}e^{ik_m^{++}J}|\chi_{k_m^{++}}\rangle \quad (J \text{ in left lead}) \tag{17.131}$$

$$|\psi_J\rangle = \sum_{p=1}^{r}\sqrt{\frac{1}{|v(k_p^{++})|}}t_{pn}e^{ik_p^{++}J}|\chi_{k_p^{++}}\rangle \quad (J \text{ in right lead}) \tag{17.132}$$

In equation (17.132) k_p^{++} is an outgoing wave vector in the right lead, whereas in equation (17.131) k_m^{++} is an outgoing wave vector in the left lead.

Similarly, if J' and J belong to the left lead and $J \leqslant J'$, the Green's function in the left lead takes the form

$$G_{LL}(J, J') = \sum_{n=1}^{r} e^{ik_n^{--}(J-J')}|\chi_{k_n^{--}}\rangle\langle\chi_{k_n^{--}}|(i\hbar V_L)^{-1} + \sum_{m,n}\sqrt{\frac{|v(k_n^{--})|}{|v(k_m^{++})|}}r_{mn}e^{ik_m^{++}J}e^{-ik_n^{--}J'}|\chi_{k_m^{++}}\rangle\langle\chi_{k_n^{--}}|(i\hbar V_L)^{-1} \quad (17.133)$$

where

$$i\hbar V_L = -\sum_n e^{-ik_n^{--}}H_{-1}^L|\chi_{k_n^{--}}\rangle\langle\chi_{k_n^{--}}| + \sum_n e^{-ik_n^{++}}H_{-1}^L|\chi_{k_n^{++}}\rangle\langle\chi_{k_n^{++}}| \quad (17.134)$$

On the other hand, if J' belongs to the left lead and J belongs to the right lead, the Green's function in the right lead takes the form

$$G_{RL}(J, J') = \sum_{p,n}\sqrt{\frac{|v(k_n^{--})|}{|v(k_p^{++})|}}t_{pn}e^{ik_m^{++}J}e^{-ik_n^{--}J'}|\chi_{k_p^{++}}\rangle\langle\chi_{k_n^{--}}|(i\hbar V_L)^{-1} \quad (17.135)$$

In equation (17.135) k_p^{++} is an outgoing wave number in the right lead, whereas (17.132) k_n^{--} is an incoming wave number in the left lead.

Hence the transmission and reflection amplitudes connecting open channels take the form

$$t_{pn} = \sqrt{\frac{|v(k_p^{+})|}{|v(k_n^{-})|}}\langle\chi_{k_p^{+}}|G_{RL}(1,1)i\hbar V_L|\chi_{k_n^{-}}\rangle \quad (17.136)$$

$$r_{mn} = \sqrt{\frac{|v(k_m^{+})|}{|v(k_n^{-})|}}\left[\langle\chi_{k_m^{+}}|G_{LL}(1,1)i\hbar V_L|\chi_{k_n^{-}}\rangle - e^{ik_m^{+}}e^{-ik_n^{-}}\langle\chi_{k_m^{+}}|\chi_{k_n^{-}}\rangle\right] \quad (17.137)$$

These are the generalisations of equations (10.88) and (10.89) to the case of a scattering region connected to arbitrary material-specific leads. As discussed at the end of section (17.9), for the special one-dimensional leads are connected to a scatterer, this reproduces the expressions for scattering amplitudes derived in chapter 9.

17.16 Summary

In this chapter, the analysis of previous chapters has been generalised to describe a scatterer connected to arbitrary leads. Expressions for scattering amplitudes were obtained both by wave function matching and by constructing the appropriate Green's functions. Expressions for the latter were obtained using superpositions of outgoing plane waves or alternatively using Dyson's equations. During the discussion, it was found that the number of incoming or outgoing channels is equal to the rank r of the inter-unit-cell coupling matrix H_1 and if r is less than the dimensionality of H_1, then a decimation procedure should be used to obtain a renormalised H_1 whose inverse exists. The generalisations presented in this chapter are necessary when using material-specific Hamiltonians and since the late 1990s [1] have underpinned the development of modern quantum transport codes [1–4], which take their mean-field Hamiltonians from self-consistent descriptions, based on density functional theory.

Appendix B contains additional technical details underpinning the above discussion and proves an important symmetry relation (equation B.46) between Green's functions, which was used in the above analysis and plays an important role in developing alternative expressions for transmission coefficients, such as equation (B.108). Consequences of time reversal symmetry are discussed in section B.9.

Appendix A Solving the H_1 problem

A.1 Decomposing a singular matrix

Consider the product $A|v_n\rangle$, where A is a $N \times N$ square matrix and $|v_n\rangle$ is a N-component column vector. Clearly $|w_n'\rangle = A|v_n\rangle$ is either a N-component vector of finite-length or a N-component vector of zero length. In the former case, if $\langle w_n'|w_n'\rangle = \lambda_n^2$, where $\lambda_n \neq 0$, then it is convenient to write $|w_n'\rangle = \lambda_n|w_n\rangle$, where $|w_n\rangle$ is of unit length and without loss of generality, to choose λ_n to be real and $\lambda_n > 0$.

That is

$$A|v_n\rangle = \lambda_n|w_n\rangle, \quad \text{where } \lambda_n \text{ is real and } \lambda_n > 0 \tag{A.1}$$

Consider also the product $A|v_m\rangle = \lambda_m|w_m\rangle$, where λ_m is real and $\lambda_m > 0$. These equations combine to yield

$$\langle v_m|A^\dagger A|v_n\rangle = \lambda_n\lambda_m\langle w_m|w_n\rangle$$

Now choose $|v_n\rangle$ to be an eigenvector of the $N \times N$ Hermitian matrix $B = A^\dagger A$.

$$A^\dagger A|v_n\rangle = P_n|v_n\rangle \tag{A.2}$$

So that $\lambda_n\lambda_m\langle w_m|w_n\rangle = P_n\langle v_m|v_n\rangle$

Since the matrix B possesses N linearly independent eigenvectors satisfying the following orthonormality and completeness conditions

$$\langle v_m|v_n\rangle = \delta_{nm} \quad \text{and} \quad \sum_{n=1}^{N}|v_n\rangle\langle v_n| = I \tag{A.3}$$

one obtains

$$P_n = \lambda_n^2 \quad \text{and} \quad \langle w_m|w_n\rangle = \delta_{nm} \tag{A.4}$$

This shows that the vectors $|w_n\rangle$ are orthonormal, but not necessarily complete, because not all λ_n are non-zero and therefore there may be less than N of them.

If the matrix A is singular, then $\det A = 0$ and only $r < N$ rows of A are linearly independent. Therefore each row of A can be written as a linear superposition of r linearly independent orthonormal vectors. This means that the rth row $\langle r_i|$ can be written

$$\langle r_i| = \sum_{m=1}^{r} a_{mi}\langle v_m| \tag{A.5}$$

and therefore

$$A = \begin{pmatrix} \langle r_1| \\ \vdots \\ \langle r_i| \\ \vdots \\ \langle r_N| \end{pmatrix} = \sum_{m=1}^{r} \begin{pmatrix} a_{m1}\langle v_m| \\ \vdots \\ a_{mi}\langle v_m| \\ \vdots \\ a_{mN}\langle v_m| \end{pmatrix} = \sum_{m=1}^{r} \begin{pmatrix} a_{m1} \\ \vdots \\ a_{mi} \\ \vdots \\ a_{mN} \end{pmatrix} \langle v_m| \tag{A.6}$$

If the normalised vector $|w_m\rangle$ is defined by the relation

$$\begin{pmatrix} a_{m1} \\ \vdots \\ a_{mi} \\ \vdots \\ a_{mN} \end{pmatrix} = \lambda_m |w_m\rangle \tag{A.7}$$

where λ_n is chosen to be real and positive, then equation (A.6) becomes

$$A = \sum_{m=1}^{r} |w_m\rangle \lambda_m \langle v_m| \tag{A.8}$$

This expression for A is called a singular value decomposition. The number r is called the rank of A and is equal to the number of non-zero eigenvalues of A. If A is Hermitian, then $|w_m\rangle = |v_m\rangle$.

A.2 Solving an eigenvalue problem involving the singular matrix A

To solve the eigenvalue equation

$$A|\psi\rangle = E|\psi\rangle \tag{A.9}$$

it is convenient to write

$$|\psi\rangle = \sum_{n=1}^{N} \phi_n |v_n\rangle \tag{A.10}$$

In this equation, the vectors $|v_1\rangle$, $|v_2\rangle$, ... $|v_r\rangle$ are chosen to satisfy equation (A.6) and the remaining $N - r$ vectors are chosen to be any convenient set of orthonormal vectors, which are orthogonal to the first r vectors.

Substituting equations (A.8) and (A.10) into equation (A.9) yields

$$\sum_{n=1}^{N} \sum_{m=1}^{r} |w_m\rangle \lambda_m \langle v_m|v_n\rangle \phi_n = E \sum_{n=1}^{N} \phi_n |v_n\rangle$$

That is

$$\sum_{m=1}^{r} |w_m\rangle \lambda_m \phi_m = E \sum_{n=1}^{N} \phi_n |v_n\rangle$$

Multiplying from the left by $\langle v_p|$, where p is an integer in the range 1 to N, yields

$$\sum_{m=1}^{r} h_{pm}\phi_m = E\phi_p \tag{A.11}$$

where

$$h_{pm} = \langle v_p|w_m\rangle\lambda_m \tag{A.12}$$

Writing

$$|a\rangle = \begin{pmatrix}\phi_1\\ \vdots\\ \phi_r\end{pmatrix},\ |b\rangle = \begin{pmatrix}\phi_{r+1}\\ \vdots\\ \phi_N\end{pmatrix},\ h_{aa} = \begin{pmatrix}h_{11} & h_{1r}\\ h_{r1} & h_{rr}\end{pmatrix} \text{ and } h_{ba} = \begin{pmatrix}h_{r+1,1} & h_{r+1,r}\\ h_{N1} & h_{N,1}\end{pmatrix}$$

yields

$$h_{aa}|a\rangle = E|a\rangle \tag{A.13}$$

and

$$h_{ba}|a\rangle = E|b\rangle \tag{A.14}$$

This means that the r-component vector $|a\rangle$ should first be obtained by solving the $r \times r$ eigenvalue problem (A.13) and then the remaining $N - r$ amplitudes contained in $|b\rangle$ are obtained from $|b\rangle = h_{ba}|\phi\rangle/E$.

A.3 Solving the Schrödinger equation for a finite-width lead with a singular H_1

Now consider the slightly more complicated equation

$$(E - H_0)|\psi_J\rangle = H_{-1}|\psi_{J-1}\rangle + H_1|\psi_{J+1}\rangle \tag{A.15}$$

where H_0, H_1, H_{-1} are $N \times N$ square matrices and $|\psi_J\rangle$, $|\psi_{J\pm1}\rangle$ are N-component vectors. If only r rows of H_1 are linearly independent, (i.e. the rank of H_1 is $r < N$), then following equations (A.8) and (A.10), it is convenient to write

$$H_1 = \sum_{m=1}^{r} |w_m\rangle\lambda_m\langle v_m| \qquad H_{-1} = H_1^{\dagger} = \sum_{m=1}^{r} |v_m\rangle\lambda_m\langle w_m| \tag{A.16}$$

and

$$|\psi_J\rangle = \sum_{n=1}^{N} \phi_n(J)|v_n\rangle \tag{A.17}$$

Substituting equations (A.16) and (A.17) into equation (A.15) and multiplying from the left by $\langle v_p|$ yields

For $p \leqslant r$

$$\sum_{n=1}^{N}(E\delta_{pn} - (H_0)_{pn})\phi_n(J) = \sum_{n=1}^{N}(h^{\dagger})_{pn}\phi_n(J-1) + \sum_{m=1}^{r} h_{pm}\phi_m(J+1) \tag{A.18}$$

where

$$(h^\dagger)_{pn} = h^*_{np} = \lambda_p \langle w_p | v_n \rangle \qquad \text{and} \qquad (H_0)_{pn} = \langle v_p | H_0 | v_n \rangle \tag{A.19}$$

For $p > r$

$$\sum_{n=1}^{N} (E\delta_{pn} - (H_0)_{pn})\phi_n(J) = \sum_{m=1}^{r} h_{pm}\phi_m(J+1) \tag{A.20}$$

In this case, it is convenient to define

$$|a(J)\rangle = \begin{pmatrix} \phi_1(J) \\ \vdots \\ \phi_r(J) \end{pmatrix}, \quad |b(J)\rangle = \begin{pmatrix} \phi_{r+1}(J) \\ \vdots \\ \phi_N(J) \end{pmatrix},$$

$$h_{aa} = \begin{pmatrix} h_{11} & \dots & h_{1r} \\ \vdots & \dots & \vdots \\ h_{r1} & \dots & h_{rr} \end{pmatrix}, \quad h_{ba} = \begin{pmatrix} h_{r+1,1} & \cdots & h_{r+1,r} \\ \vdots & \dots & \vdots \\ h_{N1} & \cdots & h_{N,1} \end{pmatrix}$$

$$h_{ab} = \begin{pmatrix} h_{1,r+1} & \cdots & h_{1,N} \\ \vdots & \dots & \vdots \\ h_{r,r+1} & \cdots & h_{r,N} \end{pmatrix}, \quad h_{bb} = \begin{pmatrix} h_{r+1,r+1} & \cdots & h_{r+1,N} \\ \vdots & \dots & \vdots \\ h_{N,r+1} & \cdots & h_{NN} \end{pmatrix}$$

$$\overline{H}_{aa} = \begin{pmatrix} (H_0)_{11} & \cdots & (H_0)_{1r} \\ \vdots & \dots & \vdots \\ (H_0)_{r1} & \cdots & (H_0)_{rr} \end{pmatrix}, \quad \overline{H}_{ba} = \begin{pmatrix} (H_0)_{r+1,1} & \cdots & (H_0)_{r+1,r} \\ \vdots & \dots & \vdots \\ (H_0)_{N1} & \cdots & (H_0)_{N,1} \end{pmatrix}$$

$$\overline{H}_{ba} = \begin{pmatrix} (H_0)_{r+1,1} & \cdots & (H_0)_{r+1,r} \\ \vdots & \dots & \vdots \\ (H_0)_{N1} & \cdots & (H_0)_{N,1} \end{pmatrix}, \quad \overline{H}_{bb} = \begin{pmatrix} (H_0)_{r+1,r+1} & \cdots & (H_0)_{r+1,N} \\ \vdots & \dots & \vdots \\ (H_0)_{N,r+1} & \cdots & (H_0)_{NN} \end{pmatrix}$$

Using the above notation, one obtains

$$\begin{aligned}(EI - \bar{H}_{aa})|a(J)\rangle + \bar{H}_{ab}|b(J)\rangle = h^\dagger_{aa}&|a(J-1)\rangle \\ &+ h^\dagger_{ab}|b(J-1)\rangle + h_{aa}|a(J+1)\rangle\end{aligned} \tag{A.21}$$

$$\bar{H}_{ba}|a(J)\rangle - (EI - \bar{H}_{bb})|b(J)\rangle = h_{ba}|a(J+1)\rangle \tag{A.22}$$

From equation (A.22),

$$|b(J)\rangle = (EI - \bar{H}_{bb})^{-1}[\bar{H}_{ba}|a(J)\rangle - h_{ba}|a(J+1)\rangle] \tag{A.23}$$

Hence equation (A.21) becomes

$$(EI - \hat{H}_0)|a(J)\rangle = \hat{H}_{-1}|a(J-1)\rangle + \hat{H}_1|a(J+1)\rangle \tag{A.24}$$

where

$$\hat{H}_0 = \bar{H}_{aa} + \bar{H}_{ab}(EI - \bar{H}_{bb})^{-1}\bar{H}_{ba} + h_{ab}^{\dagger}(EI - \bar{H}_{bb})h_{ba} \tag{A.25}$$

$$\hat{H}_1 = h_{aa} + \bar{H}_{ab}(EI - \bar{H}_{bb})^{-1}h_{ba} \tag{A.26}$$

$$\hat{H}_{-1} = \hat{H}_1^{\dagger} = h_{aa}^{\dagger} + h_{ab}^{\dagger}(EI - \bar{H}_{bb})^{-1}\bar{H}_{ba} \tag{A.27}$$

Equation (A.24) is an equation for the vector $|a(J)\rangle$ of cell J and involves invertible matrices $\hat{H}_1$ and $\hat{H}_{-1}$. The latter, along with $\hat{H}_0$ are renormalised Hamiltonians, obtained by decimating the amplitudes $|b(J)\rangle$. Clearly the terms on the right-hand sides of equations (A.25) to (A.27) are the self-energies resulting from this decimation. The decimation procedure is illustrated in figure A.1. After solving for $|a(J)\rangle$ in each cell J, the amplitudes $|b(J)\rangle$ are given by equation (A.23).

This demonstrates that the a lead composed of a periodic sequence of identical cells, coupled by a matrix H_1 of rank r, possesses $2r$ scattering channels.

A.4 An expression for inter-cell currents in terms of renormalized inter-cell couplings

Note that from equations (17.21), (A.16) and (A.17), the current carried by the state $|\psi_J\rangle$ is

$$J_{J\to J+1} = -\frac{2}{\hbar}\text{Im}\,\langle\psi_J|H_1|\psi_{J+1}\rangle = \sum_{n=1}^{N}\sum_{m=1}^{r}\sum_{p=1}^{N}\phi_p^*(J)\langle v_p|w_m\rangle\lambda_m\langle v_m|v_n\rangle\phi_n(J+1).$$

That is

$$J_{J\to J+1} = -\frac{2}{\hbar}\text{Im}\sum_{m=1}^{r}\sum_{p=1}^{N}\phi_p^*(J)h_{pm}\phi_m(J+1)$$

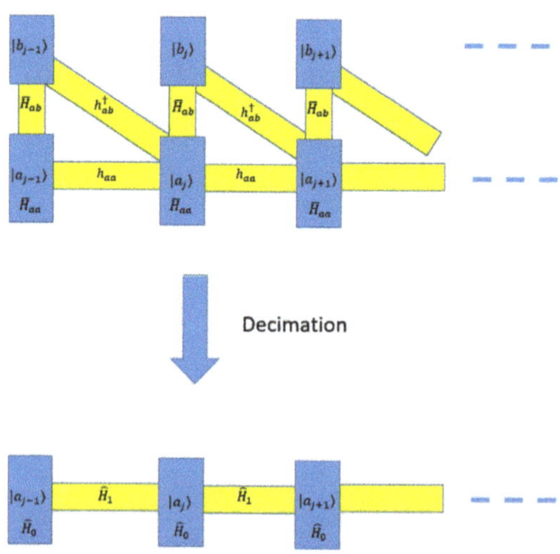

Figure A.1. The decimation procedure undertaken in proceeding from equations (A.21) and (A.22)–(A.24).

or equivalently

$$J_{J\to J+1} = -\frac{2}{\hbar}\mathrm{Im}\,\langle a(J)|h_{aa}|a(J+1)\rangle - \frac{2}{\hbar}\mathrm{Im}\,\langle b(J)|h_{ba}|a(J+1)\rangle) \tag{A.28}$$

From equation (A.23),

$$\langle b(J)| = \langle a(J)|\bar{H}_{ab}(EI - \bar{H}_{bb})^{-1} - \langle a(J+1)|h_{ab}(EI - \bar{H}_{bb})^{-1}$$

and therefore the second term on the right-hand side of equation (A.28) is

$$\begin{aligned} &-\frac{2}{\hbar}\mathrm{Im}\,\langle b(J)|h_{ba}|a(J+1)\rangle \\ &\quad = \frac{2}{\hbar}\mathrm{Im}\,\langle a(J+1)|h_{ab}(EI - \bar{H}_{bb})^{-1}h_{ba}|a(J+1)\rangle \\ &\quad - \frac{2}{\hbar}\mathrm{Im}\,\langle a(J)|\bar{H}_{ab}(EI - \bar{H}_{bb})^{-1}h_{ba}|a(J+1)\rangle \end{aligned}$$

Since $\langle a(J+1)|h_{ab}(EI - \bar{H}_{bb})^{-1}h_{ba}|a(J+1)\rangle = \langle a(J+1)|h_{ab}(EI - \bar{H}_{bb})^{-1}h_{ba}\,|a(J+1)\rangle^*$, the imaginary part of this term is zero.

Hence $J_{J\to J+1} = -\frac{2}{\hbar}\mathrm{Im}\,\langle a(J)|h_{aa} + \bar{H}_{ab}(EI - \bar{H}_{bb})^{-1}h_{ba}|a(J+1)\rangle$.

From equation (A.26), this can be written

$$J_{J\to J+1} = -\frac{2}{\hbar}\mathrm{Im}\,\langle a(J)|\hat{H}_1|a(J+1)\rangle \tag{A.29}$$

This means that the current between adjacent cells is obtained from the amplitudes $|a(J)\rangle$ using the usual formula for bond currents, provided the renormalised inter-cell coupling $\hat{H}_1$ is used.

Appendix B Translation operators, velocity operators, Green's functions, time reversal symmetry and the transmission coefficient

The aim of this appendix is to add important technical details to the discussion presented in the main body of chapter 17. It starts with a detailed definition of translation operators and wave vectors associated with an arbitrary periodic lead and then discusses the various surface Green's functions associated with a semi-infinite lead. It then derives the following fundamental symmetry relationship between various retarded and advanced Green's functions and generalised velocities

$$g_{\sigma\sigma'}^{-1} = \left[g_{p,-p'}^{-1}\right]^{\dagger} + \delta_{\sigma p}i\hbar v_{\sigma} + \delta_{\sigma' p'}i\hbar w_{\sigma'}$$

This symmetry relation was used, for example to convert equation (17.73a) to the simpler version (17.73b). One example of this symmetry is the relationship between retarded and advanced Green's functions and expresses the fact that the latter is the time reversal of the former.

In sections B.10 and B.11, these symmetry relations are used to derive the following equation for the transmission coefficient of a scatterer connected to arbitrary crystalline leads:

$$T = 4\mathrm{Tr}[\Gamma_B G_{00} \Gamma_A G_{00}^{\dagger}]$$

In section B.11, the above symmetry relation is also used to prove that the anti-Hermitian part of the self-energy of G_{00} is $\Gamma_A + \Gamma_B$.

Finally, the appendix ends with a summary of the theoretical framework developed in this book, for solving a general scattering problem.

B.1. The translation operator η

Consider a set of r linearly independent r-dimensional vectors $\{|\chi_{\theta_n}\rangle\}$, which are not necessarily orthogonal. (Here curly brackets {...} are used to denote a set of quantities.) If an rxr square matrix χ is defined to be a matrix, whose columns are formed from the column vectors $\{|\chi_{\theta_n}\rangle\}$, then

$$\chi = \left(|\chi_{\theta_1}\rangle \quad \cdots\cdots \quad |\chi_{\theta_r}\rangle \right) \tag{B.1}$$

The inverse of χ is denoted χ^{-1}, and by definition, $\chi^{-1}\chi = I$. This means that the rows of χ^{-1} can be used to define a set of row vectors $\{(\chi_{\theta_n}|\}$; i.e.

$$\chi^{-1} = \begin{pmatrix} (\chi_{\theta_1}| \\ \vdots \\ (\chi_{\theta_r}| \end{pmatrix} \tag{B.2}$$

and since equation $\chi^{-1}\chi = I$ the following orthogonality relation is satisfied $(\chi_{\theta_n}|\chi_{\theta_m}\rangle$

$$(\chi_{\theta_n}|\chi_{\theta_m}\rangle = \delta_{nm} \tag{B.3}$$

Since χ is a finite matrix, the relation $\chi^{-1}\chi = I$, implies $\chi\chi^{-1} = I$; i.e.

$$\left(|\chi_{\theta_1}\rangle \quad \cdots\cdots \quad |\chi_{\theta_r}\rangle \right) \begin{pmatrix} (\chi_{\theta_1}| \\ \vdots \\ (\chi_{\theta_r}| \end{pmatrix} = I$$

This is equivalent to the first of the following completeness conditions:

$$\sum_{n=1}^{r} |\chi_{\theta_n}\rangle(\chi_{\theta_n}| = I \quad \text{and} \quad \sum_{n=1}^{r} |\chi_{\theta_n})\langle\chi_{\theta_n}| = I \tag{B.4}$$

The second of equations (B.4) is simply the Hermitian conjugate of the first.

To each vector $|\chi_{\theta_n}\rangle$, one can associate a number θ_n, which could be real or complex and then define an rxr square matrix $\eta(J)$ via the equation

$$\eta(J) = \sum_{n=1}^{r} e^{i\theta_n J} |\chi_{\theta_n}\rangle\langle\chi_{\theta_n}| \tag{B.5}$$

In what follows, when utilising $\eta(J)$, J will be chosen to be an integer. The inverse of $\eta(J)$ is

$$\eta^{-1}(J) = \eta(-J) = \sum_{m=1}^{r} e^{-i\theta_m J} |\chi_{\theta_m}\rangle\langle\chi_{\theta_m}| \tag{B.6}$$

which can be proved by noting that $\eta(J)\eta^{-1}(J) = \left[\sum_{n=1}^{r} e^{i\theta_n J} |\chi_{\theta_n}\rangle\langle\chi_{\theta_n}|\right]$ $\left[\sum_{m=1}^{r} e^{-i\theta_m J} |\chi_{\theta_m}\rangle\langle\chi_{\theta_m}|\right] = \sum_{n,m=1}^{r} e^{-i(\theta_m-\theta_n)J} |\chi_{\theta_n}\rangle\langle\chi_{\theta_n}|\chi_{\theta_m}\rangle\langle\chi_{\theta_m}|$. Using equation (B.3), the last expression is simply $\sum_{n,m=1}^{r} e^{-i(\theta_m-\theta_n)J} |\chi_{\theta_n}\rangle\delta_{nm}\langle\chi_{\theta_m}| = \sum_{n=1}^{r} \chi_{\theta_n}\rangle\langle\chi_{\theta_n}| = I$, where in the last step, the completeness condition (B.4) was used.

To demonstrate that $\eta(J)$ is a translation operator, consider the following vector

$$|\psi_J\rangle = \sum_{n=1}^{r} A_n e^{i\theta_n J} |\chi_{\theta_n}\rangle \tag{B.7}$$

where the coefficients $\{A_n\}$ are arbitrary complex numbers and the index J is an integer.

This satisfies

$$|\psi_{J+J_0}\rangle = \eta(J_0)|\psi_J\rangle \tag{B.8}$$

which can be verified by direct substitution and making use of equations (B.3) and (B.4), in a similar manner to the above proof that $\eta(J)\eta^{-1}(J) = I$. This shows that the operator $\eta(J_0)$ acts on the vector $|\psi_J\rangle$ and translates the index J into the index $J + J_0$, i.e. $\eta(J_0)$ is a translation operator.

B.2. Book keeping for wave vectors and four useful translation operators

In section 17.8, translation operators were introduced to help solve a scattering problem by choosing the above complex numbers $\{\theta_n\}$ to be wave vectors. In sections (17.2) and (17.9), it was noted that an arbitrary lead possesses $2r$ wave vectors k_n, with labels $n = 1,2, \ldots 2r$, which can be divided into four sets. Real wave vectors k_n are labelled by integers $n = 1, \ldots M$ and $n = r + 1, \ldots r + M$, while complex wave vectors k_n are labelled by integers $n = M + 1, \ldots r$ and $n = r + M + 1, \ldots 2r$. In what follows, these sets will be denoted $\{+0\}$, $\{0+\}$, $\{-0\}$ and $\{0-\}$ and it will be convenient to introduce the following notation for sets of wave vectors.

1. The first set $\{+, 0\}$, is composed of M real wave vectors k_n, labelled $n = 1,2, \ldots M$, with positive group velocities in the direction of increasing J. In what follows, these will be denoted k_n^+, with labels $n = 1,2, \ldots M$. i.e. $k_n^+ = k_n$ for $n = 1,2, \ldots M$. The corresponding eigenvectors $\{e^{ik_n^+ J}|\chi_{k_n^+}\rangle\}$ carry

a positive current in the direction of increasing J (i.e. $I_{J\to J+1} > 0$). These will be called positive-current states. The most general eigenstate that can be constructed from these positive-current states is a superposition of the form

$$|\psi_J^{+0}\rangle = \sum_{n=1}^{M} A_n e^{ik_n^+ J} |\chi_{k_n^+}\rangle \tag{B.9}$$

where the coefficients $\{A_n\}$ are arbitrary.

2. For $M < r$, the second set $\{0, +\}$ is composed of $r - M$ wave vectors k_n labelled by integers $n = M + 1, \ldots r$. These are complex wave vectors with positive imaginary parts, for which $e^{ik_n J}$ decays exponentially in the direction of increasing J. In what follows, these will be denoted c_n^+, with labels $n = M + 1, \ldots r$. The corresponding eigenvectors are $\{e^{ic_n^+ J}|\chi_{c_n^+}\rangle\}$ and decay exponentially in the direction of increasing J. The most general vector that can be constructed from these is

$$|\psi_J^{0+}\rangle = \sum_{n=M+1}^{r} A_n e^{ic_n^+ J} |\chi_{c_n^+}\rangle \tag{B.10}$$

where the coefficients $\{A_n\}$ are arbitrary.

3. The third set $\{-, 0\}$, consists of M real wave vectors k_n labelled $n = r + 1, \ldots r + M$, with negative group velocities. In what follows, these will be denoted $k_n^- = k_{n+r}$, where $n = 1,2, \ldots M$. The corresponding eigenvectors $\{e^{ik_n^- J}|\chi_{k_n^-}\rangle\}$ carry a negative current (i.e. $I_{J\to J+1} < 0$) and the most general vector that can be constructed from these negative-current eigenvectors is of the form

$$|\psi_J^{-0}\rangle = \sum_{n=1}^{M} A_n e^{ik_n^- J} |\chi_{k_n^-}\rangle \tag{B.11}$$

where the coefficients $\{A_n\}$ are arbitrary.

4. For $M < r$, the fourth set $\{0, -\}$, consists of $r - M$ complex wave vectors k_n, labelled by integers $n = r + M + 1, \ldots. 2r$, with negative imaginary parts. In what follows, these wave vectors will be denoted denoted $c_n^- = k_{n+r}$, where $n = M + 1, \ldots r$. The corresponding eigenvectors $\{e^{ic_n^- J}|\chi_{c_n^-}\rangle\}$ increase exponentially in the direction of increasing J and the most general vector that can be constructed from these is

$$|\psi_J^{0-}\rangle = \sum_{n=M+1}^{r} A_n e^{ic_n^- J} |\chi_{c_n^-}\rangle \tag{B.12}$$

where the coefficients $\{A_n\}$ are arbitrary.

In all cases, the number of positive-current wave vectors is equal to the number of negative-current wave vectors and $\{c_n^-\} = \{(c_n^+)^*\}$. This means that if α_n and β_n are real numbers, one can write $c_n^+ = \alpha_n + i\beta_n$ and $c_n^- = \alpha_n - i\beta_n$, where β_n is positive. These sets of wave vectors and their eigenstates are summarised in table B.1.

To combine the above sets, it is useful to use standard set notation, in which the intersection of two sets $\{A\}$ and $\{B\}$ is denoted $\{A\} \cap \{B\}$ and defined to be the largest set, which contains all the elements that are common to both the sets. For example if $\{A\} = \{1,2,7\}$ and $\{B\} = \{8,3,1,9,7\}$, then $\{A\} \cap \{B\} = \{1,7\}$. Similarly the union $\{A\} \cup \{B\}$ of two given sets is the smallest set, which contains all the elements of both the sets. In this example, $\{A\} \cup \{B\} = \{2,8,3,1,9,7\}$.

From the above sets of wave vectors, one can define four other sets of wave vectors, as follows:

(a) The first set $\{++\} = \{+0\} \cup \{0+\}$ consists of wave vectors denoted k_n^{++}, which are either real, with positive-current states, or they are complex, with decaying states in the direction of increasing J. In both cases the states will be denoted $e^{ik_n^{++}J}|\chi_{k_n^{++}}\rangle$. With this notation, $k_n^{++} = k_n^+$ ($n = 1, \ldots M$) and $k_n^{++} = c_n^+$ ($n = M + 1, \ldots . r$). The most general eigenstate of energy E that can be formed from this set is of the form

$$|\psi_J^{++}\rangle = |\psi_J^{+0}\rangle + |\psi_J^{0+}\rangle = \sum_{n=1}^{r} A_n e^{ik_n^{++}J}|\chi_{k_n^{++}}\rangle \tag{B.13}$$

where $\{A_n\}$ are arbitrary coefficients.

(b) The second set $\{--\} = \{-0\} \cup \{0-\}$ consists of wave vectors denoted k_n^{--}, which are either real, with negative-current states, or they are complex and increasing with increasing J. In both cases the states will be denoted $e^{ik_n^{--}J}|\chi_{k_n^{--}}\rangle$. With this notation, $k_n^{--} = k_n^-$ ($n = 1, \ldots M$) and $k_n^{--} = c_n^-$

Table B.1. Wave vectors and their eigenstates

Symbol		Wave vectors	Channel vectors	Most generaleigenstate	Spatial dependence
$\{+0\}$	Real wave vectors, positive current	$\{k_n^+\}$	$\{|\chi_{k_n^+}\rangle\}$	$|\psi_J^{+0}\rangle = \sum_{n=1}^{M} A_n e^{ik_n^+J}\,|\chi_{k_n^+}\rangle$	
$\{0+\}$	Complex wave vectors, positive imaginary part	$\{c_n^+\}$	$\{|\chi_{c_n^+}\rangle\}$	$|\psi_J^{0+}\rangle = \sum_{n=M+1}^{r} A_n e^{ic_n^+J}\,|\chi_{c_n^+}\rangle$	
$\{-0\}$	Real wave vectors, negative current	$\{k_n^-\}$	$\{|\chi_{k_n^-}\rangle\}$	$|\psi_J^{-0}\rangle = \sum_{n=1}^{M} A_n e^{ik_n^-J}\,|\chi_{k_n^-}\rangle$	
$\{0-\}$	Complex wave vectors, negative imaginary part	$\{c_n^-\} = \{(c_n^+)^*\}$	$\{|\chi_{c_n^-}\rangle\}$	$|\psi_J^{0-}\rangle = \sum_{n=M+1}^{r} A_n e^{ic_n^-J}\,|\chi_{c_n^-}\rangle$	

($n = M + 1, \ldots. r$). The most general vector that can be formed from this set is of the form

$$|\psi_J^{--}\rangle = |\psi_J^{-0}\rangle + |\psi_J^{0-}\rangle = \sum_{n=1}^{r} A_n e^{ik_n^{--}J}|\chi_{k_n^{--}}\rangle \tag{B.14}$$

where $\{A_n\}$ are arbitrary coefficients.

(c) The third set $\{+-\} = \{+0\} \cup \{0-\}$ consists of wave vectors k_n^{+-}, which are either real, with positive-current states, or complex with states which increase in the direction of increasing J. In both cases the states will be denoted $e^{ik_n^{+-}J}|\chi_{k_n^{+-}}\rangle$. With this notation, $k_n^{+-} = k_n^{+}$ ($n = 1, \ldots M$) and $k_n^{+-} = c_n^{-}$ ($n = M + 1, \ldots. r$). The most general vector that can be formed from this set is of the form

$$|\psi_J^{+-}\rangle = |\psi_J^{+0}\rangle + |\psi_J^{0-}\rangle = \sum_{n=1}^{r} A_n e^{ik_n^{+-}J}|\chi_{k_n^{+-}}\rangle \tag{B.15}$$

where $\{A_n\}$ is a set of arbitrary coefficients.

(d) Finally the fourth set $\{-+\} = \{-0\} \cup \{0+\}$ consists of wave vectors k_n^{-+}, which are either real, with negative-current states, or complex, with decaying states in the direction of increasing J. In both cases the states will be denoted $e^{ik_n^{-+}J}|\chi_{k_n^{-+}}\rangle$. With this notation, $k_n^{-+} = k_n^{-}$ ($n = 1, \ldots M$) and $k_n^{-+} = c_n^{+}$ ($n = M + 1, \ldots. r$). The most general vector that can be formed from this set is of the form

$$|\psi_J^{-+}\rangle = |\psi_J^{-0}\rangle + |\psi_J^{0+}\rangle = \sum_{n=1}^{r} A_n e^{ik_n^{-+}J}|\chi_{k_n^{-+}}\rangle \tag{B.16}$$

where $\{A_n\}$ is a set of arbitrary coefficients. The spatial dependences of the above functions are shown in table B.2.

In summary, since real wave vectors k_n are labelled by integers $n = 1, \ldots M$ and $n = r + 1, \ldots r + M$, while complex wave vectors k_n are labelled by integers $n = M + 1, \ldots r$ and $n = r + M + 1, \ldots 2r$, the above convention means that for $n = 1,2, \ldots M$, $k_n = k_n^{+} = k_n^{+\sigma}$ and $k_{n+r} = k_n^{-} = k_n^{-\sigma}$. Also for $M < r$ and $n = 1, \ldots r - M$, $k_{n+M} = c_n^{+} = k_n^{\sigma+}$ and $k_{n+r+M} = c_n^{-} = k_n^{\sigma-}$.

Using the above notation, the most general solution to the Schrödinger equation (17.1) can be written

$$|\psi_J\rangle = |\psi_J^{++}\rangle + |\psi_J^{--}\rangle \text{ or alternatively } |\psi_J\rangle = |\psi_J^{+-}\rangle + |\psi_J^{-+}\rangle \tag{B.17}$$

The expressions in equations (B.17) are simply different notations for the same linear superposition of all possible plane waves, since they both equal to

$$|\psi_J\rangle = |\psi_J^{-0}\rangle + |\psi_J^{0+}\rangle + |\psi_J^{+0}\rangle + |\psi_J^{0-}\rangle \tag{B.18}$$

Table B.2. Spatial dependences of functions

Symbol		Wave vectors	Channel vectors	Most general eigenstate	Spatial dependence
$\{++\}$	Positive current or decaying with increasing J	$\{k_n^{++}\}$	$\{\lvert\chi_{k_n^{++}}\rangle\}$	$\lvert\psi_J^{++}\rangle = \sum_{n=1}^{r} A_n e^{ik_n^{++}J}\lvert\chi_{k_n^{++}}\rangle$	
$\{--\}$	Negative current or increasing with increasing J	$\{k_n^{--}\}$	$\{\lvert\chi_{k_n^{--}}\rangle\}$	$\lvert\psi_J^{--}\rangle = \sum_{n=1}^{r} A_n e^{ik_n^{--}J}\lvert\chi_{k_n^{--}}\rangle$	
$\{+-\}$	Positive current or increasing with increasing J	$\{k_n^{+-}\}$	$\{\lvert\chi_{k_n^{+-}}\rangle\}$	$\lvert\psi_J^{+-}\rangle = \sum_{n=1}^{r} A_n e^{ik_n^{+-}J}\lvert\chi_{k_n^{+-}}\rangle$	
$\{-+\}$	Negative current or decaying with increasing J	$\{k_n^{-+}\}$	$\{\lvert\chi_{k_n^{-+}}\rangle\}$	$\lvert\psi_J^{-+}\rangle = \sum_{n=1}^{r} A_n e^{ik_n^{-+}J}\lvert\chi_{k_n^{-+}}\rangle$	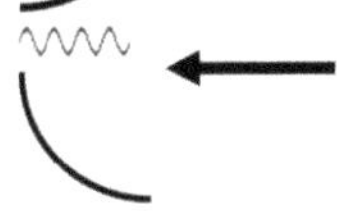

The four sets in table B.2 are particularly useful when discussing retarded and advanced Green's functions.

B.3. Translation operators for eigenstates

From the above sets of vectors, one can define four translation operators, all of which possess an inverse

$$\eta_{++}(J) = \sum_{n=1}^{r} e^{ik_n^{++}J}\lvert\chi_{k_n^{++}}\rangle\langle\chi_{k_n^{++}}\rvert,\quad \eta_{--}(J) = \sum_{n=1}^{r} e^{ik_n^{--}J}\lvert\chi_{k_n^{--}}\rangle\langle\chi_{k_n^{--}}\rvert \tag{B.19}$$

and

$$\eta_{+-}(J) = \sum_{n=1}^{r} e^{ik_n^{+-}J}\lvert\chi_{k_n^{+-}}\rangle\langle\chi_{k_n^{+-}}\rvert,\quad \eta_{-+}(J) = \sum_{n=1}^{r} e^{ik_n^{-+}J}\lvert\chi_{k_n^{-+}}\rangle\langle\chi_{k_n^{-+}}\rvert \tag{B.20}$$

These translation operators satisfy

$$\lvert\psi_{J+J_0}^{\sigma\sigma'}\rangle = \eta_{\sigma\sigma'}(J_0)\lvert\psi_J^{\sigma\sigma'}\rangle \tag{B.21}$$

where $\sigma\sigma'$ is either $++$, $--$, $+-$, or $-+$.

B.4. Left and right, retarded and advanced, surface Green's functions of semi-infinite leads

Using the above translation operators, one can define four surface Green's functions, by extending the analysis of section 17.13, where it was shown that a Green's function of the semi-infinite lead of figure B.1(a) satisfies (c.f. equation (17.113))

$$H_{-1}g(L-1, L) = I \tag{B.22}$$

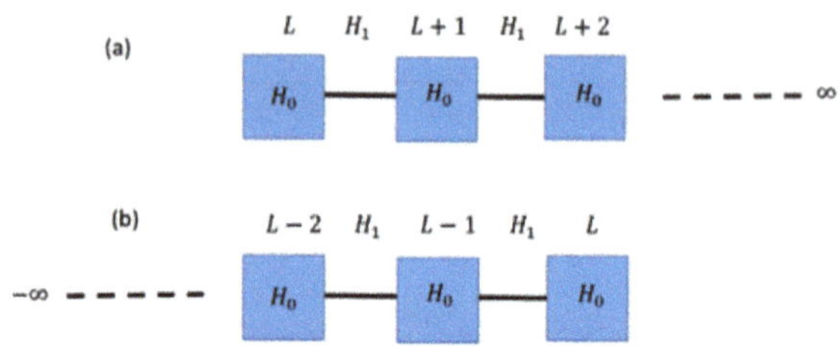

Figure B.1 Two semi-infinite leads with inter-cell Hamiltonians (a) $H_{J,J+1} = H_1$ and (b) $H_{J,J-1} = H_{-1} = H_{-1}^{\dagger}$, terminating at cell $J = L$.

Therefore (c.f. equations (17.112) and (17.114)), four alternative Green's functions, associated with the semi-infinite lead of figure B.1(a) are of the form

$$g_{\sigma\sigma'}(J, L) = \eta_{\sigma\sigma'}(J - [L - 1])(H_{-1})^{-1} \tag{B.23}$$

Equation (B.22) is a reflection of the fact that cell $L - 1$ is missing from the semi-infinite lead in figure B.1(a). Similarly, since cell $L + 1$ is missing from the semi-infinite lead in figure B.1(b) the corresponding Green's function satisfies

$$H_1\tilde{g}(L + 1, L) = I$$

Therefore, four alternative Green's functions associated with the semi-infinite lead in figure B.1(b) are of the form

$$\tilde{g}_{\sigma\sigma'}(J, L) = \eta_{\sigma\sigma'}(J - [L + 1])(H_1)^{-1}$$

In what follows, the following shorthand notation will be used for these surface Green's functions:

$$g_{\sigma\sigma'} = g_{\sigma\sigma'}(L, L),\ \tilde{g}_{\sigma\sigma'} = \tilde{g}_{\sigma\sigma'}(L, L) \tag{B.24}$$

Using the above notation, one obtains,

$$g_{\sigma\sigma'} = \eta_{\sigma\sigma'}(1)(H_{-1})^{-1} = \sum_{n=1}^{r} e^{ik_n^{\sigma\sigma'}} |\chi_{k_n^{\sigma\sigma'}}\rangle\langle\chi_{k_n^{\sigma\sigma'}}|(H_{-1})^{-1} \tag{B.25}$$

$$\tilde{g}_{\sigma\sigma'} = \eta_{\sigma\sigma'}(- 1)(H_1)^{-1} = \sum_{n=1}^{r} e^{-ik_n^{\sigma\sigma'}} |\chi_{k_n^{\sigma\sigma'}}\rangle\langle\chi_{k_n^{\sigma\sigma'}}|(H_1)^{-1} \tag{B.26}$$

Since the inverse of a product of matrices is the product of their inverses in reverse order (ie $(AB)^{-1} = B^{-1}A^{-1}$), and the inverse surface Green's functions are

$$g_{\sigma\sigma'}^{-1} = H_{-1}\,\eta_{\sigma\sigma'}(-1) = H_{-1}\sum_{n=1}^{r} e^{-ik_n^{\sigma\sigma'}} |\chi_{k_n^{\sigma\sigma'}}\rangle\langle\chi_{k_n^{\sigma\sigma'}}| \tag{B.27}$$

and

$$\tilde{g}_{\sigma\sigma'}^{-1} = H_1\,\eta_{\sigma\sigma'}(1) = H_1\sum_{n=1}^{r} e^{ik_n^{\sigma\sigma'}} |\chi_{k_n^{\sigma\sigma'}}\rangle\langle\chi_{k_n^{\sigma\sigma'}}| \tag{B.28}$$

Cartoons of Green's function $g_{\sigma\sigma'}(J, L)$ of the semi-infinite lead of figure B.1(a) are shown in figure B.2. The functions $g_{++}(J, L)$ and $g_{-+}(J, L)$ remain finite as $J \to \infty$ and as demonstrated below, are retarded and advanced Green's functions associated with sources or sinks at the surface cell of the lead. In contrast the closed channels of $g_{++}(J, L)$ and $g_{-+}(J, L)$ diverge as $J \to \infty$ and therefore these are not associated with physical sources or sinks. Similarly $\tilde{g}_{--}(J, L)$ and $\tilde{g}_{+-}(J, L)$ are retarded and advanced Green's functions of the semi-infinite lead of figure B.1(b).

It should be noted that when $H_1 \neq H_{-1}$, the surface Green's function of the semi-infinite leads of figure B.1 are not necessarily equal. As indicated in figure B.3, where the left surface of the upper semi-infinite lead is clearly different from the right surface of the lower semi-infinite lead. The relevance of the above Green's functions depends on the physical situation of interest. Since one of the semi-infinite leads can be obtained from the other by rotating through 180 degrees and exchanging H_1 and H_{-1}, in what follows, attention will be focussed on the surface Green's function of the semi-infinite lead of figure B.1(a).

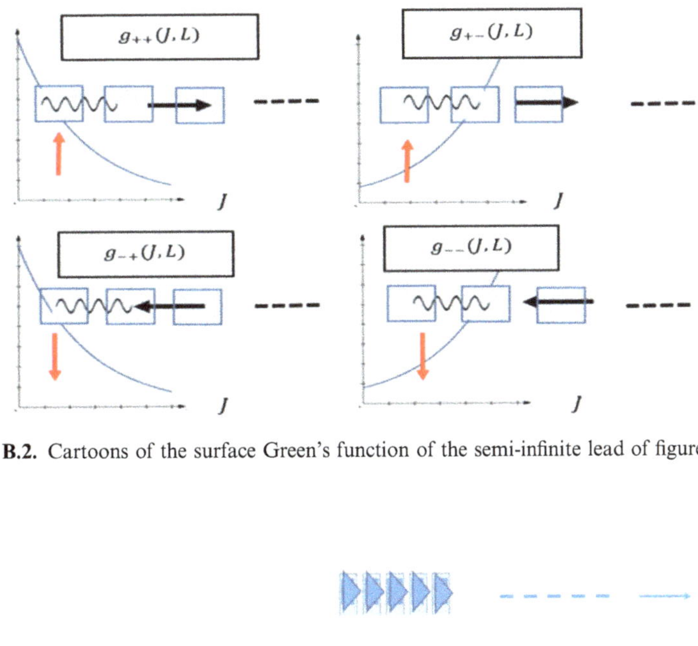

Figure B.2. Cartoons of the surface Green's function of the semi-infinite lead of figure B.1(a).

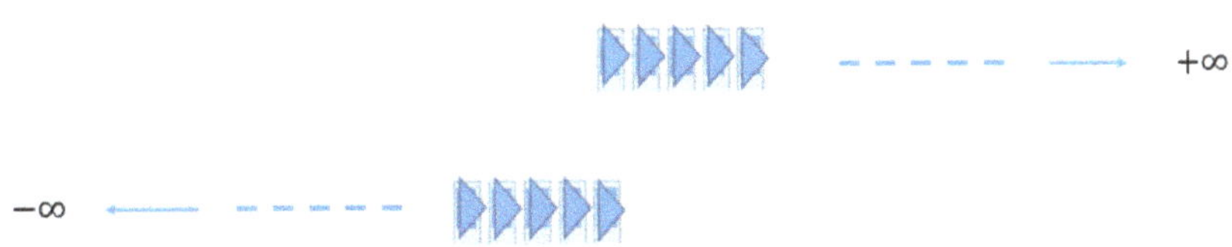

Figure B.3. Two semi-infinite leads that do not possess a centre of symmetry, because $H_1 \neq H_{-1}$. The unit cells are depicted as triangles, to indicate that the arrangement of sites within a cell do not possess reflection symmetry. Clearly, the left surface of the upper semi-infinite lead is different from the right surface of the lower semi-infinite lead.

B.5. Identification of retarded and advanced Green's functions

In this subsection, the aim is to obtain relationships between the Green's functions $g_{\sigma\sigma'}$ and in particular to prove that

$$g_{++}^{-1} = \left[g_{-+}^{-1}\right]^{\dagger} \quad \text{(B.29)}$$

Since g_{++} is a retarded Green's functions (ie a wave resulting from source at $J = L$, in figure B.1(a)), equation (B.29) demonstrates that g_{-+} is the corresponding advanced Green's function.

This result is not immediately obvious, because the Hermitian conjugate of a product of matrices is the product of their inverses in reverse order (i.e. $(AB)^{\dagger} = B^{\dagger}A^{\dagger}$) and therefore, since $H_{-1}^{\dagger} = H_1$, equation (B.27) yields

$$\left[g_{\sigma\sigma'}^{-1}\right]^{\dagger} = \sum_{n=1}^{r} e^{i(k_n^{\sigma\sigma'})^*} |\chi_{k_n^{\sigma\sigma'}}\rangle\langle\chi_{k_n^{\sigma\sigma'}}| H_1 \quad \text{(B.30)}$$

At first sight, the expression for $\left[g_{-+}^{-1}\right]^{\dagger}$ is different from the expressions for g_{++}^{-1} and therefore the additional steps provided below are needed to prove equation (B.29).

B.6. A generalised group velocity

Before proving equation (B.29) and (B.45), it will be useful to utilise equation (17.18). If θ_m and θ_n are two arbitrary wave vectors, then equation (17.18) can be written

$$\langle\chi_{\theta_m}|H_{-1}|\chi_{\theta_n}\rangle e^{-i\theta_n} = \langle\chi_{\theta_m}|H_1|\chi_{\theta_n}\rangle e^{i\theta_m^*} + \delta_{\theta_m^*\theta_n} i\hbar w(\theta_m, \theta_n) \quad \text{(B.31)}$$

where

$$i\hbar w(\theta_m, \theta_n) = \langle\chi_{\theta_m}|H_{-1}|\chi_{\theta_n}\rangle e^{-i\theta_n} - \langle\chi_{\theta_m}|H_1|\chi_{\theta_n}\rangle e^{i\theta_m^*} \quad \text{(B.32)}$$

In equation (B.31), the quantity $\delta_{\theta_m^*\theta_n}$ is an alternative notation for the quantity L_{mn} in equation (17.18). That is $\delta_{\theta_m^*\theta_n} = 1$ if $\theta_n^* = \theta_m$ and is zero otherwise.

Note that $|\chi_{\theta_n}\rangle$ is the nth channel vector and θ_n is the nth wave vector. If $|\chi_{\theta_n^*}\rangle = |\chi_{\theta_m}\rangle$, then this implies $\theta_n^* = \theta_m$. However, if $\theta_n^* = \theta_m$, then this does not imply that $|\chi_{\theta_n^*}\rangle = |\chi_{\theta_m}\rangle$, because there may be a degeneracy, with more than one channel vector belonging to the same wave vector θ_n^*. If θ_n is real and $\theta_n = \theta_m$, then following the discussion in section 17.4, linear combinations can be chosen, such that $w(\theta_m, \theta_n)$ is diagonal in the indices n, m and one can write

$$w(\theta_m, \theta_n) = \delta_{mn} v(\theta_n) \quad (\text{for } \theta_n \text{ real}) \quad \text{(B.33)}$$

where $v(\theta_n)$ is the group velocity given by

$$i\hbar v(\theta_n) = \langle\chi_{\theta_n}|H_{-1}|\chi_{\theta_n}\rangle e^{-i\theta_n} - \langle\chi_{\theta_n}|H_1|\chi_{\theta_n}\rangle e^{i\theta_n} \quad \text{(B.34)}$$

This means that for real $\theta_n = \theta_m$, $w(\theta_m, \theta_n)$ is non-zero only if $|\chi_{\theta_n}\rangle = |\chi_{\theta_m}\rangle$, even in the case of a degeneracy.

Since

$$w(\theta_m, \theta_n) = \frac{\langle \chi_{\theta_m} | H_{-1} | \chi_{\theta_n} \rangle e^{-i\theta_n} - \langle \chi_{\theta_m} | H_1 | \chi_{\theta_n} \rangle e^{i\theta_m^*}}{i\hbar} \tag{B.35}$$

$$[w(\theta_m, \theta_n)]^* = \frac{\langle \chi_{\theta_n} | H_{-1} | \chi_{\theta_m} \rangle e^{-i\theta_m} - \langle \chi_{\theta_n} | H_1 | \chi_{\theta_m} \rangle e^{i\theta_n^*}}{i\hbar} \tag{B.36}$$

Hence

$$[w(\theta_m, \theta_n)]^* = w(\theta_n, \theta_m) \tag{B.37}$$

As expected, this shows that group velocities corresponding to open channels (for which $n = m$) are real. Since positive-current channels are labelled by real wave vectors k_n^+ and negative-current channels are labelled by k_n^-, one obtains

$$v(k_n^+) > 0 \quad \text{and} \quad v(k_n^-) = -\,|v(k_n^-)| < 0 \tag{B.38}$$

B.7. Relationships between surface Green's functions

To obtain a relationship between g_{++} and g_{-+}, the completeness conditions (B.4) will be utilised, in which θ_m can sum over any one of the four sets of wave vectors defined in equations (B.13) to (B.16). From equations (B.4) and (B.27),

$$g_{\sigma\sigma'}^{-1} = \sum_{m=1}^{r} |\chi_{\theta_m}\rangle\langle\chi_{\theta_m}| g_{\sigma\sigma'}^{-1} = \sum_{m,n=1}^{r} |\chi_{\theta_m}\rangle \left[\langle \chi_{\theta_m} | H_{-1} | \chi_{k_n^{\sigma\sigma'}} \rangle e^{-ik_n^{\sigma\sigma'}} \right] \langle \chi_{k_n^{\sigma\sigma'}} | \tag{B.39}$$

Utilising equation (B.31), (with the substitution $\theta_n \to k_n^{\sigma\sigma'}$) yields

$$g_{\sigma\sigma'}^{-1} = \sum_{m,n=1}^{r} |\chi_{\theta_m}\rangle e^{i\theta_m^*} \langle \chi_{\theta_m} | H_1 | \chi_{k_n^{\sigma\sigma'}} \rangle \langle \chi_{k_n^{\sigma\sigma'}} | + \sum_{m,n=1}^{r} |\chi_{\theta_m}\rangle \delta_{\theta_m^* k_n^{\sigma\sigma'}} i\hbar w(\theta_m, k_n^{\sigma\sigma'}) \langle \chi_{k_n^{\sigma\sigma'}} | \tag{B.40}$$

$$i\hbar w(\theta_m, k_n^{\sigma\sigma'}) = \langle \chi_{\theta_m} | H_{-1} | \chi_{k_n^{\sigma\sigma'}} \rangle e^{-ik_n^{\sigma\sigma'}} - \langle \chi_{\theta_m} | H_1 | \chi_{k_n^{\sigma\sigma'}} \rangle e^{i\theta_m^*}$$

In equation (B.39), the sum over $k_n^{\sigma\sigma'}$ in the first term is simplified by using equation (B.4) and therefore

$$g_{\sigma\sigma'}^{-1} = \sum_{m=1}^{r} e^{i\theta_m^*} |\chi_{\theta_m}\rangle\langle\chi_{\theta_m}| H_1 + \sum_{m,n=1}^{r} |\chi_{\theta_m}\rangle \delta_{\theta_m^* k_n^{\sigma\sigma'}} i\hbar w(\theta_m, k_n^{\sigma\sigma'}) \langle \chi_{k_n^{\sigma\sigma'}} | \tag{B.41}$$

In the above equation, θ_m can be chosen to run over any of the four sets of the wave vectors $\{k_m^{++}\}$, $\{k_m^{--}\}$, $\{k_m^{+-}\}$, $\{k_m^{-+}\}$, with the condition that the same set must be chosen in both sums. If the chosen set is denoted $k_m^{p,-p'}$, then equation (B.39) becomes

$$g_{\sigma\sigma'}^{-1} = \sum_{m=1}^{r} e^{i(k_m^{p,-p'})^*}|\chi_{k_m^{p,-p'}})(\chi_{k_m^{p,-p'}}|H_1 + \sum_{m,n=1}^{r} |\chi_{k_m^{p,-p'}})\delta_{(k_m^{p,-p'})^* k_n^{\sigma\sigma'}} i\hbar w(k_m^{p,-p'}, k_n^{\sigma\sigma'})(\chi_{k_n^{\sigma\sigma'}}| \quad (B.42)$$

The corresponding complex conjugates $(k_m^{p,-p'})^*$ are contained in the set of wave vectors $\{p, p'\}$. Therefore the sum on the right-hand side will be non-zero only if the intersection $\{p, p'\} \cap \{\sigma, \sigma'\}$ is not empty. In other words, the sum is over all wave vectors $k_n^{\sigma\sigma'}$, which belong to $\{p, p'\} \cap \{\sigma,\sigma'\}$. In set notation, this condition is written $k_n^{\sigma\sigma'} \in \{p, p'\} \cap \{\sigma,\sigma'\}$.

Table B.3 shows the intersections between the four possible choices of $\{p, p'\}$ and $\{\sigma\sigma'\}$.

Comparing the first term on the right-hand side with equation (B.30) yields the following equation,

$$g_{\sigma\sigma'}^{-1} = \left[g_{p,-p'}^{-1}\right]^\dagger + \sum_{m,n=1}^{r} |\chi_{k_m^{p,-p'}})\delta_{(k_m^{p,-p'})^* k_n^{\sigma\sigma'}} i\hbar w(k_m^{p,-p'}, k_n^{\sigma\sigma'})(\chi_{k_n^{\sigma\sigma'}}| \quad (B.43)$$

The second term on the right-hand side can be separated into sums over n, m from 1 to M and sums over n, m from M+1 to r, and after noting equation (B.33), one obtains

$$\begin{aligned}\sum_{m,n=1}^{r} |\chi_{k_m^{p,-p'}})\delta_{(k_m^{p,-p'})^* k_n^{\sigma\sigma'}} i\hbar w\left(k_m^{p,-p'}, k_n^{\sigma\sigma'}\right)(\chi_{k_n^{\sigma\sigma'}}| &= \sum_{n=1,\, k_n^\sigma \in \{p0\} \cap \{\sigma 0\}}^{M} |\chi_{k_n^\sigma}) i\hbar v(k_n^\sigma)(\chi_{k_n^\sigma}| \\ &+ \sum_{m,n=M+1}^{r} |\chi_{c_m^{-p'}})\delta_{(c_m^{-p'})^* c_n^{\sigma'}} i\hbar w(c_m^{-p'}, c_n^{\sigma'})(\chi_{c_n^{\sigma'}}|\end{aligned} \quad (B.44)$$

Since $(c_m^{-p'})^* = c_m^{p'}$, equation (B.43) can be written

$$\begin{aligned}g_{\sigma\sigma'}^{-1} &= \left[g_{p,-p'}^{-1}\right]^\dagger \\ &+ \sum_{k_n^\sigma \in \{p0\} \cap \{\sigma 0\}} |\chi_{k_n^\sigma}) i\hbar v(k_n^\sigma)(\chi_{k_n^\sigma}| + \sum_{m,n=M+1}^{r} |\chi_{c_m^{-p'}})\delta_{c_m^{p'}, c_n^{\sigma'}} i\hbar w\,(c_m^{-p'}, c_n^{\sigma'})(\chi_{c_n^{\sigma'}}|\end{aligned} \quad (B.45)$$

Table B.3. Intersections between the four possible choices of $\{p, p'\}$ and $\{\sigma\sigma'\}$.

	{++}	{−−}	{+−}	{−+}
{++}	{++}	{00}	{+0}	{0+}
{−−}	{00}	{−−}	{0−}	{−0}
{+−}	{+0}	{0−}	{+−}	{00}
{−+}	{0+}	{−0}	{00}	{−+}

The first term is non-zero if $p = \sigma$ and the second term is non-zero if $p' = \sigma'$. Hence, one obtains the following fundamental equation

$$g_{\sigma\sigma'}^{-1} = \left[g_{p,-p'}^{-1}\right]^{\dagger} + \delta_{\sigma p} i\hbar v_{\sigma} + \delta_{\sigma' p'} i\hbar w_{\sigma'} \tag{B.46}$$

where

$$v_{\sigma} = \sum_{n=1}^{M} |\chi_{k_n^{\sigma}})v(k_n^{\sigma})(\chi_{k_n^{\sigma}}| \tag{B.47}$$

$$w_{\sigma} = \sum_{m,n=M+1}^{r} |\chi_{c_m^{-\sigma}})\, \delta_{c_m^{\sigma}, c_n^{\sigma}} w(c_m^{-\sigma}, c_n^{\sigma})(\chi_{c_n^{\sigma}}| \tag{B.48}$$

Note that from equation (B.37),

$$[w_{\sigma}]^{\dagger} = \sum_{m,n=M+1}^{r} |\chi_{c_m^{\sigma}})\delta_{c_m^{\sigma}, c_n^{\sigma}} w(c_n^{\sigma}, c_m^{-\sigma})(\chi_{c_n^{-\sigma}}| = \sum_{m,n=M+1}^{r} |\chi_{c_n^{\sigma}})\delta_{c_n^{\sigma}, c_m^{\sigma}} w(c_m^{\sigma}, c_n^{-\sigma})(\chi_{c_m^{-\sigma}}|$$

Hence, since $\delta_{c_n^{\sigma}, c_m^{\sigma}} = \delta_{(c_n^{\sigma})^*, (c_m^{\sigma})^*} = \delta_{c_n^{-\sigma}, c_m^{-\sigma}}$, one obtains

$$[w_{\sigma}]^{\dagger} = w_{-\sigma} \tag{B.49}$$

Furthermore, from equation (B.47)

$$[v_{\sigma}]^{\dagger} = v_{\sigma} \tag{B.50}$$

As an example of equation (B.46), if $\{\sigma\sigma'\}$ is chosen to be $\{++\}$ and $\{p, p'\}$ is chosen to be $\{--\}$, then one obtains

$$g_{++}^{-1} = \left[g_{-+}^{-1}\right]^{\dagger} \tag{B.51}$$

whose inverse is equation (B.29). Other relationships between Green's functions are obtained by making different choices for $\{p, p'\}$ and $\{\sigma\sigma'\}$. For example, choosing $\{\sigma\sigma'\} = \{--\}$ and $\{p, p'\} = \{++\}$ yields

$$g_{--}^{-1} = \left[g_{+-}^{-1}\right]^{\dagger} \tag{B.52}$$

Similarly choosing $\{\sigma\sigma'\} = \{++\}$ and $\{p, p'\} = \{++\}$ yields

$$g_{++}^{-1} = \left[g_{+-}^{-1}\right]^{\dagger} + i\hbar(v_+ + w_+) \tag{B.53}$$

choosing $\{\sigma\sigma'\} = \{++\}$ and $\{p, p'\} = \{+-\}$ yields

$$g_{++}^{-1} = \left[g_{++}^{-1}\right]^{\dagger} + i\hbar v_+ \tag{B.54}$$

and choosing $\{\sigma\sigma'\} = \{++\}$ and $\{p, p'\} = \{-+\}$ yields

$$g_{++}^{-1} = [g_{--}^{-1}]^{\dagger} + i\hbar w_{+} \tag{B.55}$$

Furthermore choosing $\{\sigma\sigma'\} = \{--\}$ and $\{p, p'\} = \{--\}$ yields

$$g_{--}^{-1} = \left[g_{-+}^{-1}\right]^{\dagger} + i\hbar(v_{-} + w_{-}) \tag{B.56}$$

choosing $\{\sigma\sigma'\} = \{--\}$ and $\{p, p'\} = \{+-\}$ yields

$$g_{--}^{-1} = \left[g_{++}^{-1}\right]^{\dagger} + i\hbar w_{-} \tag{B.57}$$

and choosing $\{\sigma\sigma'\} = \{--\}$ and $\{p, p'\} = \{-+\}$ yields

$$g_{--}^{-1} = [g_{--}^{-1}]^{\dagger} + i\hbar v_{-} \tag{B.58}$$

As expected, from equations (B.45) and (B.49), both g_{++} and g_{--} are Hermitian when there are no open channels (i.e. when there are no real wave vectors), because in that case, the leads are closed. In this case, equations (B.43) and (B.42) shows that g_{+-} and g_{-+} are also Hermitian.

From equations (B.51) and (B.56), $g_{++}^{-1} - g_{--}^{-1} = -i\hbar(v_{-} + w_{-})$ and from equations (B.52) and (B.53), $g_{++}^{-1} - g_{--}^{-1} = i\hbar(w_{+} + v_{+})$. Similarly, equating the right-hand sides of equations (B.57) and (B.58), taking the Hermitian conjugate of the resulting equation and noting equations (B.49) and (B.50), yields $g_{++}^{-1} - g_{--}^{-1} = i\hbar(w_{+} - v_{-})$. These equations combine to yield

$$v_{+} = -v_{-} \text{ and } w_{+} = -w_{-} \tag{B.59}$$

B.8. Retarded and advanced Green's functions

The above equations are a reflection of the intimate relationship between retarded and advance Green's functions. In chapter 14, it was noted if a wave vector $|\psi(t)\rangle$ is required to satisfy

$$i\hbar\frac{d}{dt}|\psi(t)\rangle = H|\psi(t)\rangle + |f(t)\rangle \tag{B.60}$$

where

$$|f(t)\rangle = e^{-i(\omega+i\eta/\hbar)t}|S\rangle \tag{B.61}$$

is a source which vanishes at $t = -\infty$, then

$$|\psi(t)\rangle = g_{r}(E, H)|S\rangle e^{-i(\omega+i\eta/\hbar)t} \tag{B.62}$$

In this equation, $g_r(E, H)$ denotes the retarded Green's functions given by

$$g_r(E, H) = (E - H + i\eta)^{-1} \tag{B.63}$$

where $E = \hbar\omega$ and η is a positive number, which is not necessarily small. The spatial dependence of $|\psi(t)\rangle$ is

$$|\psi(0)\rangle = g_r(E, H)|S\rangle \tag{B.64}$$

and the time dependence is such that $|\psi(-\infty)\rangle = 0$. This shows that the spatial dependence $|\psi(0)\rangle$ is determined by the parameters E and η, which control the time dependence of the source, as well as the Hamiltonian H and the shape $|S\rangle$ of the source. In the absence of the source, $|\psi(t)\rangle = 0$ for all t and in the presence of a spatially-localised source, $|\psi(t)\rangle$ is a wave that flows out from the location of the source.

Similarly, by repeating the above steps for the choice

$$|f(t)\rangle = e^{-i(\omega - i\eta/\hbar)t}|S_1\rangle \tag{B.65}$$

and denoting the corresponding solution to equation (B.60) by $|\phi(t)\rangle$, one obtains

$$|\phi(t)\rangle = g_a(E, H)|S_1\rangle e^{-i(\omega - i\eta/\hbar)t} \tag{B.66}$$

In this equation, $g_a(E, H)$ denotes the advanced Green's functions given by

$$g_a(E, H) = (E - H - i\eta)^{-1} \tag{B.67}$$

The spatial dependence of $|\phi(t)\rangle$ is

$$|\phi(0)\rangle = g_a(E, H)|S_1\rangle \tag{B.68}$$

and the time dependence is such that $|\psi(+\infty)\rangle = 0$. Since $|f(t)\rangle$ causes $|\phi(t)\rangle$ to vanish at $t = +\infty$, it is a 'sink' instead of a source. Clearly the spatial dependence of $|\phi(0)\rangle$ is completely different from that of $|\psi(0)\rangle$.

It is interesting to recall that following the discussion in appendix B, chapter 15, the definitions (B.63) and (B.67) can be used even for open systems, because for finite η the matrix H appearing in these equations can be approximated by a finite-size sub-matrix of the infinite Hamiltonian of the open system. In the limit of arbitrarily small η, they yield insight into equations such as (B.42) and (B.43), because equations (B.63) and (B.67) yield the following mathematical relationship

$$g_a(E, H) = \left[g_r(E, H)\right]^{\dagger} \tag{B.69}$$

Furthermore, since $g_a(E, H) = (E - H - i\eta)^{-1} = [(E - H^* + i\eta)^{-1}]^* = \left[g_r(E, H^*)\right]^*$, one obtains

$$\left[g_r(E, H)\right]^{\dagger} = \left[g_r(E, H^*)\right]^* \text{ and } g_a(E, H) = \left[g_r(E, H^*)\right]^* \tag{B.70}$$

The first of equation (B.70) shows that the Hermitian conjugate of the retarded Green's function of a system described by a Hamiltonian H is the complex conjugate of the retarded Green's function of a system described by a Hamiltonian H^*. In the remainder of this subsection is aimed at demonstrating explicitly that this is precisely the meaning of equation (B.51).

First note that the wave vectors K_n of a linear chain described by H^* are given by

$$(H_0^* + H_1^* e^{iK_n} + H_{-1}^* e^{-iK_n})|X_{K_n}\rangle = E|X_{K_n}\rangle \tag{B.71}$$

and the corresponding eigenstates are

$$|\phi_J\rangle = |X_{K_n}\rangle e^{iK_n J} \tag{B.72}$$

For real K_n, the current carried by such a state is

$$I^{K_n}_{J\to J+1} = -\frac{2}{\hbar}\mathrm{Im}\left[\langle X_{K_n}|H_1^*|X_{K_n}\rangle e^{iK_n}\right] \tag{B.73}$$

As in section B.2, one can define four sets of wave vectors, to produce the analogue of table B.1.

The resulting wave vectors and states are shown in table B.4. These can then be used to define four other sets of wave vectors, to produce the analogue of table B.2. The first set $\{++\} = \{+0\} \cup \{0+\}$ consists of wave vectors K_n^{++}, which are either real, with positive-current states, or they are complex, with decaying states in the direction of increasing J. In both cases the states will be denoted $e^{iK_n^{++}J}|\chi_{K_n^{++}}\rangle$. The most general eigenstate of energy E that can be formed from this set is of the form

$$|\phi_J^{++}\rangle = |\phi_J^{+0}\rangle + |\phi_J^{0+}\rangle = \sum_{n=1}^{r} A_n e^{iK_n^{++}J}|X_{K_n^{++}}\rangle$$

From equation (B.25), the retarded Green's function of H^* is

$$g_r(E, H^*) = g_{++}(E, H^*) = \sum_{n=1}^{r} e^{iK_n^{++}}|X_{K_n^{++}}\rangle(X_{K_n^{++}}|(H_{-1}^*)^{-1} \tag{B.74}$$

To compare this with the Green's function of H, note that the wave vectors of a linear chain described by H are given by

$$(H_0 + H_1 e^{ik_n} + H_{-1}e^{-ik_n})|\chi_{k_n}\rangle = E|\chi_{k_n}\rangle \tag{B.75}$$

and for real k_n, the current carried by such a state is

$$I^{k_n}_{J\to J+1} = -\frac{2}{\hbar}\mathrm{Im}\left[\langle\chi_{k_n}|H_1|\chi_{k_n}\rangle e^{ik_n}\right] \tag{B.75}$$

Table B.4. Wave vectors and states

Symbol		Wave vectors	Channel vectors	Most general eigenstate	Spatial dependence
{+0}	Real wave vectors, positive current	$\{K_n^+\}$	$\{\vert X_{K_n^+}\rangle\}$	$\vert\phi_J^{+0}\rangle = \sum_{n=1}^{M} A_n e^{iK_n^+ J}\vert X_{K_n^+}\rangle$	
{0+}	Complex wave vectors, positive imaginary part	$\{C_n^+\}$	$\{\vert X_{C_n^+}\rangle\}$	$\vert\phi_J^{0+}\rangle = \sum_{n=M+1}^{r} A_n e^{iC_n^+ J}\vert X_{C_n^+}\rangle$	
{−0}	Real wave vectors, negative current	$\{K_n^-\}$	$\{\vert X_{K_n^-}\rangle\}$	$\vert\phi_J^{-0}\rangle = \sum_{n=1}^{M} A_n e^{iK_n^- J}\vert X_{K_n^-}\rangle$	
{0−}	Complex wave vectors, negative imaginary part	$\{C_n^-\} = \left\{\left(C_n^+\right)^*\right\}$	$\left\{\vert X_{C_n^-}\rangle\right\}$	$\vert\phi_J^{0-}\rangle = \sum_{n=M+1}^{r} A_n e^{iC_n^- J}\vert X_{C_n^-}\rangle$	

For comparison, the complex conjugate of equation (B.71) is

$$\left(H_0 + H_1 e^{-iK_n^*} + H_{-1} e^{iK_n^*}\right) \vert\chi_{K_n}\rangle^* = E\, \vert\chi_{K_n}\rangle^* \tag{B.76}$$

This shows that if the wave vectors of H are $\{k_n\}$, then the wave vectors of H^* are $\{K_n\} = \{-k_n^*\}$ and the eigenstates of H^* are

$$\vert X_{-k_n^*}\rangle = \vert\chi_{k_n}\rangle^* \tag{B.77}$$

Consider a real wave vector $k_n = k_n^-$ of H and the corresponding wave vector $K_n = -k_n^-$ of H^*.

From equation (B.77), $\vert\chi_{K_n}\rangle = \vert\chi_{-k_n^-}\rangle^*$.

From equation (B.73), the current associated with a real wave vector $K_n = -k_n$ is

$$I_{J\to J+1}^{-k_n} = -\frac{2}{\hbar} Im\left[\langle X_{-k_n}\vert H_1^*\vert X_{-k_n}\rangle e^{-ik_n}\right] = -\frac{2}{\hbar}\mathrm{Im}\left[\langle\chi_{k_n}\vert H_1\vert\chi_{k_n}\rangle^* e^{-ik_n}\right] \tag{B.78}$$

i.e.

$$I_{J\to J+1}^{-k_n} = +\frac{2}{\hbar} Im\left[\langle\chi_{k_n}\vert H_1\vert\chi_{k_n}\rangle e^{ik_n}\right] = -I_{J\to J+1}^{k_n} \tag{B.79}$$

Therefore the positive current real wave vectors of H^* are

$$\{K_n^+\} = \{-k_n^-\} \tag{B.80}$$

Similarly, if $c_n^+ = \alpha_n + i\beta_n$, where α_n and β_n are real and $\beta_n > 0$, then $(-c_n^+)^* = -\alpha_n + i\beta_n$. Since this has a positive imaginary part, one obtains

$$\{C_n^+\} = \{(-c_n^+)^*\} \tag{B.81}$$

Hence the set of wave vectors $K_n^{++}$is

$$\{K_n^{++}\} = \{K_n^+\} \cup \{C_n^+\} = \{-k_n^-\} \cup \{(-c_n^+)^*\} = -\{k_n^-\} \cup \{(c_n^+)^*\} = -\{k_n^{+-}\}^*.$$

From equation (B.74), the complex conjugate of the retarded Green's function of H^* is

$$[g_{++}(E, H^*)]^* = \left[\sum_{n=1}^{r} e^{iK_n^{++}} |X_{K_n^{++}}\rangle\langle X_{K_n^{++}}|(H_{-1}^*)^{-1}\right]^*$$
$$= \sum_{n=1}^{r} e^{-i(K_n^{++})^*} |X_{K_n^{++}}\rangle^*\langle X_{K_n^{++}}|^* (H_{-1})^{-1}$$

From equation (B.78), $|X_{K_n}\rangle = |X_{-K_n^*}\rangle^*$ and since $\{K_n^{++}\} = -\{k_n^{+-}\}^*$, one obtains

$$[g_{++}(E, H^*)]^* = \sum_{n=1}^{r} e^{ik_n^{+-}} |\chi_{k_n^{+-}}\rangle\langle\chi_{k_n^{+-}}|(H_{-1})^{-1} = g_{-+}(E, H)$$

From equation (B.70), $[g_{++}(E, H^*)]^* = \left[g_{++}(E, H)\right]^\dagger$ and therefore equation (B.42) is obtained.

B.9. Time reversal

In this section, it is demonstrated that all the bond currents associated with the advanced Green's function $g_a(E, H)$ of system described by a Hamiltonian H, are equal and opposite to those associated with the retarded Green's function $g_r(E, H^*)$ of a system described by the Hamiltonian H^*.

To obtain this result, consider two distinct physical systems with the same number of sites, described by Hamiltonians A and B, whose eigenstates $|\alpha\rangle$ and $|\beta\rangle$ are given by

$$A|\alpha\rangle = E_a|\alpha\rangle \text{ and } B|\beta\rangle = E_b|\beta\rangle \tag{B.82}$$

If the amplitude of the eigenstates on site j are α_j and β_j, then in each separate system, the current from site j to site l is

$$I_{j\to l}^A = -\frac{2}{\hbar}\mathrm{Im}\left[\alpha_j^* A_{jl}\alpha_l\right] \text{ and } I_{j\to l}^B = -\frac{2}{\hbar}\mathrm{Im}\left[\beta_j^* B_{jl}\beta_l\right] \tag{B.83}$$

Since eigenstates are stationary states, these currents will not change with time. Now consider choosing the Hamiltonian B and the state $|\beta\rangle$ such that for all j and l,

$$I_{j\to l}^{B} = -\, I_{j\to l}^{A} \tag{B.84}$$

Since every bond current in system B is the reverse of every bond current in system A, it is reasonable to call system B the time reversal of system A. If z is a complex number, then $Im\, z = -\, Imz^*$ and therefore for equation (B.84) to be satisfied, one requires

$$\left[\beta_j^* B_{jl} \beta_l\right] = \left[\alpha_j^* A_{jl} \alpha_l\right]^* \tag{B.85}$$

This will be the case if $B = A^*$ and $|\beta\rangle = |\alpha\rangle^*$, in which case the second of equations (B.82) becomes

$$A^* \, |\alpha\rangle^* = E_b \, |\alpha\rangle^* \tag{B.86}$$

Comparing this with the complex conjugate of the first of equations (B.82), namely $A^* \, |\alpha\rangle^* = E_a \, |\alpha\rangle^*$ leads to the conclusion $E_a = E_b$, i.e. systems described by Hamiltonians A and $B = A^*$ possess the same energy eigenvalues. Furthermore, the eigenstates of B are $|\,\beta\rangle = |\alpha\rangle^*$; i.e. they are the complex conjugates of the eigenstates of A. With this choice of B, if system B is in an eigenstate $|\alpha\rangle^*$, and system A is in an eigenstate $|\alpha\rangle$, then all currents in system A will be the reverse of those in B. In other words, the state $|\alpha\rangle$ of A, which corresponds to the time reversal of a state $|\beta\rangle$ of B is given by

$$|\alpha\rangle = |\beta\rangle^*$$

Since the Green's function $g_r(E, A^*)$ is a state in B due to a source, the corresponding time-reversed state in A is $\left[g_r(E, A^*)\right]^*$. From equation (B.70), this is equal to $g_a(E, A)$ and therefore, all the bond currents created by the sink associated with the advanced Green's function $g_a(E, A)$ in system A are equal and opposite to those created by the source of the retarded Green's function $g_r(E, A^*)$ of B.

As an example, the Hamiltonian A describing a spin-less electron in a magnetic field is an example of a complex Hamiltonian. Furthermore the Hamiltonian of the same system in which the direction of the magnetic field is reversed, i.e. A^*. Therefore if the state of the system with a certain magnetic field is $|\beta\rangle$, then if a thought experiment is carried out in which the state is changed to $|\alpha\rangle = |\beta\rangle^*$ and the magnetic field is reversed, all bond currents will be reversed.

B.10. The velocity operator

For a doubly infinite lead, it was noted in chapter 17 that the retarded Green' function is

$g(J, J') = g^{<}(J, J')$ for $(J \leqslant J')$ and $g(J, J') = g^{>}(J, J')$ for $(J \geqslant J')$, where

$$g^{<}(J, J') = \eta_{++}(J - J')(i\hbar V)^{-1} \tag{B.87}$$

$$g^{>}(J, J') = \eta_{--}(J - J')(i\hbar V)^{-1} \tag{B.88}$$

where

$$i\hbar V = \sum_{n=1}^{r} e^{-ik_n^{++}} H_{-1}|\chi_{k_n^{++}}\rangle\langle\chi_{k_n^{++}}| - \sum_{n} e^{-ik_n^{--}} H_{-1}\,|\chi_{k_n^{--}}\rangle\langle\chi_{k_n^{--}}| \tag{B.89}$$

To show why it is reasonable to call V a velocity operator, note that from equations (B.53) and (B.54), equation (B.89) can be written i.e.

$$i\hbar V = g_{++}^{-1} - g_{--}^{-1} = i\hbar(w_+ + v_+) = -\, i\hbar(w_- + v_-) \tag{B.90}$$

where the last expression is obtained by equating the right-hand sides of equations (B.57) and (B.58), taking the Hermitian conjugate of the resulting equation and noting equations (B.49) and (B.50).

This shows why $i\hbar V$ is indeed a velocity operator, because from equations (B.47) and (B.48),

$$V|\chi_{k_n^\sigma}\rangle = |v(k_n^\sigma)||\chi_{k_n^\sigma}\rangle \tag{B.91}$$

and therefore

$$\langle\chi_{k_n^\sigma}|V|\chi_{k_n^\sigma}\rangle = |v(k_n^\sigma)| \tag{B.92}$$

For later use, it is interesting to note that from equation (B.59), $g_{++}^{-1} - g_{-+}^{-1} = -\, i\hbar v_- = i\hbar v_+$ and therefore $\left(g_{++}^{-1} - g_{-+}^{-1}\right)|\chi_{k_n^\sigma}\rangle = i\hbar|v(k_n^\sigma)||\chi_{k_n^\sigma}\rangle$. As expected, combining this with equation (B.91) yields for any choice of σ',

$$\left(g_{++}^{-1} - g_{-\sigma'}^{-1}\right)|\chi_{k_n^\sigma}\rangle = i\hbar|v(k_n^\sigma)||\chi_{k_n^\sigma}\rangle \tag{B.93}$$

Since $g_{-+}^{-1} = \left[g_{++}^{-1}\right]^\dagger$, one also obtains

$$\left(g_{++}^{-1} - \left[g_{++}^{-1}\right]^\dagger\right)|\chi_{k_n^-}\rangle = i\hbar\,|\,v(k_n^-)||\chi_{k_n^-}\rangle \tag{B.94}$$

This equality between $(g_{++}^{-1} - \left[g_{++}^{-1}\right]^\dagger)|\chi_{k_n^-}\rangle$ and $(g_{++}^{-1} - g_{-\sigma'}^{-1})|\chi_{k_n^-}\rangle$ (where $\sigma' = +$ or $-$) is the fundamental reason why the quantity $\Gamma_{k_p^+}$ in equation (B.97) of the following section, also appears in the anti-Hermitian part of the self-energy in equation (B.103).

B.11. A general expression for the transmission coefficient

As noted in the main text, substituting equation (B.94) into equation (17.73a) yields for the transmission amplitude from incoming channel k_n^- to outgoing channel k_p^+,

$$t_{pn} = i\hbar\sqrt{|v(k_p^+)||v(k_n^-)|}\,\langle\chi_{k_p^+}|g_{++}H_{10}G_{00}H_{01}g_{++}|\chi_{k_n^-}\rangle - \sqrt{\frac{|v(k_p^+)|}{|v(k_n^-)|}}\,e^{i(k_p^+ - k_n^-)}\langle\chi_{k_p^+}|\chi_{k_n^-}\rangle \tag{B.95}$$

If channels k_n^- and k_p^+ belong to physically distinct leads, then $(\chi_{k_p^+}|\chi_{k_n^-}\rangle = 0$ and

$$t_{pn} = i\hbar\sqrt{|v(k_p^+)||v(k_n^-)|}\,(\chi_{k_p^+}|g_{++}H_{10}G_{00}H_{01}g_{++}|\chi_{k_n^-}) \tag{B.96}$$

Hence

$$|t_{pn}|^2 = \hbar^2|v(k_p^+)||v(k_n^-)|(\chi_{k_p^+}|g_{++}H_{10}G_{00}H_{01}g_{++}\,|\chi_{k_n^-})(\chi_{k_n^-}\,|g_{++}^\dagger H_{10}G_{00}^\dagger H_{01}g_{++}^\dagger|\chi_{k_p^+})$$

Taking advantage of cyclic invariance under the trace yields

$$|t_{pn}|^2 = 4Tr[\Gamma_{k_p^+}G_{00}\Gamma_{k_n^-}G_{00}^\dagger] \tag{B.97}$$

where

$$\Gamma_{k_n^-} = H_{01}g_{++}\,|\chi_{k_n^-})\frac{\hbar\,|v(k_n^-)|}{2}(\chi_{k_n^-}|g_{++}^\dagger H_{10} \quad \text{and} \quad \Gamma_{k_p^+} = H_{01}g_{++}\,|\chi_{k_p^+})\frac{\hbar|v\left(k_p^+\right)|}{2}(\chi_{k_p^+}|g_{++}^\dagger H_{10} \tag{B.98}$$

It is interesting to compare these with the self-energy Σ_0, which has the form

$$\Sigma_0 = \sigma - i\Gamma \tag{B.99}$$

where

$$\sigma = \frac{\Sigma_0 + \Sigma_0^\dagger}{2} \text{ and } - i\Gamma = \frac{\Sigma_0 - \Sigma_0^\dagger}{2} \tag{B.100}$$

Since $\Sigma_0 = H_{01}g_{++}H_{10}$, Γ can be written

$$\Gamma = -H_{01}\frac{g_{++} - g_{++}^\dagger}{2i}H_{10} = H_{01}g_{++}\frac{g_{++}^{-1} - (g_{++}^{-1})^\dagger}{2i}g_{++}^\dagger H_{10} \tag{B.101}$$

From equation (B.45),

$$g_{++}^{-1} - \left[g_{++}^{-1}\right]^\dagger = i\hbar v_+ = -\, i\hbar v_- \tag{B.102}$$

Since $v\left(k_p^+\right) > 0$ and $v\left(k_p^-\right) < 0$, this yields

$$g_{++}^{-1} - \left[g_{++}^{-1}\right]^\dagger = \sum_{p=1}^{M}|\chi_{k_p^+})|v\left(k_p^+\right)|(\chi_{k_p^+}| = \sum_{p=1}^{M}|\chi_{k_p^-})|v\left(k_p^-\right)|(\chi_{k_p^-}| \tag{B.103}$$

and therefore from equation (B.98)

$$\Gamma = \sum_{p=1}^{M} \Gamma_{k_p^+} = \sum_{p=1}^{M} \Gamma_{k_p^-} \tag{B.104}$$

This shows why symmetry relations such as equation (B.102) are important, because they ensure that the the anti-Hermitian parts of the self energy (B.101) are identical to those appearing in equation (B.97).

B.12. A general expression for the transmission coefficient in the presence of two arbitrary leads

As an example of the above results, consider a scatterer connected to two leads (labelled A and B below), shown in figure B.3, chapter 17, where the units cells are depicted by triangles, to emphasise that the leads do not need to possess a centre of symmetry. For the case of two leads, equation (B.101) can be written

$$\Gamma = \Gamma_A + \Gamma_B \tag{B.105}$$

where

$$\Gamma_A = -H_{0A}\frac{g_{++}(A) - g_{++}^{\dagger}(A)}{2i}H_{A0} \quad \text{and} \quad \Gamma_B = -H_{0B}\frac{g_{++}(B) - g_{++}^{\dagger}(B)}{2i}H_{B0} \tag{B.106}$$

From equation (B.104) Γ_A and Γ_B are obtained by summing over either outgoing or incoming open channels of leads A and B, i.e.

$$\Gamma_A = \sum_{k_p^+}^{(\text{lead } A)} \Gamma_{k_p^+} = \sum_{k_p^-}^{(\text{lead } A)} \Gamma_{k_p^-} \text{ and } \Gamma_B = \sum_{k_p^+}^{(\text{lead } B)} \Gamma_{k_p^+} = \sum_{k_p^-}^{(\text{lead } B)} \Gamma_{k_p^-} \tag{B.107}$$

Finally, the transmission coefficient from lead A to lead B is

$$T = \sum_{k_p^+}^{(\text{lead}A)} \sum_{k_n^- \,(\text{real})}^{(\text{lead}B)} |t_{pn}|^2$$

i.e.

$$T = 4\text{Tr}[\Gamma_B G_{00} \Gamma_A G_{00}^{\dagger}] \tag{B.108}$$

B.13. Summary of the conceptual framework for solving a general scattering problem

This section provides an alternative derivation of equation (17.71), starting with the box below, which summarises the key steps in solving a scattering problem.

In section 17.6, for a scatterer described by a Hamiltonian h, connected to a superlead by a coupling matrix H_{01}, the following fundamental equation was derived

$$\Sigma\,|\psi_1\rangle = H_{-1}\,|\psi_0\rangle \tag{17.41}$$

where

$$\Sigma = H_{10} g H_{01} \tag{17.37}$$

and g is the Green's function of the isolated scattering region, given by

$$g = (E - h)^{-1} \tag{17.34}$$

After writing

$$|\psi_J\rangle = |I_J\rangle + |O_J\rangle \tag{17.51}$$

equation (17.64) was obtained, which is of the form

$$G_{++}^{-1}\,|O_1\rangle = -G_{-\sigma'}^{-1}\,|I_1\rangle \tag{17.64a}$$

where

$$G_{++} = [g_{++}^{-1} - \Sigma]^{-1} \text{ and } G_{-\sigma'} = [g_{-\sigma'}^{-1} - \Sigma]^{-1} \tag{17.64b}$$

In these equations $g_{\sigma\sigma'}$ is the surface Green's function of a superlead.

For a given, incident state $|I_J\rangle$, equation (17.64a) can be solved for the outgoing state $|O_J\rangle$ and then the wave function $|f\rangle$ within the scatterer can be obtained from the equation

$$g^{-1}\,|f\rangle = H_{01}\,|\psi_1\rangle \tag{17.32}$$

In section (17.8), the following alternative to equation (17.64a) was also derived

$$|O_1\rangle = g_{++}H_{10}G_{00}H_{01}g_{++}[g_{++}^{-1} - g_{-\sigma'}^{-1}]\,|I_1\rangle - g_{++}g_{-\sigma'}^{-1}\,|I_1\rangle, \tag{17.71}$$

where

$$\Sigma_0 = H_{01}g_r H_{10} \text{ and } G_{00} = [g^{-1} - \Sigma_0]^{-1} \tag{17.70}$$

From equation (17.71) expressions for transmission and reflection amplitudes were obtained.

Equation (17.64a) is a matching equation between an incoming state $|I_1\rangle$ and an outgoing state $|O_1\rangle$. The outgoing state could be a linear superposition of states carrying a positive current or states which decay with increasing J, whereas in the derivation of equation (17.64a), the incoming state was considered to be a negative current state only. For this reason, the index σ' is arbitrary and indeed the quantity $g_{-\sigma'}^{-1}$ in the above equation could be replaced by simply

$$H_{-1}e^{-ik_n^-}|\chi_{k_n^-}\rangle\langle\chi_{k_n^-}|$$

More generally however, one could consider the problem of matching wave functions in which all possible channels are allowed, in which case equations (17.53) and (17.54) would be replaced by

$$|I_J\rangle = \sum_{n=1}^{r} A_n e^{ik_n^{--}(J-1)}|\chi_{k_n^{--}}\rangle \tag{B.109}$$

and

$$|O_J\rangle = \sum_{m=1}^{r} B_m e^{ik_m^{++}(J-1)}|\chi_{k_m^{++}}\rangle \tag{B.110}$$

Since $|I_J\rangle$ now contains states with complex wave vectors, the second of equations (17.61) is replaced by

$$|I_0\rangle = \eta_{--}(-1)|I_1\rangle \tag{B.111}$$

and therefore one must necessarily choose $\sigma' = -$. For a given set of amplitudes $\{A_n\}$, equations (17.64a) or (17.71) then yield the amplitudes $\{B_n\}$.

On the other hand, this more general matching problem does not occur in the presence of a semi-infinite lead, since states in $|I_J\rangle$ of the form $e^{ic_n^- J}|\chi_{c_n^-}\rangle$ diverge as $J \to \infty$ and therefore cannot be present. In this case, since $|I_1\rangle$ is restricted to superpositions of states with real wave vectors only, then one could optionally choose $\sigma' = +$. The choice $\sigma' = -$ means that equation (17.64a) takes the form

$$G_{++}^{-1}|O_1\rangle = G_{-+}^{-1}|I_1\rangle \tag{B.112}$$

If the scattering region is closed when $H_{10} = 0$, the g is Hermitian and from equation (17.37), Σ is Hermitian. Therefore from equation (B.42)

$$G_{++}^{-1} = \left[G_{-+}^{-1}\right]^{\dagger} \tag{B.113}$$

and for a closed scattering region connected to a semi-infinite super lead, one can write

$$G_{++}^{-1}|O_1\rangle = \left[G_{++}^{-1}\right]^{\dagger}|I_1\rangle \tag{B.114}$$

Since $[g_{++}(E, H^*)]^* = \left[g_{++}(E, H)\right]^{\dagger}$ this can be written

$$G_{++}^{-1}(E, H)|O_1\rangle = [G_{++}^{-1}(E, H^*)]^*|I_1\rangle \tag{B.115}$$

In section (17.8), equation (17.71) was derived indirectly. Alternatively, it can be derived directly by noting that from equation (17.37) and the second of equations (17.64b), equation (17.64a) is of the form

$$G_{++}^{-1}|O_1\rangle = H_{10}gH_{01}|I_1\rangle - g_{-\sigma'}^{-1}|I_1\rangle \tag{B.116}$$

Since $H_{10}gH_{01} + G_{++}^{-1} = g_{++}^{-1}$, one can write the unit matrix as $I = \left(H_{10}gH_{01} + G_{++}^{-1}\right)g_{++}$ and therefore equation (B.113) can be written

$$G_{++}^{-1}|O_1\rangle = H_{10}gH_{01}|I_1\rangle - \left(H_{10}gH_{01} - G_{++}^{-1}\right)g_{++}g_{-\sigma'}^{-1}|I_1\rangle \tag{B.117}$$

i.e.

$$G_{++}^{-1}|O_1\rangle = H_{10}gH_{01}(1 - g_{++}g_{-\sigma'}^{-1})|I_1\rangle - G_{++}^{-1}g_{++}g_{-\sigma'}^{-1}|I_1\rangle \tag{B.118}$$

Hence after multiplying both sides by G_{++}, one obtains

$$|O_1\rangle = G_{++}H_{10}gH_{01}(1 - g_{++}g_{-\sigma'}^{-1})|I_1\rangle - g_{++}g_{-\sigma'}^{-1}|I_1\rangle$$

Equation (17.71) is then obtained by using the NNE

$$G_{++}H_{10}g = g_{++}H_{10}G_{00} \tag{B.119}$$

where G_{00} and Σ_0 are given by equations (17.70). (The NNE (B.119) can be proved by following the steps shown in section (A.12) of appendix A of chapter 2.)

Appendix C A note about degeneracies

In this appendix, matrix elements of the generalised velocity in the presence of degeneracies will be considered.

Case 1: First consider two degenerate real wave vectors k_n, k_m, such that $k_m = k_n$.

This means that at the chosen energy, two mini bands cross to yield the same wave vectors. The channel vectors satisfy $H(k)|\chi_k\rangle = E|\chi_k\rangle$. If two bands cross, then a given k gives the same energy E twice and a given E gives the same k twice. Since $H(k)$ is Hermitian, solving $H(k)|\chi_k\rangle = E|\chi_k\rangle$ for the k of interest shows that the channel vectors are linearly independent. If the two vectors are denoted $|\chi_n\rangle$, $|\chi_m\rangle$, with $k_n = k_m = k$, then

$$w_{mn} = \frac{\langle\chi_m|H_{-1}|\chi_n\rangle e^{-ik_n} - \langle\chi_m|H_1|\chi_n\rangle e^{ik_n}}{i\hbar} \tag{C.1}$$

Given two linearly independent vectors $|\chi_n\rangle$, $|\chi_m\rangle$ one can construct two other vectors $|\chi_1\rangle$ and $|\chi_2\rangle$, defined by

$$|\chi_i\rangle = a_n^i|\chi_n\rangle + b_m^i|\chi_m\rangle \tag{C.2}$$

The corresponding generalised velocities are

$$W_{ij} = \frac{\langle \chi_i | H_{-1} | \chi_j \rangle e^{-ik_n} - \langle \chi_i | H_1 | \chi_j \rangle e^{ik_n}}{i\hbar} \tag{C.3}$$

and satisfy

$$W_{ij} = \begin{pmatrix} a_m^{i*} & b_m^{i*} \end{pmatrix} w \begin{pmatrix} a_n^j \\ b_n^j \end{pmatrix} \tag{C.4}$$

Since the 2x2 matrix w is Hermitian, one can choose $\begin{pmatrix} a_n^1 \\ b_n^1 \end{pmatrix}$ and $\begin{pmatrix} a_n^2 \\ b_n^2 \end{pmatrix}$ to be orthonormal eigenstates of w and $\begin{pmatrix} a_m^1 \\ b_m^1 \end{pmatrix} = \begin{pmatrix} a_n^1 \\ b_n^1 \end{pmatrix}, \begin{pmatrix} a_m^2 \\ b_m^2 \end{pmatrix} = \begin{pmatrix} a_n^2 \\ b_n^2 \end{pmatrix}$. Then equation (C.4) yields $W_{ij} = W_{ii}\delta_{ij}$, which means that in the new basis, the generalised velocity W is diagonal.

Case 2: Consider two complex wave vectors k_n, k_m, such that $k_m = k_n$.

Consider also the two complex wave vectors $k_{m'} = k_m^*$ and $k_{n'} = k_n^*$. In this case, one can construct 2x2 matrices, with elements:

$$w_{m'n} = \frac{\langle \chi_{m'} | H_{-1} | \chi_n \rangle e^{-ik_{m'}^*} - \langle \chi'_m | H_1 | \chi_n \rangle e^{ik_n}}{i\hbar} = \frac{\langle \chi_{m'} | H_{-1} | \chi_n \rangle e^{-ik_m} - \langle \chi_{m'} | H_1 | \chi_n \rangle e^{ik_m}}{i\hbar}$$

$$w_{mn'} = \frac{\langle \chi_m | H_{-1} | \chi_{n'} \rangle e^{-ik_m^*} - \langle \chi_m | H_1 | \chi_{n'} \rangle e^{ik_{n'}}}{i\hbar} = \frac{\langle \chi_m | H_{-1} | \chi_{n'} \rangle e^{-ik_m^*} - \langle \chi_m | H_1 | \chi_{n'} \rangle e^{ik_m^*}}{i\hbar}$$

To diagonalise these matrices, define ν to be the 2x2 matrix with elements $w_{m'n}$ so that $\nu^\dagger$ is the 2x2 matrix with elements $w_{nm'}$. Also consider the following vectors: $|\chi_i\rangle = a_n^i |\chi_n\rangle + b_m^i |\chi_m\rangle$ and $|\chi_{i'}\rangle = a_{n'}^{i'} |\chi_{n'}\rangle + b_{m'}^{i'} |\chi_{m'}\rangle$, which allow one to construct a 2x2 matrix W with elements $W_{i'j} = \begin{pmatrix} a_{m'}^{i'*} & b_{m'}^{i'*} \end{pmatrix} \nu \begin{pmatrix} a_n^j \\ b_n^j \end{pmatrix}$.

Since the 2x2 matrix ν is not Hermitian, if $\begin{pmatrix} a_n^1 \\ b_n^1 \end{pmatrix}$ and $\begin{pmatrix} a_n^2 \\ b_n^2 \end{pmatrix}$ are chosen to be eigenstates of ν, they may not be orthogonal. Nevertheless, if $\begin{pmatrix} a_{m'}^{1'} \\ b_{m'}^{1'} \end{pmatrix}$ and $\begin{pmatrix} a_{m'}^{2'} \\ b_{m'}^{2'} \end{pmatrix}$ are chosen to be the corresponding dual vectors, then one obtains $W_{i'j} = W_{jj}\delta_{i'j}$.

References

[1] Sanvito S, Lambert C J, Jefferson J H and Bratkovsky A M 1999 General Green's-function formalism for transport calculations with spd Hamiltonians and giant magnetoresistance in Co- and Ni-based magnetic multilayers *Phys. Rev.* B **59** 11936

[2] Brandbyge M, Mozos J-L, Ordejón P, Taylor J and Stokbro K 2002 Density-functional method for nonequilibrium electron transport *Phys. Rev.* **B 65** 165401
[3] Rocha A R, Garcia-Suarez V M, Bailey S, Lambert C, Sanvito S and Ferrer J 2006 Spin and molecular electronics in atomically-generated orbital landscapes *Phys. Rev.* **B 73** 085414
[4] Ferrer J *et al* 2014 GOLLUM: a next-generation simulation tool for electron, thermal and spin transport *New J. Phys.* **16** 093029

www.ingramcontent.com/pod-product-compliance
Ingram Content Group UK Ltd.
Pitfield, Milton Keynes, MK11 3LW, UK
UKHW051908100926
8556IPUK00005B/804

9 780750 336376